British and Irish Butterflies: an Island Perspective

Supporting material is available at
http://www.cabi.org/openresources/95061.

British and Irish Butterflies: An Island Perspective

Roger L.H. Dennis

and

Peter B. Hardy

CABI

CABI is a trading name of CAB International

CABI
Nosworthy Way
Wallingford
Oxfordshire OX10 8DE
UK

Tel: +44 (0)1491 832111
Fax: +44 (0)1491 833508
E-mail: info@cabi.org
Website: www.cabi.org

CABI
745 Atlantic Avenue
8th Floor
Boston, MA 02111
USA

Tel: +1 (617)682-9015
E-mail: cabi-nao@cabi.org

A catalogue record for this book is available from the British Library, London, UK.

Library of Congress Cataloging-in-Publication Data

Names: Dennis, Roger L.H., author. | Hardy, Peter B., author.
Title: British and Irish butterflies : an island perspective / Roger L.H. Dennis & Peter B. Hardy.
Description: Boston, MA : CABI, [2018] | Includes bibliographical references and index.
Identifiers: LCCN 2017050570 (print) | LCCN 2017051602 (ebook) | ISBN 9781786394989 (ePDF) | ISBN 9781786395078 (ePub) | ISBN 9781786395061 (hbk : alk. paper)
Subjects: LCSH: Butterflies--British Isles. | Butterflies--Great Britain. | Butterflies--Ireland. | Biogeography--British Isles. | Biogeography--Great Britain. | Biogeography--Ireland.
Classification: LCC QL555.B65 (ebook) | LCC QL555.B65 D46 2018 (print) | DDC 595.78/90941--dc23
LC record available at https://lccn.loc.gov/2017050570

ISBN-13: 978 1 78639 506 1

Commissioning editor: Ward Cooper
Editorial assistant: Emma McCann
Production editor: James Bishop

Typeset by SPi, Pondicherry, India
Printed and bound in Great Britain by Bell & Bain Ltd, Glasgow

Contents

Online Supplementary Appendices

Supporting material is available at
http://www.cabi.org/openresources/95061.

This work is dedicated to all the butterfly recorders who have kindly contributed to the records and observations for the British and Irish islands, the Isle of Man and the Channel Islands.

List of Figures

Text figures

Appendix Figures

List of Plates

Island landscapes

FRONT COVER: Ramsey Island (Ynys Dewi) off St David's Peninsula, Pembrokeshire. Inset: *Pyronia tithonus*, the Gatekeeper butterfly, found on the island.
BACK COVER: The Great Orme's Head, North Wales, viewed from Penmaenmawr.

Images of British butterflies

Preface

Initially, our reason for producing this book was to provide updated records of the original work on islands by Dennis and Shreeve (1996), together with a revision of the original text to present a brief summary of research undertaken during the past 20 years. However, it soon became obvious that the original work was outdated and that readers would appreciate a wider perspective on the biogeography of the butterflies found on British and Irish islands. Also, it was felt that readers would benefit from a less technical introduction to the subject than that originally produced by the first author. Thus, the present book has a different title and, with a completely new text and wider focus, is a very different work from that of 1996. It has involved a great deal more preparation, as not only have many more appendices of additional data and other information been created, including supplementary appendices of data and analyses built for open access use, but also a much greater number and variety of illustrations have been included. Furthermore, more unusual terms are described in an extensive glossary. One thing has not changed; the prime objective remains the same: to encourage thorough observations on butterflies on the islands and to provide material for further evolutionary and ecological research to underpin the conservation of these beautiful creatures. The book, then, provides a brief overview of the fast developing research on the butterflies of the islands over the past 20 years.

In producing the book, we are aware that we are catering for different levels of expertise and knowledge on island biogeography as well as a number of other allied subject areas. Thus, the first three chapters introduce basic topics in island biogeography, past environments and aspects of butterfly biology and study. Chapters 4 to 6 drop down in scale from a broad view of butterfly geography over Britain and the European islands to finer issues affecting butterfly species richness and species incidences on the smaller offshore islands. Chapters 7 and 8 investigate changes in the faunas on islands, especially turnover and evolutionary changes, with an increasing emphasis on butterfly ecology in the changing landscapes of the Holocene. A short final chapter identifies the key features emerging from our review, summarizing the basic issues of island butterfly faunas, and then looks ahead to a new era of studies on the islands.

It will be helpful to provide a brief guide to the use of the book. It may be considered to fall into two basic sections: first, the text comprising the nine chapters; and second, a set of 15 appendices which present records for the islands and some details on more complex issues. In addition, a set of 13 supplementary appendices online provide working files of records and detailed results of analyses to facilitate research. In this way, we hope we have been able to cater for readers having different levels of expertise, interest and objectives. For those less familiar with British butterflies, we have also provided plates of all the species likely to be seen on the islands; these are not provided as an identification guide (we do not illustrate sexual dimorphism or geographical variation) but to display the extraordinary variety and beauty of these creatures; thus, to encourage enthusiasm and more concern for their plight in these days of outright habitat destruction. We have used scientific names for species, but those more familiar with (or who prefer) vernacular names will find these listed in Appendix 1 and attached to each plate.

Island biogeography is not a simple subject but covers a wide variety of academic disciplines, each having its own jargon. To facilitate reading the text, more unusual terms are noted in bold, which is an indication that they appear in the glossary. Terms applying to statistical techniques are noted in bold italics in the glossary and, to help matters further, in Appendix 8 a brief overview is given not only of essential issues in surveying islands but also the rudiments (objectives) of statistical techniques applied in the work. Although we have kept statistical jargon to a minimum, we realize that some techniques are difficult for non-specialists to understand; thus, more technical results have been stored in supplementary appendices. The reference section is extensive, the reason being to avail readers of the original sources in the 1996 booklet of information on islands, as well as sources to key publications. In addition, for easy reference, we provide the original booklet as a supplementary appendix.

In the original work of 1996, the first author was responsible for the list of records, by searching through the journals, and the text; the analysis was shared with Professor Tim G. Shreeve. In producing the present book, the work has broadly been divided as follows: Peter Hardy has undertaken the arduous process of locating all new records and of carefully building all the databases available as part of this book, including the supplementary appendices of records for general use and the scanned copy of the 1996 booklet. The island records for Britain and Ireland have been built on to the database published in Dennis and Shreeve (1996). As of 19 January 2017, records of butterflies were obtained by Peter Hardy for 393 of over 900 islands identified as being of more than 10 ha (many smaller ones too) around the coasts of Ireland, Man and the British Isles, 177 more than in 1996 (216). At the same time, 3865 species records for islands were recorded, 1614 more than the 2251 noted in 1996. The supplementary appendices also contain a file of butterflies on European islands which Peter has revised from the work of Dennis *et al.* (2008); nomenclature has been updated to match that in Kudrna *et al.* (2015) as far as possible. Many of the butterfly photographs in the book, a key attraction, have also been taken by Peter. RLHD's contribution has been to carry out new analyses and to produce the text, glossary, references, figures and appendices of analyses and results. Of the figures, Peter has produced five splendid illustrations using Alan Morton's DMAP programme (www.dmap.co.uk/).

In any work of such complexity there must be cautionary notes; several are expressed in the text. For several reasons beyond PBH's control (access to some databases, inevitable failure to locate all websites with butterfly data), it is unlikely that the work has picked up all the records available by the beginning of data analysis (1 February 2017). Furthermore, some analyses of the records are flawed if only because of biased recording, as explained in Chapter 3; how the results may be affected by biased recording is discussed at appropriate points in the text. It has to be said that, despite the detail in this book, we envisage this work as little more than a beginning for the island biogeography of butterflies on the British and Irish islands, a platform for more significant research as well as for the simple joy of discovery. With the development of techniques for tracing movements in butterflies and studies in molecular genetics (use of unique DNA signatures), we expect the knowledge of dispersal in butterflies, and of their evolutionary transformations, to be revolutionized over the next few decades. In the previous work of 1996 it proved difficult to distinguish events in ecological and geological time; that difficulty is in the process of being erased, as evident in our ability now to identify cryptic species and introgression among closely related species, revealed by the remarkable findings for both Irish *Leptidea juvernica* and British *Aricia* species.

Our hope is that this book, which focuses on broad geographical and ecological issues, will provide a guide to priorities in future research and the conservation of butterflies on British and Irish islands, the Isle of Man and the Channel Islands.

Roger L.H. Dennis and Peter B. Hardy

Acknowledgements

No academic book is entirely the product of an author's work and this book is no exception. It has depended on the generosity of many colleagues and friends for illustration, PDF files of papers, and on ideas generated by many researchers over decades. Moreover, this particular work has depended on the immense generosity of a great many butterfly enthusiasts responsible for accumulating records on the islands. We extend our grateful thanks to all the butterfly recorders who have kindly sent us their records directly or indirectly through the county and regional recorders for Butterfly Conservation UK, and who have contributed to other schemes (e.g. Big Butterfly Count).

The records have mostly been obtained from the regional coordinators for Butterfly Conservation UK's *Butterflies for the New Millennium* atlas; such is their contribution that we have listed them in a dedicated Appendix 4. The credit for these records, together with specific details of the records, lies in the original data files held by Butterfly Conservation UK, sourced by one of us (PBH) from the regional recorders; we are indebted to them all for their generosity and as such dedicate this book to those who have done the groundwork in building up the data used in this work. The contribution from Ian Rippey, who has taken great pains to accumulate records and to direct them to us, as in 1996, has been particularly important for our coverage of offshore Irish islands. Records for a detailed study of the Isle of Man and its satellites have kindly been supplied by Gail Jeffcoate, the late Dr Stephen Jeffcoate, and Garry Curtis. Selective data for UK Butterfly Monitoring Sites and the Wider Countryside Butterfly Survey have been kindly provided by Professor Tom Brereton of Butterfly Conservation UK. We thank Professor Chris Thomas, Dr Rob Wilson and Andrew Graham (regional recorder for North Wales) for data for the plot of species richness on the Great Orme's Head, North Wales. This book also provides an up-to-date supplementary appendix (Microsoft Excel file) of species for European islands; our grateful thanks to Dr Leonardo Dapporto for adding many new records for the Italian islands and to Dr Enrique Garcia-Barros for tracking down records for islands off the Galician coast. A number of colleagues have also kindly assisted us in gathering records on butterfly movements for Appendix 6; our grateful thanks to them for their experience and observations, as to all those who have contributed to this work.

We have benefited immensely from work carried out by other researchers and extend our thanks to all whose research is cited in the book whether as information, ideas, or illustrations. Professor Chris Thomas and Professor Jim Mallet kindly consented to the publication of their fascinating map of DNA variation in *Aricia*. The book includes some original work and we thank those who have helped to produce this novel material, especially Professor Richard J. Tilley. The book also summarizes work that the first author has produced on butterflies of the British Isles with other authors since 1996; on this aspect he would like to point to the outstanding original contributions to this research of his close friends and colleagues, Dr Leonardo Dapporto of Università degli Studi di Firenze, Italy, and to Professor Tim H. Sparks of Coventry University, UK, to whom he is indebted. He would also like to thank Professor Tim G. Shreeve of Oxford Brookes University, UK, for his support in producing the original book on islands in 1996 and the many publications over the years.

A number of colleagues have greatly assisted our interpretation of historical events. We convey our grateful thanks to Scott Renshaw and the British Geological Society for providing details of the borehole taken at Llandudno in 1971 by M.J.C. Nutt, and to Professor Emeritus Michael J. Tooley for a copy of his meticulous monograph on sea-level changes in the Irish Sea basin; also to Dr Philippa Tomlinson for details of Devensian and Holocene geology of the Isle of Man.

We are entirely indebted to Gail Jeffcoate, Julie Williams (Chief Executive) and Natalie Ngo (Media Officer and Assistant editor of *Butterfly*) of Butterfly Conservation UK for tracking down colleagues with images of butterflies for the book; our grateful thanks to them and all these generous contributors. Butterfly images were kindly provided by Patrick Clement, Bob Eade, Peter Eeles, Tony Hoare, Mark and Neil Hulme, Nigel Kiteley, Iain H. Leach, Tim Melling, Ian Rippey and Andy Seely; their stunning contributions are also noted in the legends to the plates. Some landscape images are also included; our grateful thanks to all contributors. The image of Bawden Rocks, Cornwall in a storm, was one of the last ever taken by an old friend (the late Dr W. Gerry Tremewan, editor of the *Entomologist's Gazette* for over 50 years and world expert of Burnet moths) shortly before he died. Gail Jeffcoate, again, kindly provided the appealing image of the Calf of Man and Matthew Oates contributed the photograph of the Isle of Wight. Wildlife film maker John Aitchison and his family generously trudged the length of Colonsay to photograph a special oak wood for us where the late Tom Dunn first found the Purple Hairstreak in 1965. A long-standing friend, Professor Andrew S. Pullin, provided the photograph of the Great Orme for the back cover, an image that impresses how close the Great Orme's Head is to becoming an island again. Steven R. Williams, Records Officer for Hilbre Bird Observatory, provided the stunning image of Hilbre from Middle Eye; Sam Whitfield, those for Flat Holm and Steep Holm during his time as Warden of Flat Holm; and Stuart Murray the glorious images of Ailsa Craig, Bass Rock, Grassholm, South Bishop Rock and Ramsay (the front cover image). Two long-standing friends, Professor Tim H. Sparks and Dr Andrew Ayres, ensured many of the above contacts, the former via Professor Mike P. Harris; Tim's daughter, doctoral student Alexandra Sparks, also provided a suite of photographs of St Kilda, all so evocative of the lost community there that it was a difficult task to select the final image! Permission to reproduce the image of Strangford Lough was provided by Mike Hartwell (Northern-eye Co. Ltd, File 776312424) and that for Lindisfarne by Graeme Peacock (D 0280; www.graeme-peacock.com). The copyright of all images credited to contributors and publishers remain with the named contributors.

Readers will find that the original book on this subject (Dennis and Shreeve, 1996) is available as a Supplementary Appendix. For its inclusion, we would like to thank Ian Johnson (Pemberley Books) and Geoffrey Morton (originally of Gem Publishing Company). In producing the book every effort has been made to contact holders of copyright to obtain permission to reproduce copyright material. However, if any have been inadvertently overlooked, the publishers will be pleased to make the necessary arrangements at the first opportunity. We would like to take the opportunity here of thanking all those publishing companies that gave permission to publish material in this book. Appropriate credit is given where reproductions are made. Copyright of illustrations is retained by authors contributing them to the book.

This work would not appear without the experience and encouragement of CABI's publishing staff: Emma McCann (Editorial Assistant), Susan Philcox (Knowledge Business Intellectual Property Manager), James Bishop (Production Editor), Victoria Davies (Marketing Assistant), Leigh-Ann Bard (Marketing Manager) and particularly Ward Cooper (Commissioning Editor); our grateful thanks to Ward for taking on the project, and to all the CABI staff for all their help and encouragement. We are especially grateful to Val Porter (Copy Editor) for hunting down the errors in the text. If the book is error-free, it is entirely owing to her painstaking scrutiny of the work.

Roger L.H. Dennis and Peter B. Hardy

1 Basics of Island Biogeography

It is one thing to make recordings or collections of butterflies from islands. Collecting has a long history but even so records for butterflies on British and Irish islands were, until recently, scattered in the journal literature (Dennis and Shreeve, 1996). However, it is another matter to make sense of the records. It was not until the early 1960s that sound scientific principles were established for island faunas. The present chapter provides an introduction to modern concepts in island biogeography, factors that affect the presence of butterflies, and other organisms, on British and Irish islands.

The Basic Model of Island Biogeography

In the 1960s, one of the most important breakthroughs in the ecological sciences was achieved; it was the publication of *The Theory of Island Biogeography* by Robert H. MacArthur and Edmund O. Wilson (1963, 1967). Although it was previously well understood that numbers of species on islands relate to island area and isolation, they demonstrated that the number of species could well reflect a dynamic equilibrium between two ongoing processes: (i) the **immigration** (viz. **colonization**) of species to an island; and (ii) their **extinction** (Fig. 1.1a); as such, it is referred to as the **equilibrium theory**. It was a theory already foreseen by Eugene Gordon Munroe (1948) working on the butterflies of the West Indies. This was an astonishingly important breakthrough because, then, even mobile organisms such as butterflies were thought to have long occupied islands – including those comprising Britain and Ireland – on which they were found; as such, overseas transfers, apart from by known long-distance seasonal migrants (e.g. *Vanessa cardui*, *Vanessa atalanta*), were considered to be rare events. This dichotomy in perception must now seem very strange to modern readers; after all, if one species can with facility make the journey in numbers, then surely all others must have some capability of achieving the journey! In fact, this very belief was one of the stumbling blocks for challenging the long-held view that British butterflies survived the last major glaciation (**Devensian**, 20 k years (20 **ka**) BP) (Dennis, 1977). Yet, even then, sufficient evidence existed in the amateur journals that very small butterflies, including many sedentary butterflies, were capable of sea crossings (Dennis and Shreeve, 1996).

The mathematical model for a single island off a mainland **source** beautifully illustrates the basis for equilibrium in species numbers on islands (Schoener, 2010); a 'steady state' in numbers of species occurs where the gross extinction rate in species equals the immigration rate of new species. In Fig. 1.1, the rates are typically shown to be curvilinear, owing to the limit in the number of species at mainland sources: as numbers on an island grow, so does the probability of extinction; similarly, the probability of immigration and colonization will decline, as there are fewer new species left in the transfer pool. This observation is supported by ecological observations, for instance by: (i) the saturation of the species

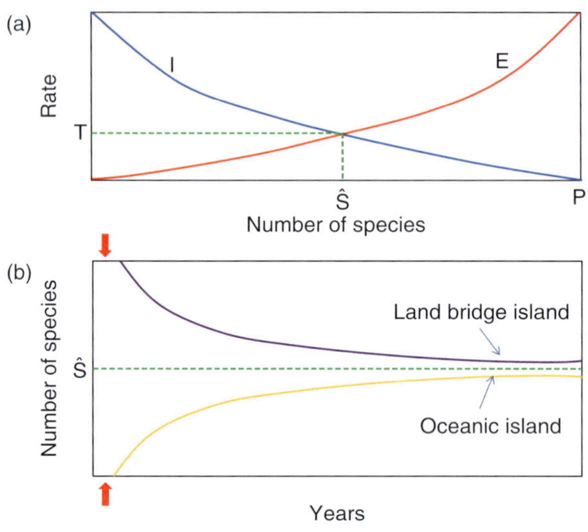

Fig. 1.1. Equilibrium island biogeography theory. (a) The equilibrium state in numbers of species for a single island based on the rate of immigration and extinction. I, immigration rate; E, extinction rate; Ŝ, equilibrium number of species on island; T, Turnover rate; P, size of the mainland source pool for the island. (b) The approach to an equilibrium for two distinct kinds of island: (i) ocean island; (ii) land bridge island. Once the number of species on the land bridge fauna have reached equilibrium, this is known as the **relaxation fauna**. The red arrows indicate the period of delay before the first species is lost (land bridge island) and gained ('oceanic' island). Oceanic islands here are used in the sense of a newly created island starting with zero species. Note how equilibrium is unlikely ever to match the number of species on the source mainland.

community with an increase in numbers of island species, leading to increased interactions (i.e. competition, predation, parasitization), thus smaller populations and extinctions (Wilson, 1969); and (ii) declining immigration rates as poorer dispersing species trail behind better dispersers (MacArthur and Wilson, 1967). Moreover, poorer dispersers are likely to be more vulnerable to extinction, owing to the link-up of colonization ability and migration capacity (Gilpin and Armstrong, 1981) (see also Chapter 3).

Equilibrium theory of numbers of species on islands, then, envisages a perpetual turnover in species; both immigration of new species and extinction of current species will have a characteristic **mean** rate and **variance** (unless conditions change significantly, i.e. through climate warming). It is perhaps useful to imagine the basic process taking place at an individual level. Consider that an island is being continually bombarded by individual butterflies and is simultaneously experiencing a continual loss of individuals (at all stages of development) (Simberloff and Wilson, 1971). Every so often an individual of a new species will arrive on the island and colonize it, or the population of a species will crash to zero and become extinct. Together, these processes describe the **relative turnover rate** (T) for an island's species:

$$T_{t1,t2} = (n_e + n_i)/(N_{t1} + N_{t2})$$

where t_0, t_1, t_2 ... t_m are time intervals (years), n_e is the number of species extinctions on an island, n_i the number of novel colonizations on an island, and N the total number of species at different times. T can be made into a percentage by multiplying by 100. Later, attention will be given to two types of island: (i) an oceanic island; and (ii) a land bridge island. Although Britain and Ireland do not have typical oceanic islands, they do have islands that simulate the conditions on a typical oceanic island, with an initial 'start-from-scratch' condition of zero species as in the case of Hawaiian islands emerging as volcanic mounds over an oceanic mantle 'hot spot' (Funk and Wagner, 1995). In the case of some British islands a number emerged offshore with deglacial isostatic rebound (see Chapter 2) of ice-laden land, their fauna wiped out by glacial tabula rasa and/or marine inundation during the last major ice advance (c. 20 ka BP). In the case of these two types of islands, very different patterns of colonization emerge, continuous gains towards an equilibrium with an 'oceanic' island and continuous losses to an equilibrium in the case of a **vicariant** land bridge island (Fig. 1.1b). These conditions describe extremes of what is found in nature; it will become evident that Britain displays more complex scenarios.

The intriguing picture of equilibrium theory is that large and small islands, near-to-source and isolated islands, have very different equilibria (Fig. 1.2); the curve for extinctions is higher for small islands than large islands, and the curve of immigration for isolated islands is lower than that for near islands. This has consequences for both the number of species and absolute turnover rates: (i) near islands (having the same area) and large islands (experiencing the same degree of isolation) have more species than far, small islands; and (ii) absolute turnover (not relative turnover) is greater for near than far islands (of the same area) and for small than large islands (of the same degree of isolation) (Fig. 1.2). Altogether, this puts isolated small islands at a huge disadvantage in terms of comparative numbers of species. In fact, the two processes of colonization and extinction have far more profound implications for their island faunas, ones that affect the communities, adaptation and evolution of their faunas (see Chapter 8).

The picture of single islands lying offshore a mainland is, of course, a simple one. Often there are archipelagos (e.g. Isles of Scilly; Outer Hebrides), and these too may be close to a mainland shore or isolated. It is at this level that one can appreciate not just how complicated island biogeography can be, but also how important equilibrium island biogeography theory has

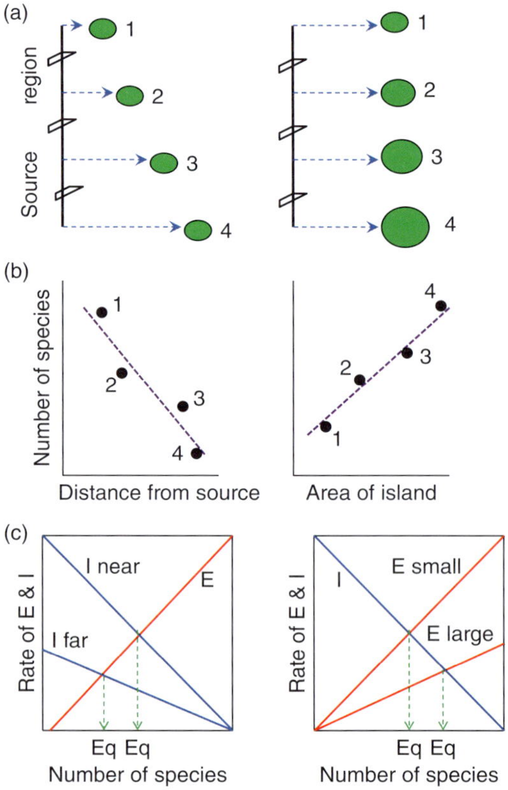

Fig. 1.2. The effect of differences in island isolation and island area on the equilibrium number of species on islands. (a) Imaginary plot of islands varying in size and isolation. (b) Scattergrams of number of species for the islands in plot (a) against isolation and area. (c) Equilibrium models for the islands shown in plot (a). E, extinction; I, isolation; Eq, equilibrium number of species. (Adapted from Dennis, 1992, 2010, using linear rather than curvilinear relationships.) Near, far, small and large refer to islands varying in isolation and size.

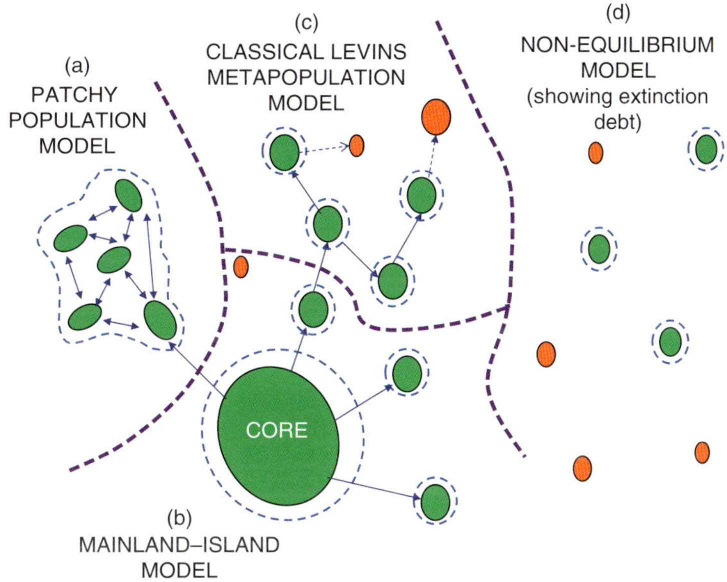

Fig. 1.3. Some basic metapopulation models for habitats on a mainland. (a) Patchy population model (e.g. *Pieris napi*). (b) Mainland–island or Boorman–Levitt model (e.g. *Aricia agestis*). (c) Classical Levins metapopulation (e.g. *Melitaea cinxia* on Isle of Wight). (d) Non-equilibrium model (showing distinct signs of extinction debt with populations becoming extinct and not replaced) (e.g. *Coenonympha tullia* in lowland UK). Green patches, colonized; red patches, vacant; dashed blue lines, boundary of a local population; arrows, transfers. Species on offshore islands which are part of archipelagos may behave in the way described by these models. No scale is provided; the scale of a metapopulation is established by the mobility of the organism. Thus islands in an archipelago may act as a patchy population to one species and as a non-equilibrium population with gradual extinction and the loss of another species. (Adapted from Dennis, 2010.)

become: archipelagos lie at the root of **metapopulation** biology (habitat islands simulating islands lacking a mainland source); the equilibrium models for them were developed by the remarkable scientist Ilkka Hanski and his team (Hanski, 1994; Hanski and Gilpin, 1997). A variety (gradient) of metapopulation models may be conceived from patchy populations to mainland–multiple island situations and extinction debt metapopulations (Dennis, 2010); all these models are relevant for different butterfly species on groups of small offshore islands in Britain and Ireland (Fig. 1.3).

A Broader View of Island Faunas

The equilibrium model, in providing such a neat picture of species numbers and turnover, is highly seductive; but it has been the subject of long and heated debate (Lomolino *et al.*, 2010). When reviewing papers that deal with this subject it is important to distinguish two related issues:

(i) the balance of immigrations (colonizations) and extinctions which defines a steady state (equilibrium) in species numbers (Fig. 1.2c); and (ii) the relationship of numbers on islands in any study region to island area and isolation, on which an equilibrium view is based (Fig. 1.2b).

The critical property of equilibrium theory is that extinction rate equals colonization rate. This has to be the case for every island in a study region (i.e. herein, British and Irish islands). It is not difficult to envisage circumstances that generate a non-equilibrium condition for a specific island or islands. Non-equilibrium may occur owing to: (i) relatively too few extinctions (i.e. **extinction debt**); (ii) a sudden extinction event affecting the island's fauna; (iii) relatively low colonization rate (i.e. an immigration backlog or lag in immigration rate; e.g. following glacial tabula rasa); (iv) a surfeit of immigration (i.e. increased rate of immigration; e.g. owing to an external stimulus such as climate change or a weather event); and (v) a relatively high species accumulation, i.e. **autochthonous speciation** (e.g. molluscs on Madeira) (Cook, 2008; Kier *et al.*, 2009).

A linear (semi-log, or log-log) relationship between species numbers on islands and island area and/or island isolation (Fig. 1.2b) is typically advanced as an indication of an equilibrium condition. In fact, the absence of a neat linear relationship is not evidence of the absence of an equilibrium, but it is evidence that island area and/or isolation are compromised by other factors affecting inter-island extinction rates and colonization rates on some or all of the islands. It is well to bear in mind that an island's 'resources' may supplant area in importance, as has been found to be the case with butterfly metapopulations (Dennis and Eales, 1997, 1999), and reflect conditions associated with geology or soils, climate, topography or land use. This issue is expressed in the simple algorithm of the **choros model** (Triantis *et al.*, 2003). Similarly, as regards immigration and colonization, some islands may be favoured by prevailing winds whilst others experience counter currents as would be the case on opposite-facing coasts of the UK mainland. Then, a number of alternative explanations for a linear species number–island area relationship, and species number–island isolation relationship, have emerged that do not necessarily equate with the factors that MacArthur and Wilson (1963, 1967) initially associated with generation of an equilibrium model. For instance, the species–distance relationship may occur owing to isolated islands having systematically fewer biotopes and lower resource potential (Lack, 1976). Similarly, the species–area effect may simply reflect a passive sampling process (i.e. a bigger 'net' to catch passing immigrants); alternatively, there may be less abiotic disturbance on large islands (Schoener *et al.*, 2001), or larger islands may have increasing biotope diversity (Ricklefs and Lovette, 1999), or indeed higher speciation rates (Losos and Schluter, 2000). Indeed, larger islands by providing larger target areas (i.e. a **target effect**) enhance colonizations and facilitate a **rescue effect** on populations that might become extinct (Lomolino *et al.*, 2005; Brown and Kodric-Brown, 1977).

It should now be evident to the reader that there is potential for great complexity in the distribution of species on islands. This is certainly likely to be the case for British islands, made up of two relatively large islands and a constellation of tiny satellites. Bear in mind that all is relative and that St Mary's, St Martin's and Tresco are relatively large compared with other islets in the Isles of Scilly, and that larger islands often have satellites and stepping stone islets (e.g. Calf and Kitterland for Isle of Man: Plate 1.1). In the case of the two large islands, mainland Britain has an entire continent as its source, whereas the current nearest source for the second largest island, Ireland, is predominantly the largest island and not the continent. However, the main historical source for Ireland would have been the continent in the immediate aftermath of the last major glaciation (Dennis, 1977). The smaller islands have one or other of the European continent (e.g. Channel Islands), mainland Britain (e.g. Isle of Wight) and Ireland (e.g. Aran islands) as their main contemporary source; but occasionally more than one source may affect the patterns of colonization on islands (e.g. Isle of Man lying between Ireland and mainland Britain). The variety of potential sources also affects the nature of processes over time. Some indication has already been given that island species may depend on more than contemporary (equilibrium) processes; geological history (**evolutionary scale** issues) can dominate current factors (**ecological scale** issues) and current processes may mimic historical ones in effect (Dennis and Shreeve, 1996). A simple process-response model of a number of these issues (Fig. 1.4) was drawn up many years ago to illustrate the interaction of factors (Dennis, 1992); it illustrates why the biogeography of the British and Irish islands is so fascinating. Perhaps the model is a little outdated now but it should still present a useful starting point for thought. Firstly, it separates historical influences from present-day ones. Then, it divides these further into initial regulators (tabula rasa), subsequent constraints (prehistory and history), island geography and island resources (ecology). Climate and human influences, in particular, have dominated the entire fauna and flora of these islands over the past 20 ka; these occupy prominent 'sides' in the figure. Admittedly, it is no easy matter to determine the part played by even this small group of factors on current island butterfly faunas, particularly when they produce similar outcomes. But investigation of island-faunal patterns exposes some prominent findings and challenges a neat equilibrium view of island biogeography. Over the long period since the Devensian glacial maximum advance, one should expect that occasional pulses of disturbances, inducing sharp changes in extinction and colonization rates, have interrupted longer periods of lower, more stable rates of turnover, if not on the catastrophic scale of Krakatoa (Bush and Whittaker, 1991).

An Integrative Island Biogeography Model

Since the development of the MacArthur–Wilson equilibrium theory, there has been a concerted attempt to build an integrative model of island biogeography, one that not only deals with immigration (colonization) and

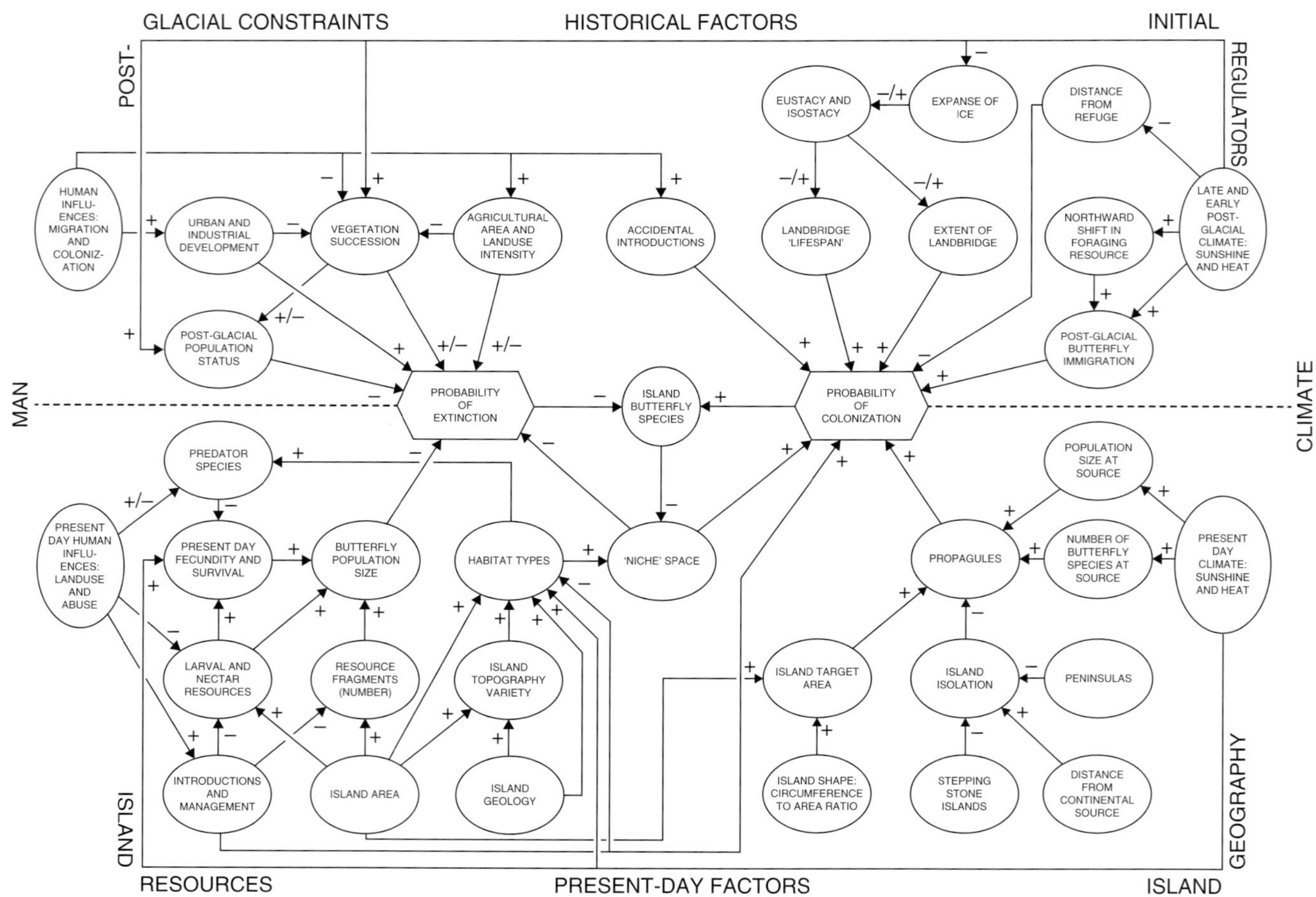

Fig. 1.4. Factors accounting for non-equilibrium species on land bridge islands as in the case of the British islands. A large number of factors may account for the number (and type) of species on islands by ultimately influencing extinction and colonization rates. (After Dennis, 1992, courtesy of Oxford University Press.)

extinction but also includes community dynamics, ecological shifts and evolutionary processes, and moreover incorporates systems interactions and feedbacks associated with changes on islands. Such a model has gradually developed with the work of numerous researchers, among them Darlington (1957), MacArthur and Wilson (1967), Losos and Schluter (2000), Lomolino (2000), Lomolino *et al.* (2010) and Whittaker and Fernández-Palacios (2007). Some of the fundamental features can be abstracted and illustrated in axes of island area and isolation (Fig. 1.5). There are several drivers of this model that can be grouped into: (i) a set of island systems' traits (i.e. area, isolation, altitude and relief complexity, age, disturbance history, latitude); and (ii) a set of species traits (i.e. crucially metabolism, size, **bauplan**, life history, **trophic** strategy). Ultimately, the outcome of area and isolation is scaled by the organisms: what is distant to a mole is but neighbourhood to a seabird like a gannet. Two key aspects characterize the model; both are in fact key to any systems approach in any science. The first is scale dependency of both island traits and species traits; the constellation of correlations linking one trait with another tends to be specific to an archipelago, an island system or subsystem, and to a taxon. The second is feedback in the system; the more complex the system, generally the more feedbacks there are linking elements within the system. The model then describes gradients in outcomes, the characteristics of which depend on the nature of organisms scaling the system and the level of interactions within communities. It is a beautifully simple way of interpreting island phenomena and takes us well beyond the turnover of species. For instance, something we are unlikely to see on British offshore islands is the development of communities that display greater diversity in particular taxa than on source mainlands, the result of **adaptive radiation** and a reminder of Wilson's (1959) extraordinary contribution in the **taxon cycle**. Rich island zones of macroevolution are well known for the Hawaiian islands (e.g. ferns, *Drosophila* flies, snails, honeycreepers; Wagner and Funk, 1995), for finches in the Galapagos islands (Grant and Grant, 2010) and for mammals in the Philippines (Heaney, 2004). As lakes may be treated as 'islands' amid a continent, then the same process has affected cichlid fish in African Rift Valley lakes (Meyer, 1993). Even so, to understand British and Irish biogeography, a systems approach and an integrative model is a crucial starting point.

Population and Ecological Changes on Islands

Offshore island faunas are typically characterized by two patterns: (i) **depleted faunas** and (ii) **disharmonic faunas**. Disharmony again describes two basic features: (i) island faunas comprise non-random assemblages of the source fauna, effectively filtered by **dispersal** and **migration**; and (ii) novel species not found on the mainland. Yet, species on islands, but present at the mainland source, may also be novel in other ways: genetically, in **phenotype**, in habitat and resource use – an issue considered in this review (see Chapter 8).

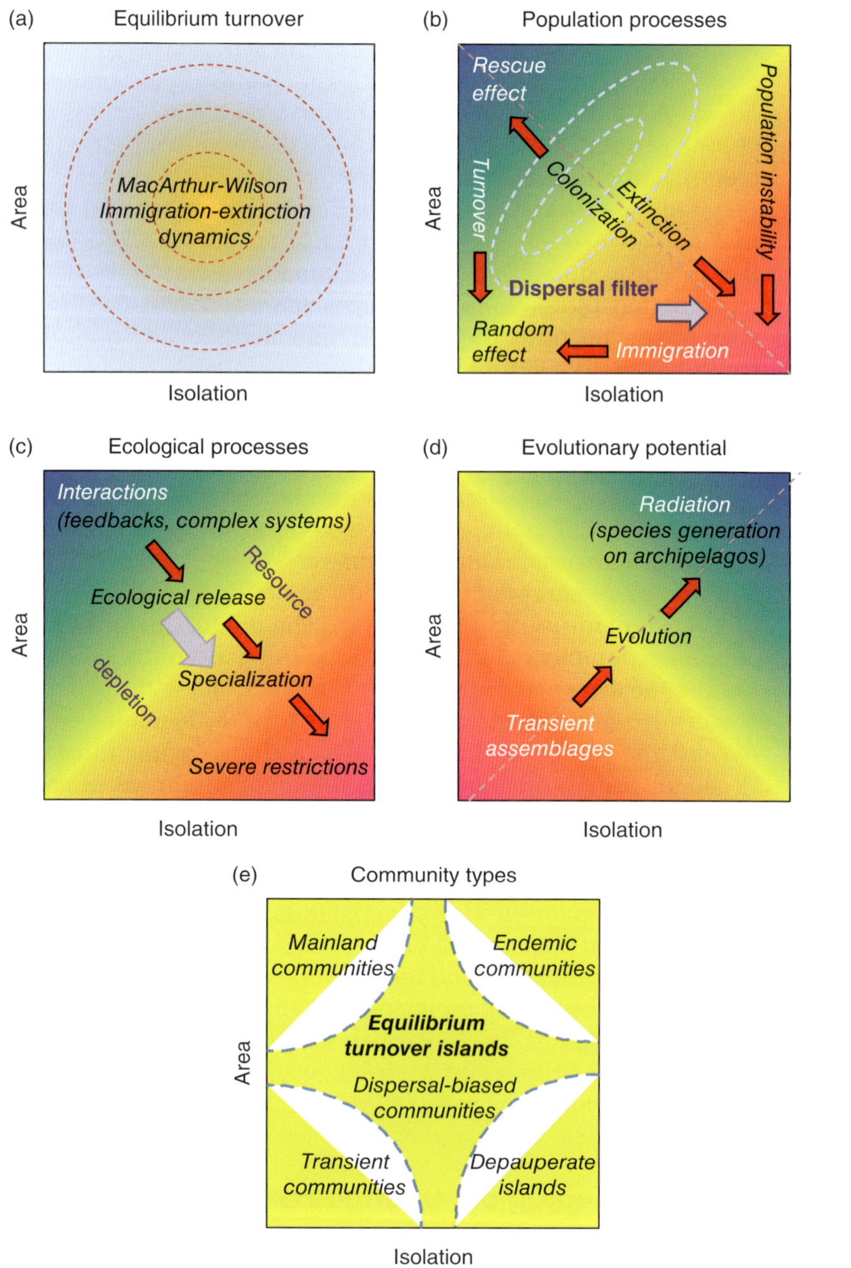

Fig. 1.5. Components of an integrative model of island biogeography based on island area and isolation. Modified and developed from a number of sources (Darlington, 1957; MacArthur and Wilson, 1967; Carlquist, 1974; Losos and Schluter, 2000; Lomolino, 2000; Lomolino *et al*., 2010; Whittaker and Fernández-Palacios, 2007; Losos and Ricklefs, 2009; Rosindell and Phillimore, 2011; García-Verdugo and Fay, 2014). (a) Typical zone (centre area) for equilibrium dynamics as envisaged by MacArthur and Wilson (1967). (b) Population processes (elliptical

Islands that have faunas most similar to those of an adjoining mainland are likely to be the depleted product of mainland communities (depleted faunas) (Fig. 1.5e) and will tend to be close to the mainland source and large in size, much as is the Isle of Wight compared with the Isles of Scilly, both of which are offshore from the British mainland coast. Populations of islands close to the mainland shore will experience low extinction rates, if only because every time they dip in number, they have the greatest chance of being 'rescued' by immigrants from the mainland source (the rescue effect) (Fig. 1.5b). But as islands become smaller, so usually does the island resource base for species, and populations also become smaller (less dense) and more unstable; moreover there is decreasing chance of the population being rescued (the island is a **sink**). A common observation is that with population instability there is an increasing likelihood of extinction. Thus, small islands will tend to have fluctuating vulnerable populations. However, as islands become increasingly isolated there is not necessarily a systematic decline in population density. With distance from a mainland shore comes **ecological release** (Fig. 1.5c); that is, with isolation a species may experience fewer enemies (i.e. parasites, predators), the populations benefiting from lower mortality. But if islands continue to decline in size and become increasingly isolated, the benefits of a lack of enemies are countered by the costs of a diminishing resource base: the syndrome of **depauperate islands** (Fig. 1.5e) is eventually encountered, with few species dependent on small fluctuating populations. Exceptions exist, of course, when a resource is favoured by suitable conditions. Small islands close to shore will also tend to have small populations but will tend be replenished as soon as lost, typically forming **transient communities** (Fig. 1.5e). In between these extremes lie the islands most typical of MacArthur–Wilson dynamics: equilibrium turnover islands. But these will be dispersal-biased communities, reflecting on an increased dispersal capacity by organisms with increasing isolation.

Ecological changes on islands affect more than population sizes and densities. All islands will tend to have a depleted resource base. As plants are affected in much the same way as animals, **heterotrophs** such as butterflies will inevitably become dependent on a depleted larval host plant and nectar resource base with decreasing island size and increasing isolation. Furthermore, the climate (temperatures, wind speed, precipitation) of islands tends to become more extreme with increasing isolation, as do conditions (geology, soils, local climate, topography and shelter from storms and salt spray) generally with declining island size and isolation (e.g. Bawden

Fig. 1.5. Continued.

isolines describe peak population density). (c) Ecological processes expected from resource availability and community interactions based on the resource-based habitat model (see Chapter 3). (d) Evolutionary potential. (e) Zonation for types of communities with equilibrium types midway in the area/isolation scaling. Not to scale, as the organisms (bauplan, mobility, etc.) themselves scale the impact of distinct processes. Conditions become more severe to the bottom right and greater potential for novel evolution towards the top right.

Rocks: Plate 1.2). The resultant flora and fauna becomes more 'polar' (Williamson, 1981), as is certainly the case with British islands (cf. Ireland with Britain) (Figure. 1.4 in Dennis, 1992) (see Chapter 5). Thus, regardless of any ecological release, specialization of species is likely to occur with a narrowing resource base. In effect the species are faced with habitat limitations in the resource base as islands become smaller and more isolated. As such, increasingly isolated islands should favour not only species with high dispersal capabilities but also those that can use a wide resource base (**generalists**). Following colonization, these will become increasingly specialized in the limiting conditions, which concomitantly results in further adaptations in life history (phenology), behaviour, resource exploitation, defences and **mobility**.

Some remarkable transformations accompany island occupancy. Typically, on islands larger animals (large vertebrates) tend towards dwarfism (**nanism**) – reflecting the reduced resources with loss of land bridge connections – and smaller animals (small vertebrates, invertebrates) tend towards **gigantism** – indicative of the reduction in numbers of enemies and consequently any need to be inconspicuous or to develop rapidly (Lomolino, 2005). A further change, highly relevant to butterflies, is a general reduction in wing size; in other insects this can lead to wing loss (beetles; Roff, 1990). Although species with larger wing expanse may be favoured in colonizing islands, they may be at a disadvantage in conditions lacking enemies and vulnerable to inadvertent dispersal unrewarded by opportunities for neighbouring land fall; smaller wing sizes are expected and this may be concomitant with reduction in body mass (see Chapter 8). But such changes are not invariant and depend on the resource base and other pressures specific to conditions on islands, such as other island occupants (i.e. seabird populations; e.g. gannet-covered Bass Rock: Plate 1.3).

Evolutionary Transformations and Novel Species

Small but significant changes in an island species' ecology and phenotype are typically referred to as **microevolutionary**; over time, with isolation and/or selection under distinctive conditions, such changes may lead to the appearance of novel species, referred to as **macroevolution**. The processes leading to genetic changes are likely to differ substantially for 'oceanic' and 'land bridge' islands, particularly as regards initial genetic variation. Oceanic islands, those emerging pristine and empty, are initially blank canvases; their populations are built up on a limited number of founders. In contrast, land bridge islands may well have substantial populations and substantial genetic variation before land severance. However, this is not always the case. In the next chapter it will become evident that for Britain prior to separation from the continent in the **Holocene**, conditions existed that markedly filtered influx of founders and subsequent populations.

For all islands, then, island populations tend to contain less genetic variation (diversity) than their source populations. Again, exceptions can occur, for instance when – following **founder events** and **bottlenecks** in island populations, perhaps owing to the destabilization of adapted genetic complexes – **introgressive hybridization** between closely related species results in greater variability (e.g. *Maniola jurtina* on West Mediterranean islands; Dapporto *et al.*, 2009). Typically, genetic differences between an island and source populations increase with isolation and a decrease in island size. Distant, small islands receive few founders, in butterflies perhaps a single gravid female; the inadvertent landing of a single male will be unproductive in the absence of a female. The larger and closer the island is to the mainland source, the more founders reach that island. Although a female butterfly is required in founding any new population, the subsequent arrival of males may contribute enormously to genetic variation, thereby enhancing adaptability. However, the smaller the island, the more likely it is that stochastic events (i.e. disturbances) will cause population bottlenecking and loss, once again, of genetic variability; this is referred to as **genetic drift**. The type of disturbance will determine the genetic outcome: when entirely random the genetic product is equally uncertain; but if deterministic, say an event of a repeated climatic pattern, survival will tend to become increasingly biased in favour of those with suitable attributes (traits) for similar disturbances or that occupy distinctive island niches (i.e. linked to cover, aspect, temporal appearance).

The genetic make-up of populations occupying islands is important, as small isolated islands are most likely to harbour unusual genetic founders from a mainland source. As most colonizing species will be similarly affected, the end product is likely to be increasingly disharmonic communities on more distant islands. This, in turn, should trigger greater divergence, because of novel resource compositions and organism interactions among the island residents (Barton, 1989). In time, this forms a platform for syndromes of distinctive island traits among residents, one attribute of which, wing reduction, was mentioned above. In butterflies, island variation in adult wing patterns is particularly distinctive; wing pattern/colour development is dependent on conflicting pressures for thermoregulation (including water balance), mate location and defences (Dennis and Shreeve, 1989; Dennis, 1993) and it is likely that balances between these pressures on islands will contrast with those found in populations at mainland sources. The degree of changes in island populations can be assessed by direct measures of genetic variation (i.e. **allozymes**, **mtDNA**), but this may not always match phenotypic changes in wing patterns, measures of **fluctuating asymmetry** (e.g. Leamy and Klingenberg, 2005) and frequency of wing abnormalities (Whittaker and Fernández-Palacios, 2007). But what has long been clear is that phenotype changes in populations can be very rapid, as in bill shape among Laysan finches in Hawaii (Conant, 1988). An earlier example in butterflies is the transformation in wing patterns and size in an isolated population of *Plebejus argus* in the Dulas Valley in North Wales from its founder population located on the **relict** or **fossil island** of the Great Orme's Head in North Wales, an equally fast event (Dennis, 1977).

One of the most difficult issues in biology is the decision whether changes in an organism have become sufficiently distinct for it to be described as a new species. For most naturalists studying butterflies at any location, generally, this does not cause a problem as the different species are recognizably distinct and easily identified; it is clear that they adhere to Ernst Mayr's (1942, 1963) long-used definition of **species**: organisms as actual or potential breeding entities determined by interactions that successfully produce viable offspring. It is a different matter when it comes to distinguishing butterflies found in different locations, especially those locations that are far apart or isolated on different islands (e.g. the Meadow Brown (*Maniola*) butterflies of the Aegean region) (Dennis *et al.*, 2000b). Unfortunately, species definitions abound (Mallet, 1995; Wilkins, 2003); as such, the problem remains of establishing pragmatic guidelines for naming a new species, which, owing to **allopatry** and manifold difficulties in confirming distinctions in the laboratory, is ultimately an arbitrary process (Whittaker and Fernández-Palacios, 2007). New molecular techniques have simplified the task of distinguishing species-level differentiation, and have been especially important for disclosing hidden (so-called cryptic) species (e.g. *Leptidea juvernica*) (Dincă *et al.*, 2011a, 2013). The fact remains that a solution to defining species is elusive: variation is multivariate, continuous and overlapping between different organisms; moreover, the amount of variation displayed by different organisms can vary enormously. The difficulties are highlighted by the amount of hybridization occurring among closely related organisms (Hausdorf, 2011) (courtship and mating are also well known even among more distantly related species; Russell, 2013) and the tendency for inflating diversity through the proclivity of naming new species (for a full discussion see Dennis, 1997a; Descimon and Mallet, 2009).

None of the issues surrounding species would be thought to raise difficulties for a review of butterflies on British islands, but it should be recalled that prior to 1977 many of the butterfly subspecies in western and northern Britain were thought to be so distinctive as to point to ancient origins and isolation in British glacial refuges. Even now, although some 17,500 butterfly species are recorded worldwide, almost 100,000 further subspecies – many of questionable status – are named (Groombridge, 1992). In this review, species and so-called subspecies/races, taken at face value, are used in different analyses to investigate distinct geographical patterns and processes. Very soon, with caution (until we have any individual's entire **genome** we do not know all we need to know), the burgeoning molecular data will allow more precise assessment of butterfly associations over Europe, applying molecular clocks and phylogenetic trees, to give a vastly improved history of origins and lineages (Grill *et al.*, 2006; Talavera *et al.*, 2013). From the standpoint of islands, especially at the European scale, their species presently may be described, and consequently their fauna distinguished, using a number of distinctive features for each species, namely: (i) their fidelity (location uniqueness) to the island (endemic or non-endemic); (ii) their age: neo (young, i.e. a sibling species) or palaeo (old, i.e. a species lacking close relatives, a distinct subgenus or higher category); and (iii) their formation: **autochthonous** (formed on the island) or **allochthonous** (formed elsewhere as on a source mainland) (Fig. 1.6). Together, these features describe eight possibilities for any species on an island.

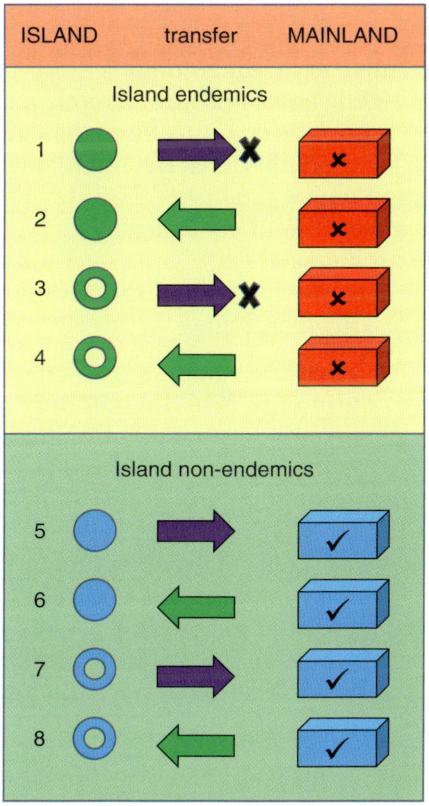

Fig. 1.6. Simple single-move relationships between an island and mainland for a single species. The model assumes endemics, where they exist, have already been created by colonization of a founder. Green circles, endemics (species restricted to island(s)); blue circles, non-endemics; whole circles, recent in origin; donut, ancient in origin; cuboids, mainland (red, unoccupied; blue, occupied); arrows, transfers (green from mainland, purple from island); crosses, barriers to movement; 1, recent endemic formed on island, and either failed to get to mainland or now extinct on mainland; 2, recent endemic formed on mainland but now extinct on mainland; 3, as for (1) but palaeoendemic; 4, as for (2) but palaeoendemic; 5, non-endemic (recent in forming) which has now spread to mainland; 6, non-endemic (recent in forming) formed and still present on mainland; 7, as for (5) but palaeo-non-endemic; 8, as for (6) but palaeo-non-endemic. Although endemics formed on islands (numbers 1 and 3), and non-endemics (numbers 6 and 8) are typical patterns of distribution and origin, all the different patterns exist in nature and more complex ones too resulting from repeated infusions from island and mainland (especially when multiple mountain peaks/ranges are interpreted as islands).

Each species, in fact, may have a different history (e.g. Tuscan islands) (Dapporto *et al.*, 2017). Without a detailed **phylogenetic analysis** (i.e. area **cladogram**) it is often not possible to determine the nature of lineages (i.e. **anagenesis**, **anacladogenesis**, **cladogenesis**) or the age or the location (i.e. whether autochthonous or allochthonous) where a species may have evolved. For instance, what may appear to be two autochthonous endemics may well have arisen by allopatry through **vicariance**, or **sympatry** through direct evolution in response to an adaptive landscape of resources such as a dichotomous host plant choice (Wright, 1932; Rosenzweig, 1995); however, one or both may have formed outside the island (e.g. owing to a past connection when eustatic sea levels were lower) and therefore be allochthonous, the mainland populations having become extinct. Through thorough survey obtaining data on all close relatives, and with due consideration of geological history, phylogenetic work hopes to unravel such problems.

From such work, fast developing (Schoener, 2010), a number of generalizations are emerging:

1. It is rare that anything remains the same for long in geological time; organisms are continually evolving not only on islands but also at mainland sources. Phylogeny has to decipher these multiple changes knowing that there is no fixed standard for reference.
2. What appears to be happening now may be quite different from what has happened in the past; again phylogenetic analysis can show that what may appear to be a recently formed endemic species could well be a relict and closer to an ancestral form (see Voda *et al.*, 2015, for complex patterns illuminated by molecular markers).
3. Phylogenetic analysis invariably aims to produce parsimonious solutions to taxonomic relationships and faunistic patterns.

This in itself is fraught with difficulties as neat solutions rarely exist. **Cladistics** relies on bifurcation of taxa in descent but the reader should consider circumstances in which this rigid assumption may fail and for a more powerful reason than the loss of **clades**, that is, multifurcation of clades (non-monophyly) (Dennis *et al.*, 1995b; Mutanen *et al.*, 2016). For instance, one might imagine an arctic species which during a deglaciation occupies different ranges of mountains in the Alpine chain of Europe. A number of clades then may exist, all evolving at much the same rate. Consider how, with multiple glacial–interglacial stages (> 40 of them are known), these taxa could then be shuffled about from the North European plain to the mountain range, back to plain and southern slopes, north and south, to produce an ever-increasingly bewildering pattern of complex associations. Phylogenetic analysis has its work cut out in deciphering this history.

Island Species as Unique Entities

Although aspects of island biogeography (i.e. biodiversity) treat different species (or subspecies) as equivalent units, it soon becomes obvious that species are not equally distributed across islands. Organisms obviously have very different capacities for migrating to islands, of colonizing them and persisting on them. This is self-evident if one compares the presence on different islands of large and small mammals, birds, plants belonging to different vegetation forms (herbs, shrubs, trees) and apterous and winged insects (Whittaker and Fernández-Palacios, 2007). Clearly some species are only capable of accessing islands when they have had land bridges, or narrow straits lacking fast currents.

A fundamental point, then, in island biogeography is the non-equivalence of species. Early studies in turnover did in fact suggest some equivalence in that the mix of species changes with island defaunation (Simberloff and Wilson, 1971); indeed, species mix changes up to and following the attainment of island equilibrium. It soon becomes apparent that just what species are found on islands is a matter of scale: scale of isolation, scale of size, and complexity of biotopes; the scaling depends on mobility of different species. The mix of species is most likely to change over moderate scales of isolation and island size, as the majority of species (within any taxon) display moderate levels of immigration capacity and colonization ability. Although distant small islands are expected to have the greatest turnover (reflecting greatest disturbance from conditions) this will tend to involve a small set of species with extensive migration and colonization abilities. Similarly, a distinctive set of species is expected to turn over on the close-to-shore large islands, in turn reflecting on limited capacity in transfers and persistence. If mobility (and turnover) is limited in a taxon, then intraspecific diversification becomes possible even in islands with close mainland sources, leading to relictuality (speciation and endemism) if there is sufficient time before major disturbance events intervene (Dapporto *et al.*, 2017).

Measures of species' functional traits, underlying measures of migration capacity and colonization ability, unavailable in 1996, have now been produced for British butterflies (Dennis, 2010; Dennis *et al.*, 2012a; Dapporto and Dennis 2013; Middleton-Welling *et al.*, 2018) (see Chapter 3). The varying movement abilities of species have been tested in the field (Cook *et al.*, 2001). As mentioned above, colonization ability and migration capacity correlate with one another – to be good (bad) at one is often to be good (bad) at the other. It has also been shown that the ability of species to transfer to, colonize and persist on new sites reflects their life histories (Shreeve *et al.*, 2001) and habitat structures (Dennis, 2010) – issues explored in some detail in later chapters. This enables us to investigate island biogeography from the standpoint of species traits; with the developments in molecular genetics it will not be long before a more complete picture can be built of the history of butterflies on British and Irish islands. The details in the following pages hopefully set the scene for those discoveries.

2 The British Islands: a Geological and Historical Outline

What is an island? It is typically accepted to be a body of land surrounded by water. Of course, there are other kinds of islands, ecological islands, such as ponds (Shropshire kettle holes; marl pits), woods and meadows, all treated as habitat islands in **metapopulation** studies. There are also **fossil islands**, once surrounded by water but now joined to the mainland by sand or shingle bars, isolated only by daily tides, or less often by spring tides. Many islands are also 'bridged' to the mainland or other islands (e.g. Seil, Skye, Foulness, Anglesey; Mersea, Hilbre and Lindisfarne: Plates 1.13 and 1.14). The problem of defining islands, and studying them, is exacerbated by the size distribution of islands, which is highly right skewed (there are vastly more small islands than large ones) (Fig. 2.1). The British Isles are dominated by two massive islands (referred to here for convenience as Britain and Ireland), a smaller number of modest sized islands (e.g. Wight, Man, Anglesey, Mull, Skye) and a very large number of small and tiny islands and islets, including sea-washed skerries, reefs and lighthouses on rocks (e.g. Eddystone, Monkstone lighthouse, South Bishop Rock: Plates 1.7 and 1.8); many are unvisited because of precarious landing conditions and perhaps regarded as less interesting for terrestrial fauna from being sea-washed by every storm (e.g. Bawden Rocks off Cornwall: Plate 1.2).

How Many British Islands Are There?

Berry (2009) quotes a maximum possible figure of 5000 British islands, but accepts that a more realistic figure lies closer to 500 (McCormick 1974a, b, c) and lists 189 in a table; some of these are lumped together as one (e.g. Colonsay and Oronsay) and others are subsumed under an

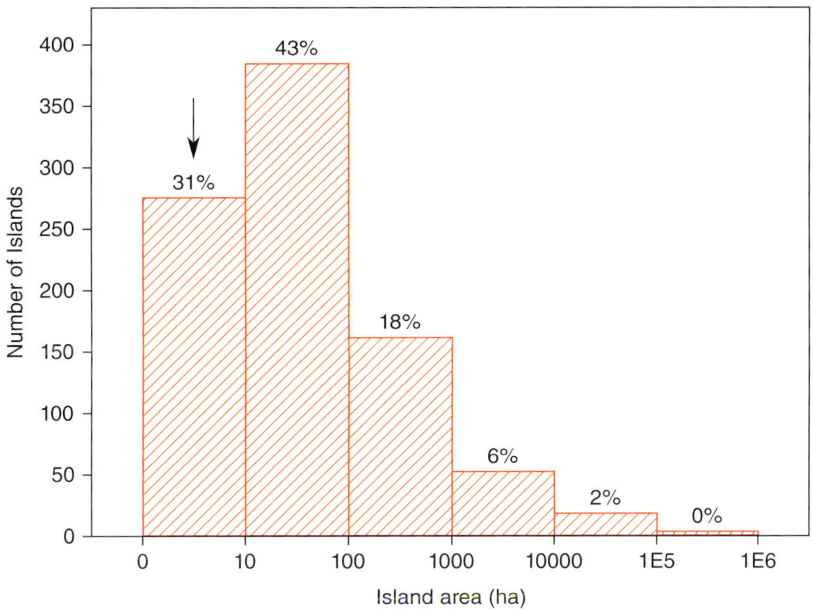

Fig. 2.1. Size distribution of British islands (*X* axis $\log_{10}$ scale). As fewer islands < 10 ha are included, the lowest class size is substantially unrepresented (vertical arrow).

inclusive label (e.g. Strangford Lough: > 90 islands occur on the 1:25,000 OSNI Activity Map for Strangford Lough) (OSNI, 2008). As it is, the Isles of Scilly alone comprise more than 140 islands and rocky islets. Owing to variable access it is inevitable that records for butterflies and other organisms are biased to larger islands; we consider the impact this may have on our understanding of island biogeography. In our study we have listed as many islands as could be located that are over 10 ha in area (including the Channel Islands off France), altogether numbering over 900 islands (see Appendix 3).

The British Islands: Composition, Context and Complexity

Not only does each species occurring on an island have distinctive attributes, so do the islands themselves. From the previous chapter it is evident that they differ in size and degree of isolation from the mainland and whether they form part of archipelagos. They also vary in position around the coasts of Britain and Ireland (see Fig. App. 3.1) and thus in climate and exposure to ambient conditions. In addition, they contrast dramatically in their composition and structure, relief, formation and cover (vegetation). Islands may be formed largely from **endogenetic processes** (i.e. extrusive volcanicity as oceanic hot spots; volcanic intrusions; plate tectonics movements as **terranes**) or **exogenetic processes** (i.e. marine, fluvial or glacial deposition and/or erosion; human and biotic agents)

(Dennis, 2010) but many have elements of both (for example, Lindisfarne is formed of the dolerite Whin Sill and wind-blown sand dunes). They may be of consolidated or unconsolidated rocks, the latter typically occurring as a superficial cover (glacial, fluvio-glacial, marine, fluvial deposits) over the solid geology. Classic examples of islands formed by deposition are the **drumlin** islands of Strangford Lough (Vernon, 1966) and those created of fluvial sands and gravels, such as the impermanent and shifting **eyots** (aits) of Whitton Island and Read's Island in the Humber estuary. Typical examples of islands formed by erosion are sea stacks left from the marine denudation of a headland; for example: the 30 m periclase and Magnesian limestone Marsden Rock, Tyne and Wear; the chalk stacks of Old Harry Rocks, Dorset (Plate 1.4); and the Needles of the Isle of Wight.

The most distinctive feature of British islands is that, depending on location, compositionally they largely mirror the south-east to north-west bands of solid geology from the London and Hampshire Basins (i.e. gently inclined Tertiary marine beds) to the Outer Hebrides (contorted Pre-Cambrian Lewisian gneiss and Torridonian sandstones); rocks forming the British Isles and its islands increase in age, formation, compaction, resistance to denudation and complexity in a north-west direction (Toghill, 2009; British Geological Survey website, undated). In many ways, this explains the general dearth of islands off the coasts of southern and eastern Britain – including their limited altitude and diversity – and the multitude of islands off the coasts of north and west Britain and Ireland, the latter made up of durable rock types and complex structures (i.e. folding, faulting), resistant to constant wave action and the high energy storms of Atlantic gales. Some islands have highly diverse lithology, comprising metamorphic, igneous and sedimentary bed rock (e.g. Anglesey; Arran, Skye) from different eras, periods and epochs; this is likely in turn to be mirrored by contrasting soils, vegetation cover and land uses. A most important distinction in island composition is based on the presence or absence of basic minerals, particularly of lime rich rocks (e.g. Jurassic Oolitic limestones, Carboniferous limestone, Cretaceous chalk), as this greatly affects the presence of particular plant species, including host plants of some notable butterflies (e.g. *Polyommatus coridon*, *Polyommatus bellargus*, *Cupido minimus*) (Dennis, 2010). Most islands comprising limestones occur off the English and Welsh coasts, but exceptions are the Outer Hebridean islands whose western shores comprise the butterfly-rich machair (shell sands).

Islands also vary enormously for their human populations (population numbers and density) and consequently the degree of 'habitat' disturbance (Berry, 2009). This feature, as with geological diversity, tends to follow island size (i.e. correlation of human population numbers with island size, **Spearman** $\boldsymbol{r_s}$ = 0.67, n = 461; population density with island area, r_s = 0.53, n = 459; both $P < 0.0001$), but also climatic conditions. Perhaps the most prominent feature of British islands is how they vary in climatic conditions. The British Isles have steep and distinctive climatic gradients; later, it will be shown how this affects gradients in butterfly species over Britain and Ireland and distorts the usual expectations of island biogeography based on island size and isolation. Dramatic climatic gradients also affect the organisms of small islands, between their west- and east-facing aspects, as long known for the fossil or relict island, the Great Orme's Head (Dennis, 1977) (see Fig. 4.3).

In many ways, then, the British and Irish islands are not uniform and this adds enormously to the excitement of visiting them to study their wildlife. In recent geological times (late Quaternary) they have also had very different histories owing to glacial–interglacial cycles, especially during the nadir of the last major ice advance (**Devensian**) c. 30–17.5 ka BP. Indeed, the time of the Devensian ice advance provides a convenient platform from which to investigate current patterns of butterfly occupancy on the islands, as the plants and animals on most islands then experienced tabula rasa (Dennis, 1977, 1992, 1993). It will also become evident that, owing to glaciation, the islands differ in status; some are land bridge islands (e.g. Britain; Ireland), whereas others have effectively risen above the waves much like oceanic islands (e.g. Tiree) (Mather *et al.*, 1974).

Glacial–Interglacial Cycles: Ice Landscapes

The tabula rasa of **biota** on mid-latitude islands during glacial–interglacial cycles has a number of potential causes: (i) long-term snow and ice cover and the gradual depletion of energy reserves for development and growth in vulnerable immobile stages (Høye et al., 2014); (ii) erosion and maceration of soils and bedrock by the ice, and consequently the destruction of biota; (iii) the burial, compression and disintegration of biota under deep till and fluvio-glacial outwash deposits; (iv) the erosion and wastage of biota during deglacial floods and mass wasting (solifluction); (v) flooding of populations in areas subjected to relative rises in sea level and by proglacial lakes (Shropshire's Lake Lapworth; southern **Doggerland**) (Clark et al. 2010); and (vi) the extinction of populations beyond the immediate zones of ice cover and deglacial flooding by a severe periglacial climate. A periglacial (**permafrost**) climate not only impacts on adult butterflies through the hazard of severe frost but prevents development and activity of stages, owing to lack of sufficient warmth and through the elimination of key resources (i.e. larval host plants; nectar flowers).

During the **Quaternary** (from 2.5 million years ago), there have been some 46 glacial–interglacial cycles (Ruddiman *et al.*, 1986; for details see Dennis, 1993), but only eight major cycles during the past 800 ka (Williams *et al.*, 1998; Lee *et al.*, 2011); typically, the major interglacial stages are short (e.g. Ipswichian 130–116 ka BP) and the major glacial stages an order of magnitude longer. The evidence for this has accumulated from stratigraphic studies, particularly from ice cores (Alley *et al.*, 1993) and sea floor sediments ($\delta^{18}O$ record), the latter leading to the labelling of stages in terms of marine isotopic stages (MIS) numbers, the **Holocene** being MIS1 (Roberts, 2014). On land, stage identification is far more difficult, much of it necessarily determined from painstaking investigations of **glacial landforms** and **fluvio-glacial landforms** (BRITICE database: Clark *et al.*, 2010), including sediments interleaved with interglacial layers characterized by biota with distinctive micro-fossil and macro-fossil markers (e.g. Coleoptera; Trichoptera; molluscs; pollen and plant remains: Godwin, 1975; Lowe and Walker 1997; Elias, 2001; Roberts, 2014). More recently, longer terrestrial records have become available (e.g. **loess** deposits: Porter and Zhisheng, 1995; pollen sequences: Tzedakis, 1993). Stratigraphic levels are typically dated by calibrated radiometric techniques (e.g. ^{14}C dating: Fairbanks *et al.*, 2005; Bronk Ramsey *et al.*, 2012) from the Devensian maximum to the present day. The calibration uses distinctive annual markers such as overlapping seasonal layers of tree rings (dendrochronology) (Becker and Kramer, 1991; Stokes and Smiley, 1996) and annually accumulating sediments (e.g. **varves**) (Sturm, 1979).

During the last glaciation (Devensian > 116–17.5 ka BP; nadir between 30 and 18 ka BP) (Bowen *et al.*, 2002; Wilson and Lord, 2014), the Fennoscandian (FIS), British and Irish (BIIS) ice sheets were confluent (Sejrup *et al.*, 2009) and likely to have had much the same impact on the biota of the British Isles as any glacial stage preceding it (Wright *et al.*, 1993; Graham *et al.*, 2011). The combined ice sheet is now known to have been twice the size (0.72 million km^2) (Clark *et al.*, 2010; Chiverrell and Thomas, 2010; Everest *et al.*, 2013) than when reconstructed in the 1970s (Bowen *et al.*, 1986) and when the origin sequence of British butterflies was last investigated (Dennis, 1977, 1992, 1993). The whole of highland Britain north of the Severn estuary, Wales and Ireland was inundated by ice, which then extended far offshore of the current coastline to beyond the continental shelf margin much as now in western Antarctica (Wilson *et al.*, 2002; Praeg *et al.*, 2015); the maximum extent occurred at slightly different times in different regions (Clark *et al.*, 2010) (Fig. 2.2). Apart from Scandinavian ice affecting the entire current eastern seaboard and Doggerland, a number of major ice caps have been identified (e.g. Scottish Highlands, Lake District and North Wales mountains), with recent additions in the south-west Peninsula (i.e. Dartmoor, Exmoor) (Patton *et al.*, 2015; Harrison *et al.*, 2015). It is highly likely from the nature of current winter snow accumulation over the southern Pennines High Peak that this region, too, was extensively covered by ice, but much of the evidence has likely been lost with erosion by deglacial meltwater (Burek, 1978) and further investigation is required. In the subsequent temporary climatic downturn of the **Younger Dryas** (11 ka ^{14}C BP or 12.9–11.7 ka cal BP), an ice sheet returned in the Scottish Highlands (i.e. the **Loch Lomond Readvance**) and corrie glaciers once again became active in upland regions of the Lake District and Wales (Roberts, 2014; Hughes *et al.*, 2012). During the Devensian, the ice is calculated to have reached thicknesses of > 1100 m over the Scottish Highlands (Boulton *et al.*, 1991) and depths of > 400 m in the lowlands (e.g. mid Cheshire) (Delaney, 2003). All the British and Irish islands covered by this ice experienced a tabula rasa dating from 17.5 ka BP.

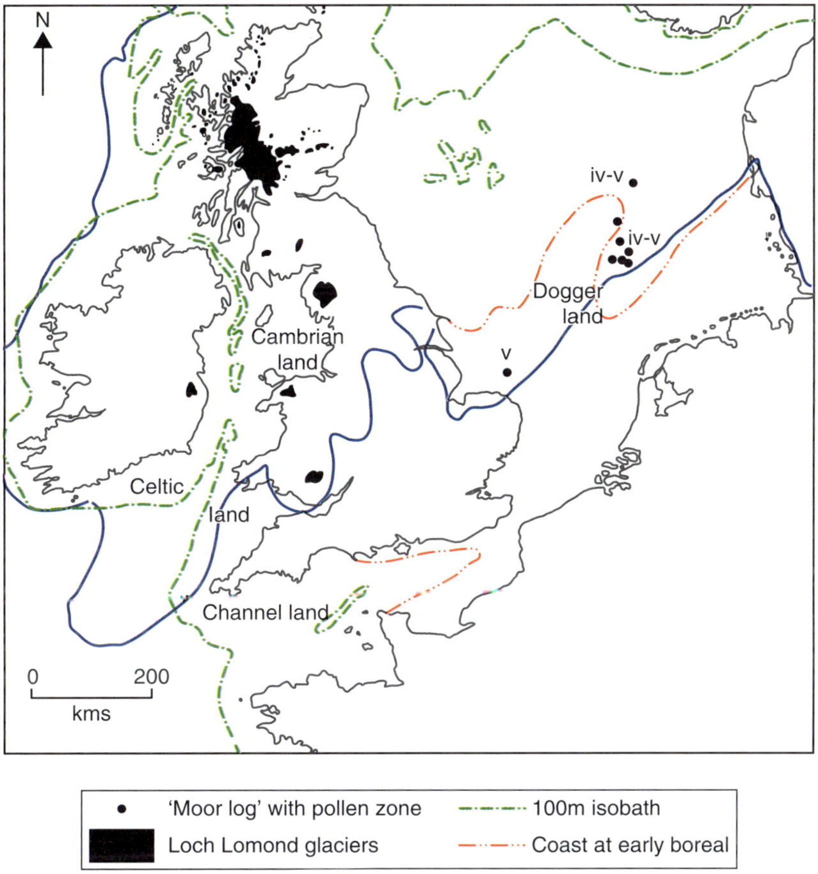

Fig. 2.2. Key features of the Devensian landscape over the British islands. Devensian glacial maximum limit (blue line); current 100 m isobaths (green line with alternating dash and dot); Loch Lomond glaciers (black areas); coast in early Holocene (red line with alternating dash and two dots); examples of moor log sites (black circles). (Redrawn from Dennis, 1992, courtesy of Oxford University Press, with Devensian limit adapted from Clark *et al.*, 2010 (*Quaternary Science Reviews* 44, p. 131, Figure 15) from the BRITICE database with kind permission of the authors and Elsevier Ltd.)

Shifting Sea Levels, Coastal Physiognomy and Island Formation

Changing sea levels throughout the late Quaternary present some of the most complex issues in reconstructing past landscapes, but acquire huge importance in determining the palaeogeography of British coastlines, specifically the formation (occurrence), timing and life span of: (i) land bridges; (ii) islands and archipelagos; (iii) potential refugia; (iv) changing land dimensions; and (v) environmental changes in developing coastal margins (e.g. dune, fen and salt marsh formation; raised beaches; transformation of river networks) (Gaffney *et al.*, 2009). An immediate point should be emphasized: the need of land bridges for the colonization of the British islands by butterflies was initially questioned in the 1970s by the first author (Dennis, 1977) and we now know a great deal more about butterfly dispersal and migration (Dennis and Shreeve, 1996; Dennis *et al.*, 1998a, b; Cook *et al.*, 2001) (see Chapter 3). It is likely that all current British butterflies are capable of crossing water barriers; the important question is how wide does a sea strait or river channel have to be to prevent crossing over a specific period of time? In fact, an organism's ability to cross barriers is likely to have a substantial **variance**; the ability of species to cross barriers is not just a matter of distance and flight endurance (which will vary among individuals), but will depend on the numbers of individuals (potential mass of migrants) attempting the crossing (Tilley and Dennis, 2017) (see Chapter 6). There is every reason to consider that butterfly populations increased dramatically with deglaciation and climate-induced range extension in the **Late Glacial** and Holocene, enhancing the ability of species to colonize outlying areas of Britain and Ireland (Dennis, 1977) (see Chapter 8).

Even so, a crucial part of island colonization depends on the extent and timing of sea barriers, as dimensions of sea barriers influence probability of transfer and subsequent success of colonization. At the empirical level of data acquired from field research, difficulties arise not only because of the modification of submarine contours (sediment aggradation, erosion, compression) (Marsset *et al.*, 1999) but owing to uncertainty of estimates for earth crustal instability (i.e. warping, tectonic movements), of ice mass and rates of isostatic adjustments (Tooley, 1978). Relative sea level rise and fall during the Devensian and Holocene has two basic components: (i) **eustacy** – sea level (volume/area) gains and losses related to ice accumulation and wastage, and outflow from pro-glacial lakes; and (ii) **isostacy** – rise and fall of the land surface in response to increased or reduced loading (by ice, water). Thus, in the vicinity of ice sheets, relative sea level is not only expected to fall as global ice builds on land areas, but the land itself is expected to be depressed with the local increase in load of ice (Fig. 2.3a). In fact, close to the ice centre, relative sea levels will be found to fall during deglaciation, rather than rise, if rebound from land depression by ice exceeds the rate of eustatic sea level recovery. If this is not complicated enough, there is likely also to be rise of the sea floor in the vicinity of ice sheet margins during glacial advances owing to: (i) reduction in load with the eustatic fall of sea level; and (ii) a developing fore-bulge from compensatory shifts within the crust and mantle (**asthenosphere**) to nearby ice sheet loading. The latter factor has long been speculative (Devoy, 1995), but firm evidence is gathering that confirms fore-bulge development off Britain (Tooley, 1978; Innes *et al.*, 2004; Roberts *et al.*, 2006; Goslin *et al.*, 2015). More certain are maximum determinations of Last Glacial Maximum eustatic falls in sea level of at least 120 m (Siddall *et al.*, 2003) (134 m in Lambeck *et al.*, 2014).

Relative sea level curves for Britain and Ireland have been accumulating since the 1960s (Tooley, 1978). These show characteristic patterns

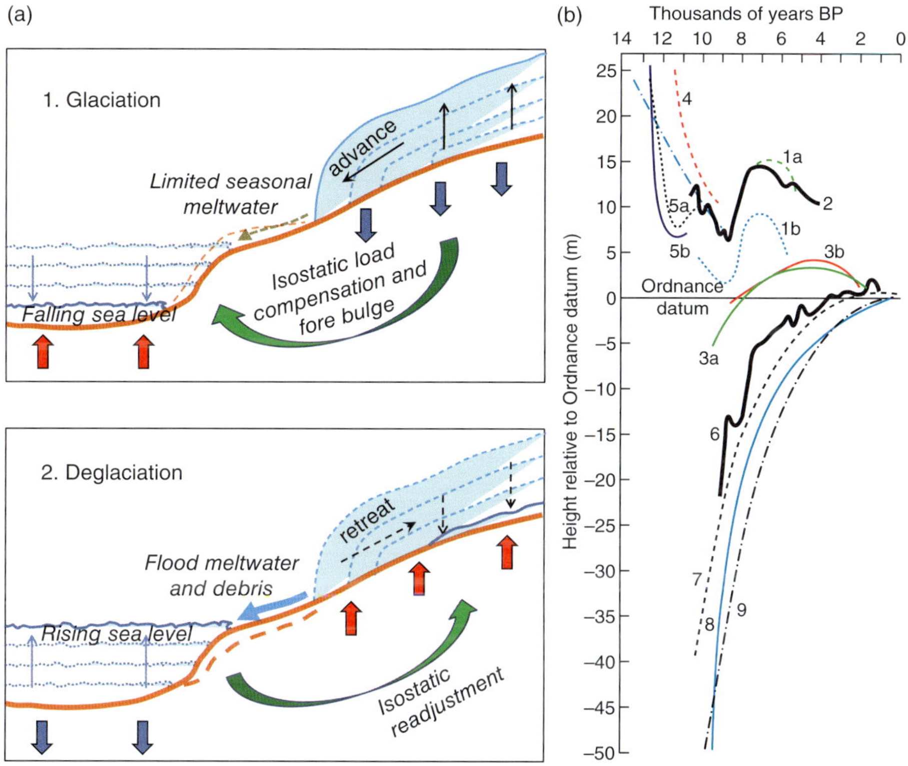

Fig. 2.3. The impact of glaciations on sea levels. (a) Simple model of relative sea level changes in the vicinity of ice sheets: 1, glaciation and ice build up; 2, deglaciation and ice melt. (b) Changes in sea level from the Late Glacial (14 ka BP) to the present around the British coastline. (Redrawn from Dennis, 1992, p. 226 Fig. 10.3, courtesy of Oxford University Press.) 1, 2, 4 and 5: north of the Central Valley of Scotland; 3: Solway Firth; 6: north-west England; 7 and 8: southern England; 9: south-west Ireland.

that follow expectations of changes over time. Basically, shorelines in northern Britain close to the ice centres confirm a net decline in relative sea level (elevation of land from isostatic recovery is relatively greater than eustatic rise in sea levels) during the Holocene (Smith *et al.*, 2012), whereas those for southern Britain register the reverse: a gradual rise in relative sea level (Tooley, 1978). Complexity typifies sea level curves near the ice centres and margins (e.g. vicinity of the Isle of Man), reflecting the changes in relative rates of land and sea level recovery following the wastage and periodic re-advances of global and local ice loads (Fig. 2.3b). Central Scotland is still rising at 1.92 mm/year whereas southern Britain is undergoing tectonic downwarping of 1.0–1.5 mm/year (Tooley, 1991). A shifting **isobase** (instantaneous mean sea level) has occurred over the Holocene millennia; north of Morecambe Bay area, raised beaches are evidence of higher relative sea levels than currently during the Holocene (Lambeck, 1995; Smith *et al.*, 2006; Brooks *et al.*, 2011). A basic problem is that empirical data are limited to the Holocene, but from these data an enormous amount has been gleaned, particularly: (i) the severing of the land bridge between Britain and the continent (Roberts, 2014); (ii) the palaeogeography of Doggerland – its occupancy during the Mesolithic (Gaffney *et al.*, 2009) – and other potential refuges during the Holocene (Bennett *et al.*, 1991); and (iii) the speed of coastal changes, including the occurrence and timing of extreme events (e.g. the discharge of Lakes Agassiz–Ojibway in North America 8470 years BP: Barber *et al.*, 1999; the **Storegga tsunami** on the Norwegian coast c. 8200 years ago: Weninger *et al.*, 2008; Smith *et al.*, 2012). However, ideally, to understand coastal developments, we require hard data on relative sea level changes since the Last Glacial Maximum. A major advance has been the detailed mapping of the North Sea and English Channel using benthic imaging and high-resolution sonar; this shows a detailed pattern of massive anastomosing channels and sediment banks (the palaeoriver system, the **Fleuve Manche**) (Bourillet *et al.*, 2006; Toucanne *et al.*, 2010), and sediment-infilled, deeply incised depressions (plunge pools) in the channel bedrock during successive glacial low sea levels, providing unprecedented insights into Devensian English Channel topography and conditions, as well as events (catastrophic flooding), leading to the formation of the English Channel itself (Gupta *et al.*, 2007, 2017).

More recently, increasingly sophisticated models have been built of relative sea level changes, coastal geography and sea floor topography (Devoy, 1995; Peltier *et al.*, 2002; Lambeck *et al.*, 2002; Lambeck, 2004). Maps of coastal palaeogeography now exist for the British and Irish Holocene at 1000-year intervals (Shennan *et al.*, 2006) and 500-year intervals (Sturt *et al.*, 2013). As the authors admit, these are still flawed owing to uncertainty over a number of parameters (e.g. modification of sea floor sediments, fore-bulge dynamics; Lekens *et al.*, 2009). However, the models have built important insights into island and archipelago formation (Sturt *et al.*, 2013). For instance, in the early Holocene, the Outer Hebrides were considerably larger than they are now and adopted their current shape after 6 k years BP; this is a very different history from that for the Inner Hebrides (e.g. Skye) (Selby and Smith, 2015) with high sea levels in the early Holocene. Britain is known to have separated from the continent c. 8 ka BP, after which Doggerland was finally submerged. Subsequently, island groups began to become isolated and fragmented (e.g. Isles of Scilly; Orkney islands); and between 8 ka and 6 ka BP low-lying coastal areas experienced inundation with marine transgressions (e.g. Fenland; Somerset and Humber levels) (Coles and Hall, 1998). There is firm evidence that the Menai Straits were active by 5.6 ka and 4.8 ka cal BP (Roberts, 2006). With the gradual assumption of current sea levels and storm wave action after 4 ka BP, the British and Irish coastlines have continued to experience modifications to the present day; then, the Isles of Scilly were finally transformed into their current cluster of islands and rocky islets. Marine processes have continued to have an impact during the last 1000 years of history; the theme has mainly been of island erosion (attenuation), but some islands have grown with spit and dune formation (Selby island) and a few have become linked to the mainland by tombolos (e.g. Great Orme's Head in North Wales; Isle of Portland) and beach accretion (e.g. Llanddwyn Island, off Anglesey).

One crucial question, the occurrence and timing of a Late Glacial and Holocene land bridge between the continent (and/or Britain) and Ireland, is still open. If the revision of Last Glacial Maximum limits, showing the entire inundation of Ireland by ice, is correct (Chiverrell and Thomas, 2010; Clark *et al.*, 2010), then most of the mammals found there must date to the Late Glacial and early Holocene (introductions). Currently, there is no firm geological evidence for a Holocene land bridge to Ireland, but biogeographical analyses of the Irish mammal fauna would point to a post-ice maximum land bridge lasting until 15 ka BP (Montgomery *et al.*, 2014). Cold-tolerant mammal species that arrived soon after the Last Glacial Maximum succumbed with the rise of temperatures in the Late Glacial; the less cold-tolerant species were lost during the Younger Dryas cold period (i.e. Loch Lomond Readvance) or soon after, with restrictions of suitable habitat conditions. Species colonizing from eastern Europe during the Holocene were prevented from crossing under their own volition to Ireland by the Irish and Celtic seas, but many mammals were introduced intentionally or inadvertently by humans. Thus, both Britain and Ireland are land bridge islands, but unlike true land bridge islands, with an initial healthy complement of species, both started with **residual** cold-tolerant biota and then experienced stringent filtering by a succession of warmer, colder and finally warmer conditions again as the ice receded and sea levels rose and closed off further access.

The situation in the Isle of Man remains equally complex and requires more research. As the island was overrun by the British–Irish ice sheet – and then subjected to higher sea levels than today from isostatic depression (Roberts *et al.*, 2006) – and the earliest specimen of *Megaloceros giganteus* there has a date of 12,920 ± 120 years cal BP, any land bridge must predate the Loch Lomond Readvance but must have been soon closed during the first millennium of the Holocene (Innes *et al.*, 2004; Roberts, 2006); the submarine topography suggests that the temporary link ran towards the Cumbrian coastline. In northern Britain, near the ice centre, more islands started as bare plates of rock, wiped clean first by ice and then by high seas, subsequently growing in size with isostatic rebound from the removal of overlying ice (e.g. Tiree) (Mather *et al.*, 1974).

A prominent feature to emerge from the changes in relative sea level and long-term coastal processes is that not only does the butterfly fauna of islands change over time, but also so does the very configuration of islands themselves: they change in size, shape and isolation from or connection to the mainland; some have fragmented into islets, a number have emerged from the sea and ice and others have disappeared or have been attenuated by erosion. Long-term consideration of islands has to take account of these changes to islands, more of which become apparent when changes to climate and vegetation are considered under the influences of human exploitation.

Climate Change During Island Formation

Tabula rasa on islands is not just caused by ice and water, but by changes in climate and, concomitantly, vegetation – thus butterfly resources such as larval host plants and nectar flowers (Dennis, 2010). The glacial/interglacial stages were associated with, in fact triggered by, massive climate changes; it is the long cold periods that were so inimical (Dennis, 1977). All in all, two critical episodes severely curtailed British butterflies: (i) the Devensian Maximum Glaciation; and (ii) the Loch Lomond Readvance (Younger Dryas) (see the integrative model in Fig. 8.2). Later, other downturns in the climate (e.g. 4–0.2 ka ^{14}C BP) must also have had a significant impact on butterfly populations and survival of species, not just regionally but over the entire set of islands (Dennis, 1993). This is evident for the simple reason that it is well known that much smaller adverse shifts in climate have affected butterfly populations, distributions and ranges over very short periods in recent history (Dennis, 1977, 1992, 1993); these observations are supported by the current impact of changes with climate warming detected in both *Butterflies for the New Millennium Atlas* records and Butterfly Monitoring Scheme data (Warren *et al.*, 2001).

The relationship between butterflies and climate can be assessed during history owing to the long period of climate records held by Britain. Temperatures have been recorded since 1659 (Manley, 1974), precipitation since 1766 (Wigley *et al.*, 1984), pressure and winds since 1864 (Lamb and Johnson, 1966). We also know that there is good association of records across Britain, between Edinburgh and central England over the period from 1764 and 1896 (Duncan, 1991). The relationship between the status of Lepidoptera populations and climate variables over this period has also been examined by Beirne (1955) and Dennis and Sparks (2007). Beyond these dates into history and prehistory, reconstruction of weather and climate changes depends on proxy variables (e.g. ice, benthic foraminifera, pollen, deuterium changes in wood cores, calcite deposits, molluscs, lichens, crop yields, etc.) (Roberts 2014; Armstrong and Bradwell, 2010). The key organisms that have provided climatic curves for Britain during the entire Devensian and Holocene are beetles (Coope in Dennis, 1977, Figure 5; Coope, 1979; Atkinson *et al.* 1987); it was shown that even beetle species of the north-east Siberian wastes were lost from Britain during the nadir of the Devensian glacial maximum advance (Coope, 1987).

Climate factors most inimical to butterfly survival are those impinging on the adult reproductive stage, including low temperatures, low insolation levels and wet conditions. Adult butterflies are ectotherms; they heat up by basking (Dennis, 1993) and are thus adversely affected by any factor that reduces radiation (i.e. cloud cover, short day length) and temperatures (i.e. air mass characteristics, wind speed, cloud, wetting). It has long been known that a drop in summer temperatures, and increased cloud and rainfall, are major factors in reducing butterfly populations (Beirne, 1955; Dennis, 1993; Pollard and Yates, 1993). Lepidoptera generally cope with cold winters (Beirne, 1955) but during the Late Devensian Maximum winter temperatures fell to as low as –50°C. However, it is the summer mean temperatures (8°C) during the Late Devensian that would have been most damaging, crucially the hazard of frequent, intermittent severe frosts; in fact, biota were subject to a regular pattern of sharp climatic switches (i.e. **Dansgaard-Oeschger (D-O) oscillations**, coinciding with **Heinrich events** from iceberg discharges) (Broecker, 1994; Hendy *et al.*, 2002) effecting massive instability at 1500-year intervals. Conditions were a little better during the Loch Lomond Readvance than the Devensian maximum glaciation (July mean temperature of c. 10°C in the Midlands) (Coope *et al.*, 1998) but would still have imposed mass extinctions.

Recent studies of Eurasian and North American butterflies support the view that a number of current species may have occupied southern Britain in the Younger Dryas (Dennis, 1993); evidence comes from field observations in Zackenberg at 74° N in Greenland (e.g. *Lycaena phlaeas*; Høye *et al.*, 2014), from Herschel Island (Leung, 2008; Burn and Zhang, 2009) and from islands in the White Sea and Siberian wastes (Bolotov, 2006; Chernov and Tatarinov, 2006). Moreover, some species, not known for long-distance migrations, have considerable capacity to disperse huge distances (e.g. *Pieris napi*) (Coulson, 2015). As **cuestas** with south-facing scarp slopes provide significantly higher temperatures (Rorison *et al.*, 1986; Bennie *et al.*, 2008), the probability for particular species occurring during the Younger Dryas on scarps and cliffs in Late Glacial Britain is increased. It is important to appreciate that substantial uncertainties still abound as to extreme conditions during the Loch Lomond Readvance; the beetle fauna indicate that conditions were dry, but this is typical of anticyclonic conditions and severe frosts, inimical to butterflies. A further consideration is that the shorter days at lower (temperate) latitudes were offset by a steeper sun angle, a higher maximum radiation load and longer snow-free period. Of course, climate alone does not restrict species, but subtle aspects of the quality and condition of resources (see Chapter 3) which may also have been impacted on by frequent severe frosts. An additional issue is the extent to which populations surviving the Younger Dryas would have been supplemented by mass immigration during the early decades of the Holocene (see below).

The absolute baseline for occupancy of British and Irish islands by any of the current British butterflies, then, outside areas covered by ice or water, was almost certainly 18 ka BP. The land was under sparse tundra and severe permafrost, and ice/water-free refuges (Channel Land; land in the current southern Irish Sea) were undoubtedly subjected to long periods of summer 'white-outs' and extreme frosts. The severe conditions in place would have forced suitable refuges for butterflies now occupying northern Europe mainly to the south of the Alpine chain, though some **cryptic refugia** of butterflies (i.e. *Erebia* species), as for trees, may well have occurred north of the fold mountains (Dennis *et al.*, 1991; Dennis, 1993; Provan and Bennett, 2008). As severe conditions returned with tho Loch Lomond Readvance, it is likely that the vast majority of western and northern British and Irish islands remained continuously unoccupied by current butterfly species before the outset of the Holocene (10 ka ^{14}C BP or 11.7 ka cal BP) (Roberts, 2014). This subject is explored more deeply in Chapter 8.

Post-Glacial Transformations to Island Landscapes

Although changes in Quaternary vegetation reflect changes in climate, there was understandably a lag in colonization owing to immigration of seeds from distant refuges and the development of soils (Ammann, 2000). Such changes describe a recurring pattern to environments through glacial–interglacial cycles. From studies in Denmark, Iversen (1958; see also Andersen, 1978; Birks, 1986) produced an appealing model of the pattern in vegetation and soil changes (Fig. 2.4). It illustrates ongoing vegetation succession and soil development to a forest **climatic climax** with deglaciation and climatic warming, followed by retrogressive succession as the climate once again cools and soil nutrient status declines (Dennis, 1992 (Fig. 10.4), 2010). Using widespread data on fossil pollen and macroscopic plant remains, it has been possible to build up the pattern of changing vegetation types over Europe through the millennia following the Devensian maximum glaciation (Huntley and Birks, 1983). Prior to the Loch Lomond Readvance, vegetation succession – starting with herbs (grasses, sedges), then shrubs (juniper, willow, crowberry) – had reached a stage of extensive tree cover, mostly birch, by the Late Glacial optimum (**Windermere interstadial**; Huntley and Birks, 1983; Fyfe *et al.*, 2009; Roberts, 2014). This was halted and reversed by the Loch Lomond Readvance, during which vegetation patterns reflected contrasts

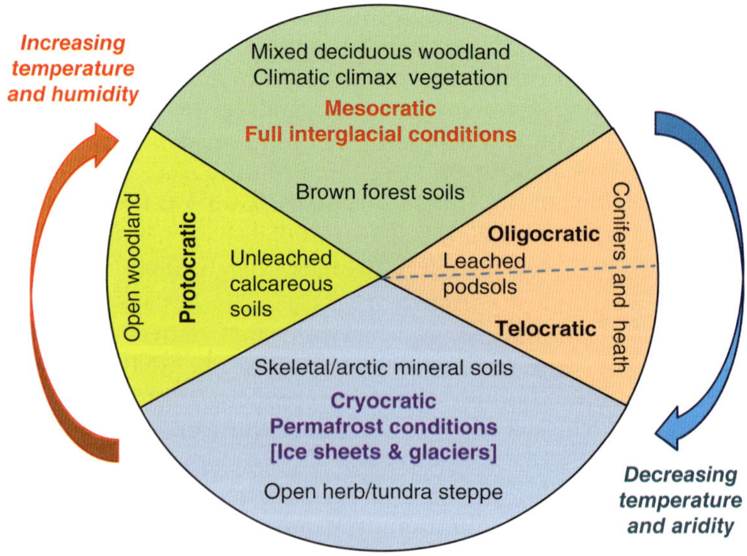

Fig. 2.4. Model of changing north European environments during a glacial–interglacial cycle.

in location and topography (i.e. aspect, slope) (Parker and Goudie, 2007). With the beginning of the Holocene began the runaway process to complete forest cover. The mid-Holocene altitudinal tree limit exceeded 600 m (Taylor, 1980); in places it reached higher levels (750–900 m; Turner and Hodgson, 1979; Turner, 1984). Bennett (1989) reconstructed the regional variation in forest tree species dominants (climatic climax vegetation) at the zenith of Holocene vegetation formation for different parts of Britain; though the scale is too coarse to illustrate open spaces on scarps and flood plains, it nevertheless impresses on us the dense forest cover 5000 years ago, which the first author (Dennis, 1977) considered to be a major isolating mechanism for many butterfly populations at that time.

It is no easy matter to gauge the impact of Late Glacial and Holocene vegetation changes on biota, especially elements lacking a detailed fossil record such as butterflies, as vegetation types may not have any modern equivalents. However, during much of the Holocene, the key control over butterfly populations would have been the degree of forest cover and open ground. Forest reached a peak between 9 ka to 5 ka ^{14}C BP and thereafter declined as human populations increased and technological innovations aided deforestation, a process at least dating from the Neolithic period and accelerating from the Norman Conquest to the modern day (Dennis, 1993). The full impact of this change on the butterfly fauna has long been well appreciated (Dennis, 1977). Although over large areas in Britain and Ireland the mid-Holocene dense forest would have been unsuitable for butterflies of open biotopes (some 39% species) (Dennis, 1993, p. 150), there were probably always open spaces caused by tree senescence and fall, lightning, fires and river floods, mass movements and the activity of large mammals (Simmons and Tooley, 1981; Grant *et al.*, 2014). One long-lived, special open space developing out of glacial erosion and outwash lakes (i.e. over-deepened depressions; **kettle holes**; ribbon lakes; **pro-glacial lakes**) is the **raised bog**, found throughout northern Britain (e.g. Whixall Moss, Shropshire; Lindow Moss, Cheshire); some species must have persisted on them throughout the Holocene (e.g. *Coenonympha tullia*) (Dennis, 1992, 1993). Larger open spaces would have existed at higher latitudes in the mountains and uplands, inland along scars and scarps, and along coastal cliffs, dunes, estuarine salt marshes and on exposed islands. In fact, owing to exposure to salt spray and higher winds, tree cover on exposed small offshore islands is more likely to have been poorly developed (i.e. **krummholz** in form) than over Ireland and Britain. Open areas would have presented a shifting patchwork for the expansive metapopulations of species dependent on such biotopes (e.g. Isles of Scilly). The key influence was the gradual opening up of the forest by humans with early beginnings in the **Mesolithic** (Simmons and Tooley, 1981). Altogether, changes in climate, rising sea level, the impact of diverse substrates and, particularly, variation in land exploitation from the **Neolithic** onwards, gradually produced an expansion of an enlarged open patchwork. The consequence has been a variety of biotopes (so-called habitats, i.e. wet and dry heath, fen, moss, moor, meadows, downland herb-rich slopes, salt marsh) (Lake *et al.*, 2015) emerging amid the shifting patterns in mid-Holocene forest cover, each accompanied by different arrays of butterfly resources (Dennis, 2010).

Historical Imperatives

In Chapter 8, focus is brought to bear upon a range of historical issues influencing the present island butterfly faunas (see Fig. 8.2; and detailed treatment in Dennis, 1993). Contemporary island biogeography tends to give these less weight; nevertheless, Holocene environmental changes are crucial for understanding present features of butterfly island biogeography, particularly adaptations to conditions and contrasts in regional butterfly communities over the islands (Dapporto *et al.*, 2015). Here, it is helpful for those unfamiliar with palaeoecology to prepare expectations by touching on two issues.

The first involves a mismatch between the responses of butterflies and their resource base; the colonization history of plants, specifically the larval host plants and adult nectar plants of butterflies, differed from that of butterfly species. Most of the plants and butterflies would have entered during the Late Glacial but the survival of butterflies (**ectotherms**, dependent on ambient heat and sunlight), unlike plants, would have been adversely affected by the Younger Dryas downturn in climate. Those butterfly species unable to survive the return of arctic conditions would then have had to face fast-diminishing land connections and widening sea straits – very likely a lack of any connection to Ireland – the main link between Britain and the continent passing to the north of the English Channel in the southern part of Doggerland. Thus, species entering from Iberia either passed over the complex network of the Fleuve Manche in the Late Glacial, subsequently surviving the Loch Lomond Readvance, or crossed over the expanding sea barriers of the English Channel and Irish Sea during the early Holocene. Survival during the Loch Lomond Readvance would have depended on warmer frost-free local climates, as butterfly adults are particularly vulnerable to summer frosts, and with mean temperatures in the vicinity of 10°C combined with clear skies, severe frosts may well have occurred frequently. Interestingly, both mechanisms of entry into the British Islands – (i) Late Glacial arrival and survival of Younger Dryas arctic conditions; and (ii) Holocene overseas migration – would have triggered severe bottlenecks in population size, leading to distinctive genetic homogeneity. Isolation during severe conditions for > 1000 years of the Younger Dryas could result in novel adaptations and the emergence of distinctive populations dating from the Late Glacial, but in the case of these Late Glacial populations there is the question of the consequences of their coming into contact with an incursion of Holocene migrants.

A second issue is the transformation of island vegetation and butterfly resources during the millennia of the Holocene. Much as the main islands of Britain and Ireland, the small islands also have a long history of land use changes (Dennis, 1993, 2010). The occurrence of butterflies on islands depends ultimately on resources comprising their habitats, which in turn reflect on natural processes such as vegetation succession, dominant sea influences (i.e. sea spray), climatic conditions and animal activity (sea bird colonies). More prominently, resources would have been affected by human land use over a long history of exploitation, as often these islands were the first places of landfall for human migrants; today some 267 British islands are still inhabited (Wikipedia, 2017; Berry, 2009, lists 114). The pattern of vegetation on even the smallest islands (e.g. in Strangford Lough, Ireland) reveals long-standing human influences (settlements, fields and their boundaries, bridges, use of fertilizers and herbicides, etc.). As on the British mainland, species colonization and extinction dynamics on islands have depended on these influences (Hodgson *et al.*, 2014), on the direct impact on habitat resources, even though it is usually too difficult to quantify the effect of these influences for island biogeography studies.

The message from this brief review is that butterfly incidence on islands is a dynamic process. It not only depends on the changing physiognomy of islands themselves but also reflects on a long history of human interference with the substrates and vegetation cover (butterfly resources), continuing today all too evidently on the largest islands. Very few islands will have escaped past human interference in the Holocene (e.g. settlement on St Kilda: Plate 1.5; granite quarrying from Ailsa Craig: Plate 1.6). During what is now referred to as the **Anthropocene**, with broadcast chemical pollution these human impacts are likely to take precedence over basic geographical parameters and historical (Pleistocene) events (Feest *et al.*, 2014; Gilburn *et al.*, 2015; Jones *et al.*, 2017; Prosser *et al.*, 2016; Hall *et al.*, 2016). The consequence is that island faunas, if dependent on mainland sources, are under threat (see Chapter 5). It is ironic, as the first author has previously made clear (Dennis, 1977, 1992, 1993), that the spread of farming which initially benefited many species of open spaces, is now proving in its modern guise entirely inimical to much of the wildlife on the British and Irish islands.

3 Island Butterflies: Understanding and Interpreting Records

In early investigations of island biogeography, species were seen very much as binary units, each undifferentiated (equivalent) in status. As research has progressed, it has become increasingly evident that each species has unique attributes that affect its probability of colonizing an island and becoming extinct on it. Butterfly species found in the British islands are listed in Appendices 1, 2 and 3. In order to understand butterfly records on islands, a first task is to appreciate what is meant by a butterfly's habitat (i.e. the resource base for colonization) and migration (the ability to transfer to an island). This chapter also examines the main sources of records for British butterflies, the nature of island records, their status, and what they can tell us about each species on an island; the detail underlies any statistical appraisal of island records. A brief guide to making records (and processing data from records) on smaller offshore islands is provided in Appendix 8.

Habitat and Biotope

Without knowledge of a species, a record of its occurrence on an island is meaningless. What gives an island butterfly record value is: (i) the status of the butterfly, i.e. its taxonomy, phylogeny, life history, gender, behaviour, population ecology and biogeography, all of which define its breeding status; and (ii) its context on the island – the environment(s) it occupies, **biotopes** in which it is found or, to be more precise, its **habitat** and the resources it requires for survival. This information provides a basis for judging the fidelity of any species (record) to an island. No species exists in a vacuum, but in a habitat; the most useful island records collect data that disclose species' habitats. Recently, closer attention has been paid to the familiar term 'habitat'; for long it was equated with what is really biotope, the environmental substrate and vegetation unit occupied (viz. bare ground, cliff, bog, meadow, wood, etc.).

Although old habits die hard, and habitat is still a convenient term for substrates with distinctive communities of species (Dennis *et al.*, 2014), it has now become clear, if only for conservation purposes, that an individual species' habitat is something more specific: it comprises an assortment of vital resources for the species, distributed in the landscape, interconnected by movements of individuals. Some of the resources are plainly evident, some invisible, to most casual observers (Dennis, 2010; Dennis *et al.*, 2003, 2006a, 2014). For a butterfly, conspicuous resources include items such as larval host plants, nectar flowers, hibernation and mate location sites; less apparent, but equally important, are sites or areas varying in radiation receipt, temperature, moisture and nutrient status that affect the quality of what are called consumer and utility resources (Fig. 3.1). In particular, butterflies are ectotherms and need to bask in sunlight to become active (Shreeve, 1992a; Dennis, 1993). A collection (mathematical set) of suitable resource zones, then, makes up a habitat; unsuitable areas within, between or beyond resource zones are referred to as **matrix** (Dennis and Shreeve, 1996). This view of habitat has come to be called the **resource-based habitat** (Dennis *et al.*, 2003; Dennis, 2010); it differs profoundly from the term habitat as casually, but conveniently, used to refer to a substrate or vegetation unit (biotope) (Dennis *et al.*, 2014). In short, it might best be summed up by the observation that habitat has explicit species-specific implications for persistence, in effect an audit of resources, whereas biotope conveys but loose inferences (implicit notions of vegetation and substrates) about the composition of a community of organisms.

A butterfly's habitat is not something static, unchanging, but a variable; that is, it varies from one location (region) to another and over time, as with climate change. Thus habitat conditions are in persistent flux on islands, their degree of instability likely to be related to island geography (viz. latitude, longitude, size, isolation, etc.). A simple example is the use by butterflies of different larval host plants in different areas of Britain (e.g. *Callophrys rubi*) (Dennis, 1992, p. 16; Dennis, 2010). A species often presents different traits (i.e. wing expanse, larval colouring, development times, broods) in different areas in relationship to changing resources, an indication of adaptation to the changing nature of resources and environmental pressures (Dennis, 1977), the shifting nature of habitats, changes all too often imperceptible to casual human observers. Thus, at any location (island), habitats of different species vary in structure, described by the composition, physiognomy and connectivity of resources (Fig. 3.2) (Dennis *et al.*, 2014); listed in Dennis (2010), these resources are discussed in Emmet and Heath (1990) and Thomas and Lewington (2010). Significantly, too, habitats for any one species vary in structure over a sample of locations (i.e. on different islands). The consequence is that species' habitats differ in their level of resource integration and compaction

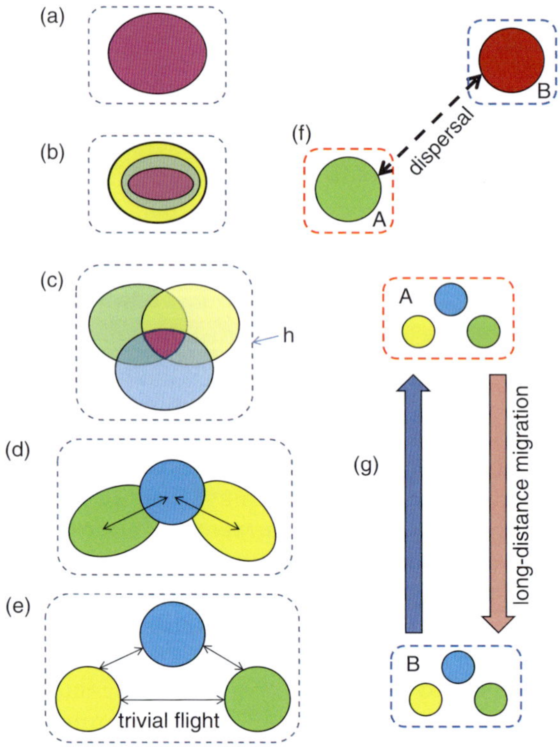

Fig. 3.1. The resource-based habitat. Basic variants of resource disposition are illustrated, each diagram illustrating just one spatial relationship between resources. (a) Equivalence and equality of resources comprising a habitat. (b) Nested intersection (equivalence, intersection and union) of resources comprising a habitat. (c) Intersection and union of resources comprising a habitat. (d) Contiguous union of resources making up a habitat. (e) Disjointed union of resources linked by 'trivial' (daily resource-seeking) flights making up a habitat. (f) Disjointed non-union (likelihood that resource spaces not used by same individuals in succeeding years) of resources linked by seasonal dispersal (movement across distinct biotopes), each part comprising a seasonal habitat. (g) Disjointed non-union of resources linked by seasonal long distance migration (movement between biomes and over large water bodies), each part comprising a seasonal habitat. Field examples of habitats will have combinations of multiple resource disposition. A maximum of three resources are illustrated in each diagram: nectar resource (blue); larval resource (green); roost sites (yellow); overwintering sites (brown); habitat boundary (h; pecked rounded squares, red summer and blue winter; habitat core purple).

(cf. habitat resource intersection scores) (Dennis *et al.*, 2014). Integration depends not just on resource intersection on the ground but also on movement, which determines connectivity in resource exploitation. Compaction is different; again, it depends on resource intersection but also on resource nestedness (much as Russian dolls) (Fig. 3.1b), resource density and the area occupied by the habitat (Dennis *et al.*, 2014). This view of habitats has vital consequences for a butterfly's existence on an

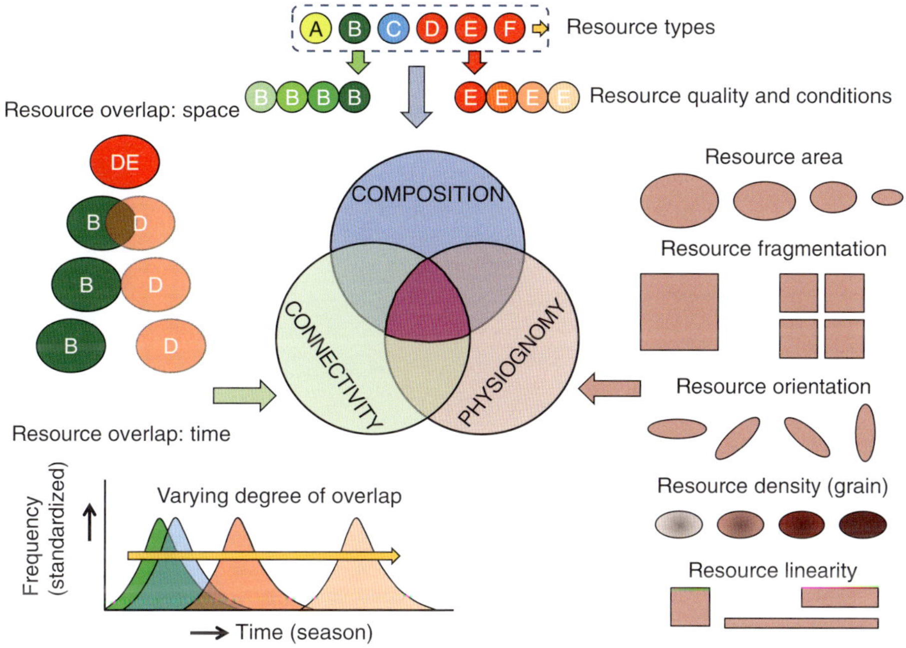

Fig. 3.2. Basic dimensionless model for habitat suitability based on three variables: resource composition, physiognomy and connectivity (Dennis *et al.*, 2006a). Each of these variables is described by a number of parameters. The purple zone where the three intersect can be regarded as the collection of conditions for which the functional habitat becomes increasingly maximized, although in fact the intersection of maximum values for the basic variables could impair function if it also increases competition and exposure to enemies. The functional habitat (resources all accessible to most individuals) depends on individual mobility in each species. Turlure (2009, p. 239) used a similar set of variables describing suitability of habitats: composition, availability and configuration; availability depends on individual resource type composition (subsets), physiognomy and connectivity, and configuration is described by resource physiognomy and connectivity.

island. From the accumulation of data on butterfly biology (Dennis, 1992, 1993, 2010; Dennis *et al.*, 2000a, 2004, 2012b) the basis for differences in butterfly habitats is now well understood; it is reflected conveniently (if still not sufficiently thoroughly) in two summary measures for British butterfly species: (i) their colonization ability; and (ii) their migration capacity (migration ability and opportunities) (Dennis *et al.*, 2012a; see also Appendix 7). Together these measures describe (and are explained by) the level of specialization (or **generalism**) in resource use among British butterflies (Fig. 3.3) (Dapporto and Dennis, 2013) and consequently their geography and vulnerability to changes in the British landscape (Dennis *et al.*, 2004, 2013; Shreeve and Dennis, 2011), including access to and persistence on islands (Dennis *et al.*, 2012a).

This section is left with this thought. Some have questioned our lack of knowledge on butterfly habitats; thus they may like to take up the

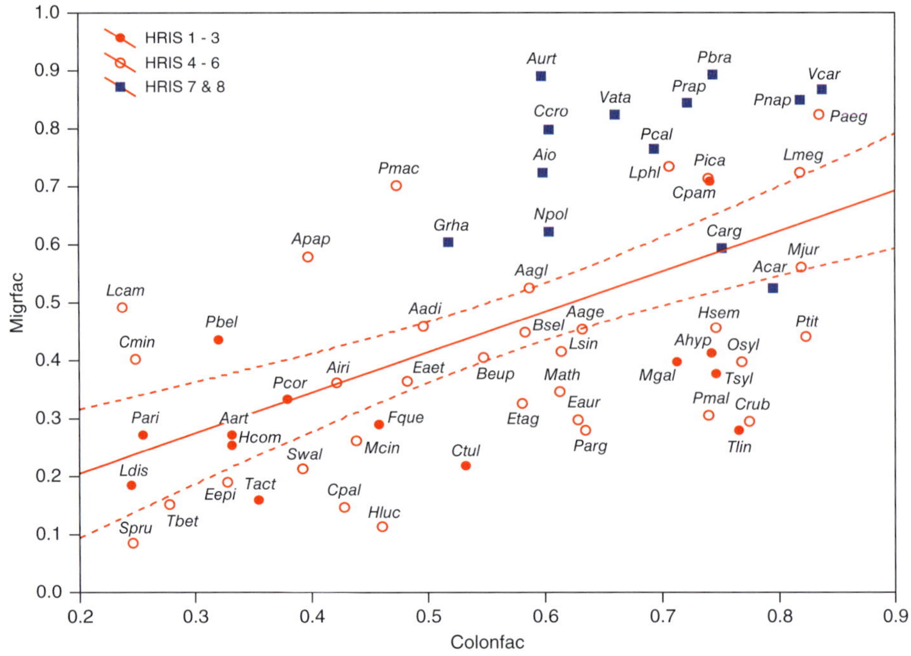

Fig. 3.3. Relationship of colonization ability and migration capacity in British butterflies. Colonfac (colonization ability) is based on ten variables and migrfac (migration capacity) on four variables (see Appendix 7 for original data and species codes). Points (species) are distinguished for habitat resource intersection scores (HRIS 1 highest, 8 lowest; see Dennis *et al.*, 2014, Fig. 5).

challenge: if you cannot explain why a species is absent from a biotope usually occupied by that particular species, it is clear you do not understand the resources and conditions that make up that species' habitat.

Habitat and Movement

An essential part of habitat, its functioning, is an organism's movement (see Appendix 6); movement by a butterfly larva and then the adult is crucial to consecutive resource use through a butterfly's life history and this defines an individual's **lifetime resource track** (Fig. 3.4) (Dennis *et al.*, 2014). But by far the greatest part of movements is achieved by adults. Baker (1978) defined migration as any movement between habitats; thus, in the earlier work of 1996 it became necessary to determine what was actually meant by a habitat (Dennis and Shreeve, 1996; Dennis *et al.*, 2003). With regard to adults, different terms have been used to describe butterfly movements, including: (i) **trivial flights** within habitats (Johnson, 1969, equates with the term **routine movements** used by Van Dyck and Baguette, 2005); (ii) **dispersal** (which can be linear and non-linear, directional and multi-directional movements outside habitats); and (iii) **migration**

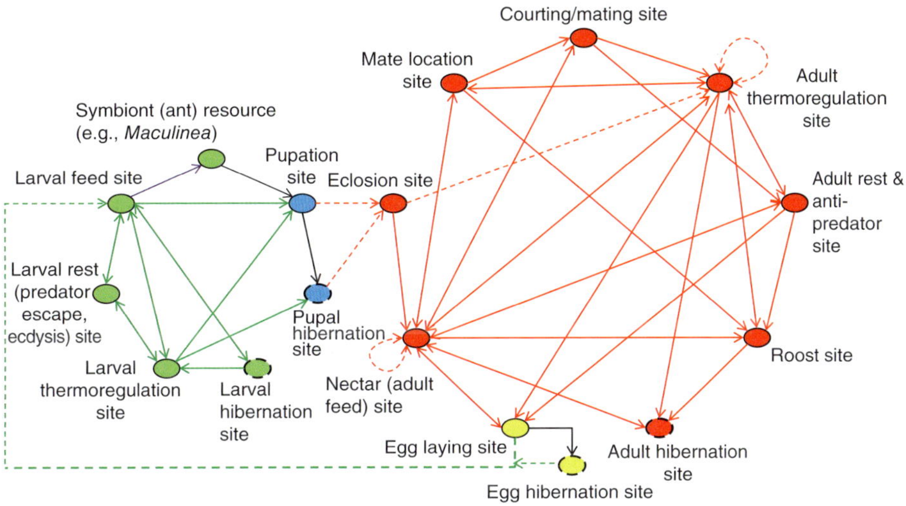

Fig. 3.4. Lifetime resource track between resource types for a single individual of a butterfly species. Movements are limited to the larval (green arrows) and adult (red arrows) stage, with one British species conveyed (transported) by a symbiont ant (purple arrow). Most adult movements are by flight, though some involve walking as well as or instead of flight (pecked line). Resource types distinguished for stages (egg, yellow; larva, green; pupa, blue; adult, red). Depending on species, hibernation stage could be any stage (pecked oval) with black arrow indicating no movement from previous site.

(long-distance seasonally reversed movements) (Dingle and Drake, 2007). Baker's (1969, 1978, 1984) research nicely illustrated the difficulties in distinguishing these terms. In fact, trivial (circuitous) movements also occur outside habitats in the matrix and dispersal-type movements occur within habitats (Dennis, 2004). For example, dispersal movements link resources used consecutively in *Gonepteryx rhamni*, which tends to have isolated, non-overlapping resources (i.e. between eclosion, nectaring and hibernating sites) (Pollard and Hall, 1980). Furthermore, there is greater difficulty in separating species on the basis of the terms dispersal and migration than perhaps is realized (e.g. *Hipparchia semele* migration in France) (Feltwell, 1976): both may involve numbers of individuals moving en masse, linear movements overlapping in distances covered and seasonal changes in movements. The term migration best distinguishes species that undergo seasonal changes in the direction of movements and whose resource-based habitats are effectively regional in scale (e.g. nymphalids such as *Vanessa cardui* and *Vanessa atalanta*; pierids such as *Colias croceus*, *Pieris brassicae* and *Pieris rapae*) (Baker, 1969); these species have very different resource-based habitat structures. One should be aware that truly migrant species, defined in this way, are also capable of engaging in dispersal and trivial flights. Similarly, it may be found that species thought only to disperse aseasonally may nevertheless engage in seasonal migrations (e.g. *Pieris napi*).

Dispersal and Migration: an Island Viewpoint

From the viewpoint of island biogeography, the focus of butterfly movements is on the end-product of dispersal and migration: transfers to islands and island colonization. It is clear, now, that species vary enormously for these movements (Dennis *et al.*, 2004). However, 20 years ago considerable differences of opinion existed on the ability of butterflies to disperse across matrix, especially open water. Readers interested in this issue are referred to the original discussion (Dennis and Shreeve, 1996, p. 43). Movement in butterfly adults is particularly difficult to study and the techniques used (i.e. **mark–release–recapture** (**MRR**), direct tracking, pollen and chemical markers) (Morton, 1982; Courtney *et al.*, 1982; Baker, 1984; Dempster *et al.*, 1986) have long been known to produce conflicting indications (Shreeve, 1992b, 1995). For example, Ford (1964, pp. 57 and 60) declared that even a hundred yards of unsuitable territory acted as a complete barrier to movement of *Maniola jurtina*. Yet this species is capable of flying through and over woods and over tall buildings, open grass and lakes (Shreeve *et al.*, 1995; Dennis, 2004). Even in the early 1990s the literature contained firm evidence of substantial movements through extensive areas of non-habitat (matrix) by most species. These matrix zones were highlighted by: (i) ex-habitat vagrants (including obvious vagrants on islands lacking resources); (ii) suburban garden records; (iii) city (especially **CBD**) records; (iv) at-sea records; (v) range expansions; (vi) frequent long-distance (reversed) migrations; and (vii) mass movements including those of mixed species (Dennis and Shreeve, 1996) (see Appendix 6).

Records of vagrants impress that there are costs as well as benefits to movements (Fig. 3.5); the frequency distributions of movements alone (numbers of movements against distance, which adopt shapes approximating either negative exponential or inverse power functions) stress the nature of selection pressures against longer-distance movements (Dennis, 2010, p. 135). Movements are essential for seeking out complementary and supplementary resources but during movements individuals become conspicuous to enemies and energy is being consumed; when movement is undertaken across extensive stretches of water the costs are high as these are hazardous (i.e. high wind speeds; minimal radiation night flights) and lack facilities for rest, and the termini may consist of resource-vacant zones. In 1996 a simple index of movement was devised: a simple sum of the binary states (yes or no) for each species based on the seven categories of movement noted above (Dennis and Shreeve, 1996: Table 7) (see Appendix 6). The scores were improved on later by weighting (summing the ranks) of these variables (Dennis *et al.*, 2004). These measures of butterfly dispersal were then tested against actual movements (vagrancy, based on host plant availability in 100 m squares) in two distinct separated environments – wooded/non-wooded and urban/non-urban areas – in the vicinity of Manchester (Cook *et al.*, 2001); the scores of the weighted index were supported by a meta-analysis on butterfly dispersal (Stevens *et al.*, 2010). More recently, a multivariate movement index has been developed (see Appendix 7) (Dennis *et al.*, 2012a; Dapporto and Dennis, 2013). Even so, this is perhaps the weakest area of our knowledge on British butterflies for two basic reasons. Firstly, data on movements differ in quality among species; for instance, movements of the high-tree species (e.g. *Satyrium w-album*, *Favonius quercus*, *Apatura iris*) are almost certainly underestimated by a mobility index based on records of observed movements and caution should be used when viewing research results on composite indices (see Appendix 7). Secondly, there is a great

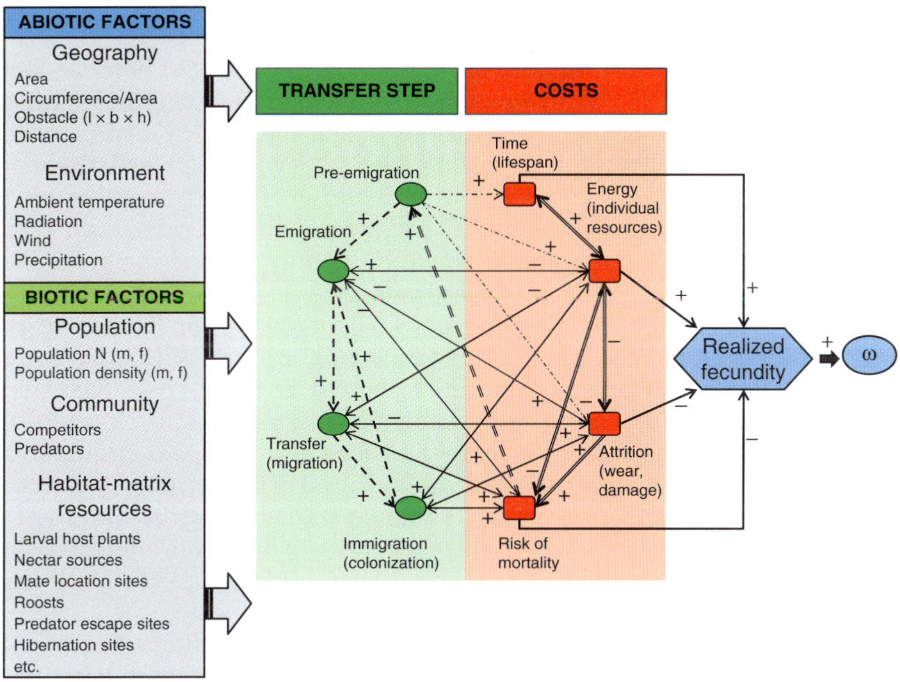

Fig. 3.5. Factors, components and costs of inter-patch dispersal affected by habitat fragmentation and resource fragmentation (of both supplementary and complementary resources constituting a habitat) (from Dennis *et al*., 2013, courtesy of Springer Netherlands). Factors (external to organisms) are divided into abiotic and biotic and affect each stage of dispersal (emigration, migration and immigration/colonization), sometimes differently (i.e. the system's response to high population density may be positive (encourage) for emigration but negative (deter) for immigration). Pre-emigration components denote evolutionary changes to populations driven by changes to demography and movements. The model illustrates the links between components of dispersal and costs and the causal impacts (+/–, correlation signs) between components of dispersal and costs. Bold pecked lines are links between dispersal components; double lines, links between cost components; double pecked line, an evolutionary feedback; fine pecked lines, changes induced by evolution that reduce costs; fine continuous lines, links between dispersal and cost components; bold continuous lines, impact of costs on realized fecundity and individual fitness (ω); l: length, b: breadth, h: height, N: numbers of individuals, m: male, f: female.

difference between migration ability, depending on a species' **bauplan**, and migration capacity that involves opportunities for movement (flight period, broods, population size, number of colonies) as well as differences in the bauplan (see Chapter 6).

A breakthrough in understanding arose when it was found that records of butterfly movements (Dennis and Shreeve, 1996) through matrix zones contradicted simple assessments of butterfly species' movements based on the dichotomy of open and closed populations (Thomas, 1984) and its modification into three scores (viz., sedentary, intermediate, wide ranging; Pollard and Yates, 1993) based on mark–release–recapture of adults (see Appendix 6). For example, it was noted that although 47 species were described as having closed populations, 36 (76.6%) of these were found as vagrants, 35 (74.5%) in suburban gardens and 22 (46.8%) in cities (e.g. *Euphydryas aurinia*: Huggins, 1972; Horton, 1977; *Polyommatus coridon* and *Melanargia galathea*: Allan, 1949; *Argynnis aglaja*: Grimwood, 1965). With Butterflies for the New Millennium (BNM) mapping by Butterfly Conservation UK, records of vagrants, garden records and city records will have become numerous, although only a fraction may be classed by recorders as belonging to these categories.

The data on range expansions (Dennis, 1977, 1993; Pollard and Yates, 1993) and repeated long-distance migrations (Williams, 1965; Baker, 1984) are also extensive and discussed in Chapter 4. Mass movements usually include typical long-distance migrants (e.g. *V. atalanta*, *V. cardui*, *C. croceus, Aglais urticae, P. brassicae* and *P. rapae*) and may include a mixed collection of species; but more surprisingly they may also include species unknown for long-distance movements. In the 1996 account (Dennis and Shreeve, 1996), several reports were cited (Dannreuther, 1933; Buckstone, 1938; Fenton, 1948; Horton, 1977; Holloway, 1980; Smith, 1993).

The most valuable records of vagrancy from the island biogeography standpoint are those for species that are unable continuously to breed and overwinter (persist) on any of the British islands (**island vagrants**). Examples include *Iphiclides podalirius* in Britain (Wilkinson, 1975, 1982; see Emmet and Heath, 1990) and continental *Papilio machaon*, *Aporia crataegi*, *Vanessa virginiensis* and *Issoria lathonia* on the Isle of Wight (Fearnehough, 1972). Such vagrants to Britain and Ireland have been recorded on several of the smaller islands (e.g. *Danaus plexippus* on Skomer, St Agnes and Tresco; *Lampides boeticus* on Hayling and the Isle of Wight; *Nymphalis antiopa* on Foula, Great Saltee and Lindisfarne) (Dennis and Shreeve, 1996). Other valuable records are of resident species on the British mainland that have also been observed on islands where clearly neither long-term breeding resources nor overwintering resources exist for them (e.g. *Argynnis paphia* on Bardsey and Inishtrahull: Darlington, 1954, and I. Rippey, personal communication; *P. machaon* at Voe, North Mainland in the Shetlands: M. Pennington, personal communication, and Baldwin, 1995; *Polygonia c-album* on Lewis and Harris: Trevor, 1994). Some care is needed in interpreting such observations of alleged island vagrants, as host plant patches may have restricted distributions on islands and be missed by recorders (e.g. *Boloria selene* on Sanday) (Campbell, 1970).

The crucial records from an island biogeography standpoint are those that indicate a capacity for species to undergo sea or lake crossings. These include records made from: (i) ships and lighthouses ('at sea' records); (ii) offshore sea-washed rocks; and (iii) islets and islands lacking species-specific resources for reproduction and survival (e.g. Bawden Rocks: Plate 1.2). Unfortunately, the days of manned lighthouses are numbered, many having been automated (e.g. Monkstone lighthouse: Plates 1.7 and 1.8). Even so, the future holds potential for automated recording of arthropod vagrants at lighthouse stations. It is not beyond our ingenuity to apply sensor-triggered cameras to non-kill variants of yellow pitfall traps that have been used to sample butterflies in the arctic (Høye *et al.*, 2014). In 1996, examination of the entomological literature disclosed that there was good evidence for some 27 of the 60 British species having made sea ventures (12 species (44%) regarded as having 'closed' populations; and seven (22%) as being sedentary). Updated records list 27 species found at sea (see Appendix 6: code SR) and 17 species having made overwater movements to islands (see Appendix 6: code OWM). It was pointed out then that this figure may significantly underestimate such movements, as the data are the product of casual recording, not compiled during any systematic survey. For 16 of the 27 species, the minimum distance travelled across the sea was found to exceed 10 km and the actual distance was thought to have been much greater (Dennis and Shreeve, 1996). Light vessel data give flight direction but this is for a single point in time and, though suggestive of direction of origin, without additional data cannot simply be assumed to indicate the direction of movement for the whole period prior to the time of observation. At the same time, it was noted that some resident species had been found much further out to sea (e.g. *Lasiommata megera*, 48 km out on Outer Dowsing light vessel off Spurn Head) (Dannreuther, 1933).

In 1996, clear evidence was also recorded of the ability of butterflies to cope with hazardous over-water conditions (Dennis and Shreeve, 1996, pp. 49 and 50):

> Out at sea, resident British species have been found flying against the wind (e.g., *Maniola jurtina* and *Lycaena phlaeas* at Sovereign light vessel on 1 July 1933 and 29 August 1933 respectively; *Coenonympha pamphilus* at Gorleston Pier, 27–31 August 1933) as well as with it (e.g., *Lasiommata megera* at Outer Dowsing light vessel, 20 August 1933) (Dannreuther, 1933). Some unusual behaviour has been observed at sea, such as the 'spiralling cloud' of 30 or more *Aglais urticae* observed midway between Troon and Lady Isle (Gibson, 1982d). A significant problem for overseas flight, compared to dispersal over land, is the lack of places for insects to rest when exhausted or during the night. Even so, several butterflies have been recorded as flying in the dark (e.g., *Pieris napi*: Heslop Harrison, 1940c; *Aglais io*: Frazer, 1939; *Satyrium w-album*: Kett, 1993; Bristow, 1994); *Vanessa atalanta*, *V. cardui* and *Aglais urticae* (Tremewan, 1953) have

frequently been recorded doing so, even in storm conditions (Sullivan, 1946). Records also exist of butterflies landing on water and taking off again (e.g., *Pieris rapae*: Dowdeswell and Ford, 1948; *Favonius quercus*: Holloway, 1980; *Coenonympha pamphilus*: Shreeve, pers. obs.), even repeatedly and sailing along in rough weather (e.g., *M. jurtina*: Heslop Harrison, 1946a). Observations exist which suggest that butterflies can also take sustenance from water bodies when they are in flight (e.g., *Vanessa atalanta*: G. Bennett, pers. comm.). What initiates these movements over water is not known, but butterflies have been observed leaving the shoreline and heading out to sea (e.g., *Hipparchia semele*: Campbell, 1952; *Limenitis camilla*: Birtley in Dannreuther, 1935; *Argynnis aglaja*: Heslop Harrison, 1939).

Several species have been recorded at Inishtrahull light vessel flying in from the south (e.g., *Pararge aegeria, Hipparchia semele*: Dannreuther. 1939).

It was also noted that mixed-species migrations have been observed to cross open sea (e.g. the swarm of pierids, nymphalids and lycaenids passing Tuskar lighthouse off Co. Wexford on 28 August 1957) (French, 1958) and, most interestingly, of Lepidoptera settling on and rising easily from water (e.g. *P. brassicae*) (Pocklington, 1872; Anonymous, 1872–73; Brown, 1887). Subsequently, additional records have been made of sea and lake crossings (e.g. *Pararge aegeria*) (Shreeve and Dennis, 2009).

Studies of butterflies on mainland Britain would suggest that there are strict limits to the capacity of each species to cross matrix. Applying logistic regression to metapopulation patches for some rarer **specialists** (*Hesperia comma, Thymelicus acteon, Plebejus argus, Satyrium pruni, Melitaea athalia*), Thomas *et al.* (1992, p. 566) concluded that 'a lack of suitable habitat patches for a few to tens of kilometres would prevent the spread of species from regions of occupied patches to those which are vacant'. However, the impact of migration ability, migration capacity and colonization ability on movements should be borne in mind, the latter two depending, respectively, on **mass effects** and the state of resources at termini of movements. It should also be considered that measurements of these species are being conducted after millennia of isolation and specialization to British environments (see Chapter 8) over short periods of survey time and may underestimate movements of these butterflies in other contexts, especially during a resource flux with deglaciation (Dennis, 1977, p. 211) and with climatic warming (Dennis, 1993).

Species Records and What They Imply

An unqualified record of a butterfly on an island may convey little about its status. Assuming that the observation is correct and not mistaken (see below), it simply indicates that the butterfly has been recorded there. Ideally, we would wish to know if the species is still resident and breeding on an island. For butterflies, this information is not easily obtained but, for example, has been achieved for *P. brassicae* in the Orkney islands and Outer Hebrides (Lorimer, 1983; Heslop Harrison, 1938a, c);

in fact, different cohorts of this butterfly occurring simultaneously have been identified, some resident, some migrant, some parasitized, others parasite-free (Heslop Harrison 1942b). Generally, records of butterfly species on islands are of adults, often a single sighting, without information on the butterflies' behaviour, their resource use or the biotopes occupied. Usually, there is little attempt to search for early stages or for other evidence of breeding and island residence. As noted previously (Dennis and Shreeve, 1996), the quality of records for butterflies on islands varies enormously, reflecting on the nature of recording on islands. The best recorders are both botanists and lepidopterists. Thus, the records obtained by Professor Heslop Harrison in the 1940s are particularly valuable; as a botanist he was particularly careful to obtain breeding records of species as well as records of adults. Even today, we would have little understanding of the ecology of Hebridean butterfly species but for his thorough surveys. In some cases, islands have had resident experienced entomologists (e.g. Canna) (Campbell, 1975a) or wardens with particular knowledge of entomology (e.g. Rhum: Wormell, 1982; Farne islands: J. Walton, personal communication). An increasing number of islands now run the **Butterfly Monitoring Scheme** transects (Skokholm; Skomer; Ramsey; Jersey) and that ensures growing experience in making observations. On other islands, there have been systematic attempts to record the wildlife, butterflies included (e.g. Shetland islands) (Harper, 1974; Harvey *et al.*, 1992; Pennington, 1993). In most cases, however, records of butterflies on islands have been undertaken as casual extensions of single or infrequent visits to the islands. These may not be thorough undertakings but nevertheless they add valuably to our knowledge.

The development of the resource-based habitat (RBH) paradigm should provide a formal basis for surveying the status of species on islands. It makes explicit reference to specific behavioural traits in relation to particular resource exploitation; in so doing, it provides the means for distinguishing residents (especially viable populations) from vagrants. Even implicit application of the RBH has proved useful, for instance in identifying vagrants of specialist species 1–2 km away from their compact habitats (e.g. *Cupido minimus* and *P. argus*) (Horton, 1977; T.G. Shreeve, personal communication; Dennis and Bardell, 1996). Although a record of individuals passing through an entire life cycle is proof of a butterfly breeding on an island, this is rarely available. Nevertheless, there are several indications of breeding status; these include evidence of reproductive behaviour, particularly mating (in copula pairs), egg laying, the development of larvae on suitable host plants, and pupae. A particularly useful indicator of residence is the presence of teneral (freshly emerged) adults, as these can only have bred *in situ*. It is important to understand that such records are not necessarily proof of the species being a long-term resident. For example, a record of egg laying indicates that the butterfly is capable of colonizing the island (e.g. *P. brassicae* on Grassholm) (Heron, 1956); it is not evidence that it is necessarily capable of persisting on it as a viable population over a succession of years. The butterfly may have laid an egg on an unusual (unsuitable even) host plant (e.g. *Polyommatus icarus* laying on *Ulex europaeus* on Tresco) (R. Sutton, personal communication), in which case certainly survival of the larva on the plant would need to be recorded.

Numbers of individuals suggest the occurrence of a population, but this is not always the case; migrant swarms give the impression of residence but may in some circumstances occur only temporarily. This conundrum is most likely to arise in species known for their long-distance mass migrations (e.g. *V. cardui* on Rhum) (Heslop Harrison, 1944). However, for species that are not known for long-distance mass migrations, if these individuals are in appropriate biotopes and involved in a range of behaviours associated with habitat use such as mate location, mating, egg laying, etc. and are not just resting and refuelling, then the combination of numbers and activities associated with habitat occupancy provide unmistakeable evidence of a resident population. Support for residence is also provided by repeated observations of species during multiple surveys, particularly when these are conducted systematically as part of long-term surveys (Butterfly Monitoring Scheme UK). Thus, it has become evident that *V. atalanta* is now a resident British species, especially in coastal areas of southern England, but perhaps as far north as North Wales (Fox and Dennis, 2010).

Conversely, lack of evidence for reproductive behaviour, or even of butterfly resources (particularly larval host plants), is not proof of vagrancy. Butterflies on British islands have been found to use a variety of host plants in different biotopes (e.g. *P. napi*) (Heslop Harrison, 1947d, 1950c, 1955b) and the same host plant in different biotopes (e.g. *P. icarus*) (Heslop Harrison, 1950c). Moreover, species may use alternative host plants in different biotopes on separate islands (Heslop Harrison, 1950c, 1955a). In the high wind-speed environments of many British islands, host plants may be found in deep rock crevices (Heslop Harrison, 1947a); this should be expected of grikes on the limestone pavements of the Aran islands off the coast of the Irish Burren. Host plants may be much reduced in height, adopting unusual (unfamiliar) conditions (e.g. **krummholz** oaks on north-east Colonsay used by *F. quercus*) (Dunn, 1965). One observation not generally appreciated is that adult females of many species have an uncanny ability for finding tiny refuges of host plants, often deeply hidden in other vegetation (Courtney, 1988; Hardy and Dennis, 2010). *V. atalanta* has been seen ovipositing on an isolated tiny leaf (< 1 cm long) of *Parietaria judaica* on a vertical wall (RLHD, personal observation, Conwy, North Wales, 5 June 2014). Surveys carried out by competent entomologists reveal that host plants and habitat resources may well be available for butterfly species on islands (e.g. Bardsey) (Morgan, 1969; Askew, 1974).

Bias and Mistaken Records

A key issue for biogeography is the validity of records. Unusual records for islands raise the spectre of error. Errors may simply be mistakes in identification; more rarely, they are deliberate (hoax or fraud). For instance, Heslop Harrison's record of *Maculinea* [now *Phengaris*] *arion* for Rhum has been suggested to have been part of a prank played on him by his students (Campbell, 1975b). Unfortunately, despite the limited nature of the British butterfly fauna, mistakes in field identification occur more frequently than atlas and monitoring scheme recorders perhaps care to admit. There are understandable problems of distinguishing some pairs of species on the wing (e.g. *Thymelicus sylvestris* and *Thymelicus lineola*; *P. rapae* and *P. napi*; *Boloria euphrosyne* and *B. selene*). In Ireland, *Leptidea sinapis* and *Leptidea juvernica* present a particular difficulty where these two species are in near contact (Nelson *et al.*, 2001; Dincă *et al.*, 2011a, 2013). Regional variation in species may also cause confusion as in the case of Hebridean forms of *P. icarus* (perhaps the mistake for *M.* [*P.*] *arion* in Campbell, 1975b) and *A. aglaja* (which may have been mistaken for *Argynnis adippe* and *A. paphia*) (Heslop Harrison, 1941d; Wilks, 1945a). Some moths are also regularly mistaken for butterfly species (e.g. *Odezia atrata* (L.) for *C. minimus* and *Euclidia glyphica* (L.) for *Erynnis tages*) (Jeffcoate, 1994; Long, 1970).

A key aspect of recording is monitoring changes in the distribution and geographical ranges of species. Owing to the distribution of recorders and recorders' interests, recording has long been affected by geographical bias (Dennis and Williams, 1986; Dennis *et al.*, 1999; Dennis and Thomas, 2000). To offset this bias, over the past three decades considerable effort has been expended by Butterfly Conservation (BC) to build interest in butterflies and attract more dedicated recorders (Asher *et al.*, 2001; Fox *et al.*, 2006). As more of the wider public are drawn into recording (Butterfly Conservation, 2017: **Big Butterfly Count**) undoubtedly more records of residents and vagrants will accumulate for all the islands. Mistakes are also bound to become more frequent, in spite of the wide availability of superbly illustrated butterfly guidebooks, if only because this activity occupying a week or so of July and August is unlikely to build expertise and familiarity with butterflies in the majority of volunteers. Even so, it is possible to screen out the more obvious errors when species are recorded well beyond their range and known distribution, and outside of their flight period. Moreover, new locations for threatened and rare species can be checked. However, the absence of a species from an island is not evidence that records for it were previously in error (e.g. records of *B. euphrosyne*, *E. aurinia*, *Erebia aethiops* on Rhum) (Heslop Harrison, 1955b; Wormell, 1982); we now know that colonizations and extinctions on small islands occur regularly (Dennis *et al.*, 2010), fitting in with experience of butterfly population fluctuations and dispersal (Pollard and Yates, 1993).

Introductions

Introductions of butterflies may result from an intentional act or unintentionally (by accident); for example, *Polygonia c-album* and *Aglais io* on

Lewis and Harris may well have come over with timber from Denmark (Trevor, 1994). *Heteropterus morpheus* may have been accidently introduced to Jersey during 1940–1945 (Long, 1970; Quinette and Lepertel, 1993). Numerous exotics in the UK have been recorded that could have been accidentally or deliberately introduced (e.g. *Melitaea didyma*, *Lasiommata maera*, *Gonepteryx cleopatra*) (Emmet and Heath, 1990). Introductions are generally unsuccessful (Oates and Warren, 1990), very likely owing to the failure to determine adequate resources (habitat) for the introduced species. Even if resources are available, these may well be insufficient for the persistence of a population; for example, the failure of substantial transfers of *M. jurtina* to Menawethan and Great Innisvouls (both < 0.5 ha) in the Isles of Scilly in 1954 (Ford, 1964; Dowdeswell, 1981). There have also been attempts to introduce several species to the Isle of Man (Oates and Warren, 1990; I. Rippey, personal communication), discussed in Chapter 7.

It is a simple matter to release large numbers of butterflies without thought for the consequences, some of them legal consequences (see Appendix 8) (see also 'Code on collecting, breeding and photography': Butterfly Conservation, 2010a). Introductions provide a valuable weapon for conservation but, when done without official approval (see 'Policy on Introductions and Re-introductions': Butterfly Conservation, 2010b), releases may result in wasted effort (if unsuccessful) and cause considerable confusion for studies focusing on butterfly conservation (if successful). The recent successful introduction of *Coenonympha tullia* on Heysham Moss – where it has been long absent – has taken careful planning over years (BBC News, 2016). Studies of butterfly DNA are now being undertaken in order to determine genetic differences and origins of British butterfly species, as for Italy (Vodă *et al.*, 2016; Dapporto *et al.*, 2017). Without the benefit of being able currently to trace an individual's genome, one may imagine the outcome of the unrecorded release of butterflies of cryptic species from distant locations on the determination of butterfly origins for islands (see Burg *et al.* 2014). This is not far-fetched: 50 *A. urticae* from the Italian Dolomites were released on Lewis in 1995 (P. Hackett, personal communication).

British and Irish Butterfly Databanks and Sources

During the years since the previous island review, a number of large databanks of records have been accumulating.

1. The New Millennium (BNM) atlas mapping records (Asher *et al.*, 2001; Fox *et al.*, 2006; Butterfly Conservation, 2015a) is an extension of the first butterfly atlas (Heath *et al.*, 1984). This project is now in its fifth 5-year recording period, records accumulating at a rate of some 400,000 a year – a total of 10 million records. Similar projects are also building mapped data for the Irish Republic (Irish National Biodiversity Data Centre, 2017) and the Isle of Man (Manx Biodiversity, 2016). These records are often available as regional atlases and their databases (see Appendix 5).

2. The Butterfly Monitoring Scheme (BMS) (Pollard and Yates, 1993) is run jointly by Butterfly Conservation in the UK and the Natural Environment Research Council (NERC) Centre for Hydrology and Ecology in partnership with other nature conservation organizations (United Kingdom Butterfly Monitoring scheme, 2017). The BMS scheme started in 1976 and monitored close to 2500 sample locations in 2015. It includes the **Wider Countryside Butterfly Survey** (WCBS) (Butterfly Conservation, 2015b), a long-awaited event since first suggested by one of us (RLHD) at a meeting at Abbots Ripton in 1997.
3. The Big Butterfly Count (Butterfly Conservation, 2017a).

The new records in this book have been obtained largely from atlas regional recorders (see Acknowledgements) and this work is a tribute to them and to the many individuals who have kindly submitted records to the building of butterfly atlas data for the islands and the basis for an increasing number of scientific papers for those with access to the data.

Selecting Islands for Study

The above details emphasize the fact that the quality of recording varies enormously from island to island. As pertinent details supporting the residence status of species are usually absent, island faunas cannot confidently be compared on the basis of so-called breeding records; in this study, as before (Dennis and Shreeve, 1996), most island comparisons use a record for a butterfly species as evidence of successful transfer from mainland (island) to island. Although information on successful breeding is not noted in the list of species for islands (see Appendix 3), even so, it is important to appreciate that islands are not expected to share equal (residence) status for butterfly species: larger islands will tend to have larger biotope areas, larger butterfly habitats and thus larger populations (see Chapter 1); therefore a record for a butterfly on a large island is more likely a record of residence than one for a smaller island, particularly as these are usually supported by multiple observations. Again, we would advise caution in assuming that a single record for a large island is evidence of breeding (see Fig. 7.3), much as we would advise caution in concluding that a record for a small island is not of a breeding population and one of long standing. Observations occur on small islands of habitual use of unusual host plants, and host plants used in unusual situations exist (e.g. *B. selene* colony confined to precipitous cliffs on Sanday) (Campbell, 1970). Amazing as it may seem, many islands off the west coast of Britain and Ireland appear to be little more than sea-washed rocks but still have breeding records (for example, *P. napi*, *P. icarus* and *A. urticae* on Oigh Sgeir (Hyskeir), a reef some 9.8 m above sea level, south-west of Canna; *A. urticae* was using *Urtica urens*) (Heslop Harrison, 1955b; Campbell, 1970). Then, conversely, even on larger islands colonizations and extinctions should be expected to occur over the years with changing (climatic and environmental) conditions (for example, *A. io* in Hebrides during the early 19th century) (Heslop Harrison, 1947b; Campbell, 1969b, 1984). Losses of species on the British mainland (see Chapter 4) warn us that this can happen easily enough when the land is used intensively, regardless of the land area.

At this juncture it is useful to point out the difficulties in island studies. The selection of islands for analysis (see Appendix 8 for study guidelines) has, of necessity, been arbitrary. For the work in 1996 (and herein), they were selected on the basis of records for the (then considered) two non-resident long-distance migrants *V. atalanta* and *V. cardui* and the two resident long-distance migrants *P. brassicae* and *P. rapae*. A study in 2010 selected islands on the basis of records for two of the three migrants *C. croceus*, *V. atalanta* and *V. cardui* (Dennis *et al.*, 2012a). These five migrant species have the highest probability of successful sea crossing to islands in large numbers and therefore of being observed on them. As the two white pierids are highly apparent and double brooded, appearing at somewhat different times of the year compared with the two nymphalids, and have the potential for colonizing inhabited islands with crucifer crops (*Brassica* spp.), they have a high probability of being recorded by resident or visiting observers on islands. Therefore, these five species, used in various combinations, provide some measure of the completeness of an island's list. In 1996, 73 islands had four of the five migrant species (Dennis and Shreeve, 1996). The current study has enumerated over 900 offshore British and Irish islands larger than 10 ha, but including many as small as 1 ha. As of 21 March 2017, four species (*P. brassicae*, *P. rapae*, *V. cardui*, *V. atalanta*) were recorded from 123 of the 910 British and Irish islands listed (see Appendix 3 and Supplementary Appendix 3.1 for British and Irish island butterfly data). The less strict selection on two of the three migrants (viz. *C. croceus* and the two *Vanessa* species) was used for a study of the incidences of individual species on islands.

Regarding these data for the islands, it is inevitable that the smallest islands (< 1 ha) will tend to be unrepresented in any analysis; small islets are not only particularly numerous but are also typically sea washed at all times, usually lacking landing and basic facilities and particularly dangerous to visit. It is important to recall that they are also islands on which any butterfly populations are expected to become frequently extinct, as evident long ago in studies of butterfly populations (Warren, 1992a). However, they could be useful as stepping stones, with adults leaving to find habitats on other islands immediately after eclosion.

Even so, geographical bias for islands varying in size affects relationships among geographical variables, a topic we return to in Chapter 5 (see Appendix 10). Such biases not only involve the relative number of small islands included in a study, but also the distribution of islands studied over the spread of the British Isles. For instance, changes in the relative number of British and Irish offshore numbers will affect estimates of species richness, if only because Ireland has fewer butterfly species than the British mainland (see Chapter 5). The scene is now set for investigating the islands' butterflies.

4 British and Irish Butterfly Species: Links and Gradients

British and Irish butterflies form but a fraction of the number listed for Europe, some 409 species (Kudrna *et al.*, 2015); these include 82 species (after 1960) that are resident (some extinct), common or infrequent migrants and a further 35 or more species regarded as rare migrants or accidental and deliberate introductions (see Appendices 1 and 2). This is a sparse fauna even when compared with the adjoining coastal margin from France to Denmark (138 species; see Supplementary Appendix 4.1). An immediate impression is that a land bridge crossing was closed before these species could cross over, but reconstruction of faunal history would tend to support the notion that these species entered Britain but were subsequently eliminated and now find it difficult to re-enter owing to a deterioration in environmental conditions, especially loss of habitat resources (Dennis, 1977). In this chapter we consider the unusual nature of the British and Irish butterfly fauna and illustrate, even for such relatively small islands, how environmental gradients have affected their ranges, distributions and changes in geography over very short periods of time.

How Unusual Are British Butterflies?

In the early 1990s, the butterfly fauna of the British islands was shown to be distinct from other parts of Europe (Dennis *et al.*, 1991, 1995b). The relationship between any two geographical areas can be determined by using elementary similarity (association) coefficients (Sneath and Sokal, 1973) (see Appendix 8). Applying several **multivariate statistical techniques** to **Jaccard coefficients** between natural areas over Europe consistently identified the British fauna as making up a distinct **faunal region** (Fig. 4.1). Jaccard coefficients simply express the number of jointly

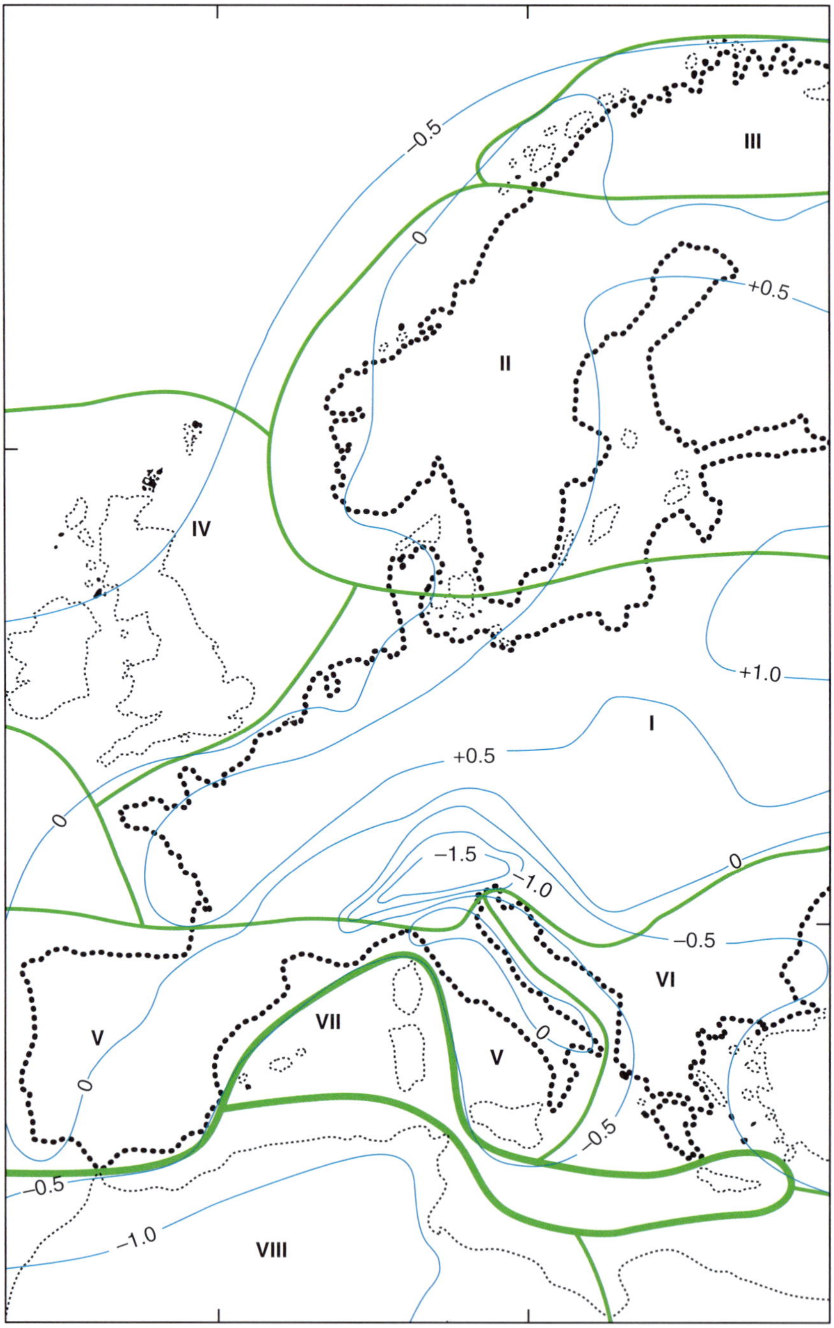

Fig. 4.1. Regions of European butterfly fauna (from a factor analysis ordination of 393 species in 85 primary areas). Green lines, regions I to VIII; blue lines, affinities (mean Jaccard coefficients) corrected for size of faunas in primary areas and their isolation from one another. (Adapted from Dennis, 1993, and Dennis *et al.*, 1995a, courtesy of Manchester University Press, the *Entomologist's Gazette* and Pemberley Books.)

occurring species in a pair of geographical areas as a decimal fraction (ranging from zero to one) of the sum of those found within the same pair of regions. Of course, a region may be distinct for several reasons. In the case of Britain and Ireland, this is not the result of the unique (novel) occurrence of species, as found on other larger European islands such as Corsica and Sardinia (Dapporto and Dennis, 2008). In fact, British species form part – a fractional or depleted part – of the fauna for the whole of Europe north of the alpine mountain chain, which the first author referred to as the **extent element** (Dennis and Schmitt, 2009). This observation was strongly supported by a detailed analysis of **affinities** between European regional faunas (Fig. 4.1) (Dennis *et al.*, 1995a). Affinity is a measure of mean similarity (e.g. average Jaccard coefficients) between any geographical area and all other areas in a study zone; the affinities were standardized for size of region and distances from other regions. The pattern of affinities for the British Isles were quite distinct from other natural regions north of the Alps (excepting the montane and arctic fringe of Scandinavia) in being negative; that is, the fauna of the British islands had less in common with surrounding continental faunas than expected from their size and proximity to other areas. Unlike Mediterranean islands, this was not because they had unique species. The British fauna had experienced significant losses of species and those remaining provide a distinct enough collection to distinguish them from other near-continental faunas.

At this point, it is useful to enumerate the evidence for the loss of species on the British mainland and the larger British islands.

1. For Britain, as with the Netherlands, there are sound 19th and 20th century records for some species, providing evidence that they have become extinct during recent history (e.g. *Aporia crataegi*, *Lycaena dispar*, *Cyaniris semiargus*; also *Phengaris arion*, now reintroduced) (Dennis, 1977, 1992; Asher *et al.*, 2001). As an adjunct to this observation, some interesting questions arise, such as why *Leptidea juvernica* occurs in Ireland but not in England (Nelson *et al.*, 2001): is this a matter of selective post-glacial immigration to Ireland or an example of yet another butterfly that has become extinct on the British mainland? The road to outright extinction on the British mainland is geographically patterned by regional extinctions, also well documented. For instance, there have been significant regional extinctions in areas once rich in butterfly species (e.g. *Cupido minimus*, *Hamearis lucina*, *Nymphalis polychloros* in North Wales; *Thecla betulae* and *Leptidea sinapis* in the Morecambe Bay region and Cumbria; many of the fritillaries in East Anglia). Regional extinctions may also be surmised from the discontinuous distributions of species, in the form of refuge populations, at the margins of the geographical ranges of species. Classic examples are the Grassington and Arnside populations of *Erebia aethiops*, the Scottish populations of *Pararge aegeria* and the west country, Kent and Essex colonies of *Melitaea athalia* (Dennis, 1977; Asher *et al.*, 2001).

2. Just over the English Channel there are a number of species that have failed to colonize the British mainland, an observation made long ago by Russell Bretherton (1968) (Table 1.2 in Dennis, 1992, lists 42 species that occur along the coast from Brest to Gröningen but are absent from Britain).
3. A number of species that do migrate here have failed to persist in Britain, though there is increasing evidence that this may change now with climatic warming (e.g. *Colias croceus*: Skelton, 2000; *Vanessa atalanta*: Fox and Dennis, 2010).
4. A number of species occurring in Britain have specialized resource demands, particularly larval host plants (Dennis, 2010) (e.g. *Papilio machaon*: Balfour-Browne, 1958), a feature highlighted as indicative of environmental constraints imposed on the fauna during the Holocene (Dennis, 1977, 1992). A detailed analysis of butterfly life history strategies (see below) in British butterflies has added further support for this hypothesis (Dennis *et al.*, 2004).
5. Gradients in species richness and modifications in **phenology** and **phenotype** are indicative of the key instrument for changes being climate, which becomes significantly cooler, windier and wetter further to the north and west (Dennis, 1977, 1993; Turner, 1986; Turner *et al.*, 1987; Dennis and Williams, 1986; Dennis and Shreeve, 1989); conditions significantly deteriorated after the climatic climax 5000 years ago (see Fig. 8.2). The loss of butterflies from the fauna was found to be predicted during the reconstruction of changing environmental conditions during the Holocene (Dennis, 1977, 1993), supporting the typical scenario of a land bridge island; that is, the loss of species once the bridge has been broken.

Island Environment–Species Gradients

One of the features supporting losses from the British fauna is the occurrence of close associations between environmental conditions and species richness; the inference is that a downturn in environmental conditions will result in the gradual tipping of species off the British mainland. In the equilibrium island biogeography theory developed by MacArthur and Wilson (1963, 1967), modelling makes the assumption of environmental homogeneity across an archipelago; but, obviously, this is rarely, if ever, the case in reality. Furthermore, conditions are highly unlikely ever to be homogeneous over a single island (the reader might like to spend a few moments mentally creating a virtual island over which no environmental differences can ever exist).

In short, species richness gradients are expected to occur on islands; this is likely to be most evident for large islands. The British mainland provides an excellent illustration of species richness gradients for a wide range of taxa (e.g. plants; mammals, birds, other arthropods, (Arnold, 1993; Eversham and Cooper, 1998; Lennon *et al.*, 2000; Preston *et al.*, 2002; Ball and Morris, 2000; Suzart de Albuquerque *et al.*, 2011) and a gradient in butterfly species has long been known for Britain (Dennis, 1977, p. 70). Although the British mainland extends over only eight degrees of latitude, there is a systematic fall off in numbers of species to the north-west (Fig. 4.2). The prime factor is climate, especially summer heat (sunshine and ambient temperatures), which declines sharply with latitude, not winter temperatures, though British butterflies are adversely affected by wet winters (Dennis, 1977, p. 75; 1992, 1993; Turner, 1986; Turner *et al.*, 1987; Pollard, 1988). In fact, summer sunshine and temperatures account for some 55–79% of the variation in butterfly species richness throughout the length of the British mainland depending on the spatial resolution (10 km to 30 km square units) of the analysis (see details in Dennis, 1993, Table 2.6, p. 81). These findings correspond with those made by Professor B.P. Beirne (1955) in his earlier work on fluctuations in British butterfly populations

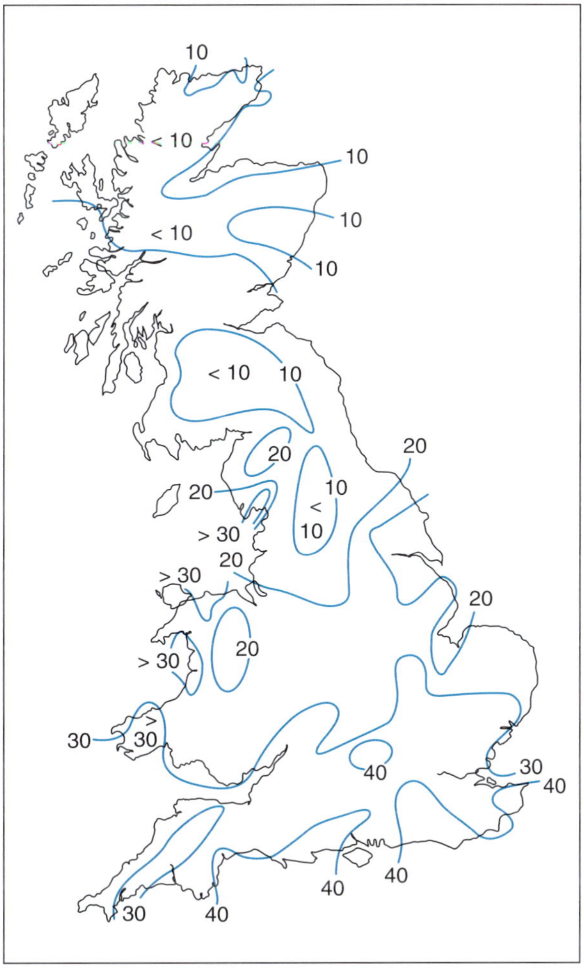

Fig. 4.2. Trends in butterfly species richness (number of species) over the mainland of Britain. (Adapted from Dennis, 1993, courtesy of Manchester University Press.)

(Dennis and Sparks, 2007). On closer inspection of environmental data, it was found that a large number of variables correlate and integrate (e.g. geology, soils, altitude, land use) with this broadly latitudinal pattern of decline in species richness (Dennis, 1993, Table 2.6, p. 80). Primarily, suitable environmental conditions for butterflies become less predictable with increasing altitude and latitude; that is, conditions become more hazardous, with the greater likelihood of unseasonable frosts, snow falls, heavy cloud cover and storms (excessive winds and precipitation) triggering population collapses and regional extinctions (Dennis, 1993).

To imagine that such gradients only affect large islands would be entirely wrong; even the very smallest of islets have highly varied environments, contrasting in sun and wind exposure, degree of shelter, substrate (soil) conditions, vegetation growth and so on. A classic case is the north-to-south gradient in species richness on the fossil island of the Great Orme's Head in North Wales (Fig. 4.3; see also Plate 1.17), now a headland marked by steep slopes, scars and a high summit (207 m). Readers may test this out for themselves: when walking around the marine drive, it will become evident that many fewer butterfly species and individuals occur in the north-east aspects of the headland than slopes facing south and south-west. Dramatic vegetation changes, indicative of environmental gradients when converted to **Ellenberg values** (Dennis, 2010), will also become noticeable: cold windy north-eastern slopes especially

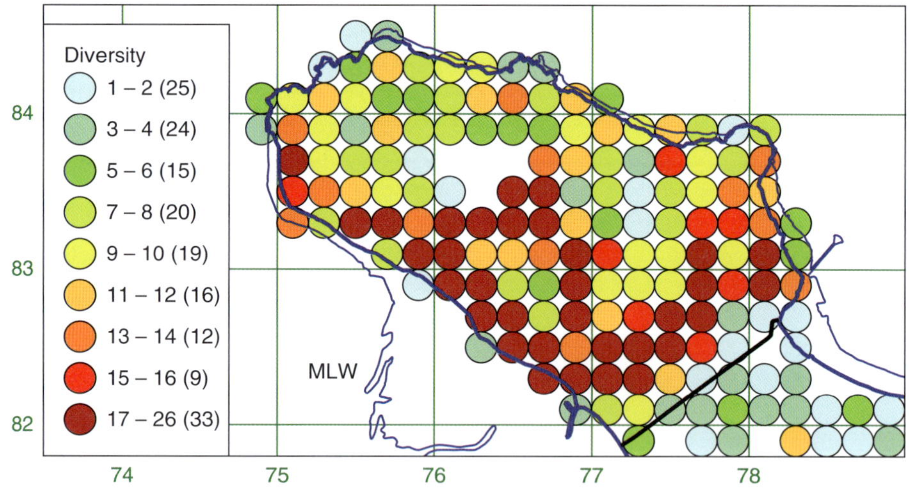

Fig. 4.3. Species richness gradients on the small ecological island of the Great Orme's Head, North Wales. Data (1996) kindly provided by Professor Chris D. Thomas and his research group (Dr Robert J. Wilson, Dr Matt Cowley, Dr Jorge L. León-Cortés and Dr David Guttiérez); also Mr Andrew Graham, the 'Butterflies for the new millennium' record coordinator for Anglesey, Caernarvonshire and Merionethshire. Species richness differs for northern and southern parts of the headland (**Mann–Whitney** U, $Z = 3.94$, $P < 0.0001$; based on 200 m squares; north: **median** = 8, $n = 69$; south: median = 13, $n = 74$).

featuring cool, moisture-loving cliff cabbage, cowslip, dog's mercury and willow bushes; hot sun-baked southern slopes sporting fennel, rock rose, valerian, cotoneaster and juniper, more familiar in Mediterranean climes. Differences are apparent on much lower and smaller islands still, as on Hilbre, an island in the Dee estuary (Cheshire and Merseyside) where species collect in the depressed sheltered track on the southern side (Appendix 2b in Dennis and Shreeve, 1996).

These climate associations have two basic impacts: (i) on the condition of critical resources (habitat conditions) for the maintenance of butterfly populations; and (ii) on the butterflies themselves as **ectotherms**, in their ability to remain active and exploit resources (Dennis, 2010). But there is more to this association of declining species richness with cooler climates; it does not explain why some species are more tolerant of testing conditions than others. Below, the different responses of species are broached; it is an area of fast-developing research.

A Recording History of Gains and Losses on the British Mainland

As outlined in Chapter 2, environmental conditions are no more stable than they are evenly distributed. From island biogeography theory, over time, it is expected that new species will colonize the islands and others will be lost from them. Basically, the smaller the island and the longer the period of time, the greater will be the number of changes expected. Such changes have affected even the largest of the islands – Britain and Ireland – during the relatively short history of butterfly recording since the 17th century.

A major downside to studying butterflies is the deficiency of a fossil record. Tertiary butterfly fossils have been found in the UK (e.g. *Nymphalites zeuneri* and *Lithopsyche antiqua* in 1980) (see Dennis, 1992, p. 222), but no fossils from the Pleistocene have yet been found of species currently occupying the islands. Regarding origins, apart from reconstructions based on geological models (Dennis, 1977), all we have is a history of first records (see Appendix 1) (Fig. 4.4a); on the British mainland discoveries date from 1634 but other sources (illuminated manuscripts) (Nazari, 2014) would suggest the possibility of earlier 'find' dates for some butterfly species (e.g. *Aglais urticae*). The record illustrates that most currently known species were found prior to 1900, but a trickle of species continues to be discovered into the 21st century, the latest a 'cryptic' species *L. juvernica* (initially referred to as *Leptidea reali*) (Nelson *et al.*, 2001). With developing research into butterfly DNA, there is the real prospect of discovering more 'cryptic' species in Britain and Ireland – exciting work currently being undertaken by Professor Tim G. Shreeve and his colleagues at Oxford Brookes University.

With the rapidly developing programmes for monitoring butterflies in the UK, Ireland, the Isle of Man and other island groups (e.g. Jersey; Isles of Scilly) there is increasing likelihood of witnessing new colonists. In recent years, well known migrants that usually fail to survive the winter have

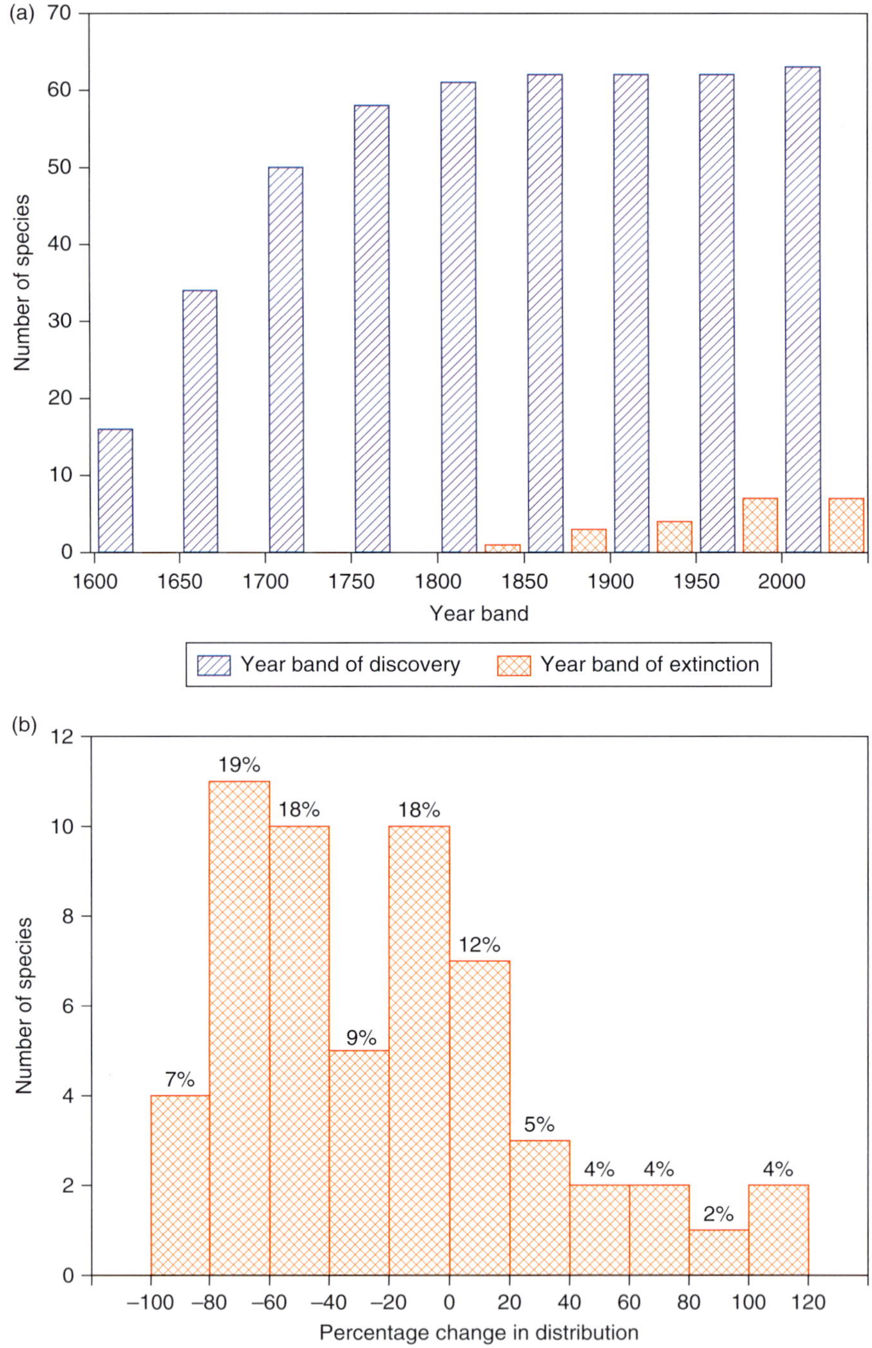

Fig. 4.4. Recorded changes in British butterflies. (a) Cumulative graph of first records and extinctions (loss of English *C. palaemon* treated as an extinction; data from Dennis, 1977; Salmon, 2000). (b) Changes (losses/gains) in the distribution of British butterflies between 1976 and 2014 (data from Fox *et al.*, 2015).

been found to establish temporary colonies. Examples include: *C. croceus* (Dorset, 1998/99: Skelton, 2000), *Lampides boeticus* (Fox, 2014), *Issoria lathonia* (Minsmere, Suffolk in 1995/96, Sussex: Pratt and Hulme, 2010); there are also tentative reports of *N. polychloros* re-establishing itself (Marren, 2015, p. 234). One well known migrant, *V. atalanta*, seems now to be a resident (Fox and Dennis, 2010) and there is perhaps the prospect of *Vanessa cardui* becoming one (Hale and Hicks, 1995), though March immigrations can occur, as witnessed in 1985 (Chesil Beach, Dorset; RLHD, personal observation). Anthropogenically enhanced climatic warming is implicated in recent colonizations of Lepidoptera (Warren *et al.*, 2001; Sparks *et al.*, 2007), but perhaps fewer butterfly species than expected from the incursion of additional moth species (Sparks *et al.*, 2005; Waring and Townsend, 2009). A great difficulty will be surviving increasingly wet British conditions predicted with climatic warming, but a more pressing factor is the diminution and downgrading (unsuitability) of resources providing habitats for species (Dennis, 2010; Hodgson *et al.*, 2014).

If the historical record fails to provide origins of species on the main islands of Ireland and Britain, it is useful in providing evidence for recent historical extinctions (Fig. 4.4a). Five species, at least, have been lost from the British mainland since 1864 (i.e. *A. crataegi* 1925; *L. dispar* 1864; *C. semiargus* 1877; *P. arion* 1979; *N. polychloros* c. 1990) (Dennis, 1977; Heath, 1983; Emmet and Heath, 1990) and perhaps a further species, *Carterocephalus palaemon* in 1976 (Collier, 1986; Ravenscroft and Warren, 1996) if the Scottish variant turns out to be a distinct clade and a 'cryptic species' compared with the lost English population. In the case of *P. arion* it has been possible to track the historical decline of the fragmented populations (Spooner, 1963; Warren, 1992b, p. 248, Fig. 11.2). The loss of these species has been fully documented (Dennis, 1977; Emmet and Heath, 1990); other species, for which there are tentative records, may also indicate losses (e.g. *Lycaena virgaureae*: Allan, 1956; perhaps also: *Lycaena hippothoe*; *Erebia ligea*: Pelham-Clinton, 1964; Dennis, 1977). In Ireland there has been loss of two species, *Erebia epiphron* (found first in 1854 by E. Birchall, the last record in 1901) and *M. athalia*, apparently abundant in Killarney in 1865 (Birchall, 1866) and not seen since (Hickin, 1992). There are also two intriguing records for *Ochlodes sylvanus* from a meadow in Lord Kenmare's demesne, Killarney and the Morrough (Wicklow) in South (1941), the butterfly since remaining unconfirmed in Ireland.

The overwhelming finding is that extinction has been the consequence of dramatic declines in, and loss of, habitats, involving finer processes of fragmentation and degradation of resources, the latter from changes in land management (such as end of coppicing, land drainage, outright hedge removal, reduction in grazing). In particular, there has been the broadcast application of industrial chemicals in agriculture, which effectively act as poisons to the natural world (Hodgson *et al.*, 2014; Jones *et al.*, 2017; Prosser *et al.*, 2016). This process, which accelerated fast from 1939, has subsequently been shown by Warren (1992b, p. 252) to be affecting a number of butterfly species (e.g. *Hesperia comma*, *Polyommatus bellargus*,

Plebejus argus, M. athalia, Argynnis adippe); the loss of fritillaries from eastern England, dominated by cereal agriculture, and nation-wide changes in woodland management, has been particularly dramatic (Dennis, 1992, p. 10; Warren, 1992b, p. 264; Hodgson *et al.*, 2014). Thus, in the face of inadequate support for nature – despite the gallant efforts of Butterfly Conservation UK – severe declines in distributions (Fig. 4.4b) likely presage the extinction of additional species over Britain.

Predicting British Species

Arising out of the reconstructions of the origin of British butterflies is the question of which species are expected to occur in the British Isles and, of those currently resident in Britain and Ireland, which are its least likely occupants.

The answer to this question is founded on the distribution of species within Europe: (i) on their locations (geography); (ii) the environments they occupy (i.e. biomes; weather and climate regimes); and (iii) the resources at the locations where they are found (resource-based habitat components). The first two are relatively easy to assess, the last very difficult, primarily because our knowledge of species habitats, for all the texts now available, is limited (Dennis, 2010). Techniques typically involve a mixture of routines in **multivariate analyses** to seek structures amongst the complex of geographical and environmental variables, and regression analysis (usually **logistic regression** (but also **discriminant function analysis**) to assess the relationship of species occurrences at sites (locations) for specific time periods, against key original variables or **'dummy' variables** from any preliminary multivariate analyses.

A simple initial view of expectations was provided by Russell Bretherton (Bretherton, 1968, p. 14), who pointed to the occurrence of a number of butterfly species (n = 38) in northern France, not found in Britain, which might be expected to occur in the British Isles but for the existence of the English Channel; the prevailing view in the 1960s was that a sea barrier of the Channel crossing was sufficient to prevent these French species occurring in Britain. Expectations of species range changes in the 21st century have led to a more sophisticated assessment based on species climatic affiliations (**Climatic Risk Atlas** (**CRA**), Settele *et al.*, 2008). Part of this assessment was to determine current expectations of distribution across Europe for each species that could be successfully modelled using climatic data; a map for each of the species (292 of 409 mapped in Kudrna *et al.*, 2015) indicated its likely occurrence in Britain and Ireland as well as elsewhere over Europe. The work was based on 22 climatic variables boiled down by multivariate analysis to four key climatic variables, with species ranges depicted within a four-dimensional niche space in which accumulated growing degree-days until August and soil water content (upper soil horizons) are subsumed within major axes of: (i) annual temperature range; and (ii) annual precipitation range (Settele *et al.*, 2008).

For the current work, we have attempted a different assessment based purely on nine geographical variables for species (**Contemporary Geography Study (CGS)**) (see Supplementary Appendices 4.1 and 4.2). A **principal components analysis (PCA)** revealed that these boil down to just two factors; the second factor, lesser by far and based primarily on a sole variable (latitude of a species' southern boundary in Europe), is not associated with species occurring in Britain (resident species or all visitors) or with species predicted to occur in Britain in the climatic risk assessment by Settele *et al.* (2008). As it is, the climatic risk assessment failed to determine the southern range boundary of many species that extend into North Africa. Species present (or absent) in Britain were then related to a primary component, which was closely correlated (high **loadings**) with the original remaining geographical variables (see Appendix 8 for PCA and Supplementary Appendices 4.1 and 4.2).

The findings of both studies are of great interest. The CRA predicted the presence of 70 butterfly species in Britain, 19 additional species to 51 of the 58 resident species occurring in Britain and Ireland (*Leptidea* treated as a single species) since 1960, a significant difference from expectations ($\chi^2_{(1)}$ = 162.5, $P < 0.0001$); seven current species in Britain were not predicted to occur on the island in this climatic assessment (see Appendix 9).

The CGS predicted the existence of 55 resident species in Britain. These include 13 additional species not currently in Britain (see Appendix 9), including some that have become extinct in Britain (e.g. *A. crataegi*, *C. semiargus*), some thought once to have been in Britain (e.g. *L. hippothoe*, *L. virgaureae*), a species that has begun to colonize Britain (i.e. *I. lathonia*) and a species long thought capable of colonizing Britain but which as yet has apparently failed to overwinter (i.e. *Nymphalis antiopa*) (see Supplementary Appendix 4.2). However, this analysis identified 16 currently resident species unlikely to occur in Britain on the basis of geography alone. Many of these are typically vulnerable specialists (e.g. *P. arion*, *P. bellargus*, *Polyommatus coridon*) and species with marginal distributions in Britain (e.g. *Thymelicus acteon*, *Melitaea cinxia*, *E. epiphron*) (see Appendix 9); but the 16 species also include two unexpected entries: species in northern France, with wide distributions in Britain and expanding their ranges on the British mainland (i.e. *Thymelicus sylvestris* and *Pyronia tithonus*) (see Appendix 9). *L. dispar* (coded in the analysis as absent in Britain), a Fenland butterfly now long extinct in Britain, was not predicted to occur in the islands. In a separate analysis of all British species (residents and rare migrants), 14 of the 16 species were predicted to occur in the British Isles, the two exceptions being the *Erebia* spp.

A final comparison can be made between the predictions from the CRA and CGS (see Appendix 9). Thirty-nine species were predicted to occur by both analyses; 31 by climate but not by geography, and 16 by geography but not by climate (see data in Supplementary Appendix 4.1).

In all predictions the reader needs to be aware of assumptions made and the levels of predictions (probabilities). If nothing else, these predictions for species warn us about the limitations of models. It is important that these two distinct analyses based on climate and geography are viewed purely as models; as such, they provide solutions based on limited sets of criteria. If they do not fit the facts (i.e. the CRA predicts the presence of British species in British locations where they are known not to occur), then it is clear that some variable(s) is/are missing from the models or that the methods are inadequate (i.e. the CRA does not account for **spatial autocorrelation** between species locations). What they do encourage are improvements in modelling and understanding. For instance, in the case of the CRA, species distributions were only processed for Europe and not other parts of species ranges (e.g. north Africa). The fact that 16 residents were not predicted by the CGS does not mean that they should not be here, but it may suggest that acceptable conditions are broader than modelled for those species (or that species adaptabilities have changed in Britain or the continent since their entry into Britain (Dennis, 1993, p. 146) (see Fig. 8.2 for Holocene reconstruction of conditions). Crucially, it is important to appreciate that the resource-based habitats of these species are not static entities; the species have been adapting to changing conditions all the time. Evidently, species doing well in Britain now, but predicted not to be here, have been able to adjust to the cooler, damper conditions in western Europe over the latter part of the Holocene. The failure to predict the occurrence in Britain of some species in the CGS, especially *T. sylvestris* and *P. tithonus*, is perhaps surprising and counterintuitive but it should alert us to divergent trajectories in the ecology of species in Britain and on the continent during the Holocene, in which isolation has a significant part to play. It is interesting that the recent discovery of errors in recording *T. sylvestris* may indicate that its distribution has been overestimated in Flanders (north Belgium) (Vantieghem *et al.*, 2017).

Ranges, Distributions and Species Diversity: Detecting Deeper Causes

Above, it was shown how species richness over Britain changes with the heat environment: species numbers decline north-westwards as the regional climate cools and northern islands are likely to have fewer species than southern ones. This does not explain why some species extend further north than others, though it is evident that different butterfly species have varying tolerances of cooler conditions: (i) individual species heat up and maintain thermal equilibria differently depending on size, basking mode, heat-susceptible wing areas exposed and substrates used for thermoregulation (Shreeve, 1992a; Dennis, 1993; De Keyser *et al.*, 2015); (ii) the heat environment of a locality, the local and microclimate, also affects species resources and life history stages differently; and (iii) larvae of species contrast in growth rates and development time, not always intuitively related to their differences in size (Dennis *et al.*, 2012b). At the root of this observation is that butterfly species contrast markedly for life history traits, ranges of tolerances and in resource use described by a continuum in resource generalism/specialism (Dennis, 2010; Dennis *et al.*,

2012a; Dapporto and Dennis, 2013); the reader may like to consider that the life histories of butterflies belonging to different families are in effect as different as gulls are to owls as to finches in the world of birds. A further word of caution: it should not be forgotten that embedded in distributions are the imprints of historical processes (Dennis, 1977); a sophisticatedly innovative recent study of butterfly diversity over the British mainland by Dr Leonardo Dapporto confirmed such patterns and reflected on the long-term differences in susceptibility of species to environmental changes (Dapporto *et al.*, 2015) (see Chapter 8).

Since 1996, a set of studies investigated the impact of life history traits and resource use on British butterfly geography. Two of these took phylogeny into consideration to account for the unevenness of 'physiological distances' between species (Dennis *et al.*, 2000a, 2004). The first study, in 2000, examined the impact of 21 resource and life history traits on five geographical variables describing ranges, distribution cover and island incidences. The spatial variables were found to be closely related to one another, and to species ecology, confirming the 1996 findings for offshore islands (Dennis and Shreeve, 1997). The geographical range of species was found to be most closely linked to the number of biotopes occupied and flight period ($\boldsymbol{R}^2$ = 61%) but the prominence of 'explanatory' factors depended on which species were included (migrants or not); and migration ability (a movement index based on records) was found also to be a prominent influence. Species with wide geographical ranges tend to have high movement indices, a wide variety and ubiquity of host plants and long flight periods and they are often multi-brooded. Those with narrow geographical ranges tend to have low movement indices, sparse and limited host plant resources and short flight periods and they are often univoltine. A follow-up study (Dennis *et al.*, 2005) found that butterfly distributions were largely accounted for (81%) by resource and life history variables linked to numbers of host plants, the unique contribution of host plant number varying only between 14% and 33%. It was shown that specialist feeders have other outlets to compensate for the lack of host plant variety.

What became clear is that many variables seemed to be operating together, some synergistically, in a complex of trait suites, to affect distributions and ranges. Each, however, has some degree of independent action on species ranges and distributions. Two detailed studies clarified the situation (Dennis, 2010): (i) an ordination of 136 non-biotope binary state ecological/life history variables (Shreeve *et al.*, 2001); and (ii) a complex ordination of larval host plants strategies (competitor, stress-tolerant and ruderal strategies = **CSR**) on 86 life history, resource, demographic, physiological, geographical and other variables (Dennis *et al.*, 2004). The degree to which host plants are ruderals, competitors or stress tolerators hugely affects species' geographical expanse, including their incidence on islands. Butterfly species classed as ruderals on their host use are best able to occupy islands and those classed as stress tolerators are least able to occupy islands – including, it is now known, habitat islands in the broad landscape of the British mainland (De Palma *et al.*, 2016).

Recently, this complex of variables has been simplified into two correlated indices of generalism/specialism, **colonfac** (colonization ability) and **migrfac** (migration capacity) (Dennis *et al.*, 2012a; see Appendix 7). Studies are now beginning to focus on the determinants of the crucial link variable between biology and geography, viz. population abundance, as in Curtis *et al.* (2015b) which investigated 1111 populations for 27 species over 54 BMS sites. Accounted variation in abundance between sites was of much the same order as the decade-earlier spatially implicit study on geographical range (i.e. 36%) (Dennis *et al.*, 2005); again, contrasts between specialists and generalists emerged strongly in the results. These studies demonstrate that species biology and ecology underlie not just geographical patterns on mainland Britain but also incidences on offshore islands.

On small islands, species will often be found to be restricted to distinct areas, owing to their resource distributions. This is neatly illustrated for the fossil island of the Great Orme's Head, where *Pieris brassicae* and *Pieris rapae* have a south-eastern to northern bias dependent on garden crucifers, nasturtium and cliff cabbage, quite different from the resource distributions of *H. semele* and *P. argus*, which are primarily located on the south and south-west of the headland (see Fig. 8.7).

Range Changes in Recent Times

In island biogeography research there is a danger that processes are conceived as being homeostatic, in a dynamic equilibrium, lacking sudden pulses of migration or extinction. The history of Krakatoa (Whittaker and Fernández-Palacios, 2007) should warn us against assuming stable states on islands; indeed, sudden changes do not require such catastrophes, just subtle shifts in climate and habitats. From the relationship of butterfly species richness with the temperature and radiation environment, we expect changes in ranges and distributions with changes in the heat environment.

Since the mid 19th century, British butterflies have been known to undergo substantial changes in distributions and geographical ranges (Dennis, 1977). With atlas mapping, this process also became obvious in the late 1980s with the sudden shift northwards of *T. sylvestris*, *P. tithonus* and *P. aegeria* (Hardy *et al.*, 1993) and the reverse: the losses of fritillaries. One butterfly, *Lasiommata megera*, which had been progressively expanding its range until 1984, very suddenly underwent a reversal for the next 30 years (Van Dyck *et al.*, 2015). Such pulses, whether gains or losses, can make for equally sudden changes to an island's fauna, not just because they may affect the island itself, but also because such processes affect the number of potential migrants to islands from nearby sources. A classic case of the former was the dramatic, if temporary, expansion of indigenous populations of both *H. semele* and *P. argus* on the fossil island of the Great Orme's Head during the drought of 1995 (Dennis and Bardell, 1996); this is matched by the recent astonishing expansion of *Aphantopus hyperantus* there, limited to a small area in the south-east in 1993–1996 but found to have covered both the western side and northern end near the lighthouse in substantial numbers by 2015 (M.P. Dennis, personal communication).

Since 1996, butterfly range and distribution changes have been modelled using 'climatic response surfaces' (Atkinson *et al.*, 1987). A study of *P. aegeria* showed that range expansion could be modelled successfully on three climatic variables (Hill, J.K. *et al.*, 1999). Mismatches, where the butterfly failed to occupy the ground expected, could be related to lack of woodland biotope (Hill *et al.*, 2001). Further research on other species showed two processes at work simultaneously: expansions related to climate changes and retractions owing to habitat loss (Warren *et al.*, 2001; Hill *et al.*, 2002). At that stage the implications of resource-based habitat were not fully appreciated (Dennis *et al.*, 2003) though modelled in the 1996 book on British offshore islands (Dennis and Shreeve, 1996, p. 44). In fact, with this knowledge we now know that range and distribution changes can be far more complex; the distribution of any butterfly species may be simultaneously expanding, stable or contracting in one small region, such as north-west England, reflecting a suite of influences, the balance of which differs between localities (Hardy *et al.*, 2014).

Such subtle changes may also be expected on small islands, especially where a gain from climate is directly opposed by habitat loss. Furthermore, habitat loss does not have to be the removal or replacement of an entire biotope, but just some subtle change in land use affecting a key resource, such as a change in grazing regime, or scrub development (e.g. *Maniola jurtina* on Tean, Isles of Scilly, in 1951; Ford, 1964; Dennis, 2015a).

Sources and Range Oscillations of Long-distance Migrants

One basic problem of island biogeography is to determine actual sources of immigrants and colonists. This may seem a simple matter, but it is not; as evidence accumulates for longer-distance movements of species, all the surrounding mainland and island landscapes become potential sources. For most British butterflies, the most likely source is still regarded as the nearest mainland or island, but this is unlikely to be the case for long-distance migrant species that undergo seasonal reversals in movements. For these species, their journey to British and Irish offshore islands may have started on the European or African continents (e.g. *V. cardui*) (Stefanescu *et al.*, 2013). Where migrants involve closely related and physically indistinguishable sibling species, the source populations are particularly difficult to determine; for example, Bath Whites found in the UK have traditionally been attributed to *Pontia daplidice*, but recent evidence would suggest that they are the more eastern sibling *Pontia edusa* (Geiger *et al.*, 1988; Wagener, 1988). Swallowtails (*P. machaon*) found on offshore islands (e.g. the Isle of Wight) originate from the continent rather than from the resident populations on the Norfolk Broads (Fox *et al.*,

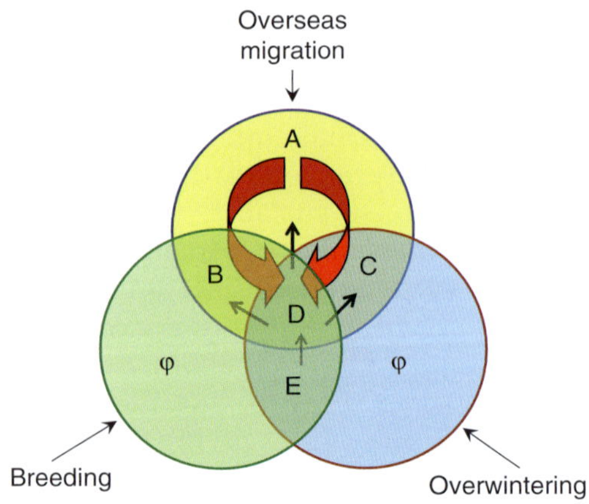

Fig. 4.5. Simple classification of migrant and resident British butterfly species as elements of three intersecting sets: species that migrate, breed and/or overwinter. A, Migrants that neither breed nor overwinter; B, migrants that breed but do not overwinter; C, migrants that overwinter but do not breed in Britain; D, migrants that are also residents, breeding and overwintering; E, residents (breed and overwinter) but whose numbers are not increased by migrants; φ, empty set. The bold red arrows suggest changes with climate warming with an increase in species; the black arrows indicate likely changes with extinctions.

2006, p. 25). To make matters more complex, the nearest sources may additionally supply immigrants (annual second broods) to the smaller islands for those long-distance migrating species that are able to breed in Britain (e.g. *V. atalanta*, *V. cardui*, *C. croceus*) and increasingly, as we have seen, some of these exceptional migrant species are able to overwinter and are beginning to take on the status of residents (Fig. 4.5).

In Britain, we are fortunate to have a number of developing valuable data sets which accumulate records on migrants (e.g. the Butterfly Monitoring Scheme: Biological Records Centre, 2006; Migrant Watch: Butterfly Conservation, 2017). The latter records the annual immigration of *V. cardui* and the Humming-bird Hawk Moth *Macroglossum stellatarum* (L.) since 2009. For *V. cardui* the data are (numbers seen/recorders): 2009 (12,800/unrecorded); 2010 (630/unrecorded); 2011 (611/539); 2012 (353/289); 2013 (862/583); 2014 (621/501); 2015 (3252/1965). There is also an older unique data set for nine long-distance migrant butterfly species covering 113 years (Williams, 1965, 1971; see Sparks *et al.*, 2005 and Appendix 1). From these, a number of observations can be established:

1. The migrants do not occur in the same numbers each year. Occasionally there are huge pulses of immigrants (e.g. *V. cardui* in 2009); often, the migrant species may be virtually absent or scarce for a number of years and then will arrive in large numbers. Recently, there has been an incursion of the rare Scarce or Yellow-legged Tortoiseshell *Nymphalis xanthomelas* (mainly east England, 2014) (Bowles and Fox, 2014).

2. Different species tend to have their 'migration years' (Williams, 1965); when the patterns are closely inspected it is found that those occurring in the same year share the same general source and UK status (breeding/overwintering) (Fig. 4.6). The long-term trend in the number of migrants has been tracked down to climatic factors (seasonal weather patterns) (Sparks *et al.*, 2005). Despite the often temporary nature of these species on offshore islands they are, nevertheless, fascinating components of the butterfly fauna that visitors to islands may expect to see.

The details in this chapter should impress on the reader that the butterflies on the British and Irish islands are unusual; from the vantage of the entire Holocene, they are a depauperate fauna and the loss of species has resulted in a unique regional fauna, different enough to emerge as a distinct faunal region in an analysis of butterfly distributions across Europe (Dennis *et al.*, 1991, 1995b). At much finer time scales of decades, different dynamics are evident. Unfortunately, in recent decades, losses of populations, and the threat of losing more species, still dominate concerns of Butterfly Conservation and the Centre for Ecology and Hydrology.

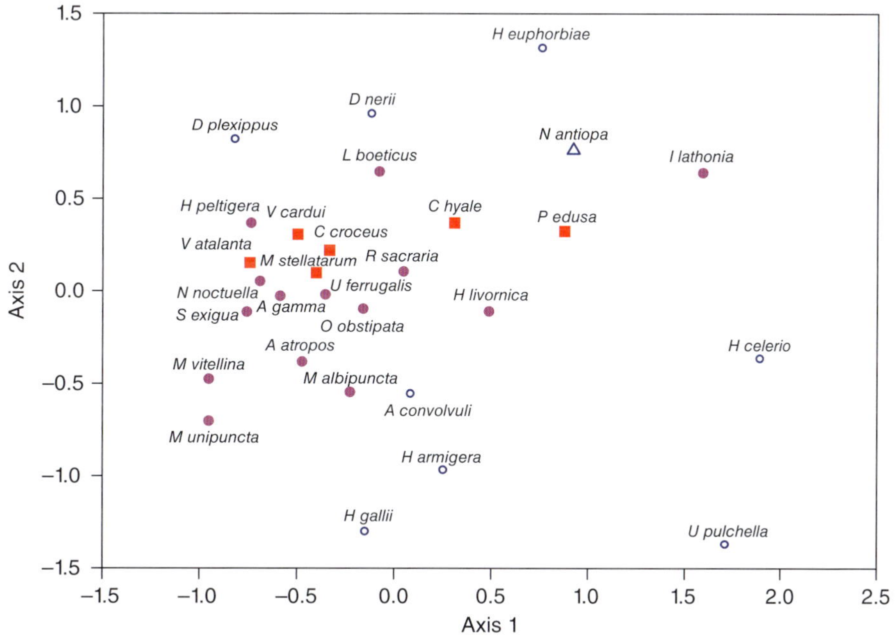

Fig. 4.6. Non-metric multidimensional scaling plot of migrant species (butterflies and moths) to the British Isles (based on correlations for standardized data for 113 years of records; Kruskal S = 0.19). Distinct life history groups of species are detected whose dates of migration are similar. Open circles, rarely able to breed or overwinter in Britain; open triangle, able to overwinter but do not breed in Britain; pink circles, breed but generally do not overwinter in Britain; red squares, both breed and overwinter in Britain. (Redrawn from Sparks *et al.*, 2005, courtesy of John Wiley & Sons Ltd.)

However, other species are expanding their ranges (e.g. *T. sylvestris*, *P. aegeria*, *P. tithonus*) (Fox *et al.*, 2015). If we really understood the habitats of our species we would be able to explain why each species behaves in exactly the way it does: whether it remains stable, expands or contracts. From year to year, brood to brood, different dynamics are evident; the status of source populations for offshore islands on mainland Britain and Ireland, as well as island populations themselves, are in continual flux. It is at this level that we must seek to understand how habitat components for species affect their destinies. Notably there are (i) fluctuations and (ii) distinct trends in species populations, often linked to weather and climate events (Pollard, 1988; Isaac *et al.*, 2011a). It is the variation within and among sites that holds clues as to the interaction of habitat components with weather and climate conditions and that will help us to understand better each butterfly's demands for resources and conditions. With the pressures on the land, pessimism for British and Irish butterflies is understandable. Even so, with a view to the future, much as residents may be lost, novel species may also appear, as expected from the increase in migrant species with climatic warming (Sparks *et al.*, 2007). This chapter points to some of the butterfly species that, hopefully, may arrive on the islands in the near future. The lesson from later chapters, which relate butterflies on offshore islands with their source populations on the British mainland and Ireland, demonstrates that this will depend, in turn, on how habitats are treated on the continent of Europe for the potential new colonist species.

5 Island Associations and Species Affinities: Geographical Influences

It is clear from the foregoing chapters that geography is a strong predictor of the species occurring throughout the British mainland. In this chapter, as a first step, we present a brief overview of all European islands. These present an island biogeography context for the butterfly faunas of the British mainland and Ireland, the two micro-continents, which act as key sources of species for the host of surrounding satellite islands and islets. Then, we examine the geographical factors influencing differences between the British mainland and Ireland in species richness and incidences. A platform is thus constructed for investigating patterns in numbers of species over the smaller islands: initially from an island viewpoint by focusing on associations among islands for their butterfly faunas; subsequently from a species viewpoint by considering the associations among them for their occurrences on islands.

European Islands: the Wider Context

It is useful, initially, to present the current study of British islands in a broader context of islands found around the European continent, as the latter display a much wider array of island biogeographic processes than found in Britain and Ireland (Dennis *et al.*, 2008). A study of 440 European butterfly species revealed that 279 of them (63.4% of the whole) occurred on 564 islands which comprise approximately 6.2% of European land area (629,263 km^2 of 10.2 million km^2). This study has now been updated with additions for islands from the Aegean and Atlantic seaboards, and more prominently by extensive fieldwork carried out subsequently in the west Mediterranean, especially islands in the Ligurian and Tyrrhenian Seas (Dapporto *et al.*, 2014, 2017; Vodă *et al.*, 2016), and the Adriatic

(Croatian islands) (Withrington and Verovnik, 2008; Verovnik, 2011; Koren *et al.*, 2015). The data are also supplemented by studies of Russian islands (i.e. the Barents Sea) (Bolotov, 2006; Bolotov and Shutova, 2006) (for European island data, see Supplementary Appendix 5.1a). Much of the total area of European islands studied in 2008 is taken up by eight very large islands (British mainland, Ireland, Iceland, Corsica, Sardinia, Sicily, Crete and Cyprus: 495,400 km^2) occupying 78.7% of whole island area; even so 259 (58.8%) species were found on the remaining smaller islands. Most species were discovered to occupy, proportionately, very few islands (69% on < 5% of islands; 67% corrected for the 'size' of island groups, that is, for the number of islands in each group), though a number, mainly long-distance migrants, are found on over 50% of islands (Fig. 5.1) (Dennis *et al.*, 2008).

In Dennis *et al.* (2008) analysis focused on 300 islands considered to have been sampled adequately (using migrants as markers of sample adequacy) (for European island geographical data, see Supplementary Appendix 5.1b). Species richness shows only a weak locational pattern determined by latitude and longitude, and is related as expected mainly to isolation from the continental source (**coefficient of determination**, $\boldsymbol{R}^2$ = 48%) and island area (R^2 = 11%) (both $P < 0.0001$). More interesting from the European viewpoint is the distribution of species **rarity** and species endemicity. **Endemism** and rarity can be defined from widely different vantages. Rarity was measured using indices based on species

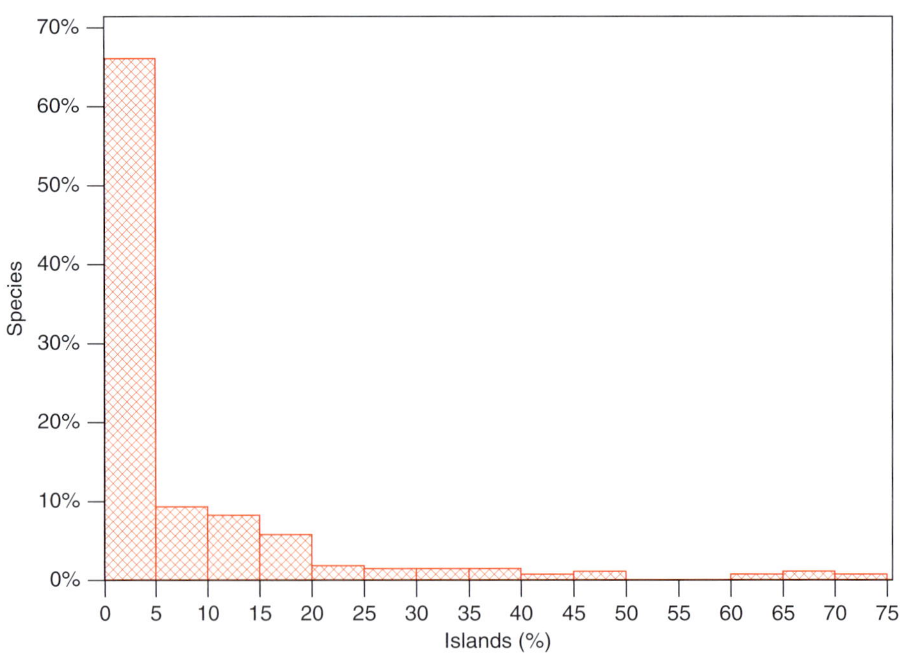

Fig. 5.1. The distribution of island occurrences (number of islands occupied) for European butterfly species (540 islands, 440 species) (Dennis *et al.*, 2008).

abundances in relation to spatial cover (i.e. the inverse of the mean proportions) for distributions of individual species over 44 countries and basic recording fields in the European butterfly atlas (Kudrna, 2002). In the study by Dennis *et al.* (2008), endemicity was measured in two ways: from a European (European endemics, Ee) and individual island (island endemics, Ei) viewpoint. Both types of endemicity have distinctive geographical bias to lower latitudes (i.e. southern Europe and Macaronesia), the additional key variables being island area for European endemics, and longitude, isolation and island elevation for island endemics (all $P < 0.0001$). The endemic species occur on islands now under increasing pressure from climate change and, increasingly, intensive human exploitation (Dapporto *et al.*, 2017). The low-latitude islands provide potential refuges for European species during each glacial advance when the most northern islands, including those in northern Britain, are swept clean of butterfly species (see Chapter 8). Latitude (low latitude) is also the key variable for rare species (accounting for R^2 = 58–62% of variation), isolation and area contributing small amounts to overall variation in rarity (R^2, 2% to 8% each) (Dennis *et al.*, 2008).

Britain does not have any island endemics, but was recorded as having four European endemics (viz. *Aricia artaxerxes*, *Polyommatus coridon*, *Erebia epiphron*, *Hipparchia semele* – species restricted to Europe west of the Urals and excluding Turkey) by Dennis *et al.* (2008); this figure may change as more cryptic species are found; for example, the discovery that '*Leptidea reali*' in Ireland is actually *Leptidea juvernica* (Dincă *et al.*, 2011a).

Comparing Butterfly Richness Profiles of Britain and Ireland

The position of Ireland in relation to the European continent and the larger British mainland is that it has had (and continually does have) the potential of being a recipient of species from the two larger more easterly sources. Ireland (35 species) has recorded 34 of the resident species and three migrants found on the British mainland (see Appendix 1 and Supplemenary Appendix 5.2); however, it is clear that one of them, *Papilio machaon*, is a **vagrant** (Fox *et al.*, 2006). The list of species has not remained unchanged. During the past two decades, since the early biogeographical review by Dennis (1992, pp. 1–21), a second Wood White (*Leptidea*) has been discovered (see below); also two *Thymelicus* species and *Polygonia c-album* have been added to the fauna (Phelan, 2008; Regan *et al.*, 2010; Irish National Biodiversity Data Centre, undated). The first Comma butterfly was recorded in 2000 in County Wexford (O'Donnell and Wilson, 2009). A *Thymelicus* species was initially reported by J. Goodwin in 2006 (Wilson *et al.*, 2007). Some confusion seems still to exist on the presence of the two very similar British *Thymelicus* species in Ireland; in fact, both now have been reported to be present, with *T. sylvestris* in Co. Kildare and *T. lineola* in Co. Wexford (Irish National Biodiversity Data Centre, 2017; Harding and Jacob, 2013; Wikipedia, 2017).

Using a dispersion diagram, Dennis (1992, p. 8) showed that species occurring in Ireland were found further north on the British mainland than those not occurring in Ireland, nicely supporting the contention of Williamson (1981) that island faunas tend to be more 'polar' than those of their sources. He also demonstrated that many species occurred on the British mainland with northern range margins that were not found in Ireland; that is, they have range margins further north than some species found in Ireland (e.g. *Thecla betulae, Leptidea sinapis, Celastrina argiolus*). A number of the absentees from Ireland were noted to lack host plants or to have very restricted host plant distributions (e.g. *Thymelicus acteon, P. machaon* (except as a vagrant), *Polyommatus bellargus, P. coridon, A. agestis and Aricia artaxerxes*) (Dennis, 1992).

These patterns are here updated (Fig. 5.2a). There is a significant difference in the latitude on the British mainland of the northernmost records for species currently found in Ireland compared with those absent from Ireland (Mann–Whitney $U = 104.5$, $Z = -4.57$; *L. juvernica* and three migrants *Colias croceus, Vanessa cardui, Vanessa atalanta* excluded); this is matched by data for southernmost records on the British mainland ($U = 129.0$, $Z = 4.15$) and species latitudinal ranges ($U = 86.0$, $Z = -4.88$) (all $n = 25, 30$; $P = 0.0001$) with Irish species having larger geographical ranges on the British mainland than those exclusive to the mainland (Fig. 5.2b). Nevertheless, for the 30 Irish species concerned, the range boundaries and range sizes in Ireland correlate significantly with those on the British mainland (Spearman r_s: northern boundary 0.67, southern boundary 0.77, latitudinal range size 0.78; $P < 0.0001$). Another interesting view of Irish versus non Irish species is given by a plot of all species in coordinates of British mainland range size (north–south) versus the latitude of the most northerly record. These two variables are highly, but imperfectly, correlated ($r^2 = 78\%$) (Fig. 5.2b). It is very noticeable that a small group of (currently) strictly British mainland species, with limited southern boundaries into England, 'fall off' the regression line (i.e. *E. epiphron, Erebia aethiops, A. artaxerxes, Carterocephalus palaemon). Coenonympha tullia* has a similar pattern but is a characteristic Irish species of raised bogs and there are old Irish records of *E. epiphron. C. palaemon* is the most extreme of the four species but it has become so only in recent history with the extinction of the English populations.

These observations are highly relevant to determining the reasons for differences between British and Irish butterflies. There are two basic potential scenarios: (i) some species never reached Ireland; and (ii) they colonized Ireland but have become extinct during the Holocene. The dramatic recent losses of populations of *C. palaemon* in England, the abundance of *C. tullia* in Ireland and the 'ghost' records of Irish *E. epiphron* should warn against assuming that these and other butterflies never managed to colonize post-Devensian Ireland but the probability of Holocene colonization in Ireland decreases for butterflies with small geographical ranges on the British mainland and confined to southern Britain. Referring to one of the finer points of distributions, Dennis (1992, p. 9) noticed that two species extended further north in Ireland than on the British mainland. At that stage *L. sinapis* was thought to cover the whole of Ireland; in fact

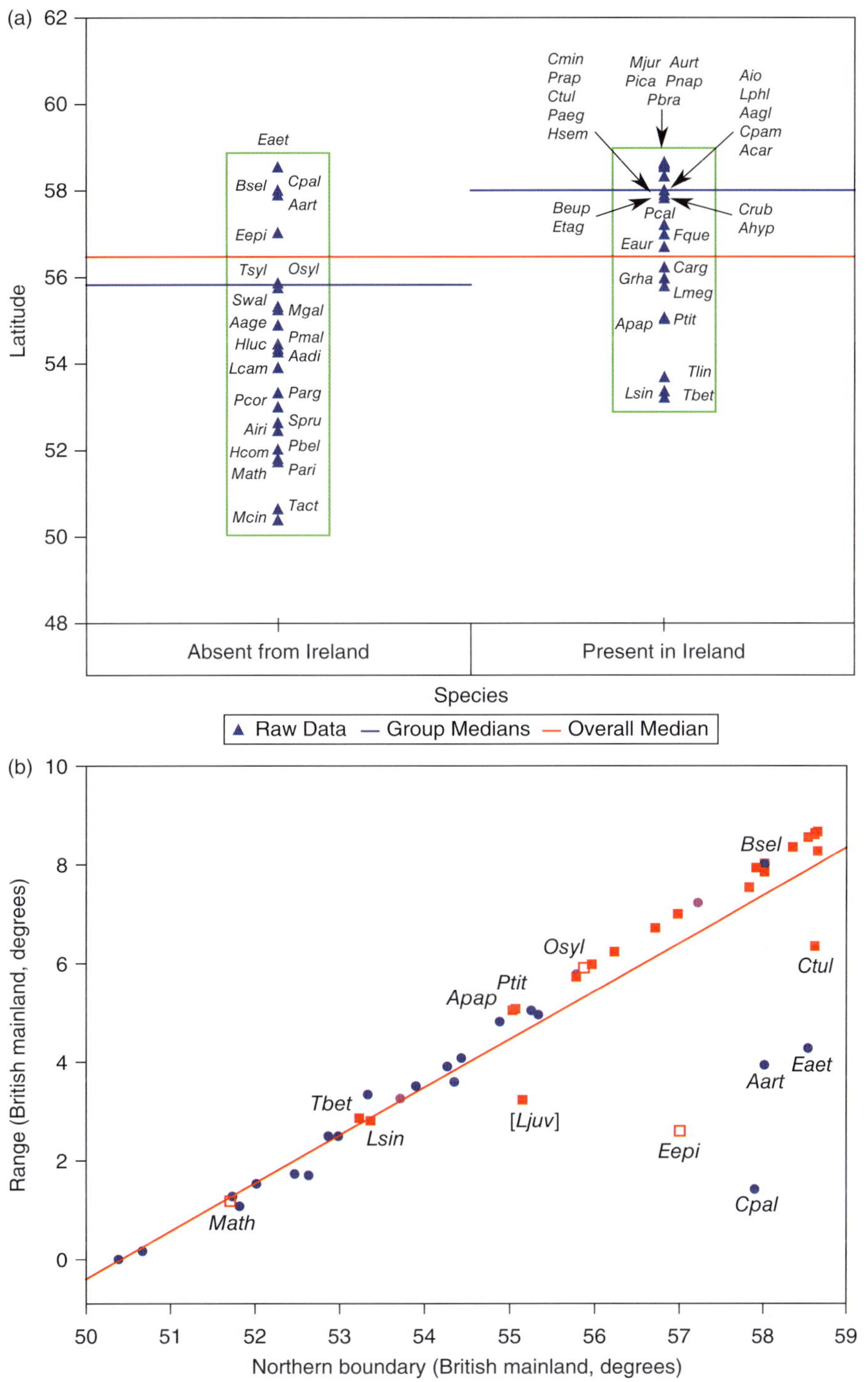

Fig. 5.2. Comparison of current Irish and British species for geographical range (latitude). (a) Dispersion diagram of species currently present and absent from Ireland for latitudinal extent on the British mainland. Species labelled by first generic letter and first three specific letters (see Appendix 1 for list). Red line, overall

it is *L. juvernica* in Ireland that extends further north than *L. sinapis* on the British mainland. The other species is *Argynnis paphia* and the question remains: why is it that *A. paphia* should do better in Ireland than in mainland Britain? Is there yet another cryptic species or does *A. paphia* have a wider habitat range (tolerance of conditions) in Ireland than in mainland Britain?

Ireland's Unusual Case of the Cryptic Wood White

In one important respect, the butterfly faunas of Britain and Ireland differ significantly: Ireland has the two *Leptidea* species (Wood Whites) to Britain's one. The species found throughout the British mainland is referred to as *L. sinapis* (Wood White); this once had a wide distribution far north into Cumberland (now Cumbria) but there has been a systematic decline of populations over the entire country during the 20th century (Asher *et al.*, 2001). Ireland has two species: (i) *L. sinapis* with its headquarters on the ecological island of the Burren limestone; and (ii) *L. juvernica* (Cryptic Wood White) found elsewhere throughout Ireland and, so far, found only on one of the smaller offshore islands: Great Saltee.

The situation in Ireland has had an interesting history. Initially it was thought that all Irish Wood Whites belonged to a single distinct subspecies, *L. sinapis juvernica* (Williams, 1946), named after a specimen from Kildare having darker olive-green ventral hindwing markings (for a detailed description, see Dennis, 1977, p. 144). The homogeneity of the butterfly's **phenotype** in Ireland was initially questioned by Heal (1965) when he found the butterfly expanding along Irish rail embankments. Peter Russell (2015) neatly summarized the subsequent history. The first significant advance in understanding was made on the discovery of a new Wood White in the Pyrenees, *Leptidea lorkovicii* (Réal, 1988), an invalid name changed to *Leptidea reali* (Reissinger, 1989; Lorković, 1994). The first indication of a different species in Ireland was by Nelson and colleagues (2001, 2011) who reported *L. reali* as occurring in Ireland. This 'new' species was subsequently found to occupy much of Ireland, with the exception of the Burren, where *L. sinapis* is incumbent. However, the subspecies term *juvernica* (originating as noted above in Kildare) could no longer be linked to *L. sinapis* on the Burren (*pace* Riley, 2007).

Fig. 5.2. Continued.
median; blue line, medians for Ireland and Britain; green boxes, quartile ranges. (b) Distribution of species for geographical range and northern limit on the British mainland. Blue dot, absent from Ireland; pink dot, arrivals in Ireland during the current millennium; open squares, species with old records in Ireland; red squares, currently present in Ireland. Data for *L. juvernica* taken from Ireland.

Confusion was potentially increased when Vlad Dincă and colleagues (2011a) discovered that *L. reali* actually comprised two species, a discovery made possible by the development of novel bar coding techniques. It is the new species, the Cryptic Wood White, that is found in Ireland. As the description by Williams (1946) applied to this species, it was named *Leptidea juvernica*. Thus, Ireland has the honour of having a unique species on its soil and first described there. Apparently, neither *L. juvernica* nor *L. reali* occur on the British mainland, but the authors would not be surprised if this finding was eventually reversed with the finding of traces of them, 'ghost' DNA, occurring in British mainland Wood Whites, much as Neanderthal DNA has been found in modern humans. At the time of writing, such a finding eludes us (Verovnik and Glogovčan, 2007; Solovyev *et al.*, 2015), as it depends on examining the entire, or huge sections of, genomes of individuals as carried out for the genetic map of the British people (Leslie *et al.*, 2015). With rapid advances in DNA technology, no doubt this process will become a standard one in due course. A question remains: was the extinct Cumbrian *Leptidea* species *L. sinapis* or *L. juvernica*?

The Influence of Geography on Island Species Richness

Over the past five decades there have been several attempts to determine the factors responsible for butterfly species richness on the British and Irish offshore islands (i.e. Dennis, 1977, 1985b; Hockin, 1981; Reed, 1985; Dennis and Shreeve, 1996, 1997; Dennis *et al.* 2008, 2012a). The early attempts were based on few islands and uncertain records. In the Dennis and Shreeve (1996) study some 73 islands were considered to have sufficient data to investigate patterns. Using the same criteria (records for four regular migrants *Pieris rapae*, *Pieris brassicae*, *V. atalanta*, *V. cardui*) (see Chapter 3), this number is now 123, though geographical data have been listed, and records sought, for some 910 islands off British and Irish shores whose area is 10 ha or greater (see Appendix 3 and Supplementary Appendix 3.1). In fact, the present study has nevertheless picked up a substantial number of islands under 10 ha (n = 261), the smallest being 0.7 ha.

A set of clearly defined geographical variables is now well known to be prominent in accounting for species richness (S) on the offshore islands: together with island area (A) and isolation from sources (I1, distance to nearest mainland source of Britain or Ireland; I2, distance to nearest larger source in area or in species numbers) and the size of faunal sources (FS1, FS2, species at sources corresponding with I1 and I2, respectively), these include island maximum elevation (ELEV), latitude (LAT) and longitude (LONG) of islands, and a simple measure of the number of surrounding islands or satellites (SAT) (Fig. 5.3; see also Appendix 10). For a number of islands in the three provinces of Ireland, Britain and the Isle of Man (PROV) in this study we also have measures of human population size (POPNO) and population density (POPDENS).

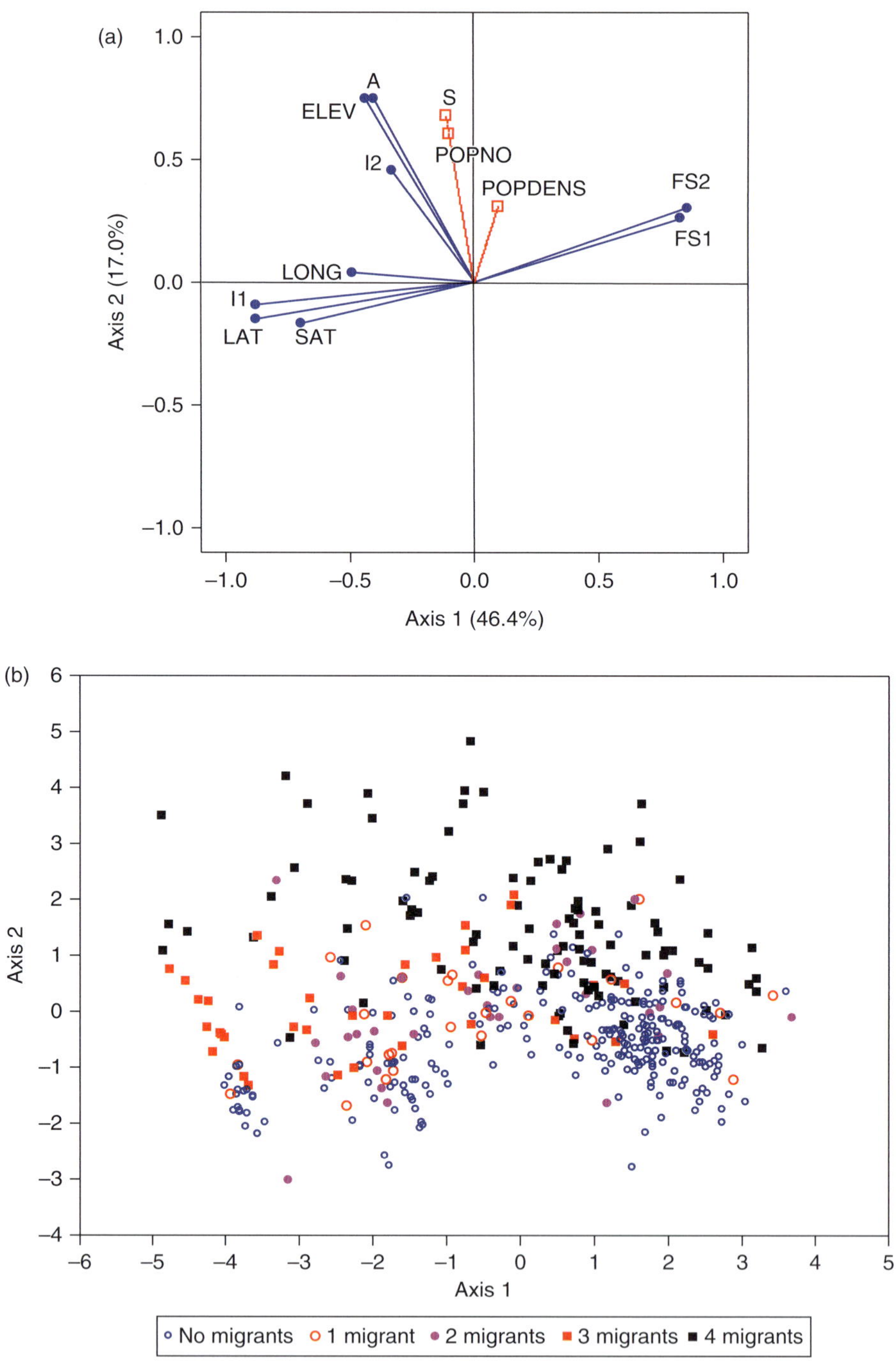

Fig. 5.3. Ordination (PCA, unrotated axes) of geographical variables for the British and Irish islands. (a) Relationships among variables for listed islands (see Appendix 3) in the first two axes (% variance): blue dots, active variables; red squares, supplementary variables; LAT, island latitude; LONG, island longitude;

At the outset of this section, it is vitally important for readers to understand what it is currently possible (and not possible) to disclose about the factors that control species richness on the British and Irish offshore islands. As an equitable sampling strategy has not been available for determining the number of species on islands, the data have inbuilt biases. Firstly, the records for islands are the product of cumulative accounts made over many years; ideally we would wish for systematic sampling of biotopes on islands throughout the same years/seasons using techniques allied to the Butterfly Monitoring Scheme (see Appendix 8). Secondly, over time, as more islands are added to the list it is inevitable that: (i) these include ever smaller islands; and (ii) the species list for individual islands will grow in number though there may be little information on butterfly population numbers and species status (i.e. turnover owing to colonizations and extinctions). Smaller islands are unlikely ever to be as well sampled for species as larger islands, if only because larger islands are likely to have resident human populations (potential recorders; places where recorders can sustain long-term stays; convenient transport, etc.) which small islands typically lack; moreover, landing on smaller islands is an altogether more hazardous affair or restricted (e.g. Little Skellig bird reserve off south-west Ireland) and consequently stays are likely to be of short duration. As there are so many islands, those that are systematically surveyed are of necessity a small subset unrepresentative of different aspects of island geography. A fundamental issue is that, if sampling is biased in any way in terms of island geography (i.e. size or proximity to sources) then, as the number of islands increases, so will the relationships change between variables outlined above and, in turn, their impact on species richness. The question is, in the case of British and Irish islands, in just what way are relationships affected? These points have consequences for any island investigation; even so, this study of British and Irish offshore islands is valuable if only in demonstrating the complexity and pitfalls in island biogeography work. In fact, it has revealed much more than that, as subsequent chapters disclose: primarily, the precariousness of faunas on our offshore islands.

In this section, we focus attention on the contrasts in findings for the 1996 study and the present study for the same island selection criteria as regards: (i) differences in island geography; (ii) changes in the relative and absolute importance of variables affecting species richness; and (iii) changing relationships (interactions) among variables describing island

---

Fig. 5.3. Continued.

A, island area; I1, island isolation (mainland); I2, island isolation (nearest source); ELEV, island maximum elevation; FS1, mainland faunal source species numbers; FS2, nearest faunal source species numbers; SAT, island satellites (Satellites 0, 0; 1,1; 2, 2 to 10; 3, 11+); S, island species richness; POPNO, human population numbers; POPDENS, human population density. (b) Scatter of islands in the first two PCA axes distinguished for numbers of four migrant markers (*Pieris rapae*, *P. brassicae*, *Vanessa cardui* and *V. atalanta*).

geography. A simple way of showing how geography of the islands in a study is likely to change as more islands are included is to compare correlation matrices for different samples of islands (see Appendix 10 part B). Plotting the relationships for island geography in reduced axes (**principal components** ordination) gives an instantaneous visual impression of variable associations (Fig. 5.3a); so, equally, does a plot of islands in these axes having different numbers of four migrant markers (two *Pieris* species and two *Vanessa* species) impress on variable sampling effects (Fig. 5.3b). The first plot (Fig. 5.3a) demonstrates the close relationship among specific geographical variables when determined for all islands listed in Appendix 3 (data in Supplementary Appendix 3.1). As might be expected, the pattern of correlations (Appendix 10 part B) shows that large islands (A) are also the highest islands (ELEV), and the size of nearest larger faunal sources (FS2) relates closely to the size of nearest mainland sources (FS1). Perhaps less expected is that northern islands (LAT) are more isolated from mainland sources (I1), and that the largest islands (A) have more satellites (SAT). Island isolation from the nearest mainland (I1) also shows a different pattern of correlations from isolation from the nearest larger source (I2). The effect of increasing the number of islands studied can be gleaned from comparing the summary statistics and correlations for the complete data file and that for the 123 islands with records of all four migrant markers (Appendix 10 parts A, B).

A direct comparison of the geographical parameters for islands in the 1996 study and the present one reveals differences for five of the six geographical variables (i.e. island latitude, area, elevation, both measures of isolation; see Appendix 10 part A). This has consequences for their impact on island species richness. Changes in the determination of factors influencing species richness on islands are also likely to be affected by the share of offshore islands included from around Britain and Ireland, respectively, as Britain has more butterfly species than Ireland. The British and Irish offshore islands have overlapping but distinct frequency distributions for geographical variables. When islands in the entire file are compared, Irish and British offshore islands differ for latitude, longitude, elevation, isolation from the mainland and for the number of satellites, but not for island area or isolation from the nearest larger source (see Appendix 10 part C). However, these differences disappear, except for longitude, when islands are selected (recorded as having four migrant butterflies), even though the proportion of Irish islands has increased from 13% (9 of 68 islands) in 1996 to 17% (21 of 123 islands) in this study.

Before directly comparing the factors influencing species richness, it is first useful to recall the ploy and findings of the 1996 study. An inspection of simple correlations revealed that island species richness was dominated by island and source latitude and the size of the faunal sources, such that for species richness S, total variation was accounted for in the descending order of FS2 > FS1 > L1 > L2 > I1, with I2 and A emerging as non-significant. The aim then was to determine the contribution of island dimensions and isolation to species richness; the method used was

partial regression analysis which first abstracted the variation owing to faunal sources and latitude prior to discerning the amount of **residual** variation subsequently determined by island area and isolation. A model of the chain of relationships is illustrated in Fig. 5.4. Previously, in the 1980s, it had already been determined that the declining species richness gradient was closely related to latitude, decreasingly suitable conditions for butterfly development and activity found towards the north and west in Britain (see Chapter 4). Thus, faunal sources for islands were also likely to decline with latitude. These controls proved dominant to island characteristics (area and isolation). Some 63% of the variation was found to be accounted for by island latitude and the size of nearest larger sources FS2 (more than by latitude and the nearest mainland sources FS1, on which nevertheless nearest larger sources depend). The two presumed prime island biogeography variables (island area and isolation) were

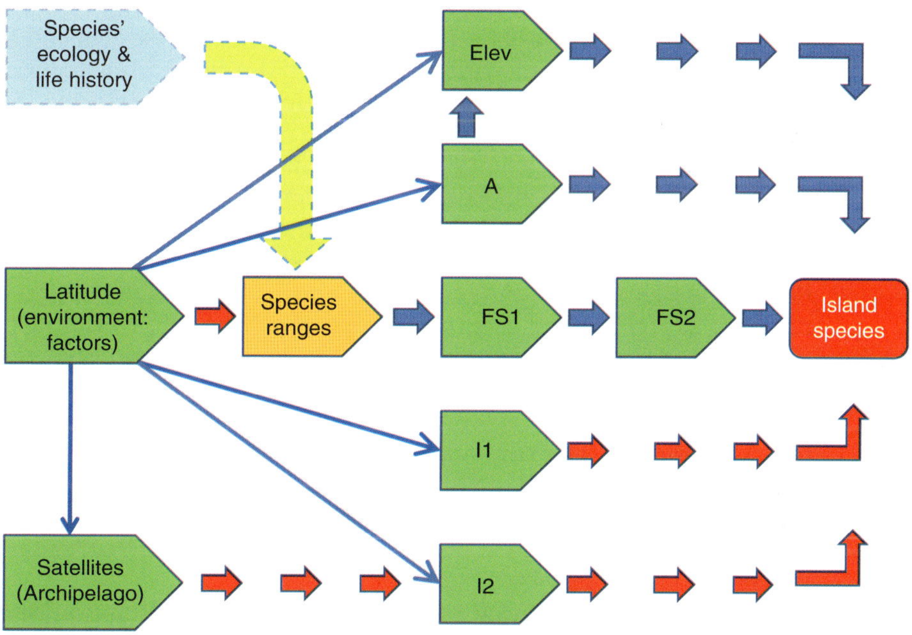

Fig. 5.4. A simple model of basic geographical factors influencing numbers on British islands. A, island area; Elev, island elevation; I1, distance (isolation) to nearest mainland of Britain or Ireland; I2, distance to nearest potential larger source (either larger island or island with more species); FS1, number of species at nearest mainland source; FS2, number of species at nearest larger source. Satellites describe an archipelago effect, islands surrounding target islands. Arrows: red, negative relationship; blue, positive relationship. Thick arrows: main impact through the effect of latitude (climate, resource availability) on butterfly geographical range size, influencing the number of species at sources for islands. Thin arrows, geographical associations imposed by latitude. Numbers and incidences of species at sources also depend on species ecology and life history (migration capacity, colonization ability).

found to account for very limited amounts of the residual variation in island species richness (R^2: A, 26–37%; Isolation I1, 17%) and less still of the total variation in island species richness (R^2: A, 14%; I1, 9.8%).

Multiple regression (forwards stepwise entry of variables) provides very different solutions for the 1996 data file (n = 68 islands; Channel Islands excluded) and the current one (n = 123 islands, the enlarged file) (Fig. 5.5a, b). For the 1996 data, the total variation accounted is 75.4%; the variables contributing significantly to this total are nearest larger source (FS2, 47%), island area (A, 20%), latitude (LAT, 6%) and isolation (I1, 2%). In the current study, the total variation is some 10% less at 65.0%, shared out among the nearest faunal source (FS2, 30%), island area (A, 31%) and island latitude (LAT, 4%). By increasing the number of islands being investigated, not only has there been a loss of explanatory power, but also the order of variables has almost been overturned, area becoming slightly more prominent than faunal source size. Isolation explains little and is no longer significant. A glance at the simple correlations (see Appendix 10 part B) – where the correlation between species' richness (S) and isolation (I1) is much higher than with island area (A) – would seem to contradict this result; it is explained by the observation that both latitude (LAT) and faunal source (FS2) correlate far more highly with island isolation than with island area, rendering isolation redundant in a multiple regression based on stepwise entry of decreasing contributions to total variation in species richness. This relationship is evident in Fig. 5.3a, which describes the pattern of associations among variables for all islands (listed in Appendix 3). Perhaps less obvious than the findings above is that relationships between numbers of species and geography are also determined by the distribution and range of geographical measures relative to the capacity of organisms to migrate and colonize islands, a subject ill understood. This issue is expanded on in Chapter 6. Although in the 1996 study the mean (± standard error) isolation to the nearest mainland source was 28.7 ± 5.0 km, the mean distance to nearest equivalent or larger source was 7.0 ± 1.1 km, with 66% of islands less than 5 km from such a potential source. This distance of 7 km is well within the transfer capabilities of most British butterflies, whereas 30 km is more testing.

Readers are now bound to have at least one question on their minds: what if all the 910 islands listed should have all four migrants and we could include them in such an analysis? The answer is that the total explained variation will likely again dip and that the order of relationships will change. Why so? A number of reasons should be immediately apparent from having read Chapter 1 but one stands out. As more small islands are added to the file the relationships (correlations) between geographical variables will continue to change. Furthermore, despite the imaginary occurrence of long-distance migrants on smaller islands, these records will be largely of passing vagrants discovered on occasional brief visits; few other species are likely to be found on the small islands, if only because potential for persistent colonization is extremely low and survey time is restricted. On the other hand, larger islands will continue to be more intensively studied. Consequently, the disparity between records for

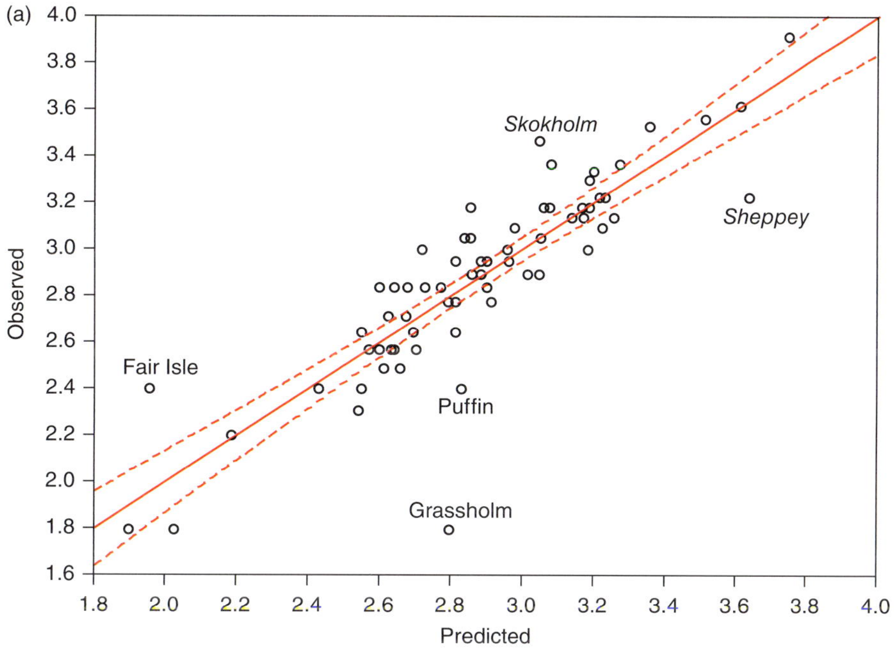

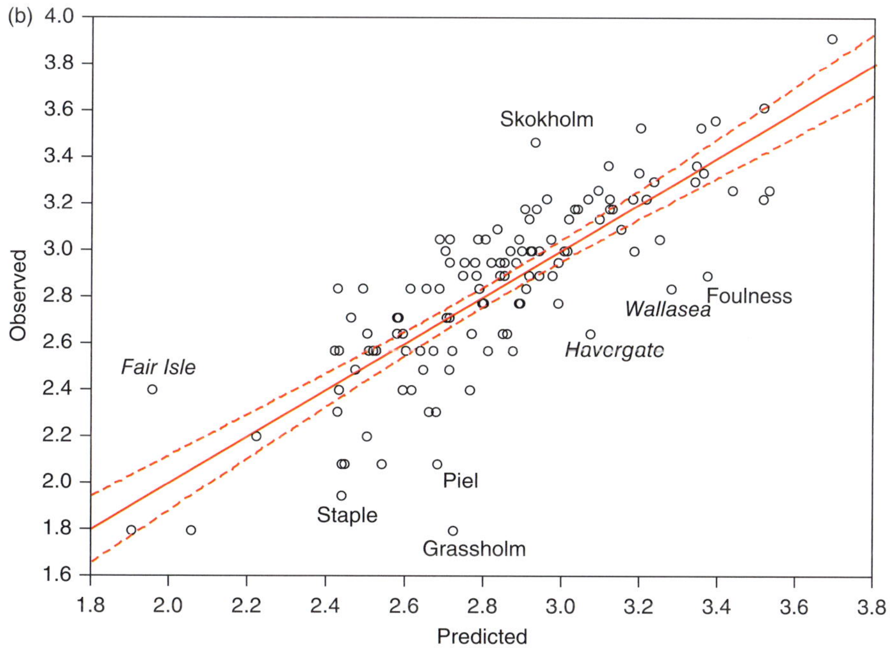

Fig. 5.5. Observed versus predicted distribution for species richness on British and Irish offshore islands. (a) 1996 study islands (n = 68) based on nearest faunal sources, island area, latitude and isolation from the mainland source. (b) 2017 study islands (n = 123) based on nearest faunal sources, island area and latitude. Regression line with 95% confidence limits. Islands marked are outliers, those in italics approaching that status.

large and small islands will continue to increase. This issue is highlighted when we discuss the findings for the Isles of Scilly.

How perhaps relationships among dependent and predictor variables in such island biogeographical analyses will change may be gleaned from the simple correlations of isolation and area with latitude and the size of faunal source (see Appendix 10 part C); note, for instance, that the correlation of latitude and faunal source increases with isolation from the mainland (I1) but decreases with isolation from the nearest larger source (I2). We would expect the isolation measure from the nearest larger source (I2) to become more important than isolation from the mainland (I1). The scatter of 'points' around the regression lines (Fig. 5.5a, b) will increase. This points to the accumulation of other, confounding (unique) influences affecting different islands; consider, for instance, the situation where much of the island is covered in a gannet colony (e.g. Bass Rock: Plate 1.3; Grassholm: Plate 1.9). In recent years there has been a growing focus on what has become known as the **small-island effect** (viz. the hypothesis that species richness below a certain island area threshold varies independently of island size) (Woodroffe, 1986; Dengler, 2010); much of this is likely to be triggered by an increasing range in specific (unique) island influences of smaller islands on island faunas.

Here, a short diversion may be useful to warn of technicalities in regression analysis and we apologize to those readers who may find this less palatable reading. The whole point of multiple regression analysis is to sift among the covariances of predictor variables in the search for ordered primacy of influential factors of species richness. But this process brutally selects out dominant variables; where variables are highly correlated it is not possible, without more stringent modelling (i.e. replacing the original variables with ordinated new variables in principal components analysis), to determine the unique contribution of individual variables. Also this approach summarizes conditions over a set of 'cases' (islands); undoubtedly there are individual islands where the effect of isolation is more prominent. To illustrate this point we consider a case of two islands in the Bristol Channel in the next chapter. In the meantime, it is instructive to examine the distribution of islands in the scattergrams for the two periods studied above (Fig. 5.5a, b); a distinctive feature of both plots is a number of **outliers**, islands that fit the model least well and have the largest residuals from the regression lines. Clearly, some islands have more species than expected (e.g. Skokholm), while others have many fewer species than expected. For example, Grassholm (Plate 1.9), formed of **keratophyre**, fails predictions for both sets of data; it is an island exposed to the full blast of gales and the fact that it houses 39,000 pairs of gannets (2009 figure, 10% world population) (Murray *et al.*, 2015) is not an insignificant detractor for butterflies. Yet, a range of factors can be responsible for the 'lack of fit' for islands having larger residuals from the predicted regression line; those having too few species may simply have poor resources for butterflies (for example, they may be precipitous with dense seabird colonies; or covered in urban and industrial development and toxic waste) while those with excess species may be particularly blessed with varied substrates, suitable biotopes and sheltered terrain affording woodland. Inevitably, differences in study intensity can produce a similar effect.

Modelling has a further advantage in that it also provides opportunities for predicting numbers of species on islands that have been inadequately studied or rarely visited, a subject explored in the next chapter.

Geography of Archipelagos: the Isles of Scilly

It should now be evident that the influence of particular geographical variables on species richness (or species incidences; see Chapter 6) is a matter of scale, that is, the scale of focus. This is no better illustrated than for archipelagos within the broader context of British and Irish offshore islands. Where islands are clustered in a mass, and the archipelago is substantially isolated from a mainland source region, the largest islands may act much as a mainland source. This is the situation occurring in the Isles of Scilly which provides a valuable example of potential interactions occurring among a number of islands (Dennis *et al.*, unpublished analyses); it is an archipelago of some 140 small islands and rocky islets (see Fig. 7.1), dominated by three relatively large islands (St Mary's, Tresco and St Martin's) and with human populations on only six of the islands (the largest three, Bryher, St Agnes, Gugh). The interactions (transfers) between such islands – and all islands can potentially contribute to transfers – are most effectively modelled as a single metapopulation; the classic case (version) for the Isles of Scilly would be the mainland–island or Boorman–Levitt model (see Fig. 1.3b). When dealing with metapopulations, then, there is no one single source of possible migrants but many; moreover, the contribution of any one island to any one target island will vary directly in proportion to its size and (inversely with) its isolation. Two simple models were devised by the late Professor Ilkka Hanski and his colleagues, the greatest contributor to our understanding of inter-island (inter-patch) transfers and metapopulation persistence, that is, the survival time for any species over an entire metapopulation (or archipelago) (Hanski *et al.*, 1994). For any island, these isolation measures are:

$$HI_1 = -\Sigma exp\left(-D_{ij}\right)$$

and

$$HI_2 = -\Sigma exp\left(-D_{ij}\right)A_j$$

where D is the shore-to-shore distance between islands and A is island area. The first of these algorithms (HI_1) is thus based purely on the distance between an island (i) and every other in the archipelago (j); the second is much the same but each distance is weighted by the area of the successive islands (Dennis and Eales, 1999). The negative exponent for distance describes the rapidly decreasing probability of transfer as distance increases. Both these measures have been calculated for 38 islands > 1 ha in extent in the archipelago.

In attempting to exploit this measure for Isles of Scilly butterflies, there is a problem. The Cornish database records a greatly imbalanced number of visits to the islands, which understandably coincides with island area ($r = 0.88$, $P < 0.0001$, $n = 13$). Even so, when numbers of species are regressed (stepwise, forwards entry of variables) against number of visits to islands, island area, and the isolation measure HI_2 (all log transformed; islands selected that have at least one species in Appendix 3), one variable alone is significant: island area ($F_{1,15} = 91.39$, $R^2 = 85.9\%$, $P < 0.0001$); the results are virtually identical for just those islands recorded as visited ($n = 13$) in the Cornish database (data source in Appendix 4). This in itself is an interesting result, as island area is more highly correlated with the measure HI_2 ($r = -0.94$, $P < 0.0001$) than with number of visits ($r = 0.88$, $P < 0.0001$). The reader is once again cautioned about the bludgeoning nature of simpler regression models. But the inference is that if an even (systematic) sampling programme were carried out over the Isles of Scilly, then island area, or its near inverse in the weighted isolation measure (HI_2), would account for species richness (and by extension incidence of species) over the islands. This is certain because when isolation on its own (unweighted by area, HI_1) is used as a single predictor, or substituted for HI_2 in the equation model including number of visits and island area, it has a low and non-significant level of determination ($R^2 = 5\%$; $P = 0.38$).

Clearly, the Isles of Scilly present many opportunities for studies into island biogeography. Later, in Chapter 7, we look at the role they have played in understanding butterfly evolution on islands.

Island Assemblages of Butterfly Species

Another way of looking at island butterflies is to compare islands for their collections of species. This is done using one or more of a selection of **association coefficients**, describing similarity or dissimilarity between locations (islands). In previous work, we have used percentage dissimilarity or disagreement (Dennis, 1977) and **Jaccard's coefficient S_J** (Dennis and Shreeve, 1996) (see Appendix 8). Jaccard's coefficient has the advantage of being directly applicable for studies in diversity and allows the separation of **species turnover** from species **nestedness** at different sites (locations) (Baselga, 2012). Other coefficients have different properties that are sometimes useful in determining relationships among faunas, particularly when joint absences of elements (viz. species) between sites are important to an investigation (Sneath and Sokal, 1973; Quinn and Keough, 2002).

In 1996, associations were examined using **ordination (non-metric scaling, NMMS)** and **cluster analysis** (algorithms: single linkage, maximum linkage, UPGMA) between 73 islands and six source units (European continent, Ireland and four latitudinal bands of the British mainland) for their butterfly faunal content (see Appendix 8 and Fig. 8.3 for a similar analysis on a reduced set of islands). Some clear patterns emerged in the two-dimensional non-metric scaling display. Despite the moderate distortion of components (island placements) (measured by **Kruskal stress** = 0.102), the plot revealed the following:

1. A clear geographical ordering of the sources and islands from south to north; distinct ordering also occurred on the second dimension in aggregating islands in archipelagos, as for example the Isles of Scilly and the Inner Hebrides.
2. A distinct shift in islands 'north' in the plot from their respective sources (viz. equivalent to the increased 'polar effect' in species richness first noticed for island organisms by Williamson (1981)).
3. The latitudinal shift or displacement in islands also affected source regions of northern England and Ireland.
4. Small islands were mainly forced into a ray at the periphery of the plot; with decreasing number of species on islands, the plot tended to splay out away from the main sources, indicating changing affinities in different regions of Britain and Ireland related to the availability of **faunal elements** (species) at nearby sources (Dennis and Shreeve, 1996, 1997).

With the increase in the numbers of islands explored in 1996 compared with 1977, especially the inclusion of small islands affected more by stochastic colonization events, the result was not as neat a reproduction of island geography. However, the same features (viz. latitudinal trend; polar displacement) occurred, indicative of the main influence of latitude over species ranges, species richness and the availability of particular species at faunal sources.

To assess the impact of island isolation and the combined impact of island size and faunal sources on island faunas, island affinities were regressed against mean island isolation and, additionally, against affinities based on the computation of virtual faunas built from the random selection of species from 'mainland' sources. For the latter, the number of species selected for each island was constrained to the number observed for each island (see Appendix 3). Three virtual faunas were established, founded respectively on: (i) the entire species pool of the study; (ii) the regional source; and (iii) a local source of the nearest 50 km square. **Affinities** are mean Jaccard coefficients $\hat{S}j$ (= $\Sigma_{i,k.n}$ S_j) taken over the n–1 islands (i.e. excluding the island under consideration). The relationship of actual island affinities was strongest with mean isolation (r = –0.54, $P < 0.001$), less with affinities based on random selection of species from the immediate source (r = 0.50, $P < 0.001$) and regional source (r = 0.45, $P < 0.001$). There was no association with affinities established randomly from the entire species pool (r = 0.17, NS). These correlations strongly implicate geographical (north–south) trends throughout the islands as the main influence on island affinities, as lower affinities for islands reflect on isolation from the core southern source of species, a feature also emerging prominently in the virtual faunas created from neighbouring sources.

It is noticeable that the virtual faunas explain little more than a quarter of the variation apparent in actual island affinities (i.e. $r^2 < 0.25$): this observation points to island recruitment of species *not* being a random process. It also suggests that the British and Irish island faunas form nested subsets of source faunas which in turn are increasingly nested, much like Russian dolls, into the richer faunas of southern Britain (Dapporto *et al.*, 2013, 2015). We now know that a nestedness pattern is not absolute and that species turnover has a share in mainland faunal diversity (see Chapter 4) (Dapporto *et al.*, 2015).

Butterfly Associations on British and Irish Islands

A view into the non-randomness of butterfly recruitment on islands is provided by the relationships (associations) among species for their presence and absence on the suite of islands (Fig. 5.6). In the earlier study

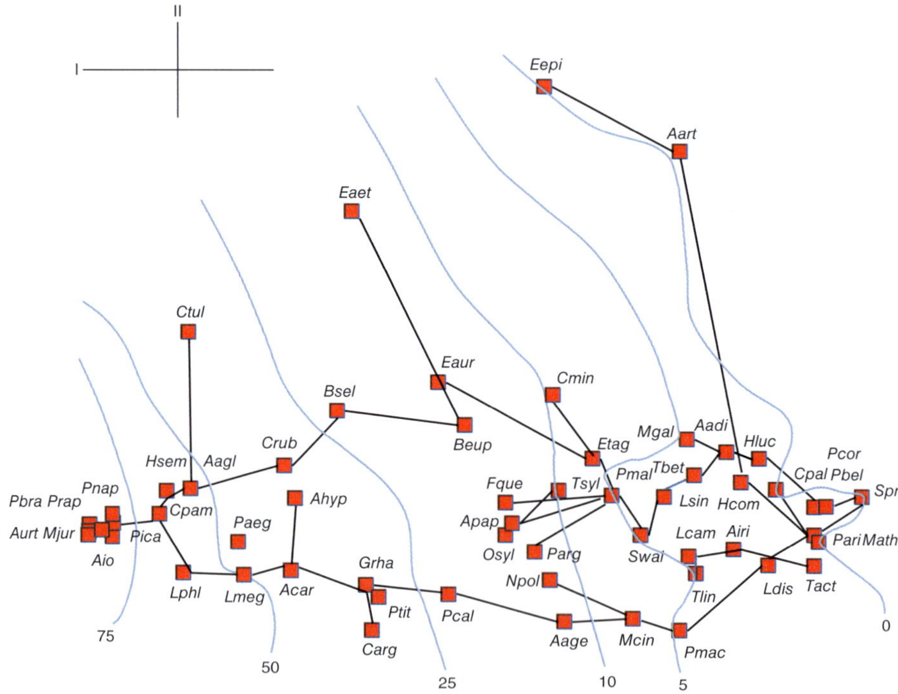

Fig. 5.6. Associations among species for their joint occurrences on British and Irish islands, including the Isle of Man and the Channel Islands. A non-metric scaling plot based on Jaccard similarities for 73 islands and eight mainland regions (North France; Holland; Belgium; four latitudinal belts of Britain; and Ireland); overlay of isolines show % occurrences on islands (Kruskal stress [phi] = 0.092). (Redrawn from Dennis and Shreeve, 1996.)

(Dennis and Shreeve, 1996) this was carried out for 57 butterfly species (migrants *C. croceus*, *V. atalanta* and *V. cardui* omitted) over 73 islands. Non-metric scaling and clustering (nearest neighbour, complete linkage, unweighted pair-group average) were carried out on the species matrix of S_J coefficients (Kruskal stress = 0.092). The islands were distinctly ordered by the two NMMS axes:

1. Axis 1 described a declining frequency of species on islands from those found on almost every island to those that have not been found on any of the British and Irish offshore islands (e.g. *C. palaemon*, *Phengaris arion*, *Satyrium pruni*, *Melitaea athalia*, *E. epiphron*). A compact cluster appearing in all analyses confirmed a kernel of ten species found, often together, on over 75% of islands (i.e. *P. brassicae*, *P. rapae*, *P. napi*, *Polyommatus icarus*, *Aglais urticae*, *A. io*, *Argynnis aglaja*, *H. semele*, *Maniola jurtina*, *Coenonympha pamphilus*).
2. Axis 2 produced a division of species into those more characteristic of western and northern islands compared to those with southern ranges in Britain limited to southern islands; these two groups were neatly highlighted by two branches of a nearest neighbour (minimum) spanning tree, respectively linking the members of the two groups. Classic cases of species that have only ever been recorded on southern islands are *T. acteon* and *Melitaea cinxia* (e.g. Isle of Wight: Plate 1.16).

The pattern of species over offshore islands once again implicates the nestedness of species over Britain and Ireland. Moving from the north and west to southern Britain, additional species are steadily recruited to islands that are unavailable further north. There are, however, clear exceptions in the frequent occurrence of *E. aethiops* and *C. tullia* in the Hebrides, the latter also extending into the Orkney and Shetland group. The ecology of these species indicates that they may long have persisted on the larger islands such as Mull and Skye (Dennis and Shreeve, 1997; see Chapters 7 and 8).

6 Assessing the Status of Individual Species on Islands: from Geography to Ecology

Island geography not only lies at the root of understanding species richness on islands; by extension, it provides the means for predicting species richness on poorly surveyed islands. This forms the starting point for this chapter, which then focuses on the presence (probability) of individual species on islands. Of course, determination of species richness using geographical variables does not determine *which* species are expected on specific islands; nevertheless, over the past three decades especially, the techniques for doing so (**regression analysis**) have evolved to the next step: predicting the presence (or absence) of species on islands. The methods used for assessing the incidence of butterfly species on European islands were first applied to *Hipparchia semele* (Dennis *et al.*, 1998a, b), techniques that have important implications for the conservation of habitat islands (Dennis and Eales, 1997, 1999). Subsequently, at a finer resolution still, attention is given to determining the probability of an individual butterfly transferring to an island and then colonizing it; Professor Richard Tilley has recently modelled novel and unforeseen perspectives on these processes.

An important issue emerging in more recent years is that to understand the presence of species on islands, greater attention needs to be paid to species life histories and ecology: measures of island geography, in themselves, provide valuable but insufficient predictors of the occurrence of different species' on particular islands. Moreover, resource and life history traits of species occurring on islands (community or faunal ecology) should reflect the impact of island geography on those species. In this chapter, then, a move is made from geography to ecology; in particular, attention is focused on two indices of migration capacity and colonization ability and their implications for butterfly island occupancy.

Using Geography to Predict Species Richness

For most naturalists, the pressing questions in visiting an island are always: 'How many species, and what species, are to be found on it?' In the previous chapter we have seen how island geography provides a sound basis for assessing the basic factors influencing the number of species found on islands; also, it was mentioned how it is possible to predict the number of species likely to be encountered on an island, given sufficient opportunity to explore it. In 1996, this allowed predictions to be made of species numbers for some 139 islands off Britain, Ireland and France, in addition to the 73 islands having records adequate for building predictive equations (Dennis and Shreeve, 1996). Predictions were based on the size of the mainland faunal source, isolation (distance) from that faunal source and island area. In fact, the close relationship among predictor variables ensured that different variable suites provided similar solutions. The current analysis of the enlarged databank (see Appendix 10 and Supplementary Appendix 3.1) has used records for 123 islands with the same migrant species markers (*Pieris rapae, Pieris brassicae, Vanessa cardui* and *Vanessa atalanta*). Predictions have been based on three significant variables (size of nearest faunal source, island area and island latitude) and are presented in Supplementary Appendix 6.2 part 1. It should be borne in mind that although the model (equation) accounts for a substantial amount of variation in species richness (R^2 = 65%) of the 123 islands that have been studied, it still fails to account for 35% of the variation; thus, estimates of species richness for individual islands, even when accompanied by error bounds, are at best an approximation of expected numbers.

A second set of predictions (Supplementary Appendix 6.2 part 1) has been produced using a different ploy, including, this time, 191 islands that have at least two of the three longest-distance migrants (viz. *Colias croceus, V. cardui* and *V. atalanta*). As it is clear that some of the independent variables are highly correlated, initially four components (**PCA, varimax rotation**) were extracted, each of which firmly describes one or more of the original nine geographical variables (see Appendix 10 part C for details). Regression analysis (stepwise entry) was then based on these four factors, two of which are significant, describing all geographical variables except island longitude, and isolation from the nearest larger source. Compared with the set of 123 islands selected for the four migrants (two *Pieris* and two *Vanessa* species), the amount of explained variation has declined by 10% to 55.6%; moreover, the number of species predicted for islands is significantly lower when paired data for islands are compared (**Wilcoxon Matched Pairs Test** Z = 24.54, P < 0.0001, n = 910 islands). This finding confirms the notion that by relaxing the criteria for selecting islands, more islands are included with fewer species, part of which may relate to their being smaller, but also (importantly) points to the likelihood of their having been less adequately surveyed.

Statistics can be seductive but, with regard to specific islands, it is always useful to get back on to firm ground and match predictions against experience in butterfly field ecology (realistic expectations). This issue becomes more critical when considering species incidences on islands.

Predicting the Incidence of a Butterfly Species on an Offshore Island: the Grayling on Grassholm

Geographical variables not only allow predictions for numbers of species found on an island but also predictions of which species are most likely to occur on it. The 1996 work (Dennis and Shreeve, 1996) did not investigate species incidences on islands, but here we use island geography for that very purpose.

The first attempt at determining the geographical factors responsible for the presence (or absence) of a butterfly species on British islands focused on *H. semele* (Dennis *et al.*, 1998a, b). This butterfly had then just been recorded on Grassholm 12.8 km off the Pembrokeshire coast (G. Thompson in Dennis, 1997b) (Plate 1.9). This was a much greater distance than determined by mark–release–recapture of adults (viz. 2.8 km) (Nieminen, 1996) but it was by no means the most isolated island on which the butterfly was known to have been recorded (e.g. Ailsa Craig; Lundy). The only other butterflies that had been found on Grassholm were all capable of long-distance migration (i.e. *P. brassicae*, *P. rapae*, *V. atalanta*, *V. cardui* and *Aglais urticae*). The Grayling butterfly is a strong flier, but for all that it is absent from the Isles of Scilly (Dennis and Shreeve, 1996). This is unexpected as the butterfly thrives on coastal biotopes (viz. cliffs, sand dunes, bare substrates with well drained grassy slopes). The issue was not just what geographical factors influenced the incidence of this butterfly on British and Irish offshore islands, but also the probability of the butterfly occurring on each island.

The study of incidence (presence, absence) is undertaken using **logistic** (and logit) **regression** analysis and **discriminant function analysis** (DFA). These techniques investigate the binary outcome of an event against a set of independent predictor variables (see Appendix 8) and have been successfully used in studies of butterfly metapopulations (Thomas *et al.*, 1992; Dennis and Eales, 1997). Logistic regression provides measures (parameters) of the success of the model (equation) whereas DFA searches for the best combination of variables that correctly maximizes known group membership. Both provide probabilities of incidence for the cases (islands) in the model and those excluded from it, for whatever reason, but deemed worthy of investigation. The study involved three collections of islands (all available islands, $n = 77$; islands within the British mainland latitudinal range of the species, $n = 69$; and a smaller group still excluding a further three well surveyed islands whose records of the butterfly were considered to be vagrants, $n = 66$). The variables included much the same selection as for species richness assessments: A, area; I, isolation (but this time sea and land distance combined), source population size records within 50 km of the nearest mainland or larger island record. Two models were again considered for mainland (FS1) and largest island (FS2) potential population sources for island colonists as for species richness; the second again proved more powerful, this time as a predictor of incidence.

A number of interesting features emerged in this study of *H. semele*:

1. Isolation was found to be a more successful predictor than island area, the opposite result of the 1996 species richness investigation (Dennis and Shreeve, 1996, 1997).
2. Although it was found that the subgroups (islands out of geographical range of the species, and the three islands with alleged vagrants) did not affect the significance of parameter estimates for geographical predictors, they did affect the classification of individual islands as potential sites for the butterfly.
3. When modelled on all three geographical variables, the incidences of the butterfly for both subgroups of islands was shown to be improbable; however, islands beyond the geographical range were predicted to have the species solely on island area, and islands with supposed vagrants were predicted to have the butterfly on the basis solely of isolation.
4. Correct classification of the islands was high (74–83% for the different subsets). Some ten islands of the 77 investigated were predicted to have the butterfly, but at the same time a further eight that have records of the butterfly were predicted to be marginal for it (see **marginality** in Glossary) (Shreeve *et al.*, 1996). The fact that the butterfly has been recorded on them belies this result, but it could well point to limiting conditions for persistence on the islands concerned (see Chapter 7 for turnover issues).

The inference of the study is that the butterfly is excluded from the two small subsets of islands by environmental conditions affecting resource availability (scarcity) or conditions (viz. severe climate for the out-of-range subset; impact of waves, salt spray, or even bird colonization on the small island vagrancy subset), and indicates that conditions are in some way marginal for the species. The reader is reminded that resources and conditions (utility and consumer) are fundamental elements of the resource-based habitat (Dennis, 2010) (see Chapter 3). These subsets of islands are distinctly marginal in the plot of geographical variables (Fig. 6.1). What is clear is that distance does not limit the butterfly's access to islands isolated by 13 km of sea. But there is evidently a limit to movement preventing the butterfly gaining a foothold on the Isles of Scilly (probability of incidence = 16%), as the habitat conditions there (bare rock substrates), and the metapopulation context of the archipelago, should facilitate persistence following initial colonization. Below, we look into the probability of an individual butterfly migrating to islands and we shall see that limits to successful transfers are clearly apparent, though involving a complex of variables (see Appendix 6).

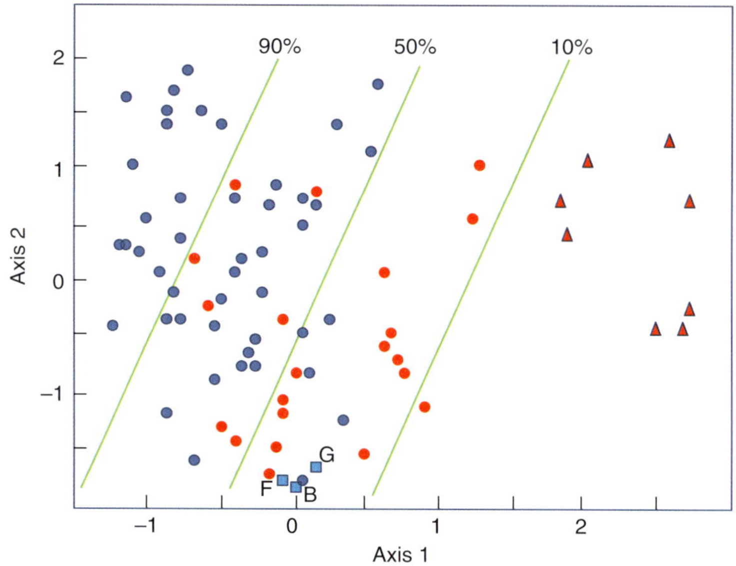

Fig. 6.1. The incidence of *Hipparchia semele* on British islands. A principal components plot: axis 1, island isolation and nearest source mass (records within 50km radius); axis 2, island area. Overlaid probability estimates from discriminant function analysis were drawn using trend surface analysis (see Dennis *et al*., 1998b, for details). Blue dots, occupied islands; red dots, unoccupied islands; red triangles unoccupied out-of-range islands; blue squares, islands with known vagrant records; B, Brownsman; F, Inner Farne; G, Grassholm. (From Dennis *et al*., 1998b, courtesy of John Wiley & Sons, Inc.)

Predicting Incidences of Butterfly Species on the British and Irish Offshore Islands

With the current data on 910 islands we are able to investigate the incidence on islands of 33 species of the current list of 60 species. As parameter estimates in logit regression models are likely to be unreliable for species that occur infrequently on the offshore islands, we have excluded 24 species that occur on fewer than ten of the 191 islands on which the models are based; the assumption is made that they are absent. Three migrant markers are used to select islands (*C. croceus*, *V. cardui* and *V. atalanta*), all regarded as likely to be recorded there in due course; modelled islands ($n = 191$) have at least two of these species. The immediate focus of attention has been on the comparative efficacy of geography and potential source mass (populations at nearest sources) in accounting for species incidences on offshore islands. Two sets of logit regressions have been performed for each species on: (i) five factors (from principal components analysis PCA; see Appendix 11 and Supplementary Appendix 6.1a) describing seven geographical variables; and (ii) the larger of two source mass influences (10 km square records for each species within 50 km of the nearest mainland and larger island sources (FS1, FS2)). The geographical variables were replaced by PCA factors to ensure independence among the geographical variables. The detailed results are available in Supplementary Appendix 6.1a, with a summary of results in Appendix 11 and the predictions for islands tabulated in Supplementary Appendix 6.2 part 2.

The key findings are as follows. As is expected, there is consistency in the sign of influence (positive or negative effect) of variables on species, including reversals in sign for species that have southern range margins (i.e. for latitude on northern species such as *Erebia aethiops* and *Coenonympha tullia*) (see Appendix 11). The highest significance levels are obtained by the two source mass variables (Fig. 6.2a); the source mass variables, as expected, are highly correlated across species. Even so, predictions for species incidences for 910 islands from geography factors correlate significantly with those of the better of the two source mass variables (r^2 = 69.6%, P < 0.0001). A single source mass variable predicts a higher level of island incidences for species than the geography factors (**Student's *t*-test**, $t_{(32)}$ = −3.17, P = 0.003). However, great care is required in interpreting such predictions from logit regression; predictions (i.e. probability of incidence) are imperfect for most species and are undoubtedly improved by combining source mass with geographical factors (see below). Numerous tests are available for logit regression, but a common sense test of its effectiveness for this island work is the degree (%) to which correct predictions are made of recorded incidences (rather than for gross presences and absences) of species; this is the case as we can be more certain of validity of recorded species on islands than we can of apparent absences, if only because sampling of islands has not been carried out in a systematic manner. There is a close correspondence of correct predictions for known occurrences across species for geography factors and source mass (r^2 = 84.9%, P < 0.0001) but, importantly, use of geography factors is more successful in determining correct predictions than source mass ($t_{(32)}$ = 2.41, P = 0.02).

The most revealing finding of the logit regression analyses is that for the relationship across species: the level (%) of correct predictions for species on islands correlates with the actual frequency of occurrences of species on islands; the first is assessed for a sample of 191 islands, the second is determined over all 910 islands. The correlation and pattern for species is similar for geography factors and source mass (r^2 = 52.9% and 54.9%, respectively, P < 0.0001); for both, the scatter of species is greater at the lower end (Fig. 6.2b). Intriguingly, species above the regression line have compact distributions in Britain and tend to be expanding their ranges, whereas those below the line are widely scattered across Britain and Ireland and are fast losing ground ($U_{(14, 19)}$ = 44.50, Z = −3.21, P = 0.0013) (Fig. 6.2c). The inference is most concerning: the distributions of species with poor predictability are being fragmented; source mass for these species is weakening and geography has less bearing on their existence. Their continued presence will depend on habitat factors not

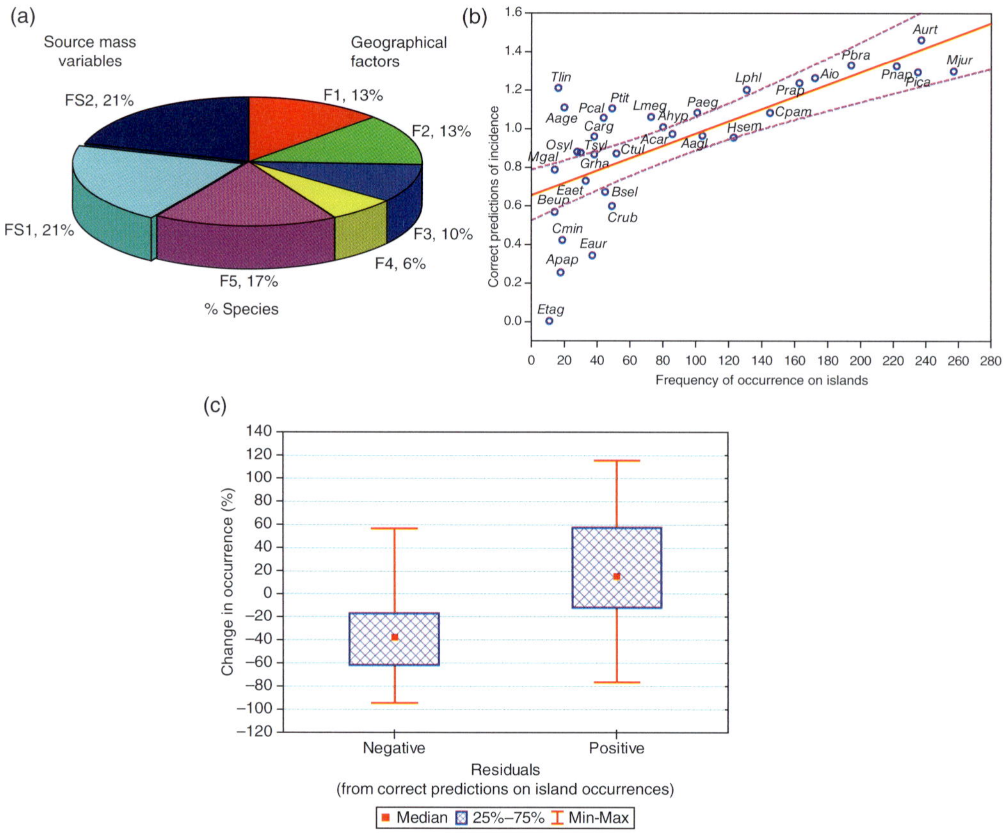

Fig. 6.2. Relationship between the correct predictions of species incidences on well sampled islands and the incidence of species over all islands. (a) Pie chart of contribution of geography and source variables in accounting for incidences of species on British and Irish offshore islands (33 species for 191 islands); F1 to F5, geographical factors; FS1 and FS2, source mass variables (see Appendix 11). (b) Scattergram of predictions from logistic regression of five geographical factors for 191 islands against frequency of occurrences (incidences on 910 islands). Correct predictions are arcsine square root transformation of proportions of recorded incidences in the sample of 191 islands. (Correlation, Pearson $r^2 = 0.51$, $P < 0.0001$). (c) Box plot of changes in occurrence (1976–2014 distribution; Fox *et al*., 2015) for negative and positive residuals from the regression line in graph in Fig. 6.2b (Mann–Whitney $U_{14,\,19} = 44.50$, $Z = -3.21$, $P = 0.0013$).

accounted in the logit regressions and, as source becomes less relevant to their existences, they carry the unhappy stigma of extinction debt.

Predictions for the 910 islands have also been made by applying discriminant function analysis to all variables (geography and source mass), the functions derived from 191 islands with at least two of the migrant markers (for detailed results, see Supplementary Appendices 6.1b and 6.2 part 3). The key feature emerging is that although the results are based on a much improved correct prediction of known island incidences, the number of predicted incidences on islands is somewhat less than for the logit equations (see Appendix 11).

The Drumlin Islands of Strangford Lough

Studies of species incidences in archipelagos provide valuable insights into transfers and colonization of habitat patches on the mainland (Fig. 6.3). An invariable problem for terrestrial (mainland) settings is the unknown extent to which findings for metapopulations are corrupted by the paucity of knowledge of potential source populations and stepping-stone resource distributions; much of the landscape is often too extensive, or inaccessible, to survey the entire areas. For islands amid water there is no doubt as to the absence of intervening opportunities for colonization and the inimical nature of the barriers.

For all the obvious limitations of Butterfly Conservation's **Big Butterfly Count** (see Chapter 3), butterfly records for repeated visits to 62 islets in Strangford Lough (2012–2016) (Enright *et al.*, 2017) have revealed some interesting features for species (see Appendix 12). This is a very different archipelago compared with the Isles of Scilly. Instead of being distant from the mainland, more than 90 islands occupy an open sea lough surrounded by mainland, including the sweeping arm of the Ards Peninsula. The islands, a drowned drumlin landscape, are mounds (mean: 308 m × 105 m, $n = 786$ drumlins) (Vernon, 1966) formed of glacial till during advance of Midlandian (Devensian, 20 ka BP) glaciers, streamlined in the direction of ice flow (south-eastwards), elliptical (lemniscate) in shape, and having a generally steeper 'stoss' end than the 'lee' (Chorley, 1959; Vernon, 1966;

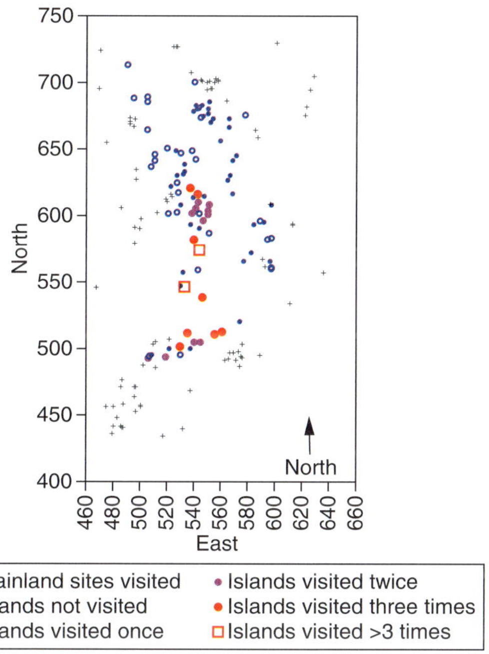

Fig. 6.3. Schematic map of islands in Strangford Lough, Ireland, illustrating number of visits to islands from 2012 to 2015. (From Enright *et al.*, 2017, courtesy of *Entomologist's Gazette* and Pemberley Books.)

Evans *et al.*, 2005) (Plate 1.10). Most have been modified by higher sea levels, tides, wave action and mass wasting; a few are little more than tidally sieved, wave-washed boulder spreads (e.g. Long Skart Rock) and unlikely ever to be occupied for more than a single brood or during a casual visit by butterfly species. Some 12 species (Fig. 6.4) were studied in relation to island isolation (mean distance from the lough margin 718 m, maximum 2625 m), connection to the mainland by a causeway (linked/unlinked), area, elevation and position (grid east and north) and number of visits (each of 15 min duration).

Despite the work being confounded by survey (sampling) issues (Fig. 6.3), a number of characteristic features of island biogeography have emerged in this Big Butterfly Count study of butterflies on Strangford Lough islands: (i) fewer species on islands than at mainland sites; (ii) greater species richness and abundance of butterflies on higher (larger) islands and lower species richness on more isolated islands; and (iii) an increasing incidence of individual species in relation to island area and elevation and decreasing incidence with isolation. Another feature emerging from the study is the expected larger populations on islands than at mainland sites, owing it is thought – in other island situations – to reduced competition (Olesen *et al.*, 2002), but which here could be owing to reduced human interference. Interestingly, the variables having the least numbers of significant correlations across all species are island isolation and 'Linked' (causeway connections) (ten significant correlations for three variables), far fewer than island area and elevation (26 significant correlations for two variables) (**Fisher Exact Test**, P = 0.009). The most interesting finding

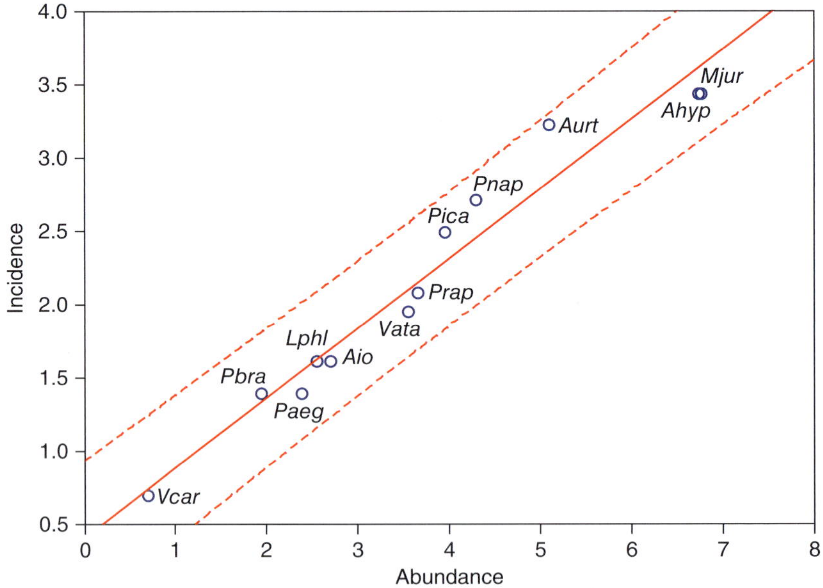

Fig. 6.4. Relationship of incidence and abundance of species on the Strangford Lough Islands. Regression line and 95% prediction band (log scales). (Courtesy of *Entomologist's Gazette* and Pemberley Books.)

was that a species such as *Aphantopus hyperantus*, generally regarded as incapable of crossing hostile matrix (Sutcliffe and Thomas, 1996), is able to cross water barriers and to build populations on tiny sites (e.g. Dunnyneill Island, < 1 ha, 1475 m from the mainland), supporting observations for that species on the British mainland (Dennis *et al.*, 2011b).

For much of the work on European islands we have had to apply surrogates of adequate sampling in the form of indicator species (long-distance migrants). If understanding of island biogeography is going to advance, then surveys of islands need to move to synoptic surveying, the timing and duration of surveys being adequate (representative) for the species likely to occupy the islands and the same for all islands. Uneven sampling presents the greatest challenge for analysis of data collected for the Strangford islands (Enright *et al.*, 2017). When the number of visits to islands is known, but unequal, then it is necessary to account for survey intensity, which the study did using two methods. As a first approach, applying logit regression equations (Table 3 in Enright *et al.*, 2017), the number of visits was entered into equations and, when acting as a significant predictor, standardized on two visits. Predictions for species were then obtained for islands lacking records by comparing probabilities with those for known occurrences on other islands using ordinal summary parameters as thresholds (mean and upper quartile probabilities of known occurrences). Successful equations were produced for four species. A second method using discriminant function analysis was based on a single visit, allowing tests of predictions for islands in the models and visited multiple times, as well as providing predictions for islands not included in the model. As *Polyommatus icarus* was the only species influenced by island isolation, there is a suggestion that isolation is more limiting for it than for *Pieris napi*, *Maniola jurtina* and *A. hyperantus*.

It is as well to appreciate, if not already evident, that, with standardization on increasingly higher numbers of visits, probabilities (predictions) of occurrence will inevitably increase. The distances to the majority of British and Irish islands are now known to be such as not to isolate them completely from occasional transfers by any British butterfly (Dennis, 2001). In simple terms it means that if an observer on the most isolated island in the lough were to be alert to every occurrence of a butterfly they would eventually record every species from the surrounding region. This raises the spectre that population structure of many species on the lough accord with a patch dynamics model rather than an orthodox (Levins) metapopulation model (see Fig. 1.3); in fact, with dramatic changes (fluctuations) in seasonal conditions we should expect the status of species to oscillate between these models.

Colonizing Islands: Modelling Individual Probabilities of Success and Failure

There are four fundamental spatial aspects to initial island colonization: (i) distance from source to island destination; (ii) island (target) area;

(iii) the expanse of source area (equating with source mass, i.e. potential population size) supplying vagrants for islands; and (iv) the disposition of essential resources on the island. That these constraints may not be so limiting, as long conceived, for initial colonization has recently been demonstrated by Richard Tilley using two probability models. First, he examined the likelihood of a single butterfly (by extension, numbers of butterflies), adopting a random direction but linear path, transferring to an offshore island of known distance and dimensions from a source population on the mainland or from another island (or islands). Although the probability is low for a single individual from any one source making a crossing to islands such as Flat Holm and Steep Holm in the Bristol Channel (c. 20 ha, 4–5 km offshore; Fig. 6.5 and Plates 1.11 and 1.12), the probability of a successful transfer is c. 100% for a population or populations of several hundred individuals emigrating (potentially in all directions of the compass) from habitats along the source shoreline (Tilley and Dennis, 2017) (see Appendix 13 part A). Then, using **search theory**, developed during the Second World War (Koopman, 1980), Tilley assessed the probability of a single individual finding a key resource (i.e. a host plant or female) on an island once it has achieved transfer to it. The likelihood of success is not only much greater than expected (see Appendix 13 part B) but also has the benefit that likelihood of success increases inversely for smaller islands even though smaller islands will tend to have fewer resources (mates, or host plants).

Tilley's model makes the assumption that distance is no object to a dispersing butterfly, at least the distances from the mainland to British near offshore islands (< 10 km). The value of Tilley's model is that it can be tested directly using data on the life history and ecology (colonization abilities and migration capacities) of each species (Fig. 6.6). When this is done for Steep Holm and Flat Holm it is found that, as may be expected,

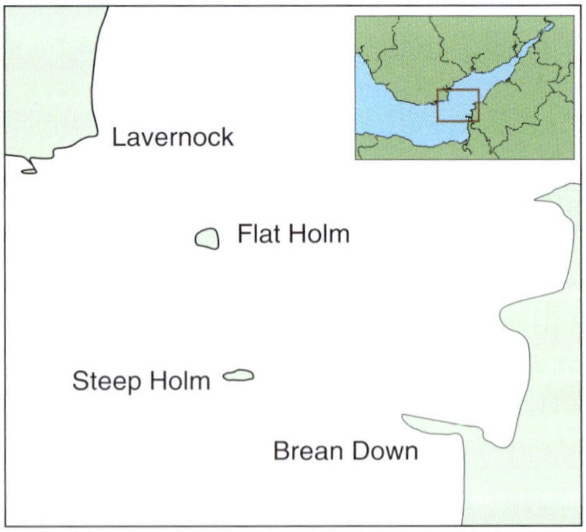

Fig. 6.5. Bristol Channel, showing the location of the islands of Flat Holm and Steep Holm. (From Tilley and Dennis, 2017, courtesy of *Entomologist's Gazette* and Pemberley Books.)

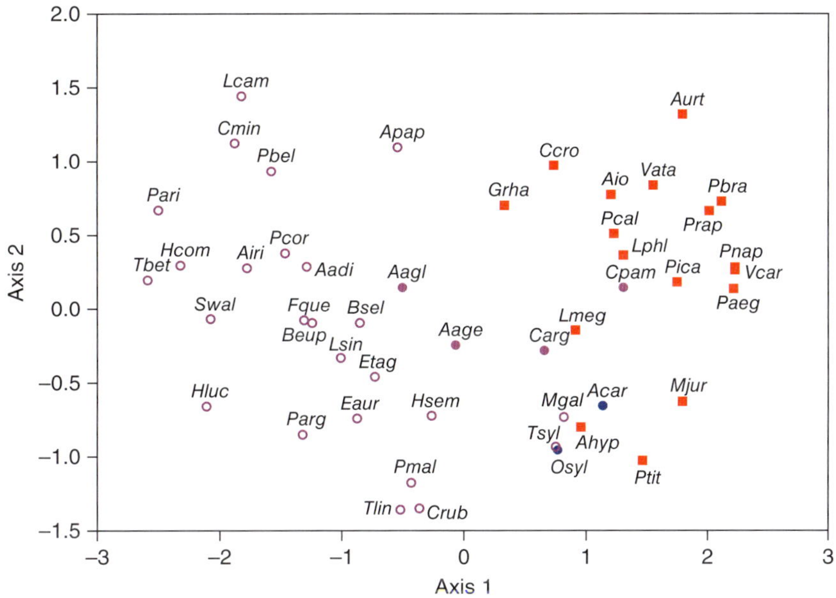

Fig. 6.6. The distribution of butterfly species on Steep Holm and Flat Holm. An ordination (PCA) for species migration capacity (migrfac), colonization ability (colonfac) and mainland source population size (source mass; 10 km records within the 100 km grid square that includes the islands); axis 1, a size component affecting all three variables equally; axis 2, distinguishing colonization ability (negative) from migration capacity (positive). Red square, species on both islands; pink dot, species only on Steep Holm; blue dot, species only on Flat Holm; open circle, species not occurring on the islands but within the 100 km square including the islands and their mainland sources; species named with first generic letter and first three species' letters (see Appendix 1).

all is not equal among species in their capacity for accessing islands, which in fact are limited by some combination of migration (**migrfac**), colonization ability (**colonfac**) and the size of populations at the faunal sources (**source mass**). The placement of species (coded as occurring or absent) on the islands in an ordination of the three variables is particularly revealing. A crucial point is that the three variables are intricately correlated on the prime ordination axis (see Appendix 13 part A); more than that, there are important functional (biological) relationships linking the variables making up colonfac, migrfac and source mass (Dennis *et al.*, 2012a; Dapporto and Dennis, 2013) (see Fig. App. 13.2). The inference is that although distance obviously influences whether or not migration (transfer) is successful, even over short distances, such as 4.5–5.0 km as in the cases of Flat Holm and Steep Holm, the main factor is likely to be the opportunity for transfers. Such opportunities are greatly affected by the number of individuals dispersing and this is not influenced just by the number of populations and their size at sources, both of which are known to be functionally associated (Mason *et al.*, 2017), but also by the length of flight periods and number of broods which increase the likelihood of coincidence with opportune weather conditions for flight. Also of great importance is that number of broods and length of flight periods have implications (in the link with generalism and resource availability) for successful colonization on arrival at islands.

At this juncture it is worth pointing out the value and limitations of models. As the eminent statistician George E.P. Box (1976, p. 791) stated, 'essentially, all models are wrong, but some are useful'. Professor Tilley has constructed valuable models that allow us to test the behaviour of different butterfly species during dispersal and resource seeking. At their core, they make us think about alternatives in butterfly behaviour. For instance, although the models treat movements of individual butterflies as being random in direction and unaffected by distant objects (potential targets) such as islands, species will tend to have preferred (mean) flight directions (Baker, 1978, 1984) and are almost certainly attracted by distant objects such as hilltops (see references in Dennis, 2010). To an unknown extent, then, change of course towards an island during dispersal may compensate for any biased movement away from the coast at source. We would expect biased movement inland at a source if only because selection will tend to operate against unsuccessful individuals taking a course over the sea. But, then, as the mainland landscape increasingly approaches a 'bare' matrix with increasing industrial use of land, this too could change.

What may not be clear from the models, but which emerges in an assessment of species found on both Flat Holm and Steep Holm, is that both islands function much as units in a regional metapopulation with habitat patch foci on the surrounding mainland. In this respect it is important to appreciate that because certain species are absent from the islands it does not mean that they are unable to transfer to them or that they have not transferred to them in the past. There will be constant turnover of species on such islands (see Hilbre and Lindisfarne in Chapter 7) and for species that have poor migration ability, weak mainland sources and demanding resource requirements, successful transfer will be occasional and persistence limited. When and where host plants are missing on islands, individuals are little more than ephemeral vagrants. The number of such species that are recorded by survey depends on how often such islands are visited and the degree of vigilance during recording. Once readers become familiar with different butterfly species they will begin to appreciate that each species will tend to have a characteristic **turnover** rate (frequency of colonization and extinction) on islands and consequently different rates of **rescue effect**. Thus, although deliberately unaccounted in the resource-seeking model (search theory) (see Appendix 13 part B), which describes post-transfer behaviour to an island, it is well to bear in mind that key resources (i.e. larval host plants) are not spatially equivalent for different species on target islands. For instance, owing to exposure and high wind speeds, small islands typically lack trees and these can be critical for some colonizing butterflies (e.g. *Favonius quercus*, *Satyrium w-album*, *Apatura iris*). However, new data accumulating on resource use on islands may provide evidence of previously unknown adaptabilities for some species (for example, egg laying by *Argynnis paphia* on hedgerow twigs, moss and conifer branches) (Heckford, 2011) and we should be prepared for some surprises.

How Well Does Island Butterfly Incidence Reflect Butterfly Ecology?

The last section hinted strongly at the importance of butterfly ecology and life history in the capacity of butterflies to transfer to and colonize islands. Today, we are well aware that species ecology and life history underlie all aspects of island biogeography, including the status of species at faunal sources, over-water migration, colonization ability and population persistence in relation to island dimensions. But only as recently as 1996 (Dennis and Shreeve, 1996) were the first steps undertaken to examine this using ecological profiles of individual British butterfly species. Initially, simple indices were used based on an earlier study of the vulnerability of butterfly species to projected climate changes (Dennis and Shreeve, 1991; Dennis, 1993). It was shown that incidences of species on islands (number of occupied islands) correlated highly (**Spearman** r_s = 0.81, $P < 0.001$) with an ecological index (called EC2) based on summed ranks of seven variables (viz. **phagy**; host plant abundance; host plant **seral stage**; biotope range; dispersal ability; **voltinism**; length of flight period). Each variable was also shown to relate significantly to island incidences in its own right. A further test was made to determine the stand-alone importance of migration ability using a movement index (MI) based on a further set of seven binary coded variables (at-sea records; ex-habitat vagrants; suburban garden records; inner-city records; documented range expansions; long-distance movements; mass movements), later applied as ranked scores (see Appendix 6); that too was found to be closely related to island incidences (r_s = 0.76, $P < 0.001$). The two indices, and their component variables, heralded a new era in studies of butterfly island biogeography; in Britain they formed the stimulus for constructing the indices of colonization ability (colonfac) and migration ability (migfrac) extracted by applying principal components analysis to a set of ecology and life history variables based on higher scales of measurement (see Appendix 7) (Dennis *et al.*, 2012a).

Initially, a wider study was made: (i) with a view to extending the list of, and quantifying, key species life history and resource variables; and (ii) to determine more broadly their relationship with species geography and status (Dennis *et al.*, 2000a) (see Chapter 4). The study included 21 life history and resource variables. Species incidences on islands (52 species on 68 islands having four migrant indicator species) were incorporated as one of the five geographical variables and found to be closely associated with species geographical ranges (Fig. 6.7), which in turn highlights the importance of faunal sources for the maintenance of island faunas. The only

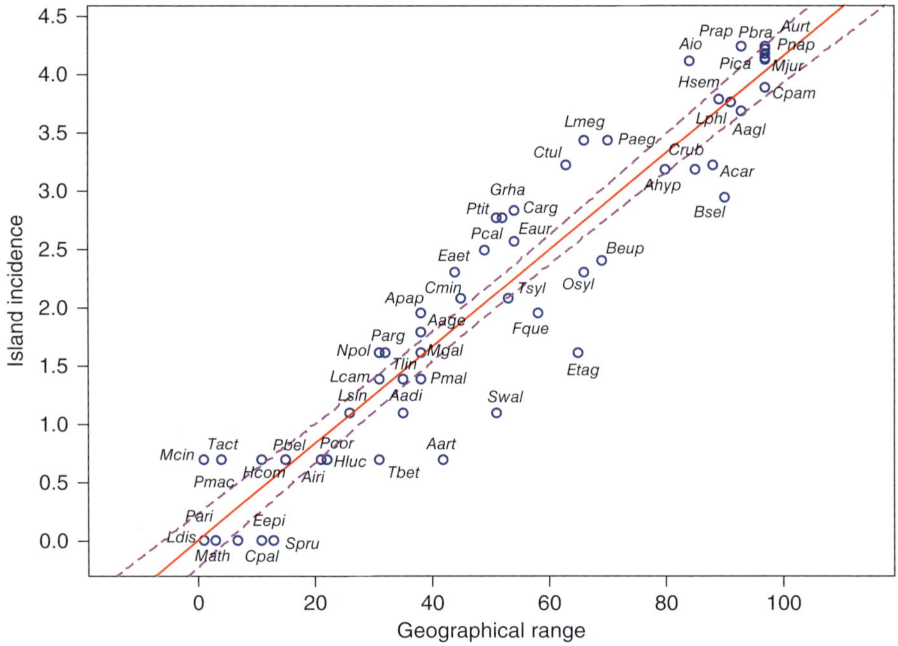

Fig. 6.7. Relationship between incidence of butterfly species on British and Irish islands (log incidences) and geographical range (10 km latitudinal bands occupied); $F_{1,58} = 472.3$, $R^2 = 89.1\%$; $P < 0.0001$; species named with first generic letter and first three species' letters (see Appendix 1). All species included (except migrants *C. croceus*, *V. atalanta*, *V. cardui*). (Modified from Dennis and Shreeve, 1996; Dennis *et al.*, 2000a.)

life history and resource variable accounting for substantial additional variation was adult flight period. Flight period is a valuable surrogate for migration capacity as it describes stochastic opportunities for successful migration from source populations. Species with long flight periods also tend to be **multivoltine** and have the capacity for building larger populations, provisioning more potential migrants. A further eight resource and life history variables were also significantly linked to species incidences on islands, cumulatively accounting for 91% of species island incidences. The inference of the strong link between island incidences and geographical ranges (distributions) is that both are determined by very much the same set of species-specific life history and ecological (resource) variables. Even so, a number of species were found to deviate in interesting ways from the regression line describing the relationship of island incidences on geographical ranges. For example, *Erynnis tages* and *Aricia artaxerxes* were found on far fewer islands than expected, whereas *C. tullia* was found on many more islands than expected. Respectively, these two groups have eastern and western distributions in Scotland, compared with which islands have a strong western bias (Dennis and Shreeve, 1996), patterns that convey insights into substantial differences in species ecology.

Species incidences on islands was then incorporated as one of 12 geographical variables in a detailed wide-ranging study investigating butterfly **life history strategies** (see **life history characteristics** and **CSR** in Glossary) (Dennis *et al.*, 2004). This involved 86 variables describing: (a) host plant and host-plant environment; (b) butterfly morphology, life history and physiology; (c) butterfly habitats and biotopes; (d) butterfly populations and mobility; and (e) butterfly geography. Species occurring more frequently on British offshore islands were found more typically to have **ruderal strategies** as opposed to being **stress tolerators** (Dennis, 2010). In detail: along with having expansive geographical ranges, wide distributions, greater site persistence (fewer population losses) and a tendency towards range expansions, species with a greater number of island incidences were found to have open populations, higher mobility, longer flight times, multiple broods, higher productivity, faster development rates and abundant and varied use of host plants and biotopes. The converse was also observed.

The 2004 study had, in effect, distinguished generalist from specialist butterfly species; in turn, the basic generalism/specialism dichotomy had identified species that were likely to differ markedly in colonization ability (colonfac) and migration capacity (migrfac). These two variables account for the ability (or lack of it) of butterflies to transfer to islands, to colonize them and then to persist on them; this was noted earlier in the short section that dealt with species on Flat Holm and Steep Holm (see Appendix 7). As might be expected, the two variables, extracted using multivariate techniques, are not entirely independent of one another (**Pearson** r = 0.57, $r^2 = 0.32$, $P = 0.001$) (Dennis *et al.*, 2012a); even so, they cumulatively contribute to the incidences of butterflies on offshore islands. The reasoning is as follows: species with a wider resource base and best able to persist on 'islands', whether offshore islands or habitat islands, may also be better able to migrate to them. A wider resource base is then expected to select for increased mobility and vice versa (Dennis and Shreeve, 1997; see Fig. App. 13.2). Separately, migrfac ($r = 0.79$) and colonfac ($r = 0.72$) correlate similarly with island incidences (Dennis *et al.*, 2012a), as noted above for Flat Holm and Steep Holm: in a multiple linear (forwards entry) stepwise regression, these two variables together were found to account for 76.1% of variation in species island occupancy ($F_{2,53} = 84.28$, $P < 0.0001$; R^2: migrfac 61.7%, colonfac 14.3%). The residuals for two species indicate substantially higher (*C. tullia*: standardized residual 2.98) or substantially lower (*Melitaea athalia*: standardized residual 2.86) occupancy than predicted (Fig. 6.8).

Using Species' Ecological Profiles to Examine Inter-island Relationships

In recent years it has become increasingly evident that the occurrence of species on islands depends on five groups of basic factors, all of which

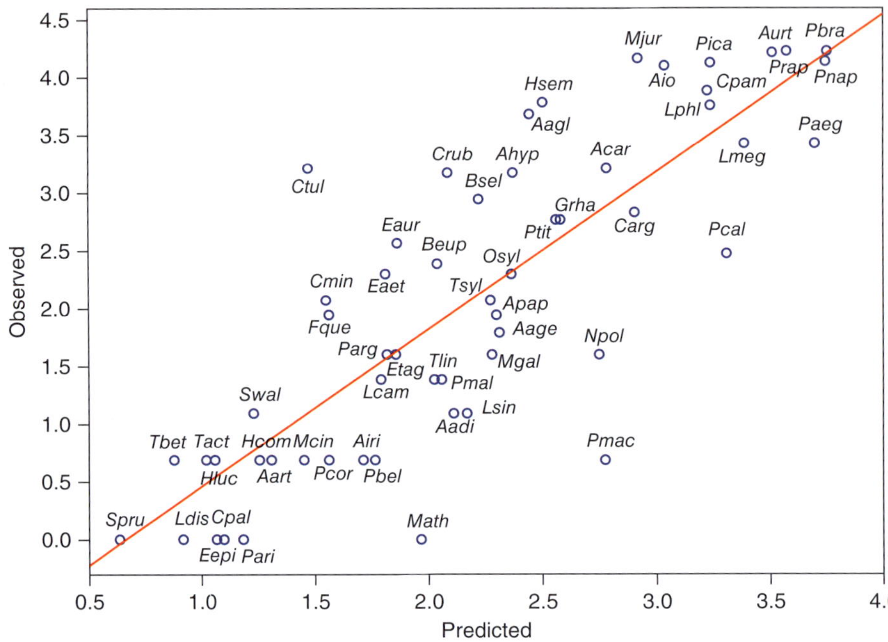

Fig. 6.8. Observed versus expected incidences on offshore islands of British and Irish butterflies from ecological measures for colonization ability (colonfac) and migration capacity (migrfac); $F_{2,54} = 60.4$, $R^2 = 69.1\%$; $P < 0.0001$); species named with first generic letter and first three species' letters (see Appendix 1).

facilitate or run counter to migration and colonization and influence extinction (viz. island geography; island heterogeneity; regional conditions; species biology; and the source pool of species) (Whittaker and Fernández-Palacios, 2007). These groups of factors can be regarded more simply as extrinsic (geographical and environmental), intrinsic (species-specific) and complex (determined by interaction between the two). Intrinsic factors incorporate the biology (**bauplan**) of species on the islands and of the regional source community, particularly their colonizing ability and migration capacity (Lomolino *et al.*, 2010).

In 1996 it was shown that species incidences on islands reflected their unique ecological profiles to the extent that it was possible to determine a species-specific propensity to migrate to, colonize and persist on islands (Dennis and Shreeve, 1996, p. 55). In the new millennium, the issue widened to a broader one, reversing the focus: whether the colonization and migration profiles (means and variances) of entire island faunas reflected the geography of islands (viz. isolation, area, marginality). From studies of butterfly species ecology (Dennis *et al.*, 2000a, 2004) the two indices of colonfac and migrfac were computed to answer this very question: colonfac and migfac profiles (**means** and **variances** and **coefficient of variation (CV)**) for island faunas should expose the history of migration, colonization and persistence of species on islands and vary in relation to island dimensions and isolation (Dennis *et al.*, 2012a). A number of features were expected:

1. As increasing island isolation should be accompanied by a greater proportion of species with increased migration capacity (and fewer species with poor migration capacity), more isolated islands should have higher migrfac means (and lower variances) than less isolated islands.
2. An increase in island area should be accompanied by lower colonfac means (higher variances) owing to colonization by (and persistence of) an increasing number of specialized species each having a limited resource base. The theoretical foundation for this is that resources – abundance, heterogeneity – and population size will tend to be greater on larger land areas, facilitating colonization, countering stochastic extinction, and providing a more stable base for more demanding specialists. Consequently, larger islands will have lower colonfac means than smaller islands.

These considerations suggest that less isolated, larger islands are likely to have more varied colonfac and migfrac values (variances) than more isolated, smaller islands. Another logical sequitur is that lower species richness is expected to correlate with higher means (and lower variances) for migrfac and colonfac values. The 2012 study (Dennis *et al.*, 2012a) investigated 103 British and Irish islands for 56 butterfly species and 12 ecological traits. Islands were selected on the basis of having at least two of the three longest-distance migrants (indicator species) (i.e. *C. croceus*, *V. atalanta* and *V. cardui*), all three if islands had fewer than five species. In the 2012 study, the mean migration capacity of island species was found to decrease with increasing island species richness (mean migrfac: ***F* statistic** $F_{1,101} = 210.50$, $R = -0.82$, $P < 0.0001$; $R^2 = 67.6\%$), but colonization ability did not significantly vary with species richness (mean colonfac: $F_{1,101} = 2.18$, $R = -0.15$, $P = 0.12$; $R^2 = 0.02\%$) (Fig. 6.9a). On the other hand, variation (using the coefficient of variation, CV) in colonization ability and migration capacity of island species increased with numbers of species on islands; again, migration capacity more so than colonization ability (CV on island numbers: colonfac $F_{1,101} = 7.88$, $R = 0.27$, $P = 0.006$; $R^2 = 7.2\%$; migrfac $F_{1,101} = 54.04$, $R = 0.59$, $P < 0.0001$; $R^2 = 34.9\%$) (Fig. 6.9b). These results suggest that changes in island faunas are more sensitive to butterfly species migration capacities, that is, to their propensity for movement, and that perhaps differences between species are less for the colonfac than migrfac measures.

Islands have two basic characteristics that distinguish them from mainland sites (habitats): (i) islands are isolated from sources by resource-vacant matrix, thus requiring some migration capacity in order to allow access; and (ii) island conditions tend to be more marginal, their resources more limited in quantity and quality compared with source regions, placing demands on species colonization. Consequently, islands present harsher conditions for migration and colonization than mainland sources; they can only be occupied by individual species if conditions match their migration

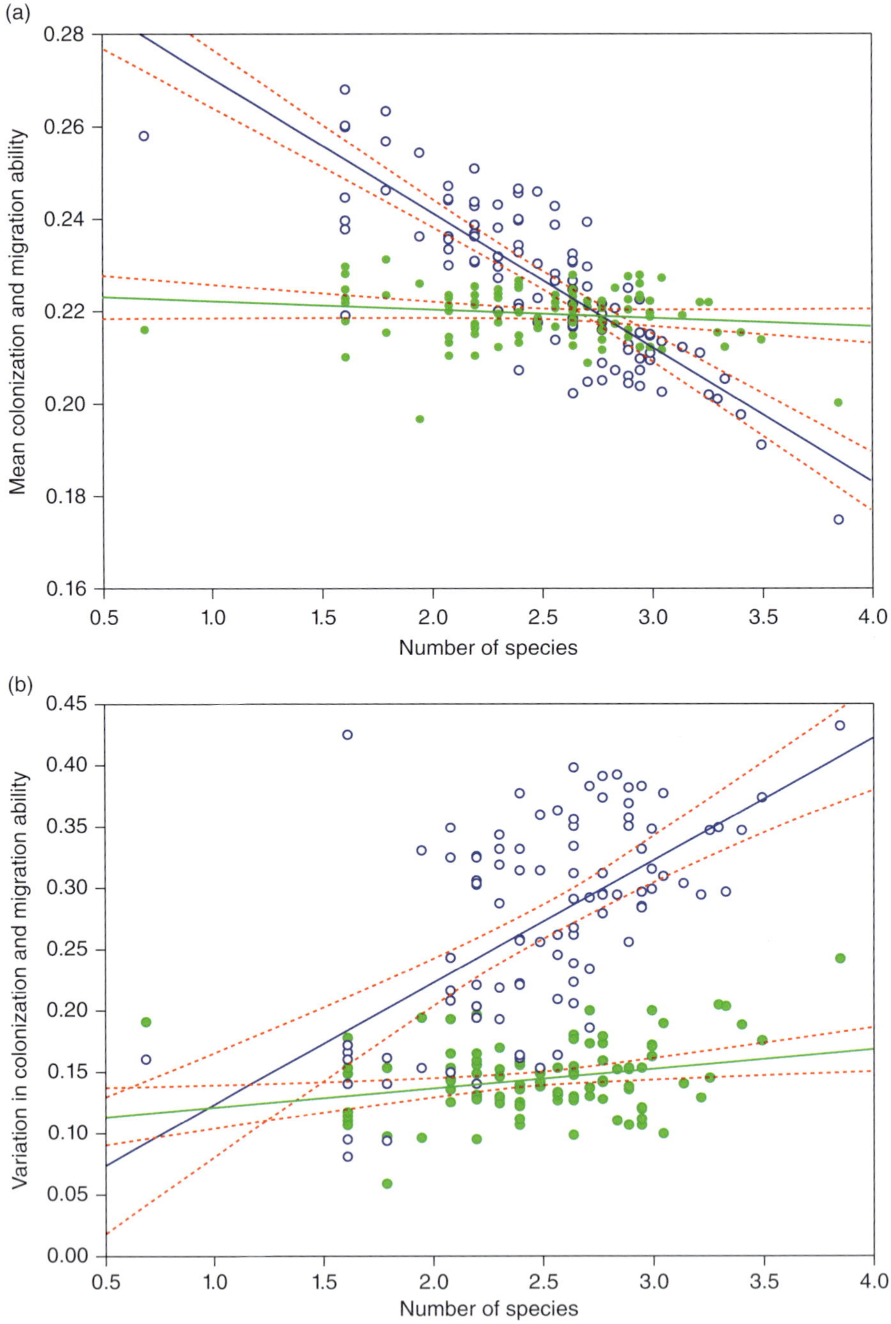

Fig. 6.9. Relationship of colonization ability (colonfac) and migration (migrfac) capacity to species richness for butterflies on islands off mainland Britain and Ireland: (a) means; (b) coefficients of variation. Open blue symbols, migration capacity; green symbols, colonization ability. (From Dennis *et al*., 2012a, courtesy of John Wiley & Sons, Inc.)

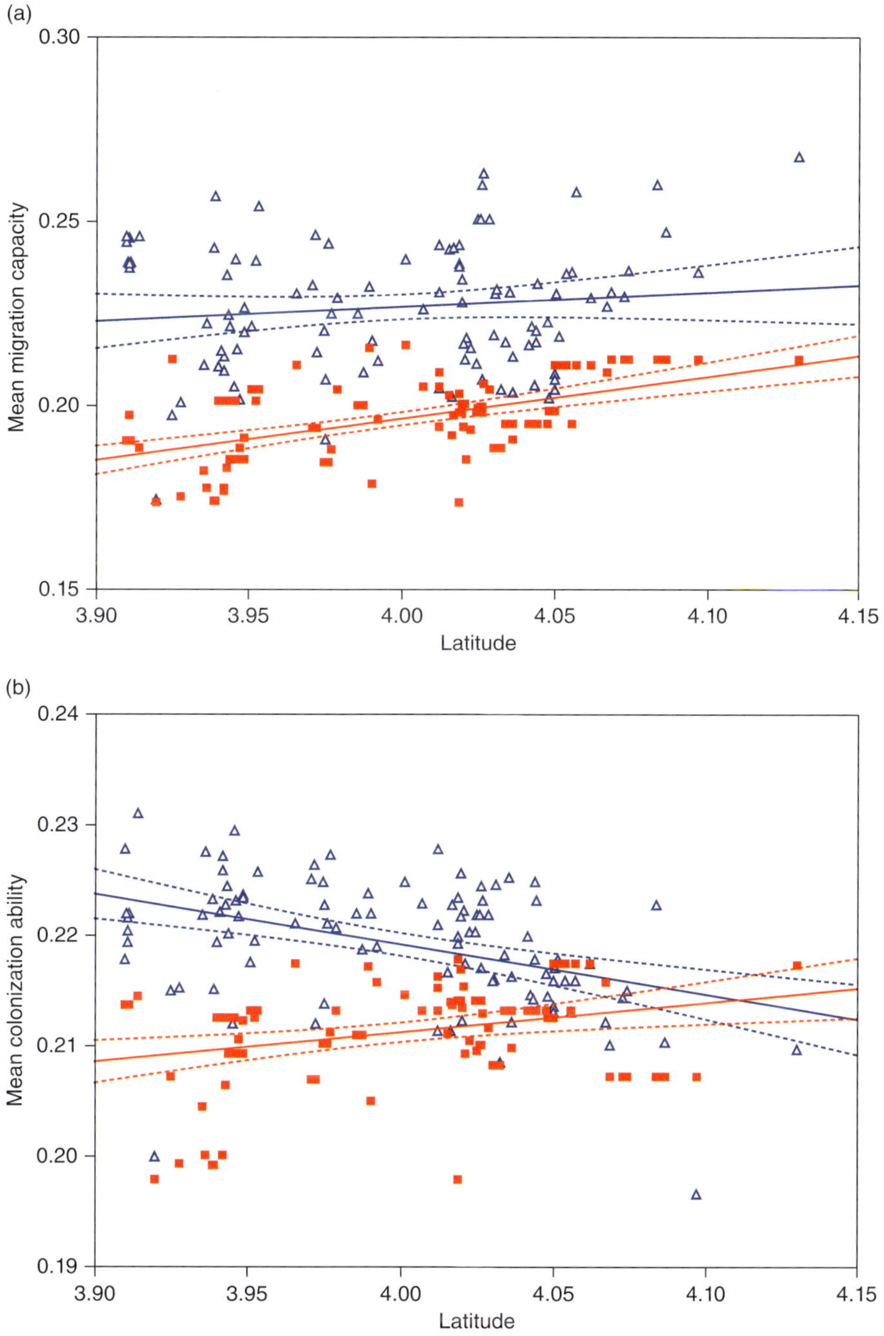

Fig. 6.10. Relationship of island and source ecological indices to latitude for butterflies occurring on islands offshore Britain and Ireland: (a) mean migration capacity (migrfac); (b) mean colonization ability (colonfac). Open blue triangles, islands; red squares, nearest mainland source (within 50 km radius). (From Dennis *et al.*, 2012a, courtesy of John Wiley & Sons, Inc.)

capacity (not just migration ability) (see Appendix 6) and colonization ability. Such conditions vary with regional geography (Dennis *et al.*, 1998a, b) and over time (Dapporto *et al.*, 2009; Fisher *et al.*, 2010). Differences between islands in both colonfac and migrfac are likely to decrease northwards if only because, with increasingly marginal conditions at higher latitudes, the species occurring at mainland sources are increasingly characterized by higher migrfac and colonfac values. In the 2012 study (Dennis *et al.*, 2012a) it was found, by applying **GLM** (full factorial **general linear models**), that mean colonfac and migrfac values were significantly higher for island faunas than for mainland source faunas (colonfac: $F_{1,101} = 30.88$, $P < 0.001$; migrfac: $F_{1,101} = 4.72$, $P = 0.03$); but, as also expected, these differences decreased with island latitude (Fig. 6.10a). The relationships are complex and interested readers are referred to the original work (Dennis *et al.*, 2012a).

A further pattern was expected following our previous work. With the existence of significant variation in species migrfac and colonfac values, the order in which species enter an island fauna is likely to be highly deterministic. As such, a pattern of nested faunas is expected on islands, with species having greater colonization ability and migration capacity occupying a larger number of islands, including all those on which species with lower migrfac and colonfac values are found; the pattern is referred to as species **nestedness** (Atmar and Patterson, 1993, 1995). Butterflies on British islands score 10.04° on a scale of 0° (complete nestedness) to 100° (absolute non-nestedness) (using the BINMATNEST programme of Rodríguez-Gironés and Santamaría, 2006). Rank order of species in the nestedness matrix was found to correlate highly with migrfac (Spearman $r_s = -0.73$) and colonfac ($r_s = -0.65$) (both $P < 0.001$). This point challenges the basis of **species turnover** that species lists will change randomly over time (Simberloff and Wilson, 1971). The greater the distinction among species in migration and colonization ability, the more likely it is that the same species migrate to and become extinct on islands. Thus, isolated smaller islands will tend to get the same occupants, owing to the species gradient in colonfac and migrfac values (see Fig. 3.3), providing support for predictions of species on islands (see Supplementary Appendices 6.1a and 6.1b). From the Dennis *et al.* (2012a) work it is suggested that nestedness among British butterflies is driven more by migration capacity than by colonization ability, but as the two variables are correlated and functionally related it would be wise to await the development of more sensitive measures of species ecology before relying too much on this suggestion. Currently, the distinctions among species for attributes describing colonfac would suggest that some degree of randomness in species turnover is expected (see Dapporto *et al.*, 2015 for details).

An interesting observation is that nestedness is not perfect, an issue we return to in the next chapter, which focuses on changes in island faunas.

7 Changing Patterns of Species Incidences on Islands

Thus far we have considered a static picture of the status of different species on islands. However, a signal feature of islands is the continual change in the community of organisms on them from the turnover of species: perpetual colonizations and extinctions. These give two kinds of patterns: (i) oscillations in abundance and incidence (presence, absence) of any species; and (ii) trends in their abundances and in community composition. The factors causing these changes are basically of two kinds: (i) stochastic (random, without any *apparent* cause); and (ii) deterministic (where causes are evident).

Community associations and shifts, as we have already seen, are typically measured using similarity (dissimilarity) coefficients such as **Jaccard's S_J** or **Sokal and Michener's S_{MM}** (Dennis and Shreeve, 1996; Dennis *et al.*, 2010) (see Appendix 8) and in the case of an island's butterfly community, the comparisons are carried across years. It has become clear that even small changes in species composition can create apparent trends in communities. For instance, a distinct (apparent) trend occurs in the community of British butterflies from AD 1600 caused purely from the variable 'find' dates for British butterfly species and then their subsequent losses (extinctions) (Dennis *et al.*, 2010). Caution is therefore required in the study of community (or species) turnover as poor sampling strategies (of species for years) may produce trends from **pseudoturnover** (inability to detect species) and/or **cryptoturnover** (irregular sampling of years) (Schoener, 2010). Important issues, then, in studying turnover are the quality of records (sampling), with all the inherent problems of bias affected by varying conspicuousness of species (Dennis *et al.*, 2006b), uniformity in surveying, and the period (length of time; timing) over which observations are made.

It is rare to have detailed records that adequately cover an island for a long period (> 20 years). However, assuming for a moment that we have adequate long-term records of an island's butterfly species, one observation is clear: the longer the period of time considered, the more certain it is that there are not just changes in a fauna but also distinct trends in community status; when considering the butterfly fauna for entire Late Glacial and Holocene Britain (> 15 ka) these changes must have been very dramatic indeed (see Chapter 8) (Dennis, 1977, 1992, 1993). This occurs because the longer the time frame the greater is the chance of increasingly higher magnitude events (pulses) occurring, in turn affecting migration and extinction (Dennis and Shreeve, 1996). Even during the past 25 years, detailed mapping by Butterfly Conservation UK's Butterflies for the New Millennium (BNM) scheme has disclosed massive changes in the ranges and distributions of species (Chapter 4), also equally dramatic contractions and regional extinctions – features already familiar to the naturalists of the 19th and early 20th centuries (Dennis, 1977). Some years stand out for extreme conditions linked to migration events in the UK, for instance 1976 and 1995 (Dennis and Bardell, 1996; De Palma *et al.*, 2016).

If changes (oscillations, trends) are almost certain to occur, does that mean there is no stability in an island's fauna? The answer to that is, of course, 'no'; there is some stability (a dynamic equilibrium) in species richness balanced by turnover in relationship to mainland source numbers, island area and isolation (see Fig. 1.5 and Chapter 5). There are also other forms of stability inasmuch as certain elements (species) retain their familiar status on islands. For instance, rare species and common species typically may remain rare and common, respectively, for long periods; we touch on these patterns briefly in this chapter as a prelude to longer-term geological changes in the next chapter.

Ecological Bases for Stable and Changing Island Faunas

Given the arrival of migrants, the basis for a species' status on an island is the existence of suitable resources in sufficient quantities for supporting a population, and distributed in a manner in which they can be found. This formula is described by the resource-based habitat (Dennis, 2010). Ease of discovery of different resources depends on resource integration and/or compaction (Dennis *et al.*, 2014) and the size of the island (Tilley and Dennis, 2017). Resources for butterflies are often equated in popular books with specific biotopes (simple or naïve habitats), particularly host plants and nectar sources, but in fact all vital resources in species' resource-based habitats tend to cut across biotopes or depend on **ecotones**, if only because with finer division in biotope categories the likelihood increases that a resource will occur in more than one of them. Even so, changes within a single biotope typically result in a shift in true habitat status for each species inasmuch as any one resource, at least, used by a species is undermined or enhanced. Under natural conditions, in the absence of human interference, terrestrial substrates will undergo vegetation succession, a shift in seres or biotopes, from bare ground at the outset of land exposure to the atmosphere, thence to short swards, tall herbs, shrubs, parkland and ultimately to woodland having a characteristic climax state (Dennis, 2010); before the Neolithic clearances the form that this woodland took differed over Britain (Bennett, 1989). With vegetation succession, different butterfly species are favoured and a shift (trend) will occur in the plant and butterfly community. Resistance to change is facilitated by island size, if only because an island's dimensions equate with substrate and topographic heterogeneity; heterogeneity has potential for forestalling wholesale change over the entire island, and species able to disperse across the island terrains tend to adapt to the average conditions of the island as a whole. This was a fundamental concept of Professor E.B. Ford and his research team on the Isles of Scilly, demonstrated by the similar frequencies in a genetic trait, hindwing margin ocellation (for example, in Plate 2.78 the female has one black hindwing spot), in *Maniola jurtina* on the three large islands (Tresco, St Mary's, St Martin's) contrasting with unique frequency patterns on five much smaller islands (Ford, 1964) (Fig. 7.1). A noteworthy corollorary of this finding is that movement between islands was insufficient to overcome the outcome of contrasting selectional regimes for wing spotting on them. However, the sole influence of island habitat heterogeneity is questionable; it should be considered how island biogeography issues (Hanski isolation) raised earlier in Chapter 5 on the Isles of Scilly and by Professor Tilley may have affected this outcome in wing spot frequency distributions.

Often island seres remain much the same over the years, at least to casual inspection, in which case succession has been arrested, as a sere 'frozen' in a **plagioclimax**, typically by abiotic conditions (weather: winds and salt wash) or biotic conditions (grazing regimes). A well documented case of stability followed by sudden change owing to a shift in land use is provided by the island of Tean, in the Isles of Scilly. Smaller exposed British islands typically lack woodland but, in the absence of grazing, may develop thick scrub. The removal of cattle from Tean, in 1950, led by 1954 not only to a shift in the location of *M. jurtina* populations from three zones on the island (from zones 1, 3 and 5 to zones 2 and 4) (Ford, 1964) (Fig. 7.1) but also to a shift in population genetics owing to changes in habitat structures and selection regimes on wing ocellation in the butterfly. Land use changes can produce wholesale changes to an island's ecology; a classic case is the isle of Canna, with the initial impact of rats on rabbit populations, and then of rabbits on the vegetation once the rats had been removed (Flux and Fullagar, 1992; Bell et al., 2011), subsequently requiring the eradication of the rabbits (BBC, 2012). One wonders what the effects would have been on the butterfly populations – another valuable research chance missed.

It is important to appreciate that unoccupied and unsuitable habitats can be converted to suitable habitats by changing conditions; one obvious factor is a change in climate, or seasonal weather conditions, which we now understand well facilitates migration. Thus, in recent years *Aphantopus hyperantus* has been able to colonize the large islands of the Isles of Scilly (I.C. Beavis, personal communication, 2004) and there are indications that this butterfly and *Anthocharis cardamines* have colonized South Uist in

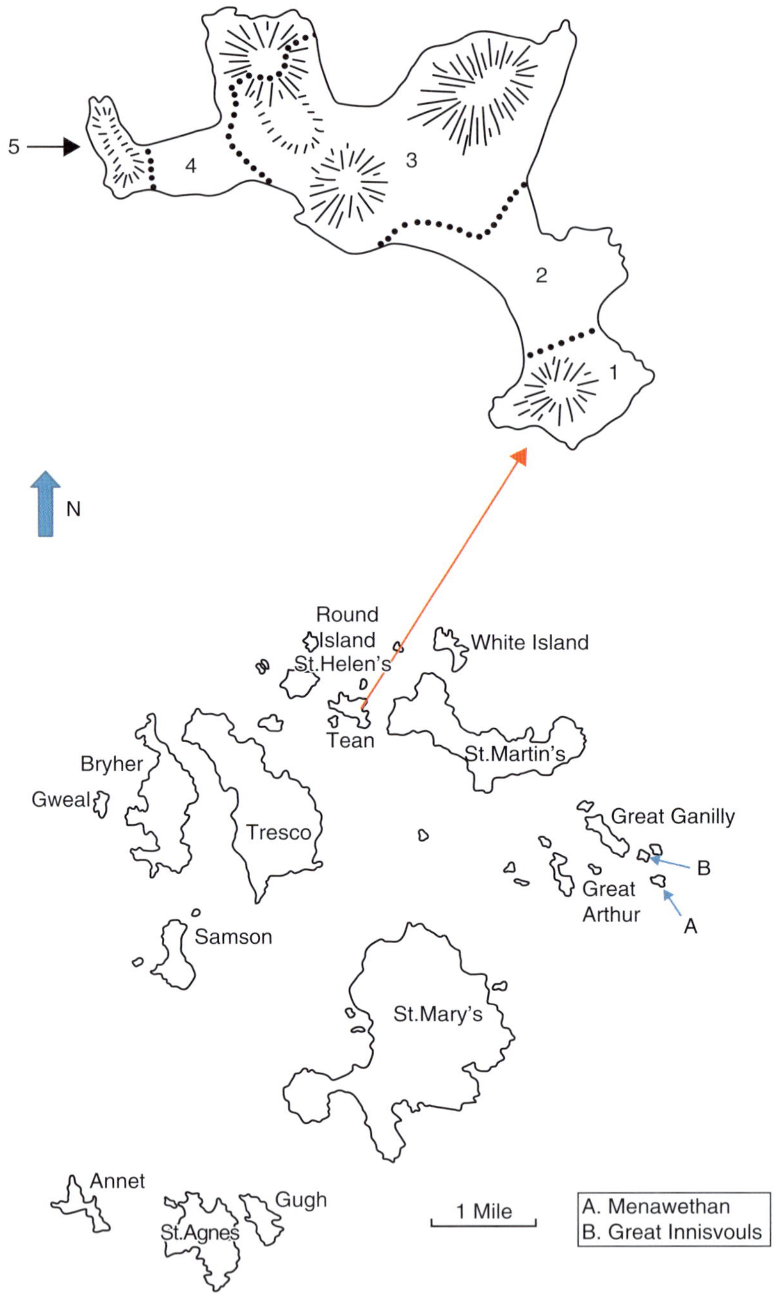

Fig. 7.1. The Isles of Scilly highlighting bounds of habitats for populations of *Maniola jurtina* on Tean. In 1950, zones 1, 3 and 5 were occupied by populations of 3000, 15,000 and 500 individuals, respectively. By 1954, after the removal of cattle in autumn 1950, the populations were restricted to zones 2 and 4 (tall grass), that previously comprised short sward avoided by the butterfly. By 1954, zones 1, 3 and 5 were covered in dense gorse, bramble and bracken (see Ford, 1964).

the Outer Hebrides (Kemp, 2014, 2015); in effect the resource base has been enhanced for these butterflies.

Historical Signals in Geographical and Ecological Data

Before the development of island biogeography theory (MacArthur and Wilson, 1963, 1967) it was thought that islands were occupied at some stage in the historical past, the plants and animals more or less surviving through to the present day. This was still a well-embedded notion for British butterflies among even the entomological scientific community in the 1970s. In circular reasoning, without thought of alternatives and the fact that potential solutions may not be exclusive, it generated the view of so-called special subspecies considered to pre-date the last glacial maximum, their distinctive phenotypes regarded as unique, a concept that in turn reinforced the view of these populations' immobility and antiquity (Ford, 1964, p.60). We are now fully aware that processes are not constant, even over short time periods (c. 10 years), as generally envisaged by the standard MacArthur and Wilson island biogeography model. Owing to ever-changing conditions, there are pulses of immigration to islands and extinction on them. Beyond that, increasingly higher-magnitude events will have characteristic recurrence intervals (Dennis and Shreeve, 1996) (see Chapter 8). Thus, the 10 June 1993 storm flood on the Great Orme's Head, North Wales, is considered to have been a <1/1000-year event; this temporarily impacted on (depressed) populations there of *Hipparchia semele* and *Plebejus argus* (Dennis and Bardell, 1996). As we have seen among some British species, extinction events occur that affect even long-established plants and animals (e.g. *Aporia crataegi*, *Phengaris arion*, *Cyaniris semiargus*) over much larger land masses.

Therefore it is of great interest to find historical signals in ecological data. In the next chapter we will consider some species that demonstrate clear ecological transformations on the British mainland over the Holocene. The question here is whether any historical signals emerge in relationships of species incidences on islands with geography or species ecology. A number of species have been noted to occur on far fewer islands than expected on the basis of either island geography or ecology (Dennis and Shreeve, 1997). The expectation is that species should occupy as many islands in accord with their migration capacity (**migrfac**) and colonization ability (**colonfac**). In the relationship of species incidences for the 73 larger islands (Dennis and Shreeve, 1996) against ecological variables, three species were noted to have particularly high anomalies (**standardized residuals**): two negative (*Papilio machaon* and *Melitaea athalia*) occurring on many fewer islands than expected, and one positive (*Coenonympha tullia*) occurring on many more islands than expected (Dennis and Shreeve, 1996) (see Fig. 6.8). Their removal led to a sharp increase in accounted variation (R^2) from 68% to 79%. Both *P. machaon* and *M. athalia* have unusual resource demands. In Britain both, in a sense, are habitat specialists; the Fenland race of *P. machaon* feeds exclusively on one host plant (*Peucedanum palustre*) (Dempster *et al.*, 1976) and *M. athalia* is one of two species known to engage, exceptionally, in site-specific monophagy on an annual plant (i.e. *Melampyrum pratense* in woodland clearings) (Warren, 1987a, b, c; Kemp *et al.*, 2008). One would not expect these species to colonize British islands and their probability of doing so is evidently overestimated by both their colonfac and migrfac values based on European-wide tolerances. In fact, swallowtails found on offshore British islands (e.g. the Isles of Thanet and Wight) (Knill-Jones, 2008; Solly, 2014) are continental vagrants not of the Fenland race.

Of much greater interest would be the unexpectedly high incidence of a species on islands; in such cases, a basic requirement would be longevity of the resource base, free of human impact. Woodland butterfly species, for the latter reason, are scarce on the British offshore islands; thus the existence of *Favonius quercus* on Colonsay's north-eastern krummholz oaks of A'Choille Mhor SSSI is possibly an early Holocene relict there (Dunn, 1965; confirmed by David Jardine in the 1980s, personal communication) (Plate 1.15). *C. tullia* is exceptional in occurring far more frequently on islands than expected: typically on Scottish islands in the Hebrides, Orkney and Shetland. In this case, colonfac and migrfac values greatly underestimate the presence of the butterfly on islands. *C. tullia* is oligophagous and a hostplant–habitat specialist on *Eriophorum–Erica tetralix* peat mosses. It formed the subject of the first study of animal metapopulations to demonstrate the primacy of patch habitat quality over patch area and isolation (Dennis and Eales, 1997, 1999). In addition to being restricted in habitat use, the butterfly is limited in voltinism and the length of the flight period. The specialized ecology of *C. tullia* certainly contrasts with other species having similar frequency of incidence on islands (e.g. *A. cardamines*, *A. hyperantus*). The distinctive feature of *C. tullia* is its confinement to the oligotrophic blanket and raised bog community which occupies a self-perpetuating climax development under the cool wet climates of northern and western Britain dating from the early to mid Holocene (< 7.5 ka BP) (Walker, 1970; Barber, 1993). Thus, it is feasible that populations of *C. tullia* have been maintained on northern archipelagos since their colonization in the wake of Late Glacial ice recession (c. 10 ka BP) (Dennis, 1993). But some care is required in assuming these island populations of *C. tullia* to be completely isolated; movement in this butterfly may be much greater than generally thought (Dennis and Eales, 1997).

In the next chapter we consider other species whose unusual features suggest they may have occupied the Scottish Isles for long periods of the Holocene, much as *Leptidea juvernica* has in Ireland.

Order and Change in Island Faunas

Over short periods of observation a general pattern builds up of the relative abundance of species on islands; a sense of continuity in status of most species becomes apparent, although suitable data for confirming such impressions are sparse. Ideally, systematically collected records for species (BMS data) standardized across all biotopes would be collected – data that currently exist for only one offshore island (viz. Jersey, 2004–2014) (Long, 2015). Even so, annual records for species are now accumulating for a range of small and large offshore islands, typically as part of Butterfly Conservation's BNM mapping project (Asher *et al.*, 2001). A basic difficulty in examining stability (or change) in such data sets arises from the considerable changes in 'sampling effort' over different years, as well as among sites (biotopes, recording squares) for larger islands. In such cases, the data require standardization among years (i.e. corrected for total numbers of records, or visits, and conditions) but even then problems of data interpretation can remain (Dennis *et al.*, 2006b; Isaac *et al.*, 2011a) (see Appendix 8).

Site faunas have long been known to have characteristic frequency distributions (**rank-abundance curve**) (Whittaker, 1965) in which there are few common species and many rare ones. Various explanations have been proffered for this recurring pattern, both descriptive (i.e. stochastic) and mechanistic or ecological (e.g. niche apportionment) (Gaston, 1994). In the case of the Isle of Wight (Fig. 7.2), the rare and common species are ones to be expected from their UK ecology profiles and population

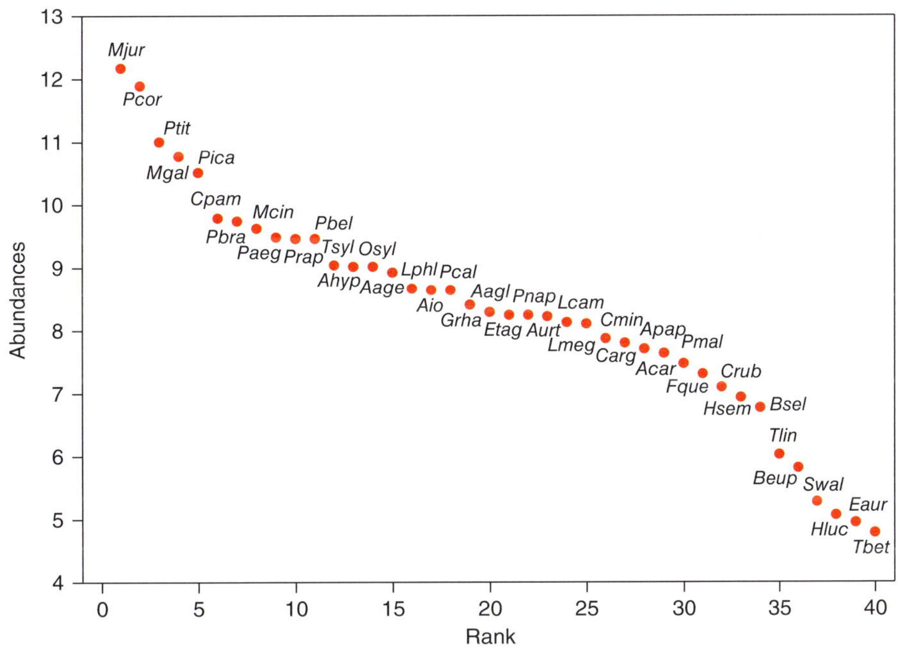

Fig. 7.2. Rank-abundance (Whittaker graph) for butterflies on the Isle of Wight (calculated for 2000–2015 records). Abundance is $\log_{10}$ records; species ranked with 1 being most abundant. Long-distance migrants removed; species named with first generic letter and first three species' letters (see Appendix 1).

declines. There is not space here to examine competing hypotheses, but a crucial aspect is the determination of population sizes. Importantly, a recent study has shown that relative population sizes are a good estimate of absolute population size in the specialist Australian butterfly *Theclinesthes albocincta* (Lycaenidae) (Collier *et al.*, 2008), as is long believed to be the case for BMS transect data (Pollard and Yates, 1993). From the resource-based habitat viewpoint this frequency distribution pattern is particularly interesting beyond the recognition that species tend to display distinctive inverse population density–mobility relationships (i.e. species existing at low density will tend to have the highest mobility) (Bink, 1992): it implies that resources for very few species can be entirely suitable or that populations for many species being recorded at a site (biotope) have significant resources and appropriate conditions that lie beyond the bounds of the target site. Within any biotope it may be expected that any butterfly species with compact (entire and highly integrated) resources will have a larger population than species having incomplete and dispersed resources. Islands, owing to their relatively greater isolation, could provide a useful comparative testing ground for this notion built on the study of mainland communities.

Regarding a community of species, a number of additional features are then perhaps expected for different elements (species) in an island's fauna:

1. Numbers will tend to remain high and low, respectively, for common and rare species over at least a short period of years.
2. The year-on-year change in status (rank) of those species will tend to be small, with rare species consistently ranked low and common species consistently ranked high (long-distance migrants excepting). The pattern in mean rank changes for species population status should approach a parabola (a polynomial function) (Fig. 7.3), and contrast with that of (virtual) communities whose populations fluctuate randomly over the years.
3. Stability in the populations of all species is likely to be significantly lower on smaller islands than on larger islands, but this will depend on the proximity of population sources and migration capacity of island residents (see Fig. 6.6).

Thus, when analyses are controlled for island geographical contexts (isolation, source mass, conditions determined by latitude and longitude), mean rank changes (standardized for n species and t years) are likely to be greater for smaller than larger islands and indicative of greater volatility. As yet, adequate data are unavailable for testing this idea.

A preliminary inspection of data for eight islands having 10 years of records and varying over two magnitudes in size and numbers of species (viz. Arran, Lindisfarne, Hilbre, Anglesey, Cape Clear, Brownsea, Wight, Jersey) supports the validity of the first two expectations, particularly the distinctive recurring parabola in mean rank changes for species when plotted against population sizes (or sums of ranks). An obvious associated feature is that year-to-year variation (coefficient of variation, CV) in standardized abundances (records) is greater for scarce than for common

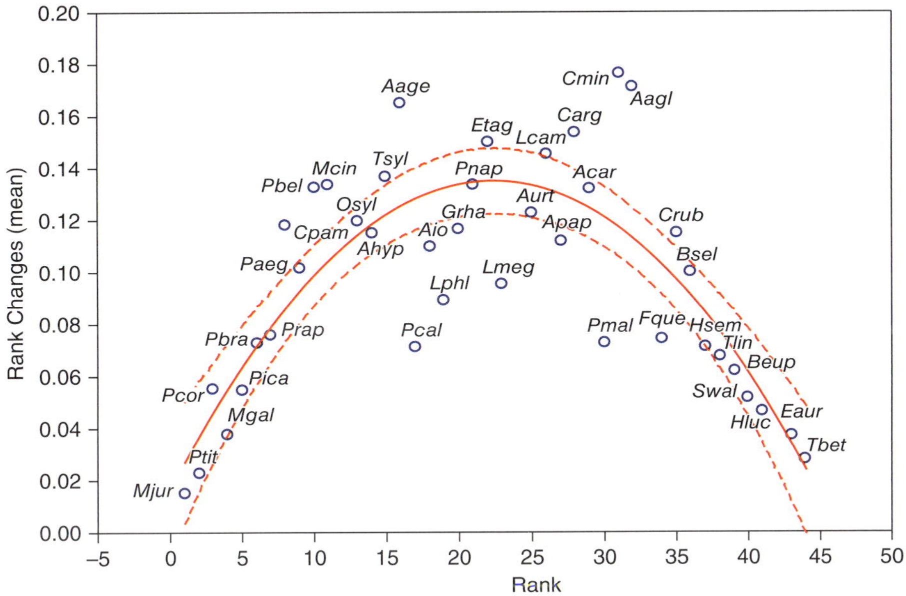

Fig. 7.3. Mean annual changes in rank order for butterflies on the Isle of Wight (migrants removed). The changes are absolute (sign removed); species ranked with 1 being most abundant. Regression line: $Y = 0.0166 + 0.0106*X - 0.0002*X^2$. Pecked lines: 95% confidence intervals. $R^2 = 64.7\%$; $F_{2,37} = 33.86$, $P < 0.0001$.

species (e.g. Wight: r = –0.60, P < 0.0001; matched by variation in CV for rank changes) reflecting on the vulnerability of rare species and the precariousness of their resource base. By comparison, the pattern produced for random population sizes in a virtual data set shows greater mean changes in ranks for species and a featureless scatter of points. The inference is that species on islands, as on sites generally, have widely varying but individually distinctive structured resources. Our understanding of these statistical patterns will only properly develop once the resource-based habitat has been widely embraced by the academic community.

Turnover and Trends: the Case of Hilbre and the Farne Islands

Relative interspecies stability in the status of some butterfly communities does not, of course, exclude the occurrence of trends in others (for example, six species show significant negative trends on the Isle of Wight: *Pieris rapae, Pieris napi, Limenitis camilla, Polygonia c-album, Lasiommata megera, A. hyperantus*). In time, sufficient data will become available for investigating the impact of turnover on oscillations and trends in species and community status over islands that differ for geography. The long-term study of small island communities presents great opportunities for distinguishing between internal (i.e. population changes, extinctions) and external (i.e. immigration, colonization) processes generated by the same stochastic and systematic (or deterministic) influences as experienced at mainland sites. Recently, this has been done for two tidal islands with relatively long recording histories (> 20 years): Hilbre (Triassic Bunter sandstone, Dee Estuary; 8 ha) and Lindisfarne (Whin Sill dolerite and dunes, off Nothumberland; 473 ha) (Plates 1.13 and 1.14); the latter is more than an order of magnitude larger than the former (Dennis *et al.*, 2010) (Fig. 7.4). Comparisons were made with the two nearest long-studied mainland sites (Leighton Moss; Wyre Forest) (Fig. 7.4). Some sparser data for the Farne Islands (Northumberland) – two clusters of small (total area 96 ha; largest island 6.5 ha) relatively isolated (2.4–6.8 km) sea-washed rocks of Whin Sill located about 10 km south-east of Lindisfarne – provided some additional useful insights. Summary butterfly records for the islands and neighbouring mainland sources can be found in Dennis *et al.* (2010, Table 1), which also has details of tests for artefacts of species conspicuousness to recorders and recorder effort.

Fig. 7.4. Outline map of the British mainland illustrating the location of Hilbre (A), Lindisfarne (B) and the Farne Islands (C) (Dennis *et al.*, 2010, courtesy of John Wiley & Sons, Inc.). Also shown are locations of two long-running mainland Butterfly Monitoring Scheme sites (LM, Leighton Moss, 54.08° N, 2.26° W, area > 250 ha, transect length 5.9 km; WF, Wyre Forest, 52.23° N, 2.14° W, area < 50 ha, transect length 2.6 km).

It might be expected that for small offshore islands the community structure would comprise random assemblages of species, if only because immigration and extinction rates are likely to be high for all species. In such situations both the so-called **near-island effect** and **small-island effect** may dominate (MacArthur and Wilson, 1967). However, this notion is likely to be simplistic; the outcome may well be affected by island factors other than area and isolation (i.e. topography, resources) but especially by species traits (i.e. migration capacity, colonization ability) (Lomolino *et al.*, 2010). Just how community trends are affected is a more complex matter, but one artefact is more certain: the longer the survey record, the more likely it is that trends occur inasmuch as environmental variability increases with time and distance (Williamson, 1987). On islands, especially small offshore ones, and without regard to variable species traits, trends are expected to be generated by the stochastics of losses and gains to the fauna more than by substantive, deterministic factors such as climate. A premonition of this reasoning is contained in an experiment carried out by Professor E.B. Ford and his team for two tiny islands in the Isles of Scilly. In 1954 (Ford, 1964, p. 66) the introductions of *M. jurtina* to Menawethan (3 ha, females $n = 120$) and Great Innisvouls (1.5 ha, females $n = 117$) both failed. On the first island the population crashed and further injection was required (females $n = 106$, 1956). These islands in the Isles of Scilly archipelago are probably **sinks** or **pseudosinks** dependent on the migration ability and the mass of individuals (cumulative population size) at larger island sources.

Hilbre and the small Farne islands, if not Lindisfarne, were expected to have much the same status as these small Scillonian islands. Total records for species on the islands were found to relate differently to source mass (mainland records as a measure of population size) and migration capacity. In the case of Lindisfarne, source mass was the sole significant predictor of the island's records ($R^2 = 57\%$, $P < 0.0002$) whereas for the Farne Islands, migration capacity was the sole predictor ($R^2 = 37\%$, $P = 0.004$). On the other hand, for Hilbre both source mass and migration capacity were discovered to contribute significantly, although source mass was almost twice as important as migration capacity ($R^2 = 71\%$, $P < 0.0001$). Unlike the case of Flat Holm and Steep Holm in the previous chapter, in this study it was possible to isolate out distinct influences of migration capacity and source mass.

Turnover of species on Hilbre and Lindisfarne has been measured as relative turnover rate (see Chapter 1) (Fig. 7.5); even if colonizations and extinctions, on which turnover is based, are overestimated owing to using nil records as absences, the lack of records is an indication of the precarious state of the populations on the islands. Turnover in butterfly species was indeed found to be linked to small population size and, as expected, was greater than at mainland sites; the patterns of both colonizations and extinctions were heterogeneous among sites. Losses were rapidly replaced by new colonizations, perhaps facilitated by low tides and temporary landing sites (i.e. for thermoregulation, feeding). Significant community

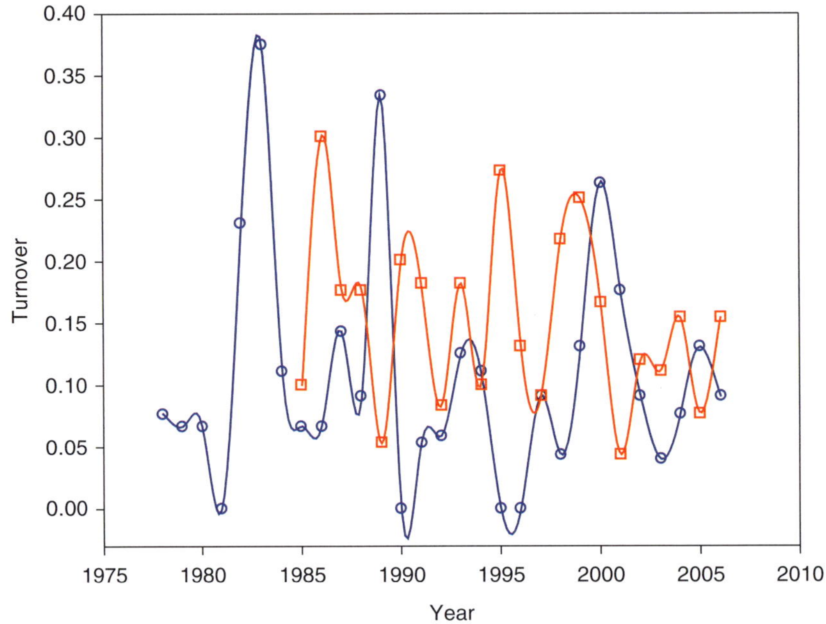

Fig. 7.5. Turnover (relative turnover rate) of butterfly species on the islands of Hilbre (red, squares) and Lindisfarne (blue, circles). Relative turnover rate (T) is $T_{t1,t2} = (ne + ni)/(N_{t1} + N_{t2})$, where t_0, t_1, $t_2...t_m$ are time intervals (years), *ne* is the number of species extinctions on an island, *ni* the number of novel colonizations on an island, and *N* the total number of species at different times. A species is considered to have colonized an island if recorded in any year following previous absences and to have become extinct if absent after a previous record.

trends over the years for both numbers of individuals recorded and incidences (presence, absence) were found at all four sites in the study and confirmed by **Mantel tests**. In fact, during the study, it became evident that time trends are universal in all butterfly communities, affected even, as indicated earlier, by the history of recording, as well as by extinctions and true arrivals (colonizations). Further to this finding, significant associations between climate and records were found for a number of different species at all sites. Ten and nine species, respectively, have significant climate signals for Hilbre and Lindisfarne, all but two relationships having the expected sign (correlation polarity) based on findings in butterfly ecology (Pollard, 1988). For instance, both *Thymelicus sylvestris* and *Pararge aegeria* became increasingly common over time on Hilbre and *Aglais io* and *L. megera* increasingly abundant on Lindisfarne.

However, only on Hilbre was a definitive climate signal (especially for temperature) found for the community trend and this was generated by the relative colonizing ability of species at nearby Wirral and Deeside sources. A warming climate has the capacity for not only building larger populations, as marginal resources become more suitable, but also of facilitating successful mainland-to-island transfers. This relationship does not describe a simple matching of island records (numbers) with raw climatic trends, but a direct matching of disparity in annual faunas with disparity in climatic data. Compared with Lindisfarne, Hilbre butterfly populations are more vulnerable to extinction, making colonizations more noticeable, but with sources on either side of it in the Dee estuary, colonizations are also more likely than on Lindisfarne. Larger Lindisfarne provides a more stable area for butterfly populations; on the other hand, species on the more isolated Farne islands are dependent entirely on the migration ability of species. For Lindisfarne, as for the two mainland sites, a greater association occurred with 'years' than any climate variable and the inference is clear: it alerts researchers to stochastic influences creating community trends. However, climate variables were noted to have an impact on the populations of a number of species at the other three sites, more than any other influences; this suggests that climate may nevertheless contribute to the trends on those sites.

The findings for Hilbre and Lindisfarne support Williamson's (1983) notion that communities are always changing. To the extent that island dimensions differ, so will their faunal content and differences in community susceptibility to stochastic influences and factors that cause directional changes. Furthermore, subtle shifts in the relative size of source populations (source mass) and of island populations (persistence) gradually alter the probability of species incidences on islands.

Desert Islands: the Perpetual Turnover of Adventives (Itinerants)

Desert islands typically present highly deleterious and/or unstable situations, preventing species from taking up long-term residence. Classical situations for British offshore islands are those regularly overwhelmed and scoured by winter storm waves (e.g. Bawden Rocks, Cornwall: Plate 1.2). Usually these are bare rocks rising little above the tides (e.g. some Strangford Lough drumlin islands) but there are larger deserted islands in the extreme climatic conditions of higher latitudes. Islets comprising unconsolidated deposits (e.g. sand bars, spits) are especially vulnerable to changes in their substrate cover from inundation and wave action, and although there are Lepidoptera capable of surviving immersion in seawater (e.g. *Luperina nickerlii leechi* on Loe Bar, Cornwall) (Spalding, 2015), no British butterfly has been recorded doing so. On desert islands crucial resources (i.e. larval host plants, nectar flowers) for butterflies may appear to be absent, but closer scrutiny may well find them. This very eventuality has been reported for Hebridean offshore islets as mentioned in Chapter 3 (e.g. Heisker, south-west of Canna). It is perhaps surprising that tiny amounts of ephemeral, exploitable resources exist on biotope 'islands' in terrestrial contexts amid agricultural monocultures and urban sprawl, yet these are located and used by butterflies (Hardy and Dennis, 1999; Dennis, 2004).

In effect, desert islands are little better than lighthouses, but even in this respect they are of immense value for understanding island biogeography processes; they provide invaluable evidence for the extraordinary migration capacity of butterflies (Dennis and Shreeve, 1996). In effect they provide stepping stones (rest sites) for migrants and are likely to be characterized by the fastest turnover rates, effectively instantaneous, of all islands. Furthermore, purely for purposes of research, these islands have a valuable role. Wang *et al.* (2015) have demonstrated in a study of snakes on Thousand Island Lake, China, why it is important for identification of the small-island effect (in which species number may vary independently of island area) to include islands that have no recorded species (see also Dennis *et al.*, 1998a, b); it does assume, however, that such islands have been adequately sampled. An ecological (resource-based) assessment is necessary to determine the resource base of small, apparently unoccupied islands.

Big Changes on Big Islands: Uncertain Visits and Problematic Extinctions on the Isle of Man

Among changes to faunal components on islands, least surprising perhaps to casual observers are those that occur on large islands. Yet, disturbingly, they do happen, as already documented for both Ireland and the British mainland (see Chapter 4). Butterfly recording history is a recent activity and much historical information has simply been lost along with the biotopes that have been transformed, as evident in the surrounds of Selborne (White, 1789). However, one exercise that can be employed is to determine how many species are likely to have been lost and specifically which species should be present. Here, we briefly provide some insights into the case for the Isle of Man just using geographical data and some simple techniques (Hardy *et al.*, 2017). Firstly, we have compared the number of species on the island with those for exactly the same sized (and shaped, where possible) contiguous areas (fourteen 10 km squares) for each Ordnance Survey grid 100 km square over Britain and Ireland, for latitude, longitude, maximum altitude and number of coastal squares. Secondly, we related the number of squares occupied historically by species on the island to records in the encircling coastal fringes of the Irish Sea (two 10 km squares inland) in Ireland, Wales, England and Scotland; the measure used is a **gravity model** (i.e. mass of records/distance) applied to eight compass sectors centred on the island, and summed. Both records and distances are standardized so as to range from 0 to 1.

These methods are not without their problems of perception, if only because butterfly distributions on the mainland, as on the Isle of Man, may have been substantially affected, perhaps differently, by wholesale changes in land use practices during the past two centuries. Such land use transformations usually lie outside individual experience, making it difficult to conceive of current landscapes as providing suitable conditions for the species allegedly observed; the physical specimens provide strong support but not always proof of the validity of past records for the sites claimed. Despite this, museum specimens and historical records prove to be invaluable indicators of past conditions. Interestingly, though frustratingly, the Isle of Man has an unusually high number of old, unconfirmed records (see Appendix 3 and Supplementary Appendix 7.1). As many as ten species may have become extinct since 1850 (Chalmers-Hunt, 1970); a further record of a southern high-forest species, in particular (i.e. *Apatura iris* from a wood in Patrick), is distinctly intriguing, raising understandable scepticism; but then it was also recorded in North Wales by W.H. Hinde in 1924 and 1944 (Asher *et al.*, 2001). At the time when Chalmers-Hunt's monograph was published it was thought that these species had long occupied this island; the losses were accepted as permanent, as is most likely to be the case for *Erebia aethiops* and *C. tullia*, neither of which are best served by a warming climate. But investigation of historical records soon afterwards furnished extensive evidence for changes in ranges and distributions (Dennis, 1977). Thus, it is not surprising that some species have 'returned' (e.g. *P. aegeria*) and novel species occur (e.g. *P. c-album*). There is a question, however, whether some of the new arrivals are inadvertent or have been assisted by introductions (e.g. *Gonepteryx rhamni*).

The first question, then, is how many species are expected on the Isle of Man, and how great is the current shortfall. Based on a regression of current numbers of species (1995–2009) for sample units (each of fourteen 10 km squares) on the mainlands of Ireland and Britain versus longitude, latitude, altitude and number of coastal squares, some 26 species are predicted to occur on the island compared with the current number of 17 species (Fig. 7.6). The deficit is very close to the number that has allegedly been lost from the Isle of Man over the past 150 years. These figures give every impression that the Isle of Man is a land bridge island gradually being depleted of its fauna (see Chapter 2). But consideration should also be given to several facts: (i) that similar areas (the size of Man) in Ireland and Britain have also steadily been losing species; (ii) that species have periodically colonized the island (e.g. *P. aegeria*) and most if not all species may well have colonized the island by overseas dispersal during the Holocene; and (iii) that with climatic warming the island still has potential to gain more species.

A closer look at the species involved is certainly interesting. Analyses of the potential source species for the Isle of Man (a logistic regression of island records against the summed and scaled gravity measure for the surrounding mainland sectors) revealed that two additional species (i.e. *Ochlodes sylvanus* and *A. hyperantus*) – three including the marginal value for *Pyronia tithonus* – are predicted to occur on the island. By contrast, two registered residents are exposed as insecure residents (i.e. *Celastrina argiolus, P. c-album* especially). Further, in a simple linear regression of occupied 10 km squares, both *C. argiolus* and *Argynnis aglaja* are indicated to be more abundant on the Isle of Man than expected from abundances in Ireland and Britain. The pattern of relationships is neatly summarized in an ordination (principal components) plot of past and present Manx records; the species are shown as varying in status on the island in a collapsed metric space described by the individual gravity measures for the eight different potential sources (Fig. 7.7). The eight sector measures proved to be highly correlated, the plot dominated by a powerful size component

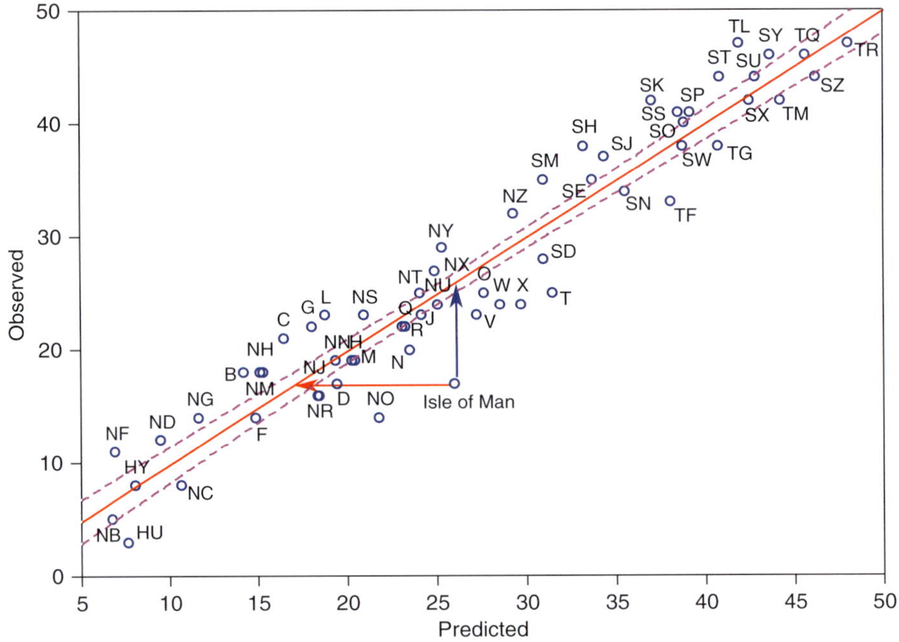

Fig. 7.6. Scattergram of observed versus predicted species richness (*n* species) for 100 km square units in Britain and Ireland on latitude, longitude, altitude and number of coastal squares ($F_{3,54}$ = 235.72, P < 0.0001, R^2 = 0.93; outliers for squares TA and S omitted). Current position of Isle of Man shown; red arrow shows its effective geographical position and blue arrow its predicted position based on all other sample squares, a loss of ten species estimated. (From Hardy *et al.*, 2017, courtesy of *Entomologist's Gazette* and Pemberley Books.)

describing the extent of range of species. Current species are clearly distinguished from extinct species and those not recorded on the island.

A further interesting vantage on the ten extinct Manx species is that collectively they have suffered a significant decline in distribution at mainland coastal sources compared with species still extant on the island (Fig. 7.8), which is indicative of their greater vulnerability to decline and extinction. Over the next decade it will be interesting to see whether *O. sylvanus*, *A. hyperantus* and *P. tithonus*, the only extinct Manx species showing increases in distribution between 1982 and 2005 on the British mainland, manage to re-establish footholds on the Isle of Man, as has *P. aegeria*. As it is, the changes in butterfly fauna on the Isle of Man have had the effect of associating the Isle of Man butterfly fauna more closely with that of Ireland than with the British mainland (Fig. 7.9).

Rare Migrants: the Future Colonists?

Conditions rarely stay the same and certainly not over long periods. In the medium-term future (AD 2050) the British and Irish islands are faced by

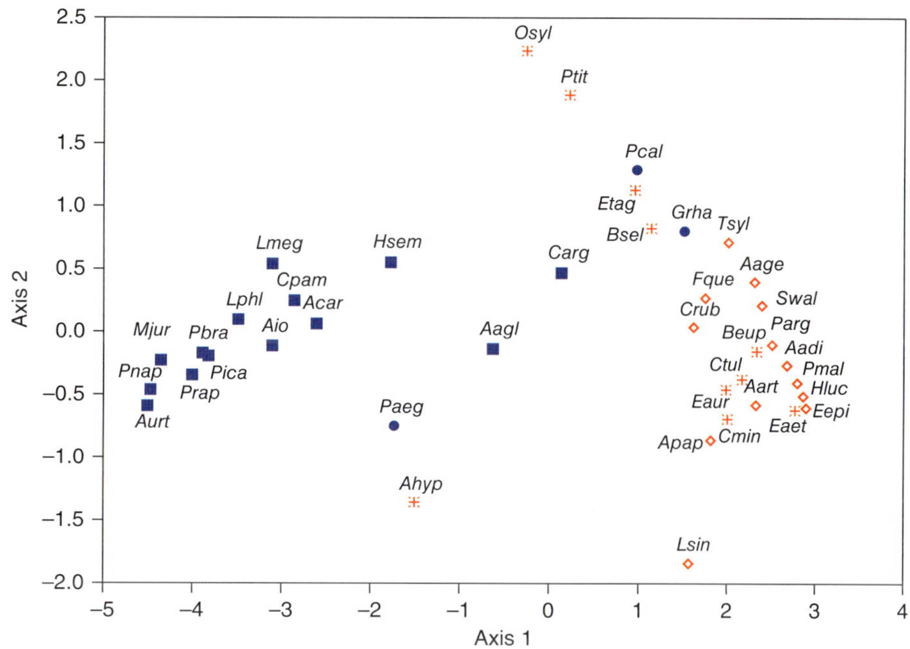

Fig. 7.7. The comparative status of Isle of Man species with those at surrounding coastal margins. Principal components plot of source species for the Isle of Man from the surrounding mainland coastal fringe of Britain and Ireland in first two dimensions based on gravity assessments of source mass (10 km records <1970 to 2009) and distance from the island for eight compass sectors. Measures of mass and distance are scaled, the former on available squares, the latter on maximum distance between sources and the island. Blue: square, extant/ubiquitous; circle, re-immigration, re-reintroduction, vagrant. Red: circle, absent; asterisk, extinct. Axis 1, 84% accounted variance; axis 2, 8% accounted variation). (From Hardy *et al.*, 2017, courtesy of *Entomologist's Gazette* and Pemberley Books.)

two vast counteracting forces: (i) biotope ('habitat') loss and degradation (Dennis, 2010); and (ii) climatic changes with enhanced greenhouse conditions (Dennis, 1993). These interact in complex ways to affect the fauna (Warren *et al.*, 2001; Hardy *et al.*, 2014). The first has serious consequences for species resources (i.e. diminution, fragmentation); the second (raised means and extremes in weather and climate variables) will undoubtedly have an impact on the quality of each species' resource base, and the conditions in which species exploit their resources. A crucial consideration in losses is that any undermining of the mainland sources (i.e. mass or populations) for potential island colonization (including coastal France for Britain) will tip the balance towards extinctions on offshore islands.

One of great delights in recent years has been the range expansion in some species. Species undergoing range expansions (e.g. *P. aegeria*) (see Chapter 4) are likely in time to generate a systematic 'diachronous' colonization of the offshore islands (e.g. *A. hyperantus* in the Isles of Scilly).

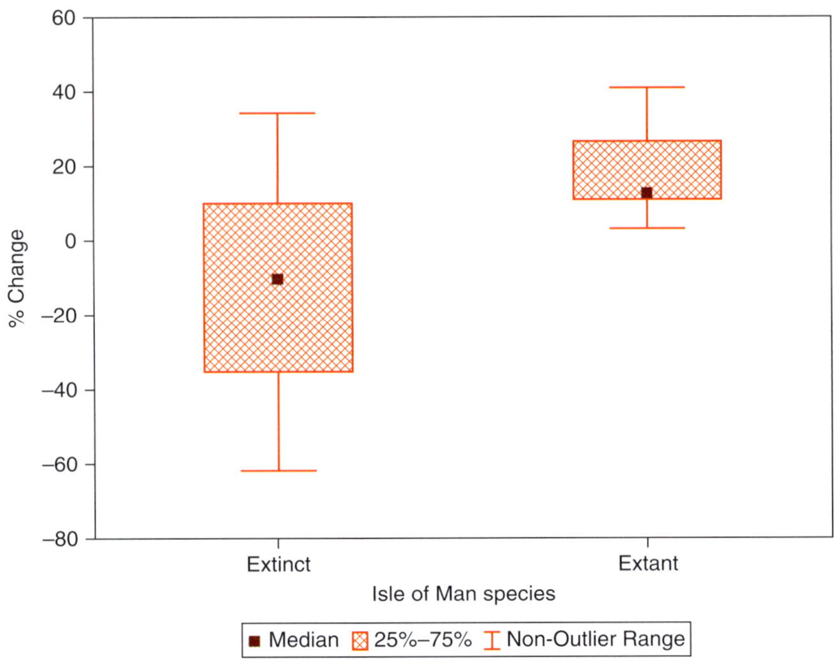

Fig. 7.8. Bar chart of distribution changes at mainland sources around the Irish Sea for extinct and extant butterfly species in the Isle of Man. A comparison is made of distributions for the periods < 1982 and > 1995 standardized to 100 from their summed cover of 10 km squares (Mann–Whitney $U_{10,17} = 28$, $Z = -2.84$, $P = 0.005$). (From Hardy *et al.*, 2017, courtesy of *Entomologist's Gazette* and Pemberley Books.)

In the long term this process has the potential to tip the balance towards more potential colonists for offshore islands, but it will almost certainly be countered by reductions in source mass for many species, that is, both the loss of populations and diminution in population size at potential sources as a result of habitat loss (Dennis, 2015a). The end product will be community changes as the species gained, in part, replace those that will be lost. Future colonists are predicted to occur in the assessment carried out by the climatic risk atlas for Europe (Settele *et al.*, 2009): some 70 additional butterfly species are regarded as potential occupants of the UK mainland by 2050, and an additional 15 species in Ireland. It should be borne in mind that this assessment is based purely on climate conditions and unrestricted migration, not the spacing, condition and availability of essential resources. It should be engraved on minds as a principle: migration is an easier matter than colonization; we observe many transfers well beyond natal habitats, but – as with attempted introductions – these rarely take root. Thus, the climatic atlas presents a highly biased viewpoint, particularly as some host plants are not available (e.g. *Aristolochia* for *Zerynthia rumina*). Even so, it provides some promise of compensation for the losses of species in the British and Irish islands, especially the

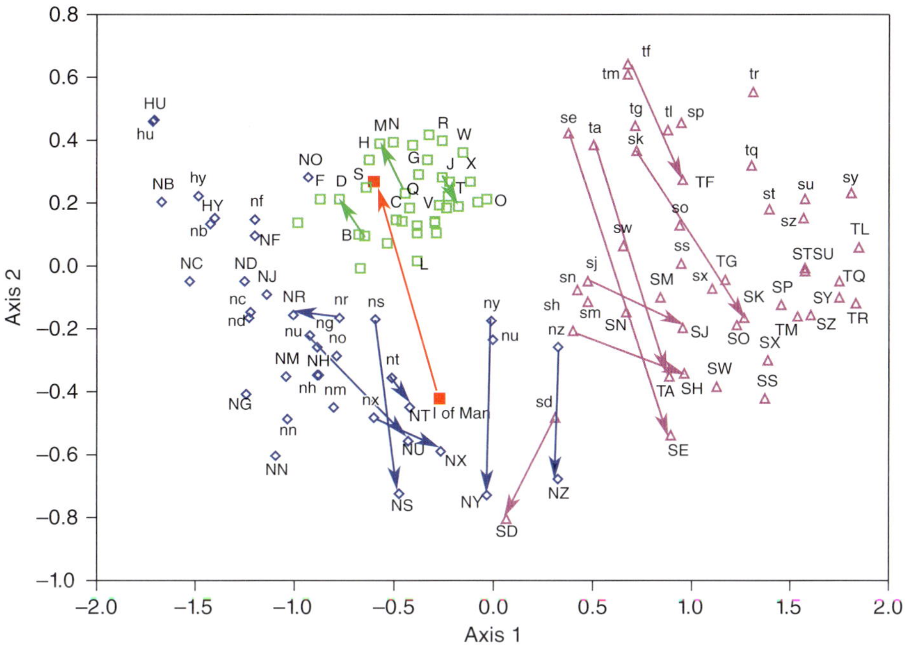

Fig. 7.9. Ordination (non-metric scaling) of relationships (percentage differences) of butterfly faunal samples (14, 10 km squares each) for 100 km grid blocks of Britain and Ireland for two periods (1, < 1982; 2, 1995–2009). Red squares: Isle of Man; circles: Britain and Ireland; blue, north of the Isle of Man; purple, south of the Isle of Man; green, Ireland. Arrows illustrate change in faunal associations for the Isle of Man during the two periods, and for units close to the island in Britain and Ireland. Letter codes (lower case: earlier period; capitals: later period). Kruskal stress 9.5%. (From Hardy *et al.*, 2017, courtesy of *Entomologist's Gazette* and Pemberley Books).

losses of those having specialist resources; there are species on the list of potential occupants that show signs of persisting on the British mainland (e.g. continental *Papilio machaon*, *Colias croceus*, *Lampides boeticus*, *Issoria lathonia*), on the Isle of Wight (e.g. *Nymphalis polychloros*) and in Ireland (e.g. *T. sylvestris*) (see Chapter 4). The outcome depends largely on the match of species' resource-based habitat demands with changing environments.

8 British and Irish Islands: an Evolutionary Vantage

An outstanding issue in the early 1970s was: how fast can evolution operate? How long does it take populations of, say, butterflies or moths to become distinct from populations found elsewhere? The answers are: unexpectedly fast and but a short time (Dawkins, 2009; Cook and Saccheri, 2013). The British and Irish islands offer exciting opportunities for the investigation of evolutionary processes in animal populations over recent geological times. This is owing to: (i) the considerable variation in size and isolation of islands and thus of populations on them; (ii) clear starting dates of species arrival or entry on the islands; and (iii) equally well defined dates for species isolation on islands. The task is facilitated by our developing knowledge of butterfly species migration capacity and colonization ability, and importantly the long period of natural history recording and data gathering over the islands. The Devensian Maximum Glaciation was responsible for the effective tabula rasa of the entire fauna over much of Britain by ice, seawater and severe climatic conditions by c. 20 ka BP (see Chapter 2). Relative sea-level rise then ensured isolation of higher ground in the creation of islands. Island size has regulated population persistence since entry: the larger the island, the greater is the likelihood of a species persisting, owing to the presence of larger and multiple local populations (metapopulations) on it; thus, we expect a species to survive longer on the British mainland and Ireland than on other British and Irish islands. On small islands, animals would have more readily become extinct, necessitating re-entry – an easy enough matter perhaps for near offshore islands but more difficult for an island like the Isle of Man over the widening sea barriers.

On an island a population facing a severe downturn in conditions cannot easily escape to a safe haven (**refuge**). For butterflies, the very nature of environmental downturns (cooling of climate, depletion of resources) counters the basic processes for survival elsewhere, viz. the ability to increase in number and to transfer (migrate) to a landmass with better resources (Dennis, 1993). A population persisting in the long term on an island faces **bottlenecks** (population, genetic) in conditions, as the longer a population survives, the greater is the likelihood that it will experience high-magnitude deleterious changes in conditions. Responses to such changes that do not result in extinction on islands almost certainly result in changes to conditions of existence, changes to the complex resource base that makes up butterfly habitats (see **resource-based habitat** in Glossary) and conditions that facilitate exploitation of resources (Dennis, 2010). These changes, in turn, result in population spatial changes (collapse in distributions and ranges), population temporal changes (i.e. narrowing of timing of appearance and activities of crucial stages, such as adult emergences, and asynchrony with the resource base), changes to life history, and evolutionary changes in a population's genotypes (DNA) and phenotypes. The latter can be extremely rapid, as was discovered in the case of the introduced colony of *Plebejus argus* in the Dulas Valley, North Wales, that had become distinct from its founding population on the Great Orme in less than 30 years (Dennis, 1977, p. 173).

It is important to appreciate that not all the changes (differences) we find in butterfly populations on islands occurred on the islands on which they are currently found. Therefore, it is valuable to examine the closely related (sibling) species belonging to *Leptidea*. The pair of species has had a prior history pre-dating the Devensian maximum advance potentially by many hundreds of thousands of years or more. Thus, key issues encapsulate their fate (extinction, refuges) during each glaciation: the location of refuges; the paths taken by species in their return northwards once the climate warmed in each interglacial; the impact of wholesale shifts in ranges on their genetics; and the changes that occurred during each period of glacial refuge. In this regard glaciations are an order of magnitude longer than interglacials.

Pulses of Loss and Gain: Return Periods and Recurrence Intervals

At the outset of Chapter 7, it was mentioned that populations of species are subject to perpetual changes in their environments caused by fluctuations and trends in conditions. In this chapter we consider the impacts of these changes over a longer time scale: such changes in conditions may have beneficial or detrimental impacts. Regarding fluctuations, increasingly higher-magnitude events have a distinctive distribution over time. Assuming stability (**stationarity**) in conditions – that is, the lack of a trend such as gradual climate change – the recurrence interval or return period

(RI) for an event over n years of magnitude m (i.e. dispersal of an individual butterfly over a distance D being equalled or exceeded) is crudely given by:

$$RI = (n+1)/m$$

The probability of the event being equalled or exceeded is then:

$$P[m] = m/(n+1)$$

Although it is important not to interpret the probabilities of **extreme environmental events** (**EEE**s) as strict predictions of their occurrence, as is often done for flood events (Wikipedia, 1 June 2017), these simple algorithms give an appreciation of the increasing scarcity of higher-magnitude events – outside personal experience – the ranking of which approximates a linear distribution in probability graphs devised by Gumbel (1958); naturally, more sophisticated methods are available for determining the probabilities (i.e. based originally on Poisson and binomial distributions; e.g. Chorley and Kennedy, 1971; Sumner, 1978).

The temporal pattern of high-magnitude events indicates that, over long periods of time, there is an increasing chance that the individuals in populations will experience sufficiently high-magnitude events that have an impact on population processes. These events may trigger expansion through population growth and dispersal (i.e. calm, hot sunny weather during the flight season) or cause population collapse by broaching limits of tolerance (i.e. floods, droughts, frost, a succession of cold cloudy days during a flight season) (Long *et al.*, 2017). Small islands and sites are particularly vulnerable to single events that may affect the entire population. It is likely that the storm surge (or tsunami) of 20 January 1607 would have had repercussions for the butterfly populations on the Bristol Channel islands, especially Denny Island (Bryant and Haslett, 2007). The 'flood' from a convectional downpour on the Great Orme's Head on 10 June 1993 (Wood and Campbell, 1995) caused high mortality in populations of *Hipparchia semele* and *P. argus*, both affected in the pupal stage (Dennis and Bardell, 1996). Deleterious events need to be larger and more broadcast to have an impact on populations occupying larger islands. However, such events clearly happen (now most frequently triggered by human activity); otherwise extinctions of species (e.g. *Lycaena dispar*, *Phengaris arion*) would not have been experienced on the British mainland.

Changes in conditions may be part of a trend (i.e. greenhouse climate warming) rather than just a random fluctuation (oscillation) in conditions. Over the 11.5 ka of the Holocene, there have been a number of highly distinctive trends in changing conditions that must have had an immense impact on butterfly populations over the British islands. In origin, these have been both natural (i.e. climate changes in temperature, moisture and cloud cover; forest cover) and human (i.e. deforestation, spread of cultivation, chemical inputs, pollution) (Dennis, 1977, 1992, 1993, 2010). Bearing this in mind, it becomes hazardous to base predictions (probability) of high-magnitude events on previous occurrences, as conditions – even over a few decades – show **non-stationarity** (Vogel *et al.*, 2011). The dangers (or benefits, depending on the direction of changes in relationship to butterfly life history) to butterfly populations are increased when high-magnitude events are imposed on growing trends; the overall effect depends on whether a high-magnitude event is in sympathy or antipathy with the trend. For instance, a wet summer on top of a cooler climate trend will have more serious consequences for butterfly populations than a drought. Thus, when an average screen temperature for July is given as 11.5°C (as for the Isle of Man for the entire Loch Lomond Readvance), cool spells and low temperature extremes (minima) during this > 1200-year period could well have caused total population collapses through aseasonal periods of severe frost when butterflies were mating and egg laying; this period was far more critical and widespread than that which occurred in warmer conditions causing the *coup de grâce* for *P. argus masseyi* in Westmorland (Wright, 1942). Ultimately, it will be a combination of such high-magnitude events on the exponential downtrend (deterioration) of 'habitats' from industrial land use that will cause the extinction of many more British Lepidoptera in the **Anthropocene** (vanderWerf, 1996; Robinson and Sutherland, 2002).

Bearing the above in mind, the likelihood is that island colonization (transfer to and colonization of islands) and extinction processes are particularly dependent on extreme events. Especially – if repeated over successive seasons – warm, calm, sunny seasonal conditions lead to population build-up at sources and increase the probability of dispersal from source populations, transfers to islands and island colonization (**migration capacity** augments **migration ability**) (see Appendix 7). On the other hand, cold, windy, cloudy conditions result in population slumps and a smaller chance of 'rescue' by immigrants (migration ability is unsupported by opportunities in migration capacity). This pattern is borne out by the strong association of (population) mass at mainland sources with measures of both migration capacity and colonization ability (see Fig. 6.6). As regards isolation, to the extent that a measure of migration capacity is dependent more on variables affecting opportunities for migration (i.e. flight period, broods, population size, numbers of colonies) rather than on a butterfly's **bauplan** (i.e. wing expanse, wing loading, wing shape, thorax musculature) which would establish definite limits to flight ability, the implication is that any event that builds population size will have the greatest impact on transfers to islands. From this observation it is reasonable to expect that any coupling of high-magnitude events with a similar trend in conditions will have an unprecedented impact on butterfly populations and migration. Thus, the immigration of butterflies into Britain in the immediate post-glacial is deemed to have been literally an invasion, unprecedented in human history (Dennis, 1977; see below). Similarly, the striking downturn in climatic conditions in the Sub Atlantic episode of the late Holocene (> 3ka BP), supported by evidence of renewed bog peat development, is likely to have caused widespread contraction in the populations (cover) of many species (Dennis, 1977).

At least 12 distinctive episodes are evident during the last 4500 years of heightened precipitation (PPT) excess over evapotranspiration (ET) (PPT » ET) and storminess (Charman, 2010; Orme *et al.*, 2015) – conditions, by the way, that would favour bog-residing *Coenonympha tullia* (Dennis and Eales, 1997).

Outlets of Change: Ecological Release and Marginality

Environmental changes, apart from having demographic consequences in population creation (i.e. colonization) or loss (i.e. extinction) can have much wider effects: population losses (metapopulation collapse) are consequent on increasing **marginality** in habitat conditions, just one of a wide array of its symptoms (viz. demographic, ecological, morphological, physiological, genetic, etc.) (Shreeve *et al.*, 1996). Population gains are associated with **ecological release**, again accompanied by a similar suite of symptoms, though with traits contrasting in polarity from marginality (Kohn, 1972; Cox and Ricklefs, 1977). The crucial trigger of change is an environmental event (or events); this leads to demographic changes via a shift in habitat components (resources and conditions of existence) and in species **niches** in the affected habitat (for wider ramifications see Dennis *et al.*, 2011a). Where environmental changes fail to effect local population extinctions, they lead to an array of adjustments (i.e. adaptations, life history and behavioural changes, variation in traits) in species within local communities. Some changes, within limits, are absorbed (contained) in the 'plastic' capacity of individuals (for example, wing phenotypes, shifts in emergence, diapause, egg-laying preferences, growth rates and development times, and dispersal, are all to a degree plastic) (Fischer *et al.*, 2003; Otaki *et al.*, 2010; Roy *et al.*, 2015; Sgrò *et al.*, 2016); others arise owing to selection for latent, uncommon or novel genotypes. The most frequently quoted example of population genetic changes is the rise and fall of f. *carbonaria* and *insularia* in *Biston betularia* (Cook *et al.*, 1999, 2004, 2011; Cook and Saccheri, 2013; Van't Hof *et al.*, 2013).

The features of marginality are not just exclusively restricted to the margin (fringe) of a species' range, but may also affect so-called core diversity areas subject to wholesale changes, such as those imposed by the onslaught of industrial land use over Britain during the past century (Shreeve *et al.*, 1996). Marginality (as also with ecological release) has temporal and spatial scales that relate to the severity and spatial extent of environmental changes. An interesting feature of increasing marginalization of a set of populations is that although declining connectivity, from reduced dispersal and gene flow, may reduce any coordinated response to change across local and regional populations, their shared gene pools in the face of shared experience may nevertheless permit a parallel response to changing pressures (Dennis, 1977, p. 193); the degree of parallel response may well increase with the magnitude of environmental change

(Shreeve *et al.*, 1996) and as a consequence of identical compromises among traits (e.g. for wing attributes; Dennis and Shreeve, 1989). Changes in the environmental conditions at local sites generate changes in communities and may cause spatial (regional) shifts in species community patterns (Dapporto *et al.*, 2015).

Glacial Refuges and the Post-glacial Invasion and Colonization Process of Britain and Ireland

The highest-magnitude event to affect the British butterfly fauna during the past 50 ka has been the Devensian Maximum Glaciation. During the long period 30–18 ka BP, retrodictions (reconstructions) of species distributions place refuges for the surviving fauna south of the Devensian polar front located at the latitude of mid Portugal (Dennis, 1977, 1992, 1993, p. 164). During each of the many glacial advances (see Chapter 2), butterfly species currently in the British Isles had refuges in the three great land blocks bounding the Mediterranean (i.e. Iberia, Italy, Balkans). Other refuges existed to the south in north Africa and the Near East (Fig. 8.1); an example is the refuges of *Polyommatus icarus* and *P. celina* (Dincă *et al.*, 2011b). More recently, though long suspected, there has been growing evidence of possible refuges for boreal fauna to the north of the Alps in eastern Europe, especially for cold-tolerant species, such as those in *Erebia*, in ice-free areas between montane ice caps and the

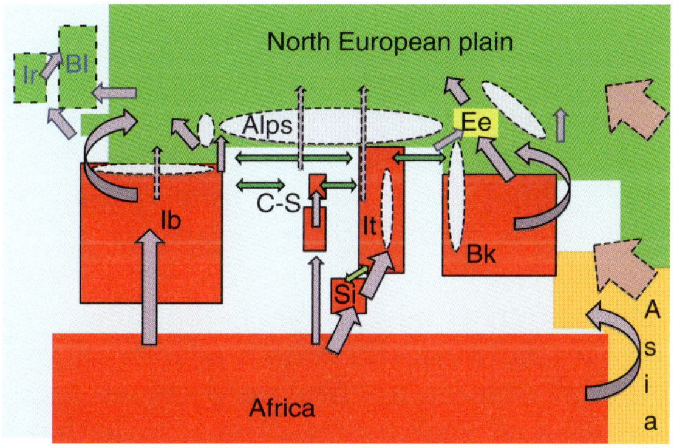

Fig. 8.1. Topological map of European and North African continental refuges for thermophilous British and north European butterfly species and other organisms during glacial stages and their tracks of recolonization immediately with deglaciation. The main refuge centres are in Iberia (Ib), Corsica–Sardinia (C-S), Italy–Sicily (It, Si) and the Balkans (Bk) but with other centres in North Africa and the Near and Middle East. For more cold-tolerant faunas there were dispersed, limited refuges north of southern limit of the alpine chain in Eastern Europe (Ee). BI, British Isles; Ir, Ireland. Purple tracks, latitudinal movements; green tracks, east–west movements. Blue bands are glaciated major mountain barriers to dispersal; orange, Asian land mass holding other refuges, especially for wide-ranging species.

Scandinavian ice sheet (e.g. *Erebia tyndarus* group) (Albre *et al.*, 2008; Schmitt and Varga, 2012). Importantly, molecular genetics research has discovered genetic markers of the repeated Pleistocene colonizations and extinctions in northern Europe in the declining intra-specific genetic diversity with latitude (Schmitt *et al.*, 2002; Hallatschek and Nelson, 2008; Habel *et al.*, 2013) and in the positive correlation between latitude and substitution rate of mitochondrial DNA sequences in five lycaenid species (Schär *et al.*, 2017).

During the start of each of the many warm periods (interglacials) several dramatic events occurred: (i) a population explosion of butterfly species; (ii) a rapid invasion of northern Europe by species from southern refuges; (iii) the shuffling of species belonging to distinct faunal elements within the refuge blocks; and (iv) the establishment of boundary (contact) zones between closely related (sibling) species that had evolved distinctions when isolated in refuges (e.g. *Maniola jurtina*: Dapporto *et al.* 2009, 2011; *Aricia* species: Mallet *et al.*, 2010) (see Fig. 8.5) during previous reshuffles generated by climate changes (Dennis, 1993, p. 179; Dennis *et al.*, 1991, 1995b, 1998c; Dennis and Schmitt, 2009; Descimon and Mallet, 2009). From the vantage of the British Isles all current species colonized the islands after c. 15 ka BP at most; owing to population collapses during the Younger Dryas, these initial colonizations were rapidly succeeded and many overtaken by subsequent reinvasions in the early Holocene (76% of species; Dennis, 1993).

The key feature of post-glacial returns of species to the British Isles is that it would have been an astonishingly dramatic process for most species emanating from all refuges. Arrival operated on a broad dispersion front from Brittany to the now drowned area of Doggerland, the latter developing marshland instrumental in the formation of the specialized biotope of extinct British *L. dispar* (Balfour-Browne, 1958; Dennis, 1977). Genetic associations provide the evidence whence British populations derived (e.g. *Euphydryas aurinia* from Iberia; Joyce and Pullin, 2001). Speed of arrival is clear for several reasons:

1. The key resources for species, herbaceous larval host plants, already had strong footholds in the British islands in periglacial ice-free tundra areas and thereafter rapidly in areas previously ice covered (a characteristic feature of most British butterflies now is that their larval host plants have larger geographical ranges and wider distributions than the species feeding on them) (Dennis and Shreeve, 1991).
2. The sharp rise in temperatures suddenly made large regions north of the alpine chain climatically suitable for temperate butterflies.
3. Selection pressures for unidirectional movements beyond refuges (migration) were maximal and unprecedented in human experience. According to Dennis (1993, p. 158):

> During the late Devensian and early Holocene return of temperate fauna from Mediterranean latitudes, the cost of leaving a current habitat would

have been much less than today (viz., shorter migration distances to suitable habitats; abundant nectar sources en route) and the benefits of moving greater (viz., hostplants occurred in Late Glacial Britain for > 70% of the present fauna; absence of parasites), whereas the cost of staying in a recently colonised site would have steadily increased (viz., resources increasingly utilised by rapidly expanding populations of fauna in the ameliorating climate; resource decline with vegetation succession; growing predator and parasite population).

These conditions – facilitating longer flight periods, longer life spans and greater ex-habitat dispersal (Bubová *et al.*, 2016) – would have led to powerful selection for increased rates of migration, the process of return northwards akin to an invasion, events occurring at a much faster rate than they can currently be reconstructed. In fact there would have been two invasions: (i) in the Late Glacial > 15 ka BP; and (ii) in the early Holocene. The arrival sequence constructed thus far has a resolution of only 500 year units (reproduced here as Fig. 8.2) (Dennis, 1993: table 4.3); however, modern evidence of butterfly migration (Williams, 1965; Baker, 1969, 1978, 1984) indicates that relatively small as well as large species can cover as much as 500 km in a single season (e.g. *Pieris napi*) (Coulson, 2015). Glacial–interglacial cycles are also considered to have enhanced, perhaps even generated, the patterns of long-distance reversed migrations in species such as *Vanessa cardui*, *V. atalanta*, *Colias croceus* and *Pieris* species (Dennis, 1993, p. 160).

The only species whose colonization northwards from refuges was retarded in the Holocene are those dependent on slower-spreading shrubs and trees (e.g. *Favonius quercus* dependent on mature oaks; *Satyrium w-album* dependent on *Ulmus* spp.) (Becker and Kramer, 1991). Although few species tend to be observed dispersing beyond their larval host plant range in Britain (an obvious exception being *Gonepteryx rhamni*), the situation would have been very different in Late Glacial and Holocene Britain. The dense populations building at the outset of the Holocene would have almost certainly had two consequences for British island biogeography:

1. A wave of colonists would have overwhelmed (genetically) any small Younger Dryas refuge populations in southern Britain.
2. The expanding English Channel – with the drowning of the Fleuve Manche that would have provided an expanse of stepping stones in the Late Glacial – is unlikely to have prevented successful transfer of all current British butterflies. Even apparently sedentary species are known to undergo mass movements when the conditions are right for dispersal (e.g. *F. quercus*) (Holloway, 1980).

Nevertheless, some species may have been latecomers; for example, there is still a question of whether *Thymelicus acteon* was accidently introduced to Dorset in recent history, much as *Heteropterus morpheus* on Jersey in the Second World War (Long, 1970, 2015).

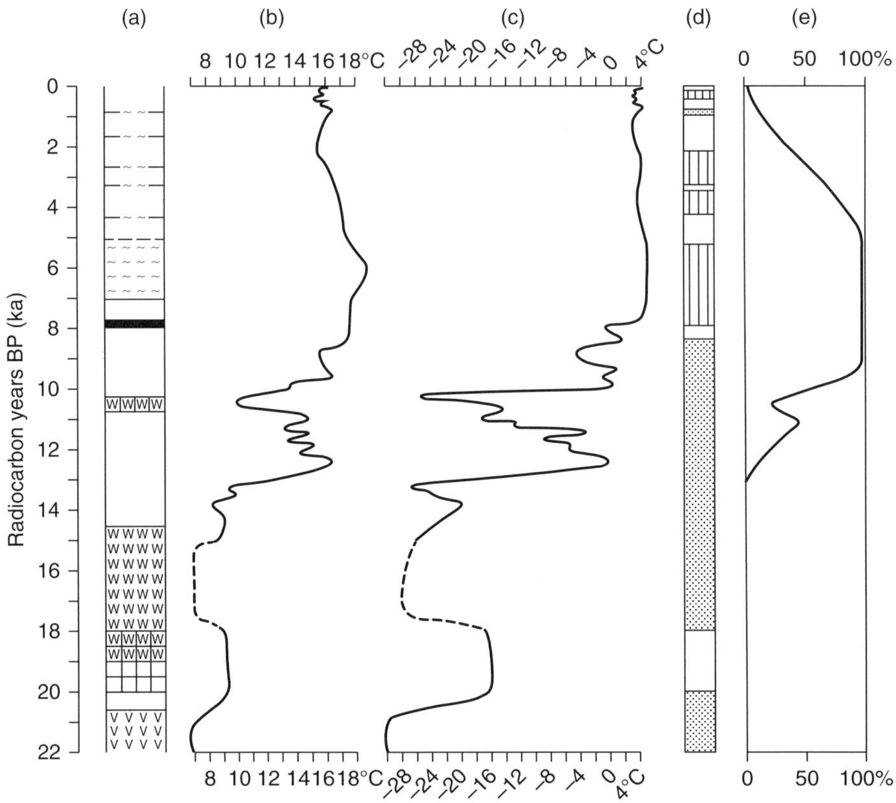

Fig. 8.2. The colonization, extinction and evolution of butterflies in Britain associated with habitat changes over the past 22,000 years (a) Geomorphological events: small squares, glacial advances in Britain; V, large ice wedges; W, small ice wedges, involutions and pingos; thick bar, severance of land connection with continent; ~~~, development and expansion of ombrogenous peat and recurrence surfaces (peat regrowth) ideal for the butterfly *Coenonympha tullia*. (b) Mean temperature of the warmest month for central England. (c) Mean temperature of the coldest month for central England. (d) Annual precipitation: grey shading, drier than period 1940–1970; vertical shading, wetter than period 1940–1970. (e) Forest cover. (f) Tolerance curves for known resident British butterflies of mean screen temperatures for the warmest month in central England: lower (thin) curve, for lowest temperature per unit time; higher (thick) curve, for the highest temperature per unit time (500-year units based on 59 species); arrows indicate direction of changes in increases or decreases in butterfly species. (g) Thermal limits for butterflies; vertical shading, tolerance of conditions in central England based on the mutual climatic range of species currently resident in lowland Britain (based on 55 species); grey shading, mean conditions for the coolest month probably critical for > 50% of the current British butterfly fauna; black, mean conditions for coolest month critical for > 90% of the current British butterfly fauna. (h) Host plant and biotope limits for the current resident butterflies within Britain: left (thin) curve, presence of habitat structural components (not symbionts), as known, for resident British butterfly species in central England; right (thick) curve, presence of host plants for resident British butterfly species within Britain. Host plants are assumed to persist for the entire pollen zone (see j) in which they are found. The bar chart shows the species vulnerable to forest climax development: stippled, butterfly species with host plants not a typical part of the field layer

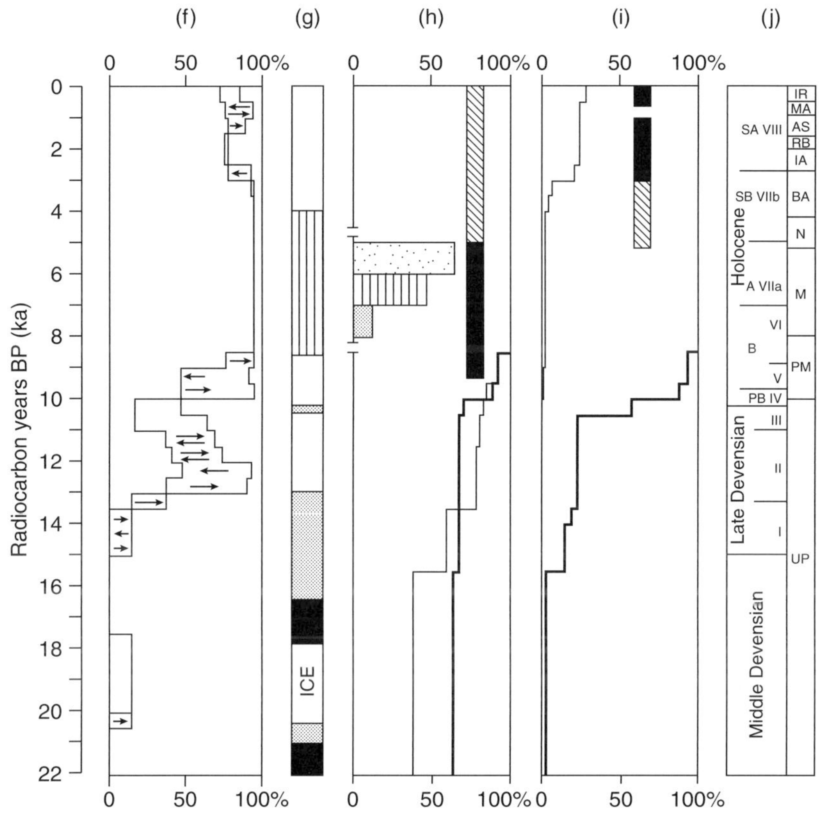

Fig. 8.2. Continued.

or high strata within forest and requiring open biotopes; vertical shading, species whose host plants attain a suitable quality in the regeneration cycle of forests but require the continuous production of open spaces; grey shading, species requiring both tree or liana host plants and biotopes. The divided bar shows the constraints for specific groups; black, period of isolation, range contraction and microevolution for species dependent entirely on open conditions; diagonal shading, population extinction and range contraction for species (n = 29 species) dependent on forest biotopes. (i) Arrival sequence and climatic constraints: thick curve, proposed cumulative earliest date for the continued presence of British butterfly fauna north of the southern English coast based on climate, host plants and biotopes (500-year periods); thin curve, cumulative curve for species experiencing upper and lower critical thresholds for mean screen temperatures of the warmest month in central England during the late Holocene (n = 59 species). The black bar shows the period of microevolutionary and ecological changes relating to the lower critical threshold summer thermal limits. This period may have started earlier (diagonal shaded zone) depending on the extent to which local and regional climates were influenced by human agents (i.e. deforestation, grazing, loss of cover, shelter, windchill). (j) Chronology and human cultures: left part, Blytt and Sernander Post Glacial periods and Godwin pollen zones (PB, Pre Boreal; B, Boreal; A, Atlantic; SB, Sub Boreal; SA, Sub Atlantic); right part, cultures (UP, Upper Palaeolithic; PM, Proto Maglemosian; M, Mesolithic; N, Neolithic; BA, Bronze Age; IA, Iron Age; RB, Romano-British; AS, Anglo-Saxon; MA, Medieval Age; IR, Industrial Revolution). (From Dennis, 1993, courtesy of Manchester University Press).

Butterfly Space–Time Responses to Changing Conditions in Post-Devensian Britain and Ireland

It would be entirely wrong to conceive of species dynamics (colonization, extinction) on British and Irish islands as conforming to a uniform stochastic model (MacArthur and Wilson, 1963) during the entire Holocene; huge changes would have occurred in distributions and no doubt many early arctic and boreal species that entered first were lost with environmental changes over the past 10 ka BP (> 11.7 ka cal BP) (Carlson, 2013; Brooks *et al.*, 2012). Although the Holocene climate in Britain was relatively stable compared with the massive climate oscillations during the Late Quaternary, wholesale changes in Holocene climates occurred that would have impinged on butterfly biogeography and island occupancy (Davis *et al.*, 2003; Liu *et al.* 2014; Rohling and Pälike, 2005): (i) range expansions and island occupation; (ii) range shrinkages and island losses; (iii) changes in overall range dimensions; (iv) fragmentation of distributions; and (v) contraction to areas with singular biotopes (Dennis, 1977, 1993). The main factors in changes were: (i) climate changes in heat and moisture (especially warm/dry versus cool/wet periods); (ii) vegetation, especially forest cover (open to closed forest conditions, then open again especially after the Norman Conquest); but also (iii) losses and development of distinctive biotopes (e.g. fen, heath, raised bog) and (iv) island formation and loss (i.e. shrinkage, archipelago formation, submergence, tombolo links).

The responses of species to changes have been assessed by applying retrodictive methods to current species tolerances of climate, host plants and habitat (biotope) categories for southern and central England (Fig. 8.2) (Dennis, 1993, table 4.1). From this it is evident that the ranges and distributions of thermophilous species would have expanded in warm dry periods and contracted in cool wet periods (vice versa for cold tolerant species). Species of high forest and of forest clearings would have expanded to the peak of forest cover after 9 ka BP, only to start losing ground after 4 ka BP, finally to be limited to woodland refuges now; conversely, those species associated with open biotopes would have spread to the far north in the early Holocene, their populations subsequently fragmented by forest cover, to expand once more as Medieval cultivation and pastures opened up the forest. Some marked changes will have occurred following the Norman Conquest as the climate shifted from warm dry conditions to the colder Little Ice Age, but the greatest changes are with us now as the land (soils and soil communities) is rapidly being poisoned by chemical inputs (Yang *et al.*, 2000; de Jong *et al.*, 2008; Firbank *et al.*, 2008; Geiger *et al.*, 2010; Hodgson *et al.*, 2014). One of the most distinctive features in butterfly species geography has been the substantial contraction of a number of species to biotope refuges, which has led to associated adaptations (see below). On offshore islands, species richness would have steadily declined as islands contracted and fragmented, and source populations declined, the exceptions being those linked up by tombolos in historical times (e.g. fossil islands: Portland, Great Orme's Head) and where rising sea levels from isostatic land recovery narrowed straits in northern Britain.

Shrinking Habitats and Ecological Islands in Holocene Britain

A contracting (refuge) phase in butterfly geography has altogether different implications from an expanding (invasive) phase; together they describe the cycles of generalism and specialism inherent in speciation (Dennis *et al.*, 2011a). Butterflies, as ectotherms, benefit enormously from warm dry summer conditions and swathes of previously unoccupied 'habitat'; these conditions lead to population explosions, migration and invasion of the new spaces. This process also 'releases' a vast amount of genetic variation within populations, as Professor E.B. Ford and his father discovered in Cumbrian *E. aurinia* populations (Ford and Ford, 1930), opening up access to wider resource use (i.e. larval host plants) and varied biotope occupancy. Cool wet summer conditions, and habitats subsequently saturated by communities of competitors and enemies, do the reverse, causing population collapses, distribution fragmentation and loss of within-population genetic variability, leading ultimately to a narrower resource base and limited biotope occupancy. The expanding phase describes a situation where the realized niche most closely approximates the potential niche of a species. A subsequent warm dry phase, or expansion in a suitable biotope, may again trigger expansion, but that depends on how long the ecological bottleneck has been in place, the extent to which a species has become adapted to (specialized on) restricted conditions, and exactly what resources become available.

These complex events characterize Holocene butterflies. Narrowing conditions (marginalization) were triggered by two processes: (i) climate stress, as the Holocene climate became cooler and wetter, typical of the present day; and (ii) shrinkage and fragmentation of key biotopes to ecological islands. The first pressure had an impact on larval growth rates, development times, survival, mating success and fecundity, therefore on the intrinsic rate of growth of butterfly populations; the second restricted the availability of resource outlets. The outcome is the increased specialization on more limited sets of larval host plants, and consequently on limited sets of other resources associated with host plants (Dennis, 2010). Two key features of British butterflies are: (i) the use of more limited sets of host plants compared with those used on the continent (Dennis, 1977) (e.g. *Polygonia c-album*: Nylin, 1988); and (ii) the restriction of a number of species to ecological islands and specialization on them (Dennis, 1977, 1992, 1993). A clue to understanding the chain of events is the occurrence of increased isolation and specialization of populations towards the range edge of species; for example, the closed nature of populations of *Pieris napi* at its northern margin compared with those in the range core (Dennis, 1993) and the increased specialization of *Erebia aethiops* on *Sesleria caerulea* at Arnside at its southern range margin (Dennis, 1982c; Kirkland, 2012).

Classic ecological refuges among species in Britain and Ireland are: (i) the *Nardus stricta* biotopes on mountains (e.g. for *Erebia epiphron*); (ii) raised bogs in lowland Britain as at Whixall and Witherslack mosses (e.g. *Sphagnum–Erica tetralix–Eriophorum* bogs for *C. tullia*); (iii) the *Phragmites communis* fenland with *Peucedanum palustre* and *Rumex hydrolapathum* in the Norfolk Broads and Cambridge fens (i.e. the remnants of the Doggerland island for *Papilio machaon* and *L. dispar*); and (iv) south-facing unstable chalk cliffs on the Isle of Wight (i.e. *Plantago lanceolata* for *Melitaea cinxia*) (Plate 1.16) (Dennis, 1977; Curtis *et al.*, 2015a). Such ecological refuges are in effect biotope islands; examples are the fens created by rising Holocene sea levels, bogs created by increasingly high water tables (high precipitation compared with evapotranspiration rates) and cliffs formed by high sea levels, marine erosion and slope mass wasting. A number have highly favourable local climates (e.g. south-facing sheltered slopes for many species), especially on permeable limestones; for example, the chalk downs are a refuge for *Polyommatus bellargus* and *Polyommatus coridon*, *Hesperia comma* and *Melanargia galathea* (Dennis, 1977; Thomas, 1993). A number of different refuges occur where northern (e.g. *C. tullia*, *E. aethiops*, *Aricia artaxerxes*) and southern (e.g. *Argynnis adippe*, *Argynnis paphia*, *Hamearis lucina*) species meet in the Morecambe Bay area (Dennis, 1992). Within Ireland, the same process is illustrated by the restriction of *Leptidea sinapis* and *Boloria euphrosyne* to the exposed Burren limestone pavements.

Ecological refuge formation is not exclusively a natural process. It is ironic that humans, from Mesolithic times, may have been responsible for creating and retaining open biotopes for butterflies dependent on them as the forest spread through the islands (Dennis, 1993); a classic example is the development of lowland dry heath so important for *P. argus* (Ravenscroft, 1990). From the early 20th century, key refuges (strings of 'nature reserves') are entirely dependent on human management for maintaining suitable forest conditions (with the demise of coppice cycles) (Rackham, 1986) and open pastures (with the loss of nearly all but a handful of Britain's meadows) (Fuller, 1987; Jefferson *et al.*, 2014). The trends experienced on the mainlands of Britain and Ireland were, of course, experienced over the millennia on every inhabited island, however small. As such, it would be a mistake to assume that biogeographical processes for butterflies and other organisms on British offshore islands have been simply little more than a monotonous process of random events, affecting stable biotopes, maintaining an equilibrium number of species throughout the Holocene. Professor E.B. Ford's (1964) work on the Isles of Scilly, disclosing the shifts and losses in populations of *M. jurtina* and *P. icarus*, and their genetic make-up, demonstrated that major changes in the status of species can occur in a few years.

Variation: Genes, Phenes and Plastic Traits

A distinctive feature of the British and Irish islands is the existence of several interwoven strands of geographical variation, not just in the diversity of adult phenotypes, which attracts most attention, but in many other traits (physiology, behaviour, phenology, etc.) (Dennis, 1993). These include: (i) distinctions among closely related species, the most intriguing being those among sibling **cryptic species** pairs (e.g. *Aricia* and *Leptidea* species) (Mallet *et al.*, 2010; Dincă *et al.*, 2011a); (ii) **subspecies**, of which > 40 are listed for butterflies on the British and Irish mainland regions and islands (e.g. *P. icarus* subspecies) (Dennis, 1977); (iii) a few island endemic 'races' (e.g. Isles of Scilly *M. jurtina cassiteridum*) (see subspecies in Glossary); (iv) gradients (**clines**) in multiple wing patterns (e.g. *C. tullia*) (Dennis *et al.*, 1984, 1986; Joyce *et al.*, 2009), as well as in many other attributes (e.g. adult emergence time, numbers of adult broods, geographical switches in resource use) (Dennis, 1992; Asher *et al.*, 2001), involving a mix of polygenic traits (e.g. *M. jurtina* wing ocellation: Ford, 1964; *Pararge aegeria*: Packer, 1984), **single nucleotide polymorphisms (SNPs)** (e.g. *A. paphia* f. *valezina*) (Brakefield and Shreeve, 1992) and plastic responses to environmental cues (Dennis, 1993; Merckx and Van Dyck, 2006; Otaki *et al.*, 2010; Kertész *et al.*, 2017); and (v) distinctive traits on ecological (biotope) islands (e.g. Fenland *L. dispar dispar* and *P. machaon britannicus*) (Duffey, 1968, 1977; Dempster *et al.*, 1976). The details for all species can be found in Dennis (1977, 1992, 1993), Dennis and Shreeve (1989, 1996), Emmet and Heath (1990) and, for *P. argus*, Thomas *et al.* (1999). All this documented variation has great complexity inasmuch as for the greater part, owing to poor sampling strategies by those who described it, we remain uncertain about: (i) its description (quantification); (ii) its space–time bounds and consistency; and, most importantly of all, (iii) the contribution of genes and degree of plasticity of the traits (Dennis and Shreeve, 1996, pp. 63–64).

It is the physical variation in wing pattern and colour that has held the greatest fascination for butterfly biologists and amateur enthusiasts, especially as this was regarded, *ipso facto*, as evidence of *in situ* formation (evolution towards new species) in British glacial refuges (Ford, 1945, 1957; Beirne, 1947; Heslop Harrison, 1950c). Distinctions on the basis of so-called subspecies have the virtual (apparent) effect of influencing the links (strengthening some, attenuating others) between islands, isolating island groups, compared with associations based simply on the presence and absence of species (Fig. 8.3) (Dennis 1977, p. 196). It should be borne in mind that subspecies associations can give the impression of fallacious ancient origins (Dennis, 1977, p. 193); the true affinities among island and regional faunas will only become clear with extensive nuclear DNA work. The antiquity of subspecies in British butterflies was first contested in 1977, following an initial assessment of factors underlying the evolution of the Great Orme's Head races (Dennis, 1972a, b, 2015b; see last section); the variation was shown to increase significantly to the north and west of

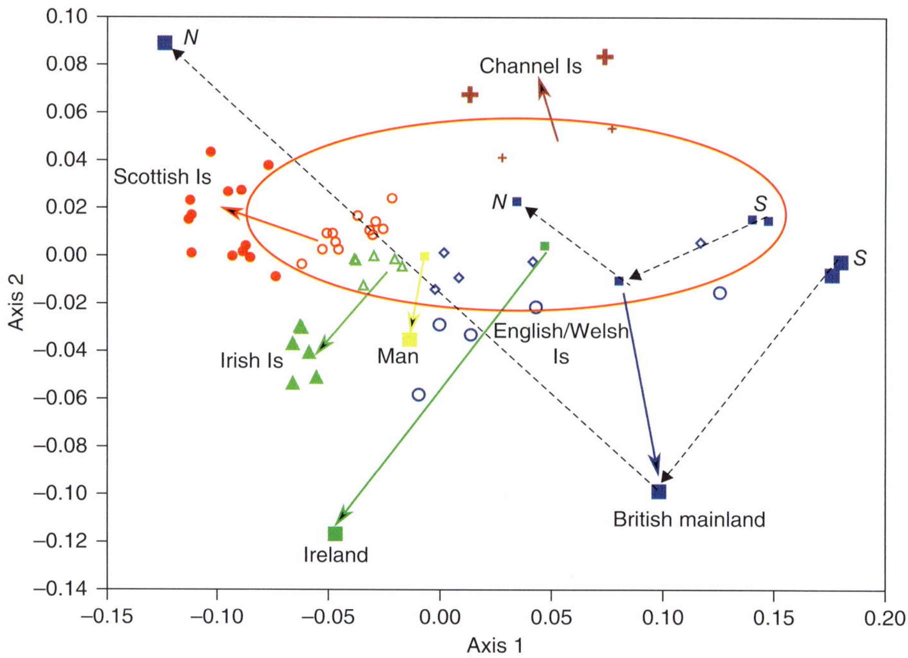

Fig. 8.3. Ordination plot (NMMS) of relationships (Jaccard coefficients) among British regions and larger islands based on species incidence and infra-species variation (subspecies; races). British regions (blue squares), English and Welsh islands (blue circles), Hebridean and Northern Scottish islands (red circles), Isle of Man (yellow square), Ireland (green square), Irish islands (green triangles) and Channel islands (crosses). British mainland (zones: < 52°, 52°–54°, 54°–56°, > 56° N) linked by pecked lines from south to north. Arrows show changes from species relationships to infra-species relationships. The red ellipse indicates the compact zone describing relationships among regions and islands for associations based only on species.

the British Isles and to correlate closely to contemporary climate, particularly to gradients in summer sunshine, temperatures, rainfall and wind speed (Fig. 8.4) (Dennis, 1977, 1993). Intriguingly, some of this phenotypic variation is as substantial as that displayed by differences between *Aricia agestis* and *A. artaxerxes*, both species occurring on the continent and known therefore to pre-date the Devensian glaciation (Mallet *et al.*, 2010), and greater than the morphological variation separating *L. sinapis* and *Leptidea juvernica*, again both species occurring on the continent and having origins dating to at least 250 ka BP on the basis of a molecular clock assessment (Dincă *et al.*, 2011a). Subsequently, it has become evident that different traits in species evolve at markedly different rates within, as well as between, species (Hennig, 1966; Cesaroni *et al.*, 1994; Otaki and Yamamoto, 2004) and that species can also undergo wing phenotypic transformations during range margin expansion (Otaki *et al.*, 2010).

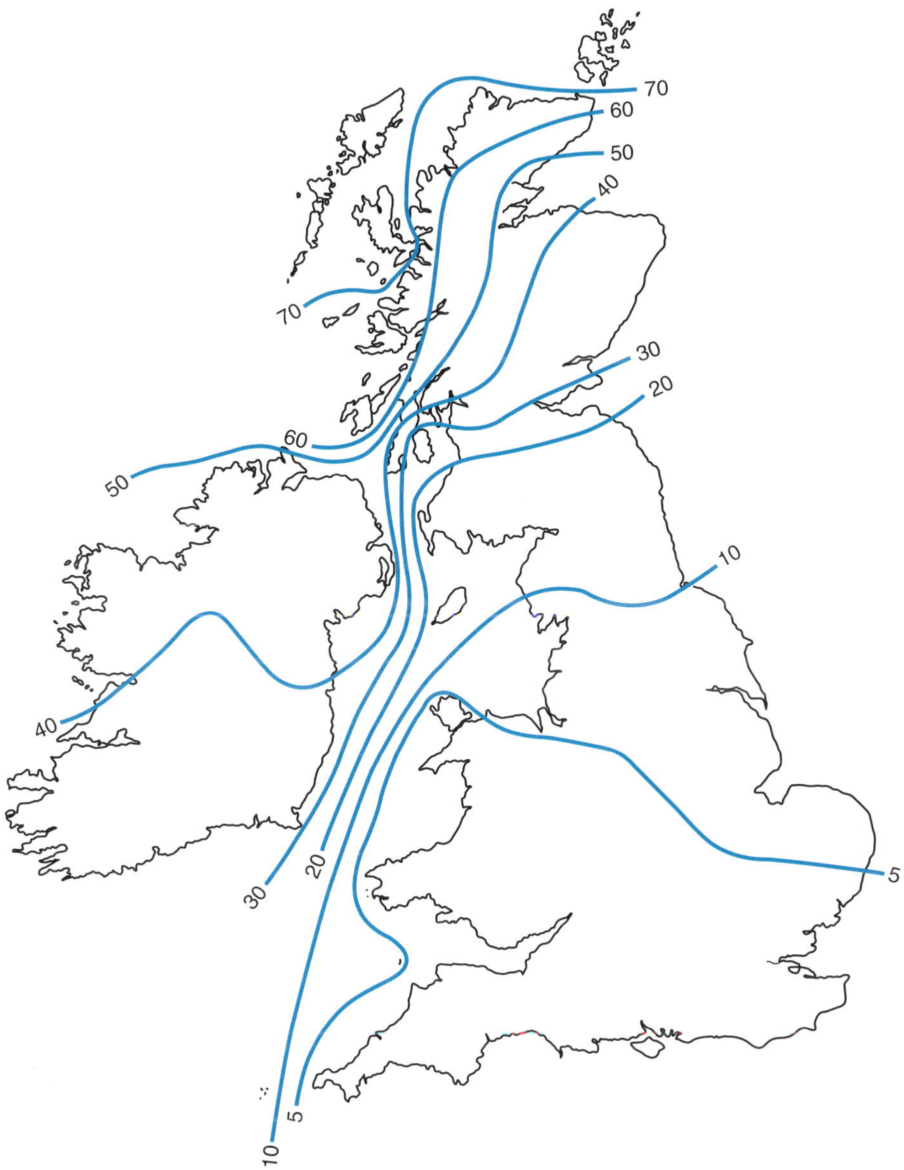

Fig. 8.4. Variation in butterfly wing patterns throughout the British islands. The lines are plotted as percentages of the total fauna for 100 km grid square intersections and island centroids, from an arbitrary base of zero in south-east England. (From Dennis, 1993, courtesy of Manchester University Press.)

Thus, for much of this variation some fundamental questions remain: (i) the contribution of genes versus plastic responses to traits; (ii) the function of genes showing differences; (iii) the time such variation takes to 'evolve'; and (iv) the external (environmental) and internal factors producing changes. Currently, despite the extraordinary advances in molecular genetics during the past two decades, only the reconstruction of an organism's entire genomics, together with an understanding of function, facilitates a *complete* understanding of the distribution of gene differences. Genetic differences vary across chromosomes, as evident from the degree of **introgression hybridization** among closely related humans and Neanderthals: three billion human nucleotide bases, of which as many as 81 million may come from Neanderthals (Green *et al.*, 2010; Prüfer *et al.*, 2014; Vernot *et al.*, 2016) are spread across the human **genome**. Similar work, we hope, is now bound to emerge for butterflies, but as there are so many species, it must await advances that greatly reduce the cost and raise the speed of such undertakings. It may well happen in the authors' lifetimes.

Models of Differentiation and Adaptation

Complicated though this variation is among butterflies on British islands, some basic elements of it can be linked to two periods and sources: (i) Devensian (or earlier Quaternary) continental refuge sources, the species colonizing the islands following deglaciation (20 ka BP); and (ii) since Late Glacial and Holocene establishment within Britain (past 15 ka). Features of the variation allow us to interpret several things: (i) the conditions in refuges (via **voltinism**, morphology, etc.) (Dennis and Shreeve, 1989); (ii) the order (and dating) of subsequent events; (iii) interactions between closely related taxa; and (iv) pressures on populations since arrival in Britain. Much of this remains surmise, but as molecular data accumulate for species the picture over the islands will become clearer. As mentioned above, one aspect of variation is evident already: different traits are known to have different evolutionary trajectories; in particular the evolutionary rates of wing pattern features have been found to differ from those of genital morphology and **allozymes** (e.g. in *Hipparchia*) (Cesaroni *et al.*, 1994). As space is limited, two cases are used here to illustrate associations, order of entry, pressures on **clades** and missing populations.

The first is the case of closely related species, classically demonstrated in *Aricia*, but with further indications from *Leptidea* and *Thymelicus*, these reflecting shifts in ranges including extinctions, and genetic changes on contact. Until recently, no evidence existed of hybridization between the units of southern *A. agestis* and northern *A. artaxerxes*. In 2010, using two gene markers (nuclear *Triose phosphate isomerise*; mitochondrial *cytB*), Mallet and colleagues (Mallet *et al.*, 2010) identified a clear hybrid zone of 150–200 km across north Wales and northern England, marked by a southern boundary for voltinism (single versus double broods) and a northern one for mtDNA haplotypes (Fig. 8.5). Although they failed to find bimodality in any traits at the boundary to support distinctions at species level, the entire genome has yet to be explored. The interesting feature is that hybridization produces new genotypes in individuals. These are subject to selection; climate change (warming) has the potential for shifting selection regimes on the units through the key attribute

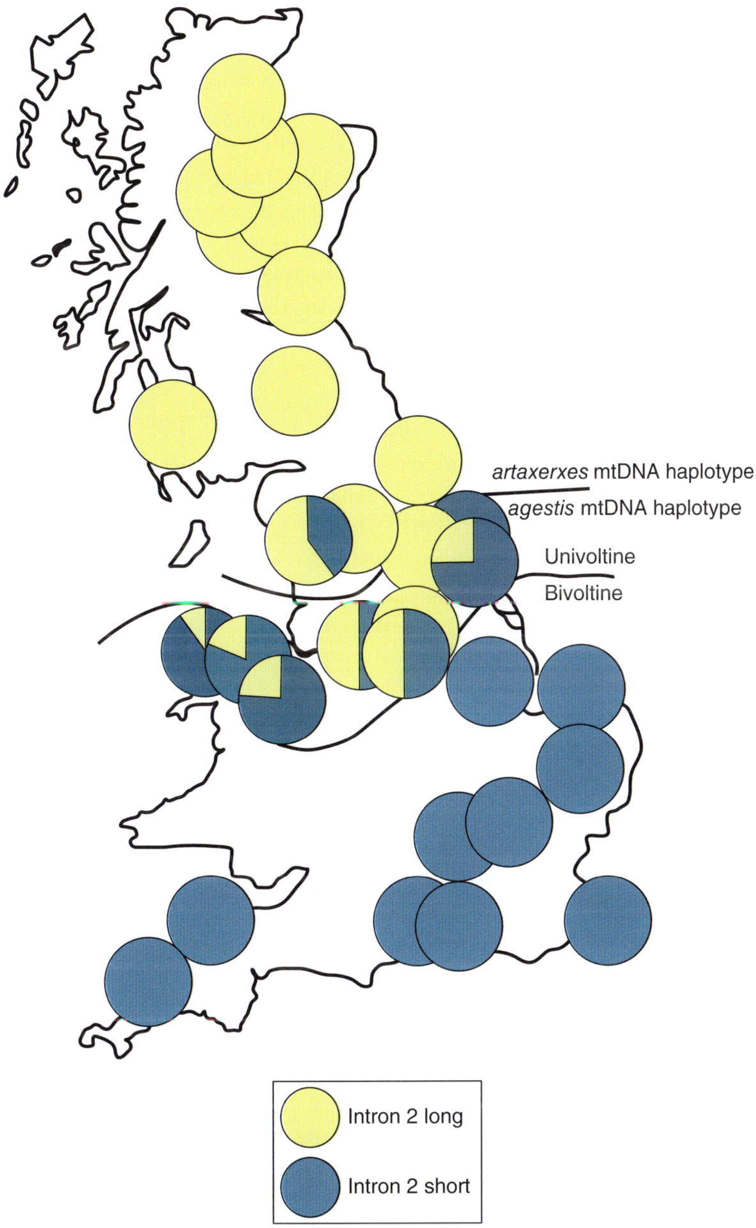

Fig. 8.5. DNA variation distinguishing two species of *Aricia* butterflies on the British mainland. Distribution of *Tpi* length variants for British populations. The pie charts show the proportions of individuals with long (*artaxerxes*-type; yellow) and short (*agestis*-type; blue) intron 2 lengths. The northern line shows the transition from *artaxerxes*-type mtDNA in the north to *agestis*-type mtDNA to the south. The southern line shows the transition in generation number, univoltine populations to the north and bivoltine populations to the south. (Courtesy of Professors James Mallet and Chris D. Thomas and John Wiley & Sons, Inc.)

voltinism (intermediate phenotypes for number of broods are likely to be strongly selected against), forcing the boundary northwards. It is likely that the two 'species' came into contact in the early Holocene and stabilized with forest development. Subsequently, withdrawal of *A. agestis* would have occurred during the cooler wetter Sub Atlantic, but not necessarily completely, as the different units could have occupied opposing east/north-facing (*A. artaxerxes*) and west/south-facing (*A. agestis*) slopes in the Peak District continuously from this period. Later, further expansions of *A. agestis* (and wider contact) would have been feasible with the opening up of the landscape during the advance of the Anglo-Saxon open field system and its expansion in Medieval times during a warming climate. The two units give some indication of conditions in their glacial refuges, bivoltinism indicative of warmer and longer seasons than found with univoltinism.

However, some features have been missed in the assessment. The autosomal recessive gene (white dorsal spot) saturates and distinguishes northern *A. artaxerxes* but not the southern part; there is a 'stress zone' (**balanced polymorphism**) along the Durham coast where this gene exists at low frequency (homozygotes 5%; Dennis, 1977, p. 175) and south of which it does not, for certain, occur. This may also represent a boundary feature suggested to distinguish a northern population isolated in Britain during the Loch Lomond Readvance and a southern one entering in the Holocene (Dennis, 1977). The significance of the 'white dorsal spot' (note: the white haloed black spot in f. *albiannulata* Harrison [1906] is not a **heterozygote**) has yet to be determined.

These sibling pairs raise another point. The coexistence of closely related species is not expected from **Gausse's exclusion principle**, but it is clear that closely related species (at least those in the same genus) can coexist (e.g. *Thymelicus sylvestris*, *Thymelicus lineola* and *T. acteon* in the same sites) (P.B. Hardy, personal observation) though their resource-based habitats (resources and tolerance of conditions) are likely to differ (Louy *et al.*, 2007; Engler *et al.*, 2014). The degree of competition and invasion introgression will depend on their genetic proximity (relatedness). *Leptidea sinapis* is more closely related to *Leptidea reali* than to *L. juvernica* (Dincă *et al.*, 2011a) and clearly on the basis of distribution in Ireland alone (as *L. sinapis* is restricted to the Burren limestone), they prefer different biotopes. The important question is: why is *L. juvernica* not in England or Wales when its post-glacial route towards Ireland is most likely to have been from the east where the bulk of this species is considered to have originated and to be located? As previously queried: is this an indication that *L. juvernica* has become extinct on the British mainland? If so, then why? Is there any evidence of Holocene introgression among English *L. sinapis*?

The second case extends the view to changes across the entire British fauna to broach the pattern of increasing wing morphology changes north and westwards illustrated earlier (see Fig. 8.4). When the pattern was first disclosed, though a clear relationship was found with climate (Dennis 1977), it was not clear how this could come about and it is entirely understandable (in hindsight) that previous researchers would think that Devensian isolation on refuges amid ice sheets covering much of the British Isles was responsible for producing such a powerful regionalization in wing phenotypes. The puzzle was that variation in colour and pattern seemed to conflict across (lacked commonality among) species. Then, with a greater understanding of the function of wing patterns things became clearer (Dennis and Shreeve, 1989). The first point to be made is that wing phenotypes of butterflies are under selection by environmental agents, including species with wide dispersal capabilities (e.g. *Pieris rapae*) (Bai *et al.*, 2015). The first author realized that patterns and colours on wings are a compromise between at least three distinct pressures: thermoregulation in these ectothermic animals, mate location (attraction) and predator escape (Fig. 8.6) (for the full explanation see Dennis and Shreeve, 1989, or Dennis, 1993). The key to changes in the wing patterns of species to the north and west is argued to be the gradual assumption of climatic gradients over Britain during the Holocene, especially during long periods when the climate has become cooler and damper (> 3 ka BP) (see Fig. 8.2). In short, conditions to the north and west of the British islands have marginalized habitats for species. No doubt a number have simply become extinct, and others have found refuges in drier, warmer biotopes (for example, butterflies on the chalk downs and limestone scars), but in a significant number of species the changes in wing patterns are one indication of the shifting balance in adaptations for thermoregulation, mate acquisition and predator escape in cooler climates.

As a cooler, wetter climate would reduce populations and movements, it is most intriguing that the patterns seem to cross islands and mainland areas. Part of this is explained by dispersal between locations (mainland and islands) during sunnier, warmer, drier periods when populations grew and more dispersal occurred. However, part is also no doubt explained by a parallel response to similar (though not identical) conditions. Currently, this is all just an intriguing theory, but it appears to have some basis in fact. It has already been tested for *P. icarus* in the Outer Hebrides (Howe *et al.*, 2007); selection pressures on wing morphology through thermoregulation in this small butterfly were found to be insignificant compared with those for mating and predator escape (De Keyser *et al.*, 2015).

A key issue is that species on islands often display a wide range of genetic differentiation. The presence of unique and variable wing patterns together in species on islands indicates an admixture of longer- and shorter-term factors and external pressures at work. Complex variation, responding to different factors and pressures, can also be found within species on islands (e.g. *M. jurtina cassiteridum* in the Isles of Scilly) (Graves, 1930; Ford, 1964; G. Thomson in Dennis, 1977; Brakefield and Shreeve, 1992, p. 199). Where differences persist in a species on an island, this suggests that dispersal is unable to counter distinctions owing either to powerful selection regimes on the island or to incompatible genetic complexes (Kisel and Barraclough, 2010). Such complex variation is more a feature of Mediterranean islands (Dennis *et al.*, 2000b; Dapporto *et al.*, 2017).

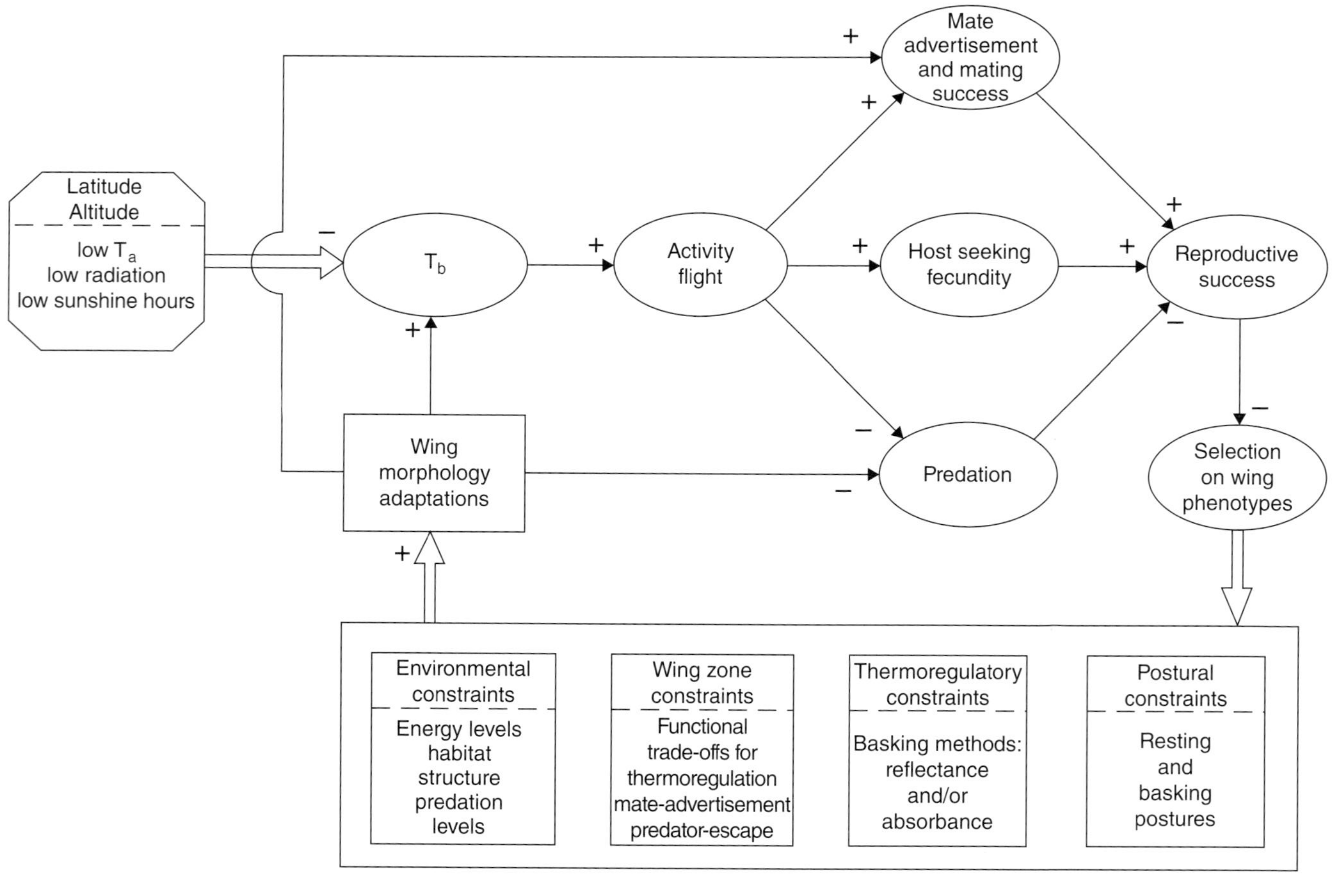

Fig. 8.6. Generalized model for butterfly wing pattern transformations in cooler climates. The type of transformation depends on a number of constraints, the most important of which are body and wing postures and thermoregulatory mechanisms (reflectance versus absorbance baskers) which have a phylogenetic basis. Octagon, zonal factors (T_a, ambient temperature); rectangle, butterfly morphology; ellipses, biological processes (T_b, body temperature); squares, constraints on morphological adaptations. (From Dennis, 1993, courtesy of Manchester University Press.)

Adjusting to Living on Small Islands

It is unusual to find distinctive changes (endemism) in the butterflies of small offshore islands, as there is greater probability of populations becoming extinct – rather than persisting to adapt to changing conditions – and to being 'rescued' by immigrants. But autochthonous endemism may occur, particularly if: (i) conditions are especially propitious for a species to persist over long time periods; (ii) there are substantial barriers to gene flow (immigrants; genetic barriers); and (iii) there are powerful selection pressures or the burgeoning of stochastic changes in genotypes with changing environmental conditions. On the mainland of Britain, there are a number of relatively compressed ecological isolates where genetic changes have occurred; classic examples are the Fenland *P. machaon* and *L. dispar* and the chalk downland species. Their populations are typically associated with use of single host plants in specific biotopes, other populations of the species in other biotopes having been lost during environmental changes during the Holocene (Dennis, 1977, 1993). Isolated populations often have reduced genetic variation compared with areas where populations are dense (e.g. *P. coridon*: Schmitt and Seitz, 2002; Krauss *et al.*, 2004; *E. aurinia*: Smee, 2011; Bulman *et al.*, 2007), increasing their vulnerability to extinction (Frankham, 2005). This may occur from one or more processes: (i) a **founder event** of few Holocene colonists; (ii) **genetic drift** (bottlenecking) from stochastic population collapses; or (iii) selection for optimal exploitation of a specialized narrow resource base coupled with the elimination of populations occurring more widely.

Even so, examples do exist of small island populations accruing unusual phenotypes and other adaptations (e.g. Hebridean *M. jurtina* and *Argynnis aglaja*; Heslop Harrison, 1938a,c, 1945c, 1953). It is appropriate to finish this account by examining one of the characteristic features of winged insects found on small islands: nanism (small size and wing expanse), if only to demonstrate how, potentially, causes may be complex. To do so, the example taken is that of two unrelated species, *P. argus caernensis* and *H. semele thyone* on the Great Orme's Head in North Wales. This is no longer an island but through its Holocene history has likely been an island; it is also a geological island and still functions much like an island, with the town of Llandudno acting as matrix (see Appendix 15). Both of the butterfly populations are dwarfed, emerge earlier than elsewhere and largely occupy the south and west scars and scree slopes of the Orme (Fig. 8.7 and Plate 1.17) (Dennis, 1977); they are similar, too, in that both overwinter in early stages, as eggs and very young larvae, respectively, and feed on young spring growth as immatures. The butterflies also differ in wing patterns from populations of the species along the North Wales coast (Dennis, 1977). There the similarity ends; the former feeds on a wide range of plants, but mainly *Helianthemum* sp. and *Lotus corniculatus* (Thomas, 1985) and is ant-attended (i.e. *Lasius alienus* more than by *L. niger*) (Diego Jordano and Chris Thomas in Thomas and Lewington, 2010), the latter largely on a grass, *Festuca ovina*. Even so, both are specialists in early stages on plant quality (young tender growth).

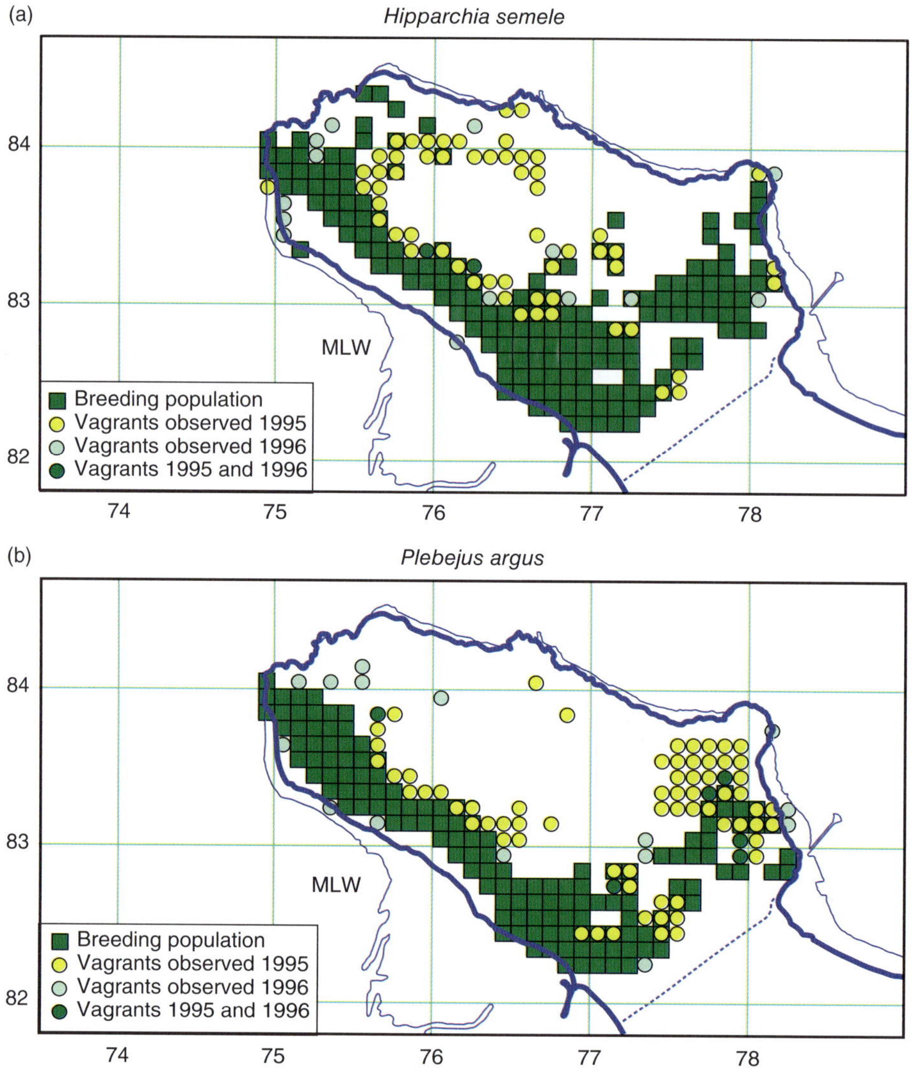

Fig. 8.7. The distribution of *Hipparchia semele thyone* and *Plebejus argus caernensis* on the Great Orme's Head, North Wales, from systematic cover of the headland during the summer of 1996. Shaded area, zones with suspected local breeding populations of (a) *Hipparchia semele* and (b) *Plebejus argus*. Circles, 100 m squares in which vagrants were observed during the summer of 1995 (yellow), during 1996 (lime green), and both years (dark green). Records absent from Pen Isa and the Parc, which have restricted access. (Redrawn using DMAP from Dennis and Bardell, 1996, courtesy of *Entomologist's Gazette* and Pemberley Books.)

There are several reasons used to explain the evolution of a similar (but much more extreme) condition called **brachyptery** (or **aptery** as in many beetles) (Kotze, 2008), including: (i) adaptation from losses to the population owing to high wind speeds; and (ii) a reduced food base, quality and/or quantity. Beirne (1947) postulated that these Orme species survived the Devensian maximum on an ice-free refuge (i.e. Cambrian Land). In fact, the entirety of this region, including the Orme, was covered by Irish, Scottish, Lake District and Welsh ice south to the Bristol Channel (Addison *et al.*, 1990). Ford (1957) considered that the genetic differences are small; it was subsequently revealed that changes in wing phenotypes can occur very rapidly indeed, within 20 years (Dennis, 1972a, 1977). Then there is the problem of why dwarfism has not been countered by immigration from surrounding populations, especially in *H. semele*, which, as we have seen, is more than capable of dispersing long distances (Dennis *et al.*,1998a). The reason was argued to be temporal isolation following a period of adaptation and spatial isolation: part of the Orme's Grayling population emerges well before the species elsewhere (Dennis, 1972b). In fact, the dwarf race has invaded the surrounding sand dunes, a biotope (elsewhere along the coast) on which occur significantly the largest individuals of the Grayling in Britain. At the same time, it was discovered that the Grayling gets significantly larger through the long flight season on the Orme.

Basically, it was argued that the answers lie in the present special conditions on the Great Orme's Head (for the Holocene history of the headland see Dennis, 1972a,b, 1977). Environmental conditions on the headland, as on many British islands, lead to an early growing season and lack of frosts (see Appendix 15, Fig. App. 15.1) but, owing to a number of factors (including geology, topography, local climate), also to host plant desiccation as the season progresses (Fig. App. 15.2). Ground temperatures are especially high owing to the high radiation load on slopes that effectively have an overhead sun (Bennett *et al.*, 2015) and reflection of the sun's rays off the sea (Ryrholm, 1988). These factors will tend to induce rapid development and fast growth rates but penalize slow development, leading to early emergences and concomitantly smaller size. In addition, there are pressures on the adult stage owing to high wind speeds in this exposed environment (Dennis and Sparks, 2006), favouring small wing expanse and penalizing larger butterflies in early summer (Fig. App. 15.3). Size has implications for mating success, fecundity and predator avoidance, and there are likely balances caused by **negative feedback**.

Explanation of these features in these two unrelated species on the Great Orme's Head still remains to be ascertained but it is highly likely that the reasons involve a complex of feedbacks between variables, the links of which are to be sought in a systems approach. As the details are perhaps too involved to examine fully here, some additional detail is provided in Appendix 15 for those interested in considering this issue further.

This final example has been discussed in some detail to warn against simplistic notions of factors operating in butterfly island biogeography: numerous factors may influence dwarfism, and other traits, on islands even through the same medium (i.e. larval growth rates) (Conover and Present, 1990). For example, Russell (2014) reported on finding (now extinct) Graylings on the shingle fossil island, Pilsey Island (now part of Thorney Island), with uniformly dark patterned hindwings and also with much reduced ventral forewing anal angle eyespots; this finding again suggests that rapid phenotypic changes are possible in populations.

It has long been the hope of the first author that the fascinating situation on the Great Orme's Head will be studied in depth. It can be added to the long list of uncertain issues raised in this book. We wish future researchers the great joy of investigation into the issues raised, and of finding solutions to protecting butterflies on Irish and British islands and on the Channel Islands and the Isle of Man.

9 Island Studies: a Glance Back and the View Ahead

The current review confirms the findings of the original work (Dennis and Shreeve, 1996), particularly the ability of species to reach (transfer to) islands and the importance of mainland and larger island source populations for the persistence of species on smaller offshore islands. Knowledge acquired since 1996 has substantially added to our understanding of the biogeography of butterflies on British and Irish islands and it may be useful to summarize some of the main features that emerge. As long ago as 1973 (Dennis, 1977) it was obvious that butterfly species richness and species incidence patterns on the British and Irish islands is dominated by environmental gradients, especially by geology, topography and climate, and consequently by basic butterfly resources (i.e. larval host plants) and living conditions. As the years have passed the cumulative and integrative effect of these environmental gradients has been highlighted in gradients of species richness (which declines northwards but with some unique northern species), distributions, and population dynamics, changes in phenology (voltinism) and phenotypes, and distinctive genetic gradients and zonation (Dennis, 1993). The foundations for these discoveries had already been laid down in a reconstruction of origins of the fauna, which was found to be far more recent – post the Devensian Maximum Glaciation – than previously believed to be the case (Dennis, 1977), and in multivariate analyses of the entire European butterfly fauna (Dennis *et al.*, 1991, 1995a,b).

The two basic influences expected to control spatial patterns across islands – island area and island isolation, so elegantly modelled by MacArthur and Wilson (1963, 1967) – were, comparatively speaking, found to be of limited influence relative to geographical position (island latitude). Even so, they were found to have an impact and their expected significant influences were teased out using regression analysis (Dennis and Shreeve, 1997). At that time, it was evident that our calculations were based on only a small sample of the offshore islands and then a sample largely biased to larger islands. As such, we expected that as the databank on islands increased, and it included a greater number of smaller islands, the pattern of relationships would change. The overall level of explanation of species richness would decline but the effect of island area and island isolation would increase relative to the influence of environmental gradients. Not only that, but the relative influences of island area and island isolation would also change. This latter shift in associations was expected simply because of the nature of correlations, fixed relationships, among the geographical variables describing the British islands (i.e. area, isolation, altitude, latitude, longitude, number of satellites) and the understanding that in adding more islands to the study these relationships would change if 'sampling' the islands happened to be geographically biased.

Therein lies the rub. One can only analyse what is available and although adjustments (selection of islands) can be made to the sample to reduce bias, at the very least there would be a relative increase in the number of islands off Ireland, for which data were originally very limited. Island data are also inevitably affected by uneven sampling, based on number of visits to them: large islands are understandably visited more often than small islands. Even so, owing to overall high levels of accounted variation, we have been able to use island geography and knowledge of species found at nearby sources to predict numbers of species for a large number of British and Irish offshore islands as well as the presence (or absence) of individual species on the islands. These findings are available in summary in the book and in data files as supplementary appendices for, at one level, those who wish to know what to expect to find on an island, and at another level, those who take on the important task of improving this work as techniques are advanced. Much as the smaller offshore islands are offshoots of Britain and Ireland, the British islands are but an appendage of the European continent. Therefore, we have set our findings on the patterns in species richness in a broader context of findings for European islands as a whole and the contrast in species found on the continent, Britain and in Ireland. This view highlights the fact that the butterfly fauna of the British islands is an impoverished version of that on the continent, especially the widespread **extent** (faunal) **element**. Even so, the reader will have found that the British fauna is distinct in many other ways.

Islands: a Pattern of Intricate and Endless Changes

Thus, we now have a good understanding of the general pattern of butterfly species over the offshore islands. However, the main advance of the current study is a much better understanding of the changes on islands over time. All the indications of the present study are that we must expect turnover of species and changes in the species communities on islands. Just how much change depends on island size (thus resource availability) and isolation: the larger the island, the more numerous and varied will be the resources and the more stable will be populations of species. The more isolated islands are, the less frequently will immigrants replace any losses on them. Greater stability of butterfly populations on near-shore larger islands is apparent on the Isle of Wight, reflected in the abundance-rank relationships of the species populations (Chapter 7). Small islands of a few hectares, however, have rapid turnover (extinction, colonization) of species, more especially those islands close to the mainland shore whence new colonists arise (e.g. Hilbre, Lindisfarne) and from which vagrants arrive that continually 'rescue' species lost owing to stochastic population changes. Further away from mainland shores, the colonists on small isolated islands will be species noted especially for their long-distance migration ability, often species that invade annually from the continents of Europe and Africa; they, too, will experience an apparent rapid turnover, particularly so as some are usually incapable of overwintering (e.g. *Vanessa cardui*, *Colias croceus*) and which therefore are like summer visiting seabirds.

One important finding in the 1996 study, and supported by this review, is that all species are capable of dispersing to the majority of islands placed a few kilometres offshore of the British and Irish coastline. Of course, there are exceptions: species that occur nowhere near a coastline and therefore lack coastal source populations for the islands (e.g. *Satyrium pruni*). Moreover, at the northern extremes of Britain, owing to steep environmental gradients, there are groups of islands that also lack coastal source populations; isolation in these cases includes both sea and land distances. Regardless of ability of a species for movement, isolation is a great filter of immigrants and with greater island isolation fewer species achieve a landfall, or at least frequent landfalls. There are distinctly isolated islands such as the St Kilda group that may well lie beyond the migration (dispersal) ability of most resident British species without a helping hand from human introductions, either by intention or by accident. Once species manage to transfer to an island, the simple model raised by Professor Richard Tilley in Chapter 6 would also suggest that colonists should not have difficulty in finding resources that do occur on islands. The great difficulty in persisting seems to arise in the quality and abundance of critical resources (Dennis, 2010); this should not come as a surprise to the many butterfly enthusiasts who are well aware that an abundance of a host plant frequently disappoints with the absence of an expected species. It is important to reflect on the simple observation that on offshore islands hazardous environmental conditions are likely to have a greater impact on the quality of resources, and therefore on their potential exploitation.

Recent studies have led to further insights on dispersal and colonization. Two indices for migration and colonization (see Appendix 7), built on life history attributes, demonstrate that species form a graded series in their ability to colonize islands and persist on them (Dennis *et al.*, 2012a; Dapporto and Dennis, 2013). From this, in turn, we may infer that the turnover of species is not just a simple random process but to a large degree deterministic, governed by each species' particular facility for migration (dispersal) and colonization. This is highly important as it ensures that the presence of species on islands is more predictable than if their occurrence were entirely random (turnover on even the smallest islands will have 'taxonomic pattern'). Even so, this review has exposed complications and flaws in our understanding of migration and dispersal. There is a world of difference between migration ability (an individual's capability or facility for moving a specific distance) and migration capacity (species' potential for movements and used in a broad sense of numbers of individuals moving a certain distance). The former is bound up much with a butterfly species' **bauplan**, the structure and 'engine' of each butterfly; the latter, additionally, with opportunities for movement and dependent more on factors that generate numbers of individuals (i.e. population size, flight periods, number of broods a year) and thus potential numbers of vagrants. The rapid fall-off in numbers migrating over distances (described by a negative exponential or inverse power function) in capture–mark–release–recapture studies suggests that isolation can be a severe filter for migrants; many travel nearby but few travel far, thus the greater number of migrants the greater is the number of opportunities for trialling distances and the greater the likelihood for a successful transfer to an island. This 'formula' explains the lack of appearance on islands of species that are rare and exist in sparse populations on a source mainland. Imagine that for each species there is a critical point, a distance to an island, beyond which rescue following extinction is ineffectual; at this point the rate of extinction exceeds the rate of immigration; beyond this point, migrants arrive but sparingly and colonization is temporary. It is important to appreciate that this point is not actually fixed in time but shifts with seasons, with environmental changes. Biogeography is not just about flexibility of species behaviour, but the flexibility of the human mind to appreciate the complexity of species behaviour in the face of shifting opportunities.

From these observations, it might be considered that the frequency of a species' incidence on British and Irish islands would be proportional to its probability of persistence, particularly as persistence (colonization ability) is correlated with migration capacity and both are functionally related to source mass, that is, the number and size of populations at a mainland or larger island source. This is evident from the study of species on Flat Holm and Steep Holm (see Chapter 6 and Appendix 13). The reader must expect that different islands (i.e. large versus small, near to shore and far offshore) experience very different genres of population dynamics; the species on them will have contrasting statuses. Thus predictions for species occurrences on islands in this study carry different implications for small and large, or near offshore and far offshore islands. There will be small isolated islands dominated by long-distance migrants whose status is primarily vagrant with ruderal life history associations. In contrast, small near-to-shore islands will receive many more specialists, as well as moderate generalists, which seem to have poorer dispersal capabilities and very limited potential for persistence. Large islands, with their more varied resource base and insulation from weather events, have the greatest potential for long-term residents, but even so those with the most stable populations on them will tend to have species with generalist resource requirements (good competitors). Specialists on the larger islands will depend on the presence of more unusual biotopes, as in the case of lycaenids (blues) on the Isle of Wight's chalk downland. This feature of butterfly persistence on islands has yet to be fully explored: the ecological basis of persistence is determined by resource availability as described by the resource-based habitat (Dennis, 2010).

Lessons from Island Studies: the Need to Conserve Sources

A key feature of the analysis of records for islands is the primacy of sources, mainland sources and larger island sources. Source numbers of species for species richness on islands, and source numbers of 10 km squares for particular species' occurrences on islands, dominate the geographical variables (island area, island isolation, island altitude). The source records in turn reflect environmental gradients the length of Britain and Ireland. The diminution (or, worse still, loss) of populations of species at sources will result, in time, in the loss of those species on dependent islands. The fact is that species on smaller islands are dependent on mainland and larger sources for their long-term persistence; they regularly become extinct and sources permit recolonization, or they are 'rescued' before becoming extinct. The only context in which small islands can be important for persistence is one in which they function as part of a metapopulation (archipelago), as in the case of the Isles of Scilly, or the Hebrides and the Orkney and Shetland groups. The message for conservation is: protect the mainland sources, especially those along the coastline, and the populations on the larger islands. Larger islands can become particularly valuable refuges, as they tend to experience less detrimental human interference than the mainland. It is very possible that some species have persisted on the large islands throughout the entire Holocene (e.g. *Coenonympha tullia* on the climatic-climax raised bog communities on Skye).

The process of loss of species on large islands is unhappily illustrated by the account of the Isle of Man (Chapter 7). This study demonstrated clearly that species absent and recorded as lost from Man are those that have also declined along the coastal fringes of Wales and Lancashire. Species that have recently colonized Man are those undergoing range expansions. In this context of providing conditions (space, biotopes, habitats) for species, there is no room for complacency. Even with the unique efforts put into butterfly conservation by the dedicated work of Butterfly Conservation UK during its lifetime (founded 1968), losses have dominated the British scene despite all the valiant efforts of a dedicated staff, whereas major gains have serendipitously been gifted by climate change. All the knowledge (scientific papers and reports) massing in the archives will be unable to reverse this invidious trend as the problem is one of outright habitat destruction, the removal and degradation of key resources on which our wildlife depends. To conserve species, it is not more papers on general trends that we need on these issues but a focus on the altogether more difficult task: determined ploys that captivate the backing of a burgeoning public opinion to contest detrimental land use practices, and to demand landscape solutions, over the wider industrially exploited landscape, linked to practical work that tests alternative measures. In this, conservation organizations need to function in concert as a single body rather than as competing entities with narrow remits. One thing is abundantly clear: for all the clever algorithms, if we cannot explain the absence of a species where it might be entirely expected, or its scarcity where we might expect abundance, we simply do not understand its habitat requirements. The starting point is to understand species' habitats, their resource-based habitats and how to cater for species that occupy the same sites but have conflicting demands. Treating biotopes as habitats may capture a community of species, and that is a good start where time is limited, when holding on to more natural biotopes is an urgent matter; but if the intention is to conserve a particular species there is no alternative but to understand all its requirements and put them in place.

Directions for Future Research of Island Butterflies

Our hope is that this work will also encourage further studies of islands at all intellectual levels, but primarily to determine their potential for conservation. Fundamentally, we need a greater understanding of butterfly dispersal, the colonization process and the ability of species to persist in limited spaces. Islands (and other offshore structures such as lighthouses) offer invaluable research 'tools' for such practical studies. Islands provide finite bounded spaces which present advantages over mainland terrestrial sites for similar work:

1. Immigration to, and emigration from, an island depend on transfers across entirely inimical matrix and can be more precisely monitored and tested using marked individuals.
2. Population parameters for any species studied relate to the island space entirely and are far less likely to be confused with the dynamics of adjoining sites (i.e. biotopes or habitats).
3. Populations and their resources can be more effectively quantified over the entire space, especially when islands are uninhabited; when owned by national bodies (e.g. National Trust, Royal Society for the Protection of Birds) rather than privately there is the prospect of greater support for the research with synoptic recording programmes such as the Butterfly Monitoring Scheme and regular, consistent vegetation surveys.

It is an unfortunate fact that ecological and biogeographical work on mainland Britain is often frustrated by well meaning but misguided attempts to spread species but that fail to document such efforts with Butterfly Conservation UK; also by changes by individuals on adjoining inaccessible land. In particular, emphasis should be given to long-term studies. An outstanding issue in the study of species turnover is the impact of changing conditions. Islands, as environments, rarely experience stable conditions for long. There will be seasons that promote colonizations, others that trigger multiple extinctions. Changing weather conditions over the offshore islands are a prominent determinant of migrations to islands (hot, sunny summers with favourable wind directions), as they are of extinctions (cold, wet, stormy summers). These events are unlikely to be picked up by research without investment in long-term studies.

A future focus on island butterfly ecology and population genetics is certainly the next step. In this and the previous work (Dennis and Shreeve, 1996, 1997) we have given much attention to the impact of the basic geography of the islands. These geographical variables do explain much, both of species richness and individual species' occurrences on islands. But island biogeography practitioners will be aware of the residuals in explanation we have left behind. Further explanation lies in the land use of islands, the ecology of species and the population genetics of species. Something as simple as an exposure of Carboniferous limestone can make the difference between a rich and a poor island, much as can a dense colony of seabirds, but most differences in species and population sizes that matter between islands are likely to be found to be more subtle. Therefore, we need a better understanding of the factors that cause oscillations in population size and persistence. To assist in these future studies there is a growing armoury of techniques and technology. Students of movements in and outside habitats are now gaining access to remote observation and tags, to radio-labelling, radar tracking and isotope analysis, drones with cameras and better 'traps' (e.g. Altizer *et al.*, 2015; Hungate *et al.*, 2016; Ivosevic *et al.*, 2017). The biological significance of variations in phenotypes across islands, which have for long presented a puzzle for entomologists (e.g. *Maniola jurtina* wing spotting on the Isles of Scilly by Professor E.B. Ford, 1964) now have a range of molecular (DNA) markers for determination (Wink, 2006; Gómez-Diaz and Gonzáles-Solis, 2007). At long last, entomologists who have for so long been interested in the regional variation among British butterflies (i.e. tension zones, clines, seasonal polymorphisms, wing pattern variation, etc.) have access to molecular biological tools that will eventually present firm explanations for it.

Above all, there is a need to understand butterfly habitats better and the extent to which resources and environmental conditions other than the simple presence of the larval host plant(s) affect population parameters, thus population persistence. Management cannot operate effectively unless we have such detailed information at our disposal. So often one witnesses broadcast management changes in sites that counter basic hard-wired scientific findings with apparent absence of thought to any balance in management practices for the benefit of a target species, let alone the species community that includes organisms other than butterflies.

A classic example is the widespread clearance of cotoneaster scrub from the upper slopes of the Great Orme's Head, when it has been demonstrated just how important such cover can be to a key species there, *Plebejus argus*, for a range of activities (e.g. resting, roosting, mating and feeding) (Dennis and Sparks, 2006) in the high-wind environment of the Irish Sea. Evidence-based methods are crucial if we are to be seen to be applying science to management and conservation (Pullin and Knight, 2003).

Plates

Plate 1. Island Landscapes

Plate 1.1. The Calf of Man and Kitterland as seen from the Isle of Man (courtesy of Gail E. Jeffcoate©).

Plate 1.2. Bawden Rocks, off Newdowns Head, north Cornwall (courtesy of the late Dr W.G. Tremewan).

Plate 1.3. Bass Rock in the Firth of Forth, Scotland (courtesy of Stuart Murray©).

Plate 1.4. Old Harry Rocks on the south side of Studland Bay, Dorset.

Plate 1.5. Hirta, the largest of the St Kilda archipelago, NW of North Uist (courtesy of Alexandra and Tim Sparks ©).

Plate 1.6. Ailsa Craig off SW Scotland (courtesy of Stuart Murray©).

Plate 1.7. Monkstone lighthouse in Bristol Channel (courtesy of Sam Whitfield©).

Plate 1.8. Bishop Rock South (Emsger) off St David's Head, Pembrokeshire coast, SW Wales (courtesy of Stuart Murray©).

Plate 1.9. Grassholm island off the Pembrokeshire coast, South Wales (courtesy of Stuart Murray©).

Plate 1.10. The Strangford Lough archipelago, Northern Ireland (courtesy of northern-eye and Mike Hartwell©).

Plate 1.11. Flat Holm and Steep Holm in the Bristol Channel (courtesy of R.J.D. Tilley©).

Plate 1.12. Flat Holm island in the Bristol Channel (courtesy of Sam Whitfield©).

Plate 1.13. Hilbre, in the Dee Estuary, Cheshire (courtesy of Steven R. Williams and Hilbre Bird Observatory©).

Plate 1.14. Lindisfarne island off the Northumberland coast (courtesy of Graeme Peacock©).

Plate 1.15. A'Choille Mhor SSSI oak woods on Colonsay, Inner Hebrides (courtesy of John Aitchison©).

Plate 1.16. Compton Bay from Shippard's Point on the south-facing cliffs of the Isle of Wight (courtesy of Matthew Oates©).

Plate 1.17. The Great Orme's Head (and west side shown below) with the tombolo linking it to the mainland, North Wales (courtesy of Andrew Mills, Bob Baxter and Mills Media Creative, Wirral©).

Plate 1.1. Satellites and stepping-stone islands. The southern tip of the Isle of Man in the Irish Sea, looking across the tiny stepping-stone island of Kitterland in the Sound to the Calf. Unusually for a small satellite island, the Calf has been well surveyed for butterflies (courtesy of Gail E. Jeffcoate©).

Plate 1.2. Exposure as an impediment to butterfly colonization. Bawden Rocks, off Newdowns Head, north Cornwall (SW 700532) present examples of small islands (35 m high and 0.5 ha in area) extremely vulnerable to the effects of storm waves, which regularly sweep over the islands (courtesy of the late Dr W.G. Tremewan).

Plate 1.3. Seabird colonies as an impediment to butterfly colonization. Bass Rock in the Firth of Forth is the site of Britain's largest gannet colony; birds are at such high density that there is little room left for colonizing butterflies and demonstrates that island area is not always a good measure of island area available for butterfly colonization (courtesy of Stuart Murray©).

Plate 1.4. Islands and islets created by erosion. The eroding chalk cliffs of Handfast Point and the stacks Old Harry Rocks on the south side of Studland Bay, Dorset. Such cliffs sometimes provide limited resources for butterflies (e.g. cliff cabbage for pierids) but are virtually inaccessible to survey.

Plate 1.5. An example of an isolated abandoned island. Hirta, the largest of the small archipelago of St Kilda, 64 km from the island of North Uist, was finally abandoned in 1930, and since 1957 has belonged to the National Trust for Scotland. The outlines of houses, fields and unique forms of stone structures, known as cleitean (*cleit* is a bothy), are still evident in this recent photograph. Old records of butterflies on once-occupied islands suggest the insects may have been given a helping hand in colonization (courtesy of Alexandra and Tim Sparks ©).

Plate 1.6. Ailsa Craig, an island exploited intensively for its granite resources, has a precipitous cliff topography that hinders recording. Aerial photographs of islands provide useful resources to determine the potential for butterfly colonists. This photograph of Ailsa Craig would suggest that the interesting record of *Coenonympha tullia*, an occupant of raised and blanket bogs, is questionable from the vantage of being even a temporary colonist (courtesy of Stuart Murray©).

Plate 1.7. Lighthouses, even remotely operated ones, provide potential sampling points for Lepidoptera in the British and Irish seaboard. The now remotely operated Monkstone lighthouse in Bristol Channel (located about 5 km east-north-east of Lavernock Point and 8 km south of Cardiff) was built in 1839 and rebuilt in 1993. It is built on underwater rock and extends 13–21 m above mean high-water mark (courtesy of Sam Whitfield©). (See Anonymous, 2017, for list of lighthouses around the British Isles.)

Plate 1.8. An example of a 'desert' island off the British and Irish coasts. Bishop Rock South (Emsger), 8 km off St David's Head, Pembrokeshire coast, SW Wales, formed of rhyolite, has a limited area for persistent butterfly colonization but potential for recording, the lighthouse having been constructed in 1839 (courtesy of Stuart Murray©).

Plate 1.9. Grassholm island, 13 km off the Pembrokeshire coast with its steep inaccessible cliffs and large gannet populations, presents much the same problems for butterfly colonization and butterfly recording as Bass Rock (courtesy of Stuart Murray©).

Plate 1.10. The Strangford Lough archipelago, Northern Ireland, a completely enclosed group of unusual islands formed in a drowned drumlin landscape. Unlike many archipelagos, the surrounding mainland provides the main source of butterfly species (courtesy of northern-eye and ©Mike Hartwell [File 776312425; N-E_040908_3207]).

Plate 1.11. Near-offshore islands in the Bristol Channel. The distinctive profiles of Flat Holm and Steep Holm (island on right) viewed from the Welsh coast at Lavernock Point; Brean Down on the Somerset coast is seen in the far distance (courtesy of R.J.D. Tilley©).

Plate 1.12. Near-offshore islands in the Bristol Channel. Flat Holm island, illustrating contrasting land uses including a lighthouse (courtesy of Sam Whitfield©).

Plate 1.13. Causeway tidal islands with known turnover of butterfly species. Hilbre, in the Dee Estuary, Cheshire, photographed from the islet of Middle Eye (courtesy of Steven R. Williams and Hilbre Bird Observatory©, www.hilbrebirdobs.blogspot.co.uk).

Plate 1.14. Causeway tidal islands with known turnover of butterfly species. Lindisfarne island off the Northumberland coast (courtesy of Graeme Peacock©).

Plate 1.15. The remarkable stunted (krummholz) oaks of north-eastern Colonsay in the Inner Hebrides, an unusual ancient wooded site (A'Choille Mhor SSSI) and potential biotope for woodland butterflies such as *Favonius quercus* and *Pararge aegeria* (courtesy of John Aitchison©).

Plate 1.16. Island refuges for butterfly species. Compton Bay from Shippard's Point on the south-facing cliffs of the Isle of Wight, biotope for *Melitaea cinxia* (courtesy of Matthew Oates©).

Plate 1.17. An old island and a potential island. An aerial view from the east of the Great Orme's Head, a headland on the North Welsh coast, and the low-lying tombolo on which stands the tourist resort of Llandudno (courtesy of Andrew Mills, Bob Baxter and Mills Media Creative, Wirral©). The second photograph illustrates the dipping carboniferous limestone strata, scars and scree slopes on the western side, biotope for *Hipparchia semele* and *Plebejus argus*.

Plate 2. Images of British Butterfly Species

Each butterfly in Appendix 1 is illustrated (Plates 2.1 to 2.83). Butterfly images were kindly provided by Patrick Clement, Bob Eade, Peter Eeles, Tony Hoare, Mark and Neil Hulme, Nigel Kiteley, Iain H. Leach, Tim Melling, Ian Rippey and Andy Seely, and Butterfly Conservation, UK, who retain the copyright to the photographs. Photographs lacking definitive citations were taken by one of the authors (Peter B. Hardy).

Plate 2.1. *Carterocephalus palaemon* Chequered Skipper (courtesy of Iain H. Leach, Butterfly Conservation).

Plate 2.2. *Heteropterus morpheus* Large Chequered Skipper (courtesy of Peter Eeles, Butterfly Conservation).

Plate 2.3. *Thymelicus sylvestris* Small Skipper.

Plate 2.4. *Thymelicus lineola* Essex Skipper.

Plate 2.5. *Thymelicus acteon* Lulworth Skipper.

Plate 2.6. *Hesperia comma* Silver-spotted Skipper.

Plate 2.7. *Ochlodes sylvanus* Large Skipper.

Plate 2.8. *Erynnis tages* Dingy Skipper (courtesy of Iain H. Leach, Butterfly Conservation).

Plate 2.9. *Pyrgus malvae* Grizzled Skipper (courtesy of Iain H. Leach, Butterfly Conservation).

Plate 2.10. *Papilio machaon* Swallowtail (courtesy of Iain H. Leach, Butterfly Conservation).

Plate 2.11. *Iphiclides podalirius* Scarce Swallowtail.

Plate 2.12. *Leptidea sinapis* Wood White (courtesy of Bob Eade, Butterfly Conservation).

Plate 2.13. *Leptidea juvernica* Cryptic Wood White (courtesy of Ian Rippey, taken at Craigavon Lakes, Co. Armagh, 24 May 2013).

Plate 2.14. *Colias hyale* Pale Clouded Yellow (courtesy of Peter Eeles and Butterfly Conservation).

Plate 2.15. *Colias alfacariensis* Berger's Clouded Yellow (courtesy of Tim Melling, Butterfly Conservation).

Plate 2.16. *Colias croceus* Clouded Yellow.

Plate 2.17. *Gonepteryx rhamni* Brimstone.

Plate 2.18. *Gonepteryx cleopatra* Cleopatra.

Plate 2.19. *Aporia crataegi* Black-veined White.

Plate 2.20. *Pieris brassicae* Large White.

Plate 2.22. *Pieris napi* Green-veined White.

Plate 2.24. *Anthocharis cardamines* Orange-tip.

Plate 2.26. *Thecla betulae* Brown Hairstreak (courtesy of Neil Hulme and Butterfly Conservation).

Plate 2.21. *Pieris rapae* Small White.

Plate 2.23. *Pontia daplidice* Bath White.

Plate 2.25. *Callophrys rubi* Green Hairstreak.

Plate 2.27. *Favonius quercus* Purple Hairstreak (courtesy of Mark Hulme and Butterfly Conservation).

Plate 2.28. *Satyrium w-album* White-letter Hairstreak (courtesy of Iain H. Leach, Butterfly Conservation).

Plate 2.29. *Satyrium pruni* Black Hairstreak (courtesy of Neil Hulme and Butterfly Conservation).

Plate 2.30. *Lycaena phlaeas* Small Copper.

Plate 2.31. *Lycaena dispar* Large Copper (courtesy of Tony Hoare and Butterfly Conservation).

Plate 2.32. *Lycaena virgaureae* Scarce Copper.

Plate 2.33. *Lycaena tityrus* Sooty Copper (courtesy of Tim Melling, Butterfly Conservation).

Plate 2.34. *Lycaena hippothoe* Purple-edged Copper (courtesy of Mark Hulme and Butterfly Conservation).

Plate 2.35. *Lampides boeticus* Long-tailed Blue.

Plate 2.36. *Cacyreus marshalli* Geranium Bronze (courtesy of Tony Hoare and Butterfly Conservation).

Plate 2.37. *Cupido minimus* Small Blue (courtesy of Patrick Clement and Butterfly Conservation).

Plate 2.38. *Cupido argiades* Short-tailed Blue.

Plate 2.39. *Plebejus argus* Silver-studded Blue (courtesy of Neil Hulme and Butterfly Conservation).

Plate 2.40. *Aricia agestis* Brown Argus (courtesy of Bob Eade, Butterfly Conservation).

Plate 2.41. *Aricia artaxerxes* f. *artaxerxes* Northern Brown Argus (courtesy of Tim Melling, Butterfly Conservation).

Plate 2.42. *Polyommatus icarus* Common Blue.

Plate 2.43. *Polyommatus coridon* Chalk Hill Blue.

Plate 2.44. *Polyommatus bellargus* Adonis Blue.

Plate 2.45. *Cyaniris semiargus* Mazarine Blue.

Plate 2.46. *Celastrina argiolus* Holly Blue.

Plate 2.47. *Phengaris arion* Large Blue (courtesy of Neil Hulme and Butterfly Conservation).

Plate 2.48. *Hamearis lucina* Duke of Burgundy Fritillary (courtesy of Iain H. Leach, Butterfly Conservation).

Plate 2.49. *Limenitis camilla* White Admiral.

Plate 2.50. *Apatura iris* Purple Emperor.

Plate 2.51. *Vanessa atalanta* Red Admiral.

Plate 2.52. *Vanessa cardui* Painted Lady.

Plate 2.53. *Vanessa virginiensis* American Painted Lady (courtesy of Tony Hoare and Butterfly Conservation).

Plate 2.54. *Aglais urticae* Small Tortoiseshell.

Plate 2.55. *Aglais io* Peacock.

Plate 2.56. *Nymphalis polychloros* Large Tortoiseshell (courtesy of Neil Hulme and Butterfly Conservation).

Plate 2.57. *Nymphalis xanthomelas* Scarce Tortoiseshell (courtesy of Neil Hulme and Butterfly Conservation).

Plate 2.58. *Nymphalis antiopa* Camberwell Beauty.

Plate 2.59. *Polygonia c-album* Comma.

Plate 2.60. *Araschnia levana* European Map (courtesy of Tony Hoare and Butterfly Conservation).

Plate 2.61. *Boloria selene* Small Pearl-bordered Fritillary.

Plate 2.62. *Boloria euphrosyne* Pearl-bordered Fritillary.

Plate 2.63. *Issoria lathonia* Queen of Spain Fritillary.

Plate 2.64. *Argynnis adippe* High Brown Fritillary.

Plate 2.65. *Argynnis aglaja* Dark Green Fritillary.

Plate 2.66. *Argynnis paphia* Silver-washed Fritillary (courtesy of Iain H Leach, Butterfly Conservation).

Plate 2.67. *Euphydryas aurinia* Marsh Fritillary.

Plate 2.68. *Melitaea cinxia* Glanville Fritillary (courtesy of Nigel Kiteley and Butterfly Conservation).

Plate 2.69. *Melitaea athalia* Heath Fritillary (courtesy of Iain H Leach, Butterfly Conservation).

Plate 2.70. *Pararge aegeria* Speckled Wood.

Plate 2.71. *Lasiommata megera* Wall.

Plate 2.72. *Erebia epiphron* Small Mountain Ringlet (Courtesy of Neil Hulme and Butterfly Conservation).

Plate 2.73. *Erebia aethiops* Scotch Argus.

Plate 2.74. *Erebia ligea* Arran Brown.

Plate 2.75. *Melanargia galathea* Marbled White.

Plate 2.76. *Hipparchia semele* Grayling.

Plate 2.78. *Maniola jurtina* Meadow Brown.

Plate 2.80. *Coenonympha pamphilus* Small Heath (courtesy of Tim Melling, Butterfly Conservation).

Plate 2.82. *Danaus plexippus* Monarch (courtesy of Andy Seely and Butterfly Conservation).

Plate 2.77. *Pyronia tithonus* Gatekeeper.

Plate 2.79. *Aphantopus hyperantus* Ringlet.

Plate 2.81. *Coenonympha tullia* Large Heath (courtesy of Tim Melling, Butterfly Conservation).

Plate 2.83. *Danaus chrysippus* Plain Tiger.

Appendices: Records of Butterflies from British and Irish Islands

Appendix 1. Checklist of British and Irish Butterflies[1]

The scientific names of species in the checklist follow Kudrna *et al.* (2015)[2]. There has been much disagreement in the past over the notation of higher taxa in the Hesperioidea and Papilionoidea (cf. Ackery, 1984; Scott, 1985; Brock, 1990; Scoble, 1992); with developments in molecular biology this eventually promises to be resolved (Wahlberg *et al.*, 2005; Kim *et al.*, 2010; Heikkilä *et al.*, 2012). Butterflies occurring as native species, residents formerly native but now extinct and common and infrequent immigrants are included in the list, as are those that may have been accidentally introduced but which breed or have bred in the islands (e.g. *Heteropterus morpheus*). Exceptionally rare immigrants, rare accidental introductions and adventives into the British islands are excluded from it (see Appendix 2 and classification in Emmet and Heath, 1989). Allocation of species to Appendix 1 or Appendix 2 is in some ways arbitrary, as the status of species can (and does) change (e.g. *Parnassius apollo*) (Morley and Chalmers-Hunt, 1959; Dennis, 1977, p. 104). The vernacular names follow Emmet and Heath (1989).

Regarding regular migrants to the British Isles, Williams (1971) summarized records of the observed numbers of nine migratory butterfly species (marked *) and 20 migratory moth species in Britain for the period 1850–1962. The migrants were overseas immigrants lacking regular resident populations; this is likely to change in future as increasing numbers of some of these species are found to overwinter (e.g. *Vanessa atalanta*) (Fox and Dennis, 2010).

Abbreviations and notes

RE, regionally extinct
CR, critically endangered

EN, endangered
VU, vulnerable
NT, near threatened
LC, least concern
LP(F), legal protection (full) under section 9 of the Wildlife and Countryside Act 1981 or the Wildlife (Northern Ireland) Order 1985 (protection against killing, injuring, taking from the wild, possessing, selling or advertising for sale any wild specimen unless in possession of a licence)
LP(P), legal protection (partial) under the above legislation (covering sale only)

R, resident species
E, resident now known to be extinct
M, frequent immigrant
I, infrequent immigrant
O, immigrant which may overwinter
B, immigrant which may breed
D, deliberate introduction
A, accidental introduction
?, status (identification or native record) uncertain; a ? following another symbol indicates uncertain data
BM, British mainland (including islands off England, Wales and Scotland) and Channel Islands
Ire, Ireland and its islands (Baynes, 1964; Emmet and Heath, 1989)
+, Species not determined, and could be *Pontia edusa* (Fabricius, 1777) (see Wagener, 1988)

Superscript numbers immediately following the name of the species indicate the form of so-called subspecies listed for islands in Appendix 3.

Photographs of each of the 83 species are recorded as Plates 2.1 to 2.83.

Superfamily HESPERIOIDEA LATREILLE, 1809

Family Hesperiidae LATREILLE, 1809

Subfamily Heteropterinae AURIVILLIUS, 1925

Carterocephalus palaemon (PALLAS, 1771) Chequered Skipper, **EN**, **LP(P)** (BM: R; E and D in England) Plate 2.1

Heteropterus morpheus (PALLAS, 1771) Large Chequered Skipper (BM: A in Jersey only, E) Plate 2.2

Subfamily Hesperiinae LATREILLE, 1809

Thymelicus sylvestris (PODA, 1761) Small Skipper, **LC** (BM and Ire: R) Plate 2.3

Thymelicus lineola (OCHSENHEIMER, 1806) Essex Skipper, **LC** (BM and Ire: R) Plate 2.4

Thymelicus acteon (ROTTEMBURG, 1775) Lulworth Skipper, **NT**, **LP(P)** (BM: R) Plate 2.5
Hesperia comma (LINNAEUS, 1758) Silver-spotted Skipper, **NT**, **LP(P)** (BM: R) Plate 2.6
Ochlodes sylvanus (ESPER, [1777]) Large Skipper, **LC** Plate 2.7
ssp. *faunus* (TURATI, 1905) (BM: R; Ire: ? E; South, 1941)

Subfamily Pyrginae BURMEISTER, 1878
Erynnis tages (LINNAEUS, 1758) Dingy Skipper, **VU** Plate 2.8
ssp. *tages* (LINNAEUS, 1758)[1] (BM and Ire: R)
ssp. *baynesi* HUGGINS, 1956[2] (Ire: R)
Pyrgus malvae (LINNAEUS, 1758) Grizzled Skipper, **VU** (BM: R) Plate 2.9

Superfamily PAPILIONOIDEA LATREILLE, [1802]

Family Papilionidae LATREILLE, [1802]

Subfamily Papilioninae LATREILLE, [1802]
Papilio machaon LINNAEUS, 1758 Swallowtail, **NT** Plate 2.10
ssp. *britannicus* SEITZ, 1907[1] (BM: R)
ssp. *gorganus* FRUHSTORFER, 1922[2] (BM and Ire: I, O, B)
Iphiclides podalirius (LINNAEUS, 1758) Scarce Swallowtail (BM: I, B) Plate 2.11

Family Pieridae DUPONCHEL, [1835]

Subfamily Dismorphiinae SCHATZ, [1886]
Leptidea sinapis (LINNAEUS, 1758) Wood White, **EN**, **LP(P)** (BM and Ire: R) Plate 2.12
Leptidea juvernica WILLIAMS, 1946 Cryptic Wood White (Ire: R) Plate 2.13

Subfamily Coliadinae SWAINSON, 1827
*Colias hyale** (LINNAEUS, 1758) Pale Clouded Yellow (BM and Ire: I, O? B) Plate 2.14
Colias alfacariensis BERGER, 1948 Berger's Clouded Yellow (BM: I, B) Plate 2.15
*Colias croceus** (GEOFFROY, 1785) Clouded Yellow, **LC** (BM and Ire: I, O, B) Plate 2.16
Gonepteryx rhamni (LINNAEUS, 1758) Brimstone, **LC** Plate 2.17
ssp. *rhamni* (LINNAEUS, 1758)[1] (BM: R)
ssp. *gravesi* HUGGINS, 1956[2] (Ire: R)
Gonepteryx cleopatra (LINNAEUS, 1767) Cleopatra (BM: I, D and A) Plate 2.18

Subfamily Pierinae DUPONCHEL, [1835]

Aporia crataegi (LINNAEUS, 1758) Black-veined White, **RE** (BM: E, I or A, D) Plate 2.19

Pieris brassicae (LINNAEUS, 1758) Large White, **LC** (BM and Ire: R) Plate 2.20

Pieris rapae (LINNAEUS, 1758) Small White, **LC** (BM and Ire: R) Plate 2.21

Pieris napi (LINNAEUS, 1758) Green-veined White, **LC** Plate 2.22

ssp. *napi* (LINNAEUS, 1758)[1] (BM: R)

ssp. *septentrionalis* VERITY, 1916[2] (BM: R)

ssp. *britannica* MÜLLER and KAUTZ, 1939[3] (Ire: R)

ssp. *thomsoni* WARREN, 1968[4] (probably = *britannica*) (BM: R)

*Pontia daplidice**+ (LINNAEUS, 1758) Bath White (BM and Ire: I, O? B?) Plate 2.23

Anthocharis cardamines (LINNAEUS, 1758) Orange-tip, **LC** Plate 2.24

ssp. *cardamines* (LINNAEUS, 1758)[1] (BM: R)

ssp. *britannica* (VERITY, 1908)[2] (BM: R)

ssp. *hibernica* (WILLIAMS, 1916)[3] (Ire: R)

Family Lycaenidae [LEACH], [1815]

Subfamily Theclinae BUTLER, 1869

Callophrys rubi (LINNAEUS, 1758) Green Hairstreak, **LC** (BM and Ire: R) Plate 2.25

Thecla betulae (LINNAEUS, 1758) Brown Hairstreak, **VU**, **LP(P)** (BM and Ire: R) Plate 2.26

Favonius quercus (LINNAEUS, 1758) Purple Hairstreak, **LC** (BM and Ire: R) Plate 2.27

Satyrium w-album (KNOCH, 1782) White-letter Hairstreak, **EN**, **LP(P)** (BM: R) Plate 2.28

Satyrium pruni (LINNAEUS, 1758) Black Hairstreak, **EN**, **LP(P)** (BM: R) Plate 2.29

Subfamily Lycaeninae [LEACH], [1815]

Lycaena phlaeas (LINNAEUS, 1761) Small Copper, **LC** Plate 2.30

ssp. *eleus* (FABRICIUS, 1798)[1] (BM: R)

ssp. *hibernica* GOODSON, 1948[2] (Ire: R)

unnamed Scottish race (Thomson, 1980)[3] (BM: R)

Lycaena dispar (HAWORTH, 1803) Large Copper, **RE**, **LP(F)** Plate 2.31

ssp. *dispar* (HAWORTH, 1803)[1] (BM: E)

ssp. *rutilus* WERNEBERG, 1864[2] (BM and Ire: D, E)

ssp. *batavus* (OBERTHÜR, 1923)[3] (BM: D; Ire: D, E)

Lycaena virgaureae (LINNAEUS, 1758) Scarce Copper (BM: ? E) Plate 2.32

Lycaena tityrus (PODA, 1761) Sooty Copper (BM: I) Plate 2.33

Lycaena hippothoe (LINNAEUS, 1761) Purple-edged Copper (BM: ? E) Plate 2.34

Subfamily Polyommatinae SWAINSON, 1827

*Lampides boeticus** (LINNAEUS, 1767) Long-tailed Blue (BM: I, B) Plate 2.35
Cacyreus marshalli (BUTLER, 1898) Geranium Bronze (BM: A) Plate 2.36
Cupido minimus (FUESSLY, 1775) Small Blue, **NT**, **LP(P)** (BM and Ire: R) Plate 2.37
Cupido argiades (PALLAS, 1771) Short-tailed Blue (BM: I) Plate 2.38
Plebejus argus (LINNAEUS, 1758) Silver-studded Blue, **VU**, **LP(P)** Plate 2.39
ssp. *argus* (LINNAEUS, 1758)[1] (BM: R)
ssp. *cretaceus* TUTT, 1909[2] (BM: R)
ssp. *masseyi* TUTT, 1909[3] (BM: E, but see Birkett, 1995)
ssp. *caernensis* THOMPSON, [1937][4] (BM: R)
Aricia agestis ([SCHIFFERMÜLLER], 1775) Brown Argus, **LC** (BM: R) Plate 2.40
Aricia artaxerxes (FABRICIUS, 1793) Northern Brown Argus, **VU**, **LP(P)** Plate 2.41
ssp. *artaxerxes* (FABRICIUS, 1793)[1] (BM: R)
ssp. *salmacis* (STEPHENS, 1828)[2] (BM: R)
Polyommatus icarus (ROTTEMBURG, 1775) Common Blue, **LC** Plate 2.42
ssp. *icarus* (ROTTEMBURG, 1775)[1] (BM: R)
ssp. *mariscolore* (KANE, 1893)[2] (BM and Ire: R)
unique island race (Heslop Harrison, 1950b)[3] (BM: R)
Polyommatus coridon (PODA, 1761) Chalk Hill Blue, **NT**, **LP(P)** (BM: R) Plate 2.43
Polyommatus bellargus (ROTTEMBURG, 1775) Adonis Blue, **NT**, **LP(P)** (BM: R) Plate 2.44
Cyaniris semiargus (ROTTEMBURG, 1775) Mazarine Blue, **RE** (BM: E, I) Plate 2.45
Celastrina argiolus (LINNAEUS, 1758) Holly Blue, **LC** Plate 2.46
ssp. *argiolus*[1] (LINNAEUS, 1758) (BM: R)
ssp. *britanna*[2] (VERITY, 1919) (BM and Ire: R)
Phengaris arion (LINNAEUS, 1758) Large Blue, **CR**, **LP(F)** Plate 2.47
ssp. *arion* (LINNAEUS, 1758)[1] (BM: D)
ssp. *eutyphron* (FRUHSTORFER, 1915)[2] (BM: E)

Subfamily Riodininae GROTE, 1895

Hamearis lucina (LINNAEUS, 1758) Duke of Burgundy Fritillary, **EN**, **LP(P)** (BM: R) Plate 2.48

Family Nymphalidae SWAINSON, 1827

Subfamily Limenitinae BEHR, 1864

Limenitis camilla (LINNAEUS, 1764) White Admiral, **VU** (BM: R) Plate 2.49

Subfamily Apaturinae BOISDUVAL, 1840

Apatura iris (LINNAEUS, 1758) Purple Emperor, **NT**, **LP(P)** (BM: R) Plate 2.50

Subfamily Nymphalinae SWAINSON, 1827

*Vanessa atalanta** (LINNAEUS, 1758) Red Admiral, **LC** (BM and Ire: M, O, B) Plate 2.51

*Vanessa cardui** (LINNAEUS, 1758) Painted Lady, **LC** (BM and Ire: M, O, B) Plate 2.52

Vanessa virginiensis (DRURY, 1773) American Painted Lady (BM and Ire: I) Plate 2.53

Aglais urticae (LINNAEUS, 1758) Small Tortoiseshell, **LC** (BM and Ire: R) Plate 2.54

Aglais io (LINNAEUS, 1758) Peacock, **LC** (BM and Ire: R) Plate 2.55

Nymphalis polychloros (LINNAEUS, 1758) Large Tortoiseshell, **RE**, **LP(P)** (BM: E, I, O, B) Plate 2.56

Nymphalis xanthomelas ([SCHIFFERMÜLLER], 1775) Scarce Tortoiseshell (BM: I, O) Plate 2.57

*Nymphalis antiopa** (LINNAEUS, 1758) Camberwell Beauty (BM and Ire: I, O) Plate 2.58

Polygonia c-album (LINNAEUS, 1758) Comma, **LC** (BM and Ire: R) Plate 2.59

Araschnia levana (LINNAEUS. 1758) European Map (BM: I in Jersey, D and A) Plate 2.60

Subfamily Heliconiinae SWAINSON, 1822

Boloria selene ([SCHIFFERMÜLLER], 1775) Small Pearl-bordered Fritillary, **NT** Plate 2.61

ssp. *selene* ([SCHIFFERMÜLLER], 1775)[1] (BM: R)

ssp. *insularum* (HESLOP HARRISON, 1937)[2] (BM: R)

Boloria euphrosyne (LINNAEUS, 1758) Pearl-bordered Fritillary, **EN**, **LP(P)** (BM and Ire: R) Plate 2.62

*Issoria lathonia** (LINNAEUS, 1758) Queen of Spain Fritillary (BM and Ire: I, B?) Plate 2.63

Argynnis adippe ([SCHIFFERMÜLLER], 1775) High Brown Fritillary, **CR**, **LP(F)** Plate 2.64

ssp. *vulgoadippe* VERITY, 1929 (BM: R)

Argynnis aglaja (LINNAEUS, 1758) Dark Green Fritillary, **LC** Plate 2.65

ssp. *aglaja* (LINNAEUS, 1758)[1] (BM and Ire: R)

ssp. *scotica* WATKINS, 1923[2] (BM: R)

unique island race (Heslop Harrison, 1945c)[3] (BM: R)

Argynnis paphia (LINNAEUS, 1758) Silver-washed Fritillary, LC (BM and Ire: R) Plate 2.66

Subfamily Melitaeinae REUTER, 1896

Euphydryas aurinia (ROTTEMBURG, 1775) Marsh Fritillary, **VU** Plate 2.67

ssp. *aurinia* (ROTTEMBURG, 1775)[1] (= *anglicana* FRUHSTORFER, 1916) (BM: R)

ssp. *hibernica* (BIRCHALL, 1873)[2] (Ire: R)

ssp. *scotica* (ROBSON, 1880)[3] (BM: R)

Melitaea cinxia (LINNAEUS, 1758) Glanville Fritillary, **EN**, **LP(P)** (BM: R) Plate 2.68

Melitaea athalia (ROTTEMBURG, 1775) Heath Fritillary, **EN**, **LP(F)** (BM: R; Ire: ? E; Lavery, 1989) Plate 2.69

Subfamily Satyrinae BOISDUVAL, [1833]

Pararge aegeria (LINNAEUS, 1758) Speckled Wood, **LC** Plate 2.70
- ssp. *tircis* (GODART, 1821)[1] (BM and Ire: R)
- ssp. *oblita* HESLOP HARRISON, 1949[2] (BM: R)
- ssp. *insula* HOWARTH, 1971[3] (BM: R)

Lasiommata megera (LINNAEUS, 1767) Wall, **NT** Plate 2.71
- ssp. *megera* (LINNAEUS, 1767)[1] (BM and Ire: R)
- ssp. *caledonia* VERITY, 1911[2] (BM: R)

Erebia epiphron (KNOCH, 1783) Small Mountain Ringlet, **NT**, **LP(P)** Plate 2.72
- ssp. *aetherius* f. *nelamus* (Ire: ? E; Redway, 1981)
- ssp. *mnemon* (HAWORTH, 1812)[1] (BM: R)
- ssp. *scotica* COOKE, 1943[2] (BM: R)

Erebia aethiops (ESPER, 1777) Scotch Argus, **LC** Plate 2.73
- ssp. *aethiops* (ESPER, 1777)[1] (BM: R)
- ssp. *caledonia* VERITY, 1911[2] (BM: R)

Erebia ligea (LINNAEUS, 1758) Arran Brown (BM: ? E) Plate 2.74

Melanargia galathea (LINNAEUS, 1758) Marbled White, **LC** Plate 2.75
- ssp. *serena* VERITY, 1913 (BM: R)

Hipparchia semele (LINNAEUS, 1758) Grayling, **VU** Plate 2.76
- ssp. *semele* (LINNAEUS, 1758)[1] (BM: R)
- ssp. *thyone* (THOMPSON, 1944)[2] (BM: R)
- ssp. *scota* (VERITY, 1911)[3] (= *atlantica* (HESLOP HARRISON, 1946) (BM: R)
- ssp. *clarensis* DE LATTIN, 1952[4] (Ire: R)
- ssp. *hibernica* HOWARTH, 1971[5] (Ire: R)

Pyronia tithonus (LINNAEUS, 1771) Gatekeeper, **LC** Plate 2.77
- ssp. *tithonus* (LINNAEUS, 1771)[1] (BM: R)
- ssp. *britanniae* (VERITY, 1915)[2] (Ire: R)

Maniola jurtina (Linnaeus, 1758) Meadow Brown, **LC** Plate 2.78
- ssp. *jurtina* (LINNAEUS, 1758)[1] (BM: R)
- ssp. *insularis* THOMSON, 1969[2] (BM: R)
- ssp. *iernes* GRAVES, 1930[3] (Ire: R)
- ssp. *cassiteridum* GRAVES, 1930[4] (BM: R)
- ssp. *splendida* WHITE, 1871[5] (BM: R)

Aphantopus hyperantus (LINNAEUS, 1758) Ringlet, **LC** Plate 2.79
- ssp. *hyperantus* (LINNAEUS, 1758)[1] (BM and Ire: R)
- unnamed Hebridean and Scottish race north of 56° N[2] (Dennis, 1977) (BM: R)

Coenonympha pamphilus (LINNAEUS, 1758) Small Heath, **NT** Plate 2.80
- ssp. *pamphilus* (LINNAEUS, 1758)[1] (BM and Ire: R)
- ssp. *rhoumensis* HESLOP HARRISON, 1948[2] (BM: R)
- unique island race (Heslop Harrison, 1950b)[3] (BM: R)

Coenonympha tullia (MÜLLER, 1764) Large Heath, **VU**, **LP(P)** Plate 2.81
ssp. *scotica* STAUDINGER, 1901[1] (BM and Ire: R)
ssp. *polydama* (HAWORTH, 1803)[2] (BM: R)
ssp. *davus* (FABRICIUS, 1777)[3] (BM: R)

Subfamily Danainae BOISDUVAL, 1833
*Danaus plexippus** (LINNAEUS, 1758) Monarch (BM and Ire: I) Plate 2.82
Danaus chrysippus (LINNAEUS, 1758) Plain Tiger (BM: I) Plate 2.83

We have not illustrated sexual dimorphism in species and would recommend those unfamiliar with the species, their appearance and biology to refer to the beautifully illustrated and informative book by Thomas and Lewington (2010). Scientific details on species are documented in Emmet and Heath (1989, paperback edition 1990) and tabulated in Dennis (2010). During the past 100 years or so, British butterflies have been in dramatic decline (Thomas and Lewington, 2010). The current situation is effectively summarized in 'A new Red List of British butterflies' (Fox *et al.*, 2011). For policies on collecting, breeding and photography, see the Butterfly Conservation UK website: http://butterfly-conservation.org/files/bc-policy_collecting-breeding-photography-2010.pdf

Note

[1] All species recorded in Appendix 1 have been recorded in Britain, Ireland and the smaller islands, including the Isle of Man and the Channel Islands.
[2] See Sattler and Tremewan (2009) for species attributed entirely to SCHIFFERMÜLLER by Kudrna *et al.* (2015).

Appendix 2. Butterflies Occurring on the British and Irish Islands as Adventives (Accidental and Deliberate Introductions)[1]

Abbreviations and notes

R, Rare immigrant
D, Deliberate introduction
A, Accidental introduction (e.g., escape from captivity)
?, status (identification and mode of entry) uncertain
Note: a number of these species possibly occur as rare immigrants.
Period noted: 17, 1700–1799; 18, 1800–1899; 19a, 1900–1949; 19b, 1950–1999; 20, 2000 to date.

Superfamily HESPERIOIDEA

Family Hesperiidae

Subfamily Hesperiinae
Hylephila phyleus (DRURY, 1773) Fiery Skipper (A, 18)
Saliana longirostris (SEPP, [1840]) Shy Saliana (A, 19a)

Subfamily Pyrginae
Carcharodus alceae (ESPER, 1780) Mallow Skipper (A, 19a)
Pyrgus armoricanus (OBERTHÜR, 1910) Oberthür's Grizzled Skipper (A, 18)

Superfamily PAPILIONOIDEA

Family Papilionidae

Subfamily Parnassiinae

Parnassius apollo (LINNAEUS, 1758) Apollo (?R but also A and D, 18, 19a, 19b)

Parnassius phoebus (FABRICIUS, 1793) Small Apollo (A, 18)

Subfamily Zerynthiinae

Zerynthia rumina (LINNAEUS, 1758) Spanish Festoon (A, 18)

Zerynthia polyxena ([SCHIFFERMÜLLER], 1775) (D or A, 18)

Subfamily Papilioninae

Papilio glaucus LINNAEUS, 1758 Tiger Swallowtail (A, 19a)

Papilio demetrius CRAMER, 1782 Black Swallowtail (A, ?)

Family Pieridae

Subfamily Coliadinae

Colias palaeno (LINNAEUS, 1761) Moorland Clouded Yellow (A, 18, 19a)

Subfamily Pierinae

Euchloe crameri (BUTLER, 1869) Butler's Dappled White (A or R, 18, 19a), or *E. ausonia* (HÜBNER, [1804]), not *E. simplonia* (FREYER) as in Emmet and Heath (1989)

Family Lycaenidae

Subfamily Theclinae

Rapala manea (ssp. *schistacea*) (MOORE, [1881]) Slate Flash (A, 19a)

Strymon melinus HÜBNER, 1818 Grey Hairstreak (A, 18)

Satyrium spini ([SCHIFFERMÜLLER], 1775) Blue-spot Hairstreak (?A, 18)

Satyrium ilicis (ESPER, 1779) Ilex Hairstreak (?A, 18)

Subfamily Lycaeninae

Lycaena alciphron (ROTTEMBURG, 1775) Purple-shot Copper (A, 18)

Subfamily Polyommatinae

Leptotes pirithous (LINNAEUS, 1767) Lang's Short-tailed Blue (A or R, 19a)

Polyommatus dorylas ([SCHIFFERMÜLLER], 1775) Turquoise Blue (? A, 17, 18, 19a)

Glaucopsyche alexis (PODA, 1761) Green-underside Blue (A or R, 19a)

Family Nymphalidae

Subfamily Apaturinae
Apatura ilia ([SCHIFFERMÜLLER], 1775) Lesser Purple Emperor (? A, 18)

Subfamily Nymphalinae
Junonia villida (FABRICIUS, 1787) Albin's Hampstead Eye (? A, 17)
Junonia oenone (LINNAEUS, 1758) Blue Pansy (A, 19b)
Junonia lemonias (LINNAEUS, 1758) Lemon Pansy (A, 19b)
Colobura dirce (LINNAEUS, 1758) Zebra (A, 19a)
Hypanartia lethe (FABRICIUS, 1793) Small Brown Shoemaker (A, 19b)
Vanessa indica (HERBST, 1794) Indian Red Admiral (A, 19b)

Subfamily Heliconiinae
Dryas julia FABRICIUS, 1775 Julia (A, 19a)
Heliconius charitonius (LINNAEUS, 1767) (A, ?)
Boloria dia (LINNAEUS, 1767) Weaver's Fritillary (A, D, 18, 19a)
Argynnis cybele (FABRICIUS, 1787) Great Spangled Fritillary (A, 18), not *A. aphrodite* as in Emmet and Heath (1989, 1991)
Argynnis niobe (LINNAEUS, 1758) Niobe Fritillary (possibly extinct resident, but also D or R, 18)
Argynnis pandora ([SCHIFFERMÜLLER], 1775) Mediterranean Fritillary (A or R, 19a, 19b)
Phalanta phalantha (DRURY, [1773]) Common Leopard (A, ?)

Subfamily Melitaeinae
Melitaea didyma (ESPER, 1779) Spotted Fritillary (A, 19b)

Subfamily Satyrinae
Lasiommata maera (LINNAEUS, 1758) Large Wall (A, 19a)
Erebia albergana (DE PRUNNER, 1798) Almond-eyed Ringlet (? A, 18, or extinct resident?)
Hipparchia fagi (SCOPOLI, 1763) Woodland Grayling (A, 19a)
Chazara briseis (LINNAEUS, 1764) Hermit (? A, 18)
Arethusana arethusa ([SCHIFFERMÜLLER], 1775) False Grayling (A or R, 19b)

Subfamily Morphinae
Opsiphanes tamarindi FELDER & FELDER, 1861 Tamarindi Owlet (A, 19a, 19b)
Opsiphanes cassiae (LINNAEUS, 1758) Cassia's Owl-butterfly (A, 19b)
Caligo illioneus (CRAMER, 1775) Illioneus Giant Owl (A, 19a)

Note

[1] Sources: Anonymous (1992, [photograph of *Junonia lemonias*]); Dennis and Shreeve, 1996; Emmet and Heath (1989, 1991); Knill-Jones and Angell (1996); Mead-Briggs and Eales (2010).
https://en.wikipedia.org/wiki/List_of_butterflies_of_Great_Britain
http://www.ukbutterflies.co.uk/taxonomy.php

Appendix 3. List of Butterflies on British and Irish Offshore Islands

The following list records the butterflies observed on British islands, Irish islands, the Isle of Man and Channel Islands as of 10 September 2017. Analyses reported in the book were conducted on the file as of 1 February 2017. No attempt has been made to distinguish breeding records from those of vagrants. The quality of data is often inadequate for this purpose and the status of insects on islands can readily change with time, especially for small islands. The islands have been placed in regional groups and references for the butterflies on islands in each group are indicated at the end of the group. These references include the personal communications listed in Appendix 4 as well as the published works tabulated in the main bibliography. Also, attention is drawn to a number of islands in each regional group which lack records and which could be usefully surveyed.

A. Abbreviations and Notes for the Island Records

Various symbols have been used to designate the status of species and records for islands:

Superscript numbers immediately following the name of the species indicate the form of so-called subspecies on islands named in the main checklist (Appendix 1); when the number is placed in parentheses the status of the form has not been formally designated. In some cases (e.g. *Lasiommata megera caledonia*) insufficient distributional details exist to designate island populations appropriately.

The letter D against a record indicates that the species has been deliberately introduced.

A question mark (?) indicates that the record is in doubt.

An asterisk (*) on its own indicates that the butterfly species has only been recorded on the island prior to 1960; some of these older records may suggest that the species no longer exists on the island or, at very least, they are in need of confirmation.

A dagger (†) on its own indicates that the species was recorded during the period up to 1995 – it was certainly recorded in the period 1960–1995 and may have also been recorded prior to 1960.

A double dagger (‡) on its own indicates that the species was recorded during the period 1996–2015.

A single and a double dagger together (†‡) indicate that the species was recorded both prior to 1996 (certainly in the period 1960–1996 and possibly also earlier) and in the period 1996–2015.

An asterisk plus a double dagger (*‡) indicate that the species was recorded prior to 1960 and also in the period 1996–2015, but not in the period 1960–1995.

Unfortunately it has not been possible to determine which species with the single-dagger symbol were also present on the island prior to 1960; the likelihood is that the majority of them were, although there will probably have been some instances of colonization since that year. The reader should be aware, after reading this book, that there is a continual turnover of species on islands; individual species come, go and return. This was not how records of butterfly species for British islands were envisaged prior to 1996.

References are given to the National Grid: two letters (one letter for Ireland) designating the 100 km squares, and two numbers the 10 km squares. Islands larger than one 10 km square are referenced by their midpoints.

Also, notes are added when butterflies on any particular island were counted under the Butterfly Monitoring Scheme (BMS) or the Wider Countryside Butterfly Survey (WCBS) and the years when this was done. The spread (distribution) of islands (midpoints) is illustrated in Fig. App. 3.1.

B. Islands off the French Coast

Channel Islands

ALDERNEY [WA50]: *P. machaon*[2]*‡; *I. podalirius**; *C. hyale*†; *C. croceus*†‡; *G. rhamni*[1]†‡; *P. brassicae*†‡; *P. rapae*†‡; *P. napi*[1]†‡; *P. daplidice*‡; *A. cardamines*[1]†‡; *C. rubi*†‡; *F. quercus**‡; *S. w-album*†‡; *L. phlaeas*[1]†‡; *L. boeticus*‡ (1 in 2010); *P. argus*[1]†; *A. agestis*†‡; *P. icarus*[1]†‡; *C. argiolus*[1]†‡; *V. atalanta*†‡; *V. cardui*†‡; *A. urticae*†‡; *N. polychloros*†‡; *A. io*†; *P. c-album*†‡; *I. lathonia*‡ (1 in 2010); *A. aglaja*[1]†‡; *A. paphia**; *M. cinxia*†‡; *P. aegeria*[1]†‡; *L. megera*[1]†‡; *H. semele*[1]†‡; *P. tithonus*[1]†‡; *M. jurtina*[1]†‡; *A. hyperantus*[1]‡; *C. pamphilus*[1]†‡. BMS (5 transects) 2008–present (4), 2011–present (1).

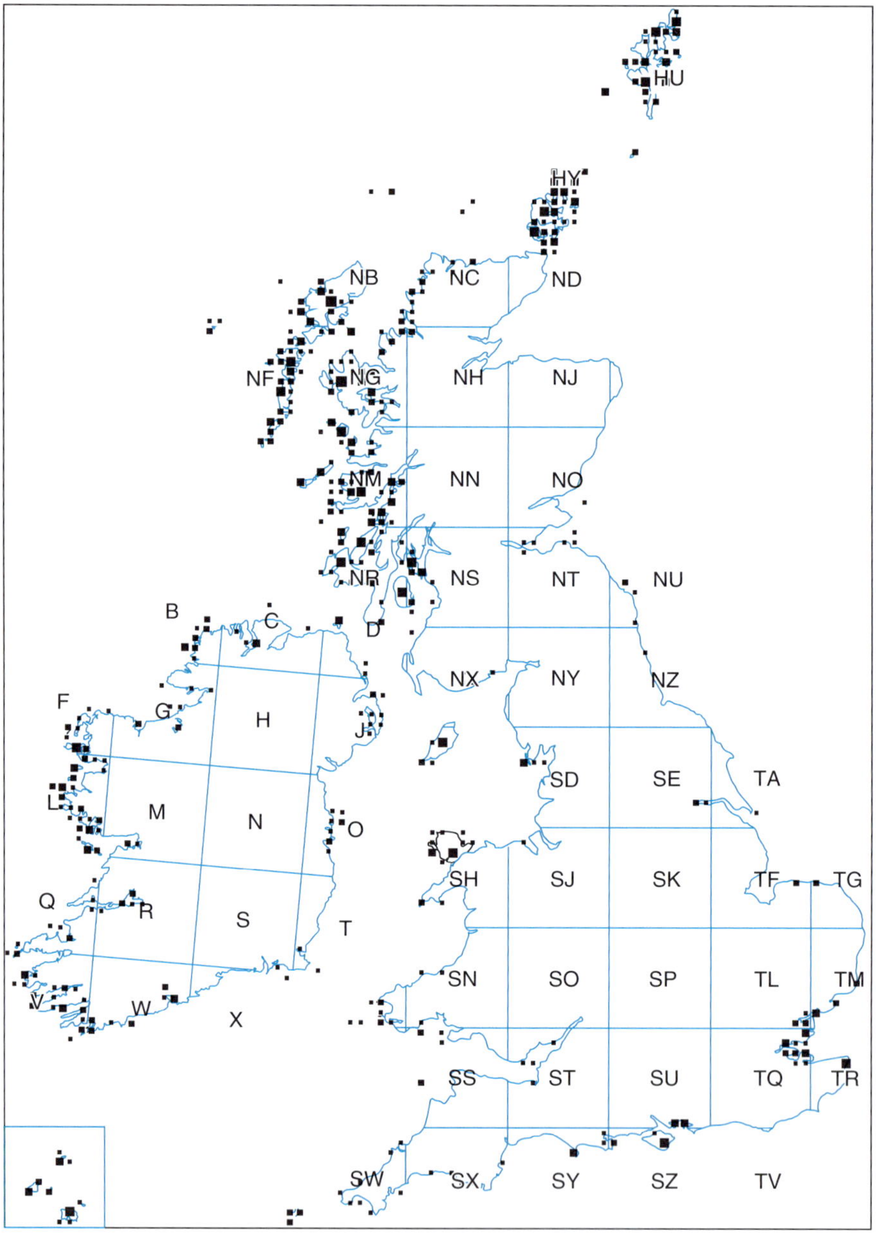

Fig. App. 3.1. The distribution of offshore islands around the coasts of the Britain, Ireland, the Isle of Man and the Channel Islands. Each dot indicates the position of an island, the symbol being relative to the size (order of magnitude) of each island. The British Isles and Ireland are divided up by their respective ordnance grid networks.

HERM [WV38]: *C. croceus**‡; *G. rhamni*[1]†; *P. brassicae*†‡; *P. rapae*†‡; *P. napi*[1]†‡; *P. daplidice*‡; *C. rubi*†‡; *L. phlaeas*[1]†‡; *P. argus*[1]†; *P. icarus*[1]†‡; *C. argiolus*[1]†‡; *V. atalanta*†‡; *V. cardui*†‡; *A. urticae*†‡; *N. polychloros*‡; *A. io*‡; *P. c-album*‡; *A. aglaja*[1]†; *M. cinxia*†; *P. aegeria*[1]†‡; *L. megera*[1]†‡; *H. semele*[1]†‡; *P. tithonus*[1]†‡; *M. jurtina*[1]†‡; *C. pamphilus*[1]†‡.

JETHOU [WV38]: *P. brassicae**; *P. rapae**; *L. phlaeas*[1]‡; *P. argus*[1]†; *P. icarus*[1]‡; *C. argiolus*[1]*; *V. atalanta*‡; *V. cardui*‡; *A. aglaja*[1]†; *P. aegeria*[1]‡; *L. megera*[1]‡; *H. semele*[1]‡; *P. tithonus*[1]‡; *M. jurtina*[1]*; *C. pamphilus*[1]*.

LES HOUMETS (OMPTOLLE) [WV38]: *P. machaon*[2]*‡.

SARK [WV47]: *P. machaon*[2]†‡; *I. podalirius**; *C. croceus*†‡; *G. rhamni*[1]†‡; *P. brassicae*†‡; *P. rapae*†‡; *P. napi*[1]†‡; *A. cardamines*[1]*; *C. rubi*†‡; *F. quercus*‡; *L. phlaeas*[1]†‡; *L. boeticus*†‡; *P. argus*[1]†‡; *A. agestis*†‡; *P. icarus*[1]†‡; *C. argiolus*[1]†‡; *L. camilla*‡; *V. atalanta*†‡; *V. cardui*†‡; *A. urticae*†‡; *N. polychloros*†‡; *N. antiopa*‡ (1 in 2007); *A. io*†‡; *P. c-album*†‡; *B. selene*[1]†; *I. lathonia**; *A. aglaja*[1]†‡; *A. paphia**; *M. cinxia*†; *P. aegeria*[1]†‡; *L. megera*[1]†‡; *H. semele*[1]†‡; *P. tithonus*[1]†‡; *M. jurtina*[1]†‡; *C. pamphilus*[1]†‡; *D. plexippus*†.

BRECQHOU [WV47]: *P. brassicae*‡; *P. rapae*‡; *L. phlaeas*[1]‡; *P. argus*[1]‡; *A. agestis*‡; *P. icarus*[1]‡; *V. atalanta*‡; *V. cardui*‡; *A. urticae*‡; *A. io*‡; *A. aglaja*[1]‡; *P. aegeria*[1]‡; *L. megera*[1]‡; *H. semele*[1]‡; *P. tithonus*[1]‡; *M. jurtina*[1]‡.

GUERNSEY [WV27]: *P. machaon*[2]†‡; *I. podalirius**; *C. hyale*†; *C. croceus*†‡; *G. rhamni*[1]†‡; *P. brassicae*†‡; *P. rapae*†‡; *P. napi*[1]†‡; *P. daplidice**; *A. cardamines*[1]†‡; *C. rubi*†‡; *F. quercus*†‡; *L. phlaeas*[1]†‡; *L. tityrus*†; *L. boeticus*†‡; *P. argus*[1]†; *A. agestis*†‡ (4 in 2014); *P. icarus*[1]†‡; *C. argiolus*[1]†‡; *V. atalanta*†‡; *V. cardui*†‡; *A. urticae*†‡; *N. polychloros*†‡; *N. antiopa**; *A. io*†‡; *P. c-album*†‡; *I. lathonia**‡; *A. aglaja*[1]†; *M. cinxia*†‡; *P. aegeria*[1]†‡; *L. megera*[1]†‡; *H. semele*[1]†‡; *P. tithonus*[1]†‡; *M. jurtina*[1]†‡; *C. pamphilus*[1]*; *D. plexippus**. WCBS 2009.

LIHOU [WV27]: *C. croceus*‡; *P. brassicae*‡; *P. rapae*‡; *P. napi*[1]‡; *L. phlaeas*[1]‡; *A. agestis* (1 in 2015); *P. icarus*[1]‡; *V. atalanta*‡; *V. cardui*‡; *A. urticae*‡; *A. io*‡; *P. aegeria*[1]‡; *L. megera*[1]‡; *H. semele*[1]‡; *P. tithonus*[1]‡; *M. jurtina*[1]‡; *C. pamphilus*[1]‡.

JERSEY [WV64]: *H. morpheus*†; *T. lineola*†‡; *O. sylvanus*†‡; *P. malvae**; *P. machaon*[2]†‡; *I. podalirius*‡; *L. sinapis*†; *C. hyale*†‡; *C. croceus*†‡; *G. rhamni*[1]†‡; *G. cleopatra*†; *A. crataegi**; *P. brassicae*†‡; *P. rapae*†‡; *P. napi*[1]†‡; *P. daplidice**; *A. cardamines*[1]†‡; *C. rubi*†‡; *F. quercus*†‡; *S. w-album*†‡; *L. phlaeas*[1]†‡; *L. dispar*[2]*(?); *L. boeticus*†‡; *C. argiades**; *A. agestis*†‡; *P. icarus*[1]†‡; *C. semiargus**‡ (1 in 2002); *C. argiolus*[1]†‡; *L. camilla*†‡; *A. iris**‡ (1 in 2014); *V. atalanta*†‡; *V. cardui*†‡; *A. urticae*†‡; *N. polychloros*†‡; *N. antiopa*†; *A. io*†‡; *P. c-album*†‡; *A. levana*‡; *B. euphrosyne**; *I. lathonia**‡; *A. aglaja*[1]*; *A. paphia**; *M. cinxia*†; *P. aegeria*[1]†‡; *L. megera*[1]†‡; *M. galathea*‡ (1 in 1999); *H. semele*[1]†‡; *P. tithonus*[1]†‡; *M. jurtina*[1]†‡; *A. hyperantus*[1]*‡; *C. pamphilus*[1]†‡. BMS (42 transects) 2004–2006 (1), 2004–2007 (5), 2004–2008 (2), 2004–2009 (4), 2004–2010 (2), 2004–present (12), 2005–present (2), 2007–2010 (1), 2007–present (1), 2010–present (1) and 2015–present (11).

No records for Burhau, Raz and Les Casquets [WA51] or smaller islands around Alderney; Crevichon near Herm [WV38]; L'Islet, St. Aubin's [WV65], La Motte [WV64], Ile de Guerdain [WV54], Les Ecréhous [WV76] and Les Minquiers [WV63] near Jersey.

References

Austin (2016), Baker (1900), Blyth (1901), Brereton (2016), Bryan (1987), Coney (1904), Freeman (1976), Higgs (1986), Hull (1996), Kaye (1922), King (1922, 1923), Lawlor (2011), Le Quesne (1945, 1946, 1947, 1973), Long (1970, 1987, 2010, 2016), Long and Austin (1986), Lowe (1904, 1911), Luff (1873, 1874, 1882, 1886, 1890a, 1890b, 1893a, 1893b, 1898, 1900a, 1900b, 1905a, 1907), McGowan (2017), Moon (1994), Newman (1931), Peet (1987), Piquet (1873), Riley (1975), Shayer (1964, 1967), Vicary (1994), Walker (1888, 1892, 1897), Wedd (2016).

C. Islands off the British mainland

Northumberland

LINDISFARNE [NU14]: *T. sylvestris*‡ (1 in 2005); *C. croceus*†‡; *P. brassicae*†‡; *P. rapae*†‡; *P. napi*[4]†‡; *A. cardamines*[2]†‡; *L. phlaeas*[1]†‡; *P. icarus*[1]†‡; *V. atalanta*†‡; *V. cardui*†‡; *A. urticae*†‡; *N. antiopa**?†; *A. io*†‡; *P. c-album*‡; *A. aglaja*[1]†‡; *P. aegeria*[1]‡; *L. megera*[(2)]‡; *H. semele*[1]†‡; *M. jurtina*[2]†‡; *A. hyperantus*[1]†‡; *C. pamphilus*[1]†‡. BMS 1977–present.

NORTHERN HARES [NU23]: *P. napi*[4]†; *V. atalanta*†.

LONGSTONE [NU23]: *V. atalanta*†.

NORTH WAMSES [NU23]: *V. atalanta*†; *A. urticae*†.

SOUTH WAMSES [NU23]: *A. cardamines*[2]†; *A. urticae*†.

BROWNSMAN [NU23]: *P. brassicae*†; *P. rapae*†‡; *P. napi*[4]†; *A. cardamines*[2]‡; *L. phlaeas*[1]‡; *P. icarus*[1]†; *V. atalanta*†; *V. cardui*†; *A. urticae*†; *A. io*†; *A. aglaja*[1]‡; *L. megera*[(2)]‡; *H. semele*[1]†; *M. jurtina*[2]†.

STAPLE [NU23]: *P. brassicae*‡; *P. rapae*†‡; *P. napi*[4]‡; *V. atalanta*†‡; *V. cardui*†‡; *A. urticae*†; *A. io*†‡.

INNER FARNE [NU23]: *P. brassicae*†‡; *P. rapae*†‡; *P. napi*[4]†‡; *A. cardamines*[2]‡; *L. phlaeas*[1]‡; *P. icarus*[1]*; *V. atalanta*†‡; *V. cardui*†‡; *A. urticae*†‡; *A. io*†‡; *P. c-album*‡; *A. aglaja*[1]†‡; *P. aegeria*[1]‡; *L. megera*[(2)]†‡; *H. semele*[1]†; *M. jurtina*[2]†‡; *A. hyperantus*[1]‡. BMS from 2016.

WEST WIDEOPEN [NU23]: *V. atalanta*†; *V. cardui*‡; *A. urticae*†.

COQUET [NU20]: *V. atalanta*‡; *V. cardui*‡.

ST MARY'S or BAIT [NZ37]: *T. sylvestris*‡; *P. brassicae*‡; *P. rapae*‡; *P. napi*[4]‡; *A. cardamines*[2]‡; *L. phlaeas*[1]‡; *P. icarus*[2]‡; *V. atalanta*‡; *V. cardui*‡; *A. urticae*‡; *A. io*‡; *P. c-album*‡; *P. aegeria*[1]‡; *L. megera*[(2)]‡; *M. jurtina*[2]‡; *A. hyperantus*[1]‡; *C. pamphilus*[1]‡.

No records from other Farne islands such as Knivestone, Big Harcar, Megstone, Crumstone and East Wideopen [NU23]

References

Castle-Russell (1932), Clynes (1996), Dunn (1986), Dunn and Parrack (1986), Hodges (1996), McElwee (1996), Parrack (1986, 1994), Perkins (2016), Stringer and Wight (1994), Tanner (1995), Walton (1995), Woof (1958).

Yorkshire

SPURN POINT [TA41] ("fossil island"): *T. sylvestris*‡; *T. lineola*‡; *O. sylvanus*‡; *C. croceus*‡; *G. rhamni*[1]‡; *P. brassicae*‡; *P. rapae*‡; *P. napi*[2]‡; *A. cardamines*[2]‡; *C. rubi*‡; *L. phlaeas*[1]‡; *A. agestis*‡; *P. icarus*[1]‡; *C. argiolus*[2]‡; *V. atalanta*‡; *V. cardui*‡; *A. urticae*‡; *A. io*‡; *P. c-album*‡; *P. aegeria*[1]‡; *L. megera*[1]‡; *P. tithonus*[2]‡; *M. jurtina*[2]‡; *A. hyperantus*[1]‡; *C. pamphilus*[1]‡.

No records for Whitton Island [SE82] and Read's Island [SE92] in the Humber Estuary.

References

Clough (2017), Davison (2017), Spence (2008).

Norfolk, Suffolk, Essex and Kent

SCOLT HEAD [TF84]: *T. sylvestris*‡; *T. lineola*‡; *O. sylvanus*‡; *C. croceus*‡; *G. rhamni*[1]‡; *P. brassicae*‡; *P. rapae*‡; *P. napi*[2]‡; *C. rubi*‡; *L. phlaeas*[1]‡; *A. agestis*‡; *P. icarus*[1]‡; *C. argiolus*[2]‡; *V. atalanta*‡; *V. cardui*‡; *A. urticae*‡; *A. io*‡; *P. c-album*‡; *A. aglaja*[1]‡; *P. aegeria*[1]‡; *L. megera*[1]‡; *H. semele*[1]‡; *P. tithonus*[2]‡; *M. jurtina*[2]‡; *A. hyperantus*[1]‡; *C. pamphilus*[1]‡. BMS 2009–present.

BLAKENEY [TG04]: *T. sylvestris*‡; *T. lineola*‡; *G. rhamni*[1]‡; *P. brassicae*‡; *P. rapae*‡; *P. napi*[2]‡; *L. phlaeas*[1]‡; *A. agestis, P. icarus*[1]‡; *V. atalanta; V. cardui; A. urticae*‡; *A. io*‡; *P. c-album*‡; *A. aglaja*[1]‡; *L. megera*[1]‡; *H. semele*[1]‡; *P. tithonus*[2]‡; *M. jurtina*[2]‡; *A. hyperantus*[1]‡. BMS 2008–present.

HAVERGATE [TM44]: *T. lineola*‡; *P. brassicae*‡; *P. rapae*‡; *P. napi*[2]‡; *L. phlaeas*[1]‡; *P. icarus*[1]‡; *V. atalanta*‡; *V. cardui*‡; *P. c-album*‡; *P. aegeria*[1]‡; *L. megera*[1]‡; *P. tithonus*[2]‡; *M. jurtina*[2]‡; *C. pamphilus*[1]‡.

ORFORDNESS [TM44]: *T. sylvestris*‡; *T. lineola*‡; *G. rhamni*[1]‡; *P. brassicae*‡; *P. rapae*‡; *P. napi*[2]‡; *C. rubi*‡; *L. phlaeas*[1]‡; *A. agestis*‡; *P. icarus*[1]‡; *V. atalanta*‡; *V. cardui*‡; *A. io*‡; *P. c-album*‡; *P. aegeria*[1]‡; *L. megera*[1]‡; *H. semele*[1]‡; *P. tithonus*[2]‡; *M. jurtina*[2]‡; *A. hyperantus*[1]‡; *C. pamphilus*[1]‡.

PEWIT (1) [TM22] *P. tithonus*[2]‡.

SKIPPER'S [TM22]: *T. lineola*‡; *O. sylvanus*‡; *G. rhamni*[1]‡; *P. brassicae*‡; *P. rapae*‡; *P. napi*[2]‡; *L. phlaeas*[1]‡; *P. icarus*[1]‡; *C. argiolus*[2]‡; *V. atalanta*‡; *V. cardui*‡; *A. urticae*‡; *A. io*‡; *P. c-album*‡; *P. aegeria*[1]‡; *L. megera*[1]‡; *P. tithonus*[2]†; *M. jurtina*[2]‡; *A. hyperantus*[1]‡; *C. pamphilus*[1]‡.

MERSEA [TM01]: *T. sylvestris*‡; *T. lineola*‡; *O. sylvanus*‡; *C. croceus*‡; *G. rhamni*[1]‡; *P. brassicae*‡; *P. rapae*‡; *P. napi*[2]‡; *A. cardamines*[2]‡;

C. rubi‡; *F. quercus*‡; *L. phlaeas*[1]‡; *A. agestis*‡; *P. icarus*[1]‡; *C. argiolus*[2]‡; *L. camilla*‡ (1 in 2010); *V. atalanta*‡; *V. cardui*‡; *A. urticae*‡; *A. io*‡; *P. c-album*‡; *A. paphia*‡ (1 in 2013); *P. aegeria*[1]‡; *P. tithonus*[2]‡; *M. jurtina*[2]‡; *A. hyperantus*[1]‡; *C. pamphilus*[1]‡.

RAY [TM01]: *C. rubi*‡; *A. urticae*‡.

NORTHEY [TL90]: *T. sylvestris*‡; *P. brassicae*‡; *P. rapae*‡; *P. napi*[2]‡; *V. atalanta*‡; *A. io*‡; *P. c-album*‡; *P. aegeria*[1]‡; *P. tithonus*[2]‡; *M. jurtina*[2]‡.

OSEA [TL90]: *P. brassicae*‡; *P. icarus*[1]‡; *M. jurtina*[2]‡.

WALLASEA [TQ99]: *T. sylvestris*‡; *O. sylvanus*‡; *C. croceus*‡; *P. brassicae*‡; *P. rapae*‡; *P. napi*[2]‡; *A. cardamines*[2]‡; *L. phlaeas*[1]‡; *A. agestis*‡; *P. icarus*[1]‡; *V. atalanta*‡; *V. cardui*‡; *A. urticae*‡; *A. io*‡; *P. tithonus*[2]‡; *M. jurtina*[2]‡; *C. pamphilus*[1]‡.

FOULNESS [TR09]: *T. lineola*‡; *O. sylvanus*‡; *P. brassicae*‡; *P. rapae*‡; *P. napi*[2]‡; *A. cardamines*[2]‡; *L. phlaeas*[1]‡; *A. agestis*‡; *P. icarus*[1]‡; *V. atalanta*‡; *V. cardui*‡; *A. urticae*‡; *A. io*‡; *P. aegeria*[1]‡; *L. megera*[1]‡; *P. tithonus*[2]‡; *M. jurtina*[2]‡; *C. pamphilus*[1]‡. WCBS 2007–2008.

TWO TREE [TQ88]: *T. sylvestris*‡; *T. lineola*‡; *O. sylvanus*‡; *C. croceus*‡; *G. rhamni*[1]‡; *P. brassicae*‡; *P. rapae*‡; *P. napi*[2]‡; *A. cardamines*[2]‡; *L. phlaeas*[1]‡; *A. agestis*‡; *P. icarus*[1]‡; *C. argiolus*[2]‡; *V. atalanta*‡; *V. cardui*‡; *A. urticae*‡; *A. io*‡; *P. c-album*‡; *P. aegeria*[1]‡; *L. megera*[1]‡; *M. galathea*‡; *P. tithonus*[2]‡; *M. jurtina*[2]‡; *A. hyperantus*[1]‡; *C. pamphilus*[1]‡. BMS (2 transects) 1976–2015 and 2003 (only).

CANVEY [TQ78]: *T. sylvestris*‡; *T. lineola*‡; *O. sylvanus*‡; *P. malvae*‡ (1 in 2010); *P. machaon*[2]‡ (1 in 2013); *C. croceus*‡; *G. rhamni*[1]‡; *P. brassicae*‡; *P. rapae*‡; *P. napi*[2]‡; *A. cardamines*[2]‡; *C. rubi*‡; *L. phlaeas*[1]‡; *A. agestis*‡; *P. icarus*[1]‡; *C. argiolus*[2]‡; *V. atalanta*‡; *V. cardui*‡; *A. urticae*‡; *A. io*‡; *P. c-album*‡; *P. aegeria*[1]‡; *L. megera*[1]‡; *M. galathea*‡; *P. tithonus*[2]‡; *M. jurtina*[2]‡; *A. hyperantus*[1]‡; *C. pamphilus*[1]‡. BMS 2015–present.

SHEPPEY [TR06]: *T. sylvestris*†‡; *T. lineola*†‡; *O. sylvanus*†‡; *C. croceus*†‡; *P. brassicae*†‡; *P. rapae*†‡; *P. napi*[1]†‡; *A. cardamines*[2]†‡; *L. phlaeas*[1]†‡; *A. agestis*‡; *P. icarus*[1]†‡; *C. argiolus*[2]†‡; *V. atalanta*†‡; *V. cardui*†‡; *A. urticae*†‡; *N. polychloros**; *A. io*†‡; *P. c-album*†‡; *A. aglaja*[1]†, *P. aegeria*[1]†‡; *L. megera*[1]†‡; *M. galathea*†, *P. tithonus*[2]†‡; *M. jurtina*[2]†‡; *A. hyperantus*[1]‡; *C. pamphilus*[1]†‡. BMS 1977–1978; WCBS 2007–2010.

THANET [TR36] ("fossil island"): *T.sylvestris*‡; *T. lineola*‡; *O. sylvanus*‡; *P. machaon*[2]‡; *C. croceus*‡; *G. rhamni*‡; *P. brassicae*‡; *P. rapae*‡; *P. napi*[1]‡; *A. cardamines*[2]‡; *S. w-album*‡; *L. phlaeas*[1]‡; *L. boeticus*‡; *A. agestis*‡; *P. icarus*[1]‡; *C. argiolus*[2]‡; *V. atalanta*‡; *V. cardui*‡; *A. urticae*‡; *A. io*‡; *P. c-album*‡; *P. aegeria*[1]‡; *L. megera*[1]‡; *M. galathea*‡; *P. tithonus*[2]‡; *M. jurtina*[2]‡; *C. pamphilus*[1]‡.

No records for New, Horsey, Hedge-end and Sandcreek Salts [TM22], Rat, Pewit (2), Bonner's Saltings and Cobmarsh [TM01], Sunken and Great Cob [TL91], Pewet [TL90], Potton and Rushley [TQ98] and Hard [TQ88] in Essex, nor for Hoo Ness [TQ77], Nor and Millfordhope [TQ86], Oakham Marsh, Bee Ness, Darnet Ness, Burntwick and Greenborough/Slayhills [TQ87] and Fowley [TQ96] in Kent.

References

Brazil (2016), Easterbrook (2016), Fletcher (1901), Parker (2015), Philp (1993), Smith (2013, 2015), Solly (2014).

Sussex and Hampshire

THORNEY (including Pilsey island) [SU70]: *V. atalanta*‡; *A. io*‡; *P. aegeria*‡; *H. semele*[1]†.

HAYLING [SU70]: *T. sylvestris*†‡; *T. lineola*†‡; *H. comma*†; *O. sylvanus*†‡; *P. machaon*[2]‡ (1 in 2013); *C. croceus*†‡; *G. rhamni*[1]†‡; *P. brassicae*†‡; *P. rapae*†‡; *P. napi*[2]†‡; *A. cardamines*[2]†‡; *C. rubi*†; *F. quercus*†‡; *L. phlaeas*[1]†‡; *L. boeticus**; *C. minimus*‡; *A. agestis*†‡; *P. icarus*[1]†‡; *P. coridon*‡ (2 in 2009); *C. argiolus*[2]†‡; *L. camilla*†‡; *V. atalanta*†‡; *V. cardui*†‡; *A. urticae*†‡; *N. polychloros*‡ (1 in 2007); *A. io*†‡; *P. c-album*†‡; *A. aglaja*[1]†‡ (1 in 1996); *P. aegeria*[1]†‡; *L. megera*[1]†‡; *M. galathea*†‡; *H. semele*[1]†‡; *P. tithonus*[2]†‡; *M. jurtina*[2]†‡; *A. hyperantus*[1]‡; *C. pamphilus*[1]†‡. BMS 1989–2004.

PORTSEA [SU60]: *T. sylvestris*‡; *T. lineola*‡; *O. sylvanus*‡; *C. croceus*‡; *G. rhamni*[1]‡; *P. brassicae*‡; *P. rapae*‡; *P. napi*[2]‡; *A. cardamines*[2]‡; *C. rubi*‡; *L. phlaeas*[1]‡; *C. minimus*‡; *P. icarus*[1]‡; *C. argiolus*[2]‡; *V. atalanta*‡; *V. cardui*‡; *A. urticae*‡; *A. io*‡; *P. c-album*‡; *P. aegeria*[1]‡; *L. megera*[1]‡; *M. galathea*‡; *H. semele*[1]‡; *P. tithonus*[2]‡; *M. jurtina*[2]‡; *C. pamphilus*[1]‡.

WIGHT [SZ58]: *T. sylvestris*†‡; *T. lineola*†‡; *T. acteon**; *O. sylvanus*†‡; *E. tages*[1]†‡; *P. malvae*†‡; *P. machaon*[2]†‡; *I. podalirius*‡ (1 in 2003, 1 in 2011); *L. sinapis**; *C. hyale**‡ (1 in 2013); *C. croceus*†‡; *G. rhamni*[1]†‡; *A. crataegi**; *P. brassicae*†‡; *P. rapae*†‡; *P. napi*[2]†‡; *A. cardamines*[2]†‡; *C. rubi*†‡; *T. betulae*†‡; *F. quercus*†‡; *S. w-album*†‡; *L. phlaeas*[1]†‡; *L. boeticus*‡; *C. marshalli*‡ (1 in 2010); *C. minimus*†; *P. argus*[1]†; *A. agestis*†‡; *P. icarus*[1]†‡; *P. coridon*†‡; *P. bellargus*†‡; *C. semiargus**; *C. argiolus*[2]†‡; *H. lucina*†‡; *L. camilla*†‡; *A. iris*†‡; *V. atalanta*†‡; *V. cardui*†‡; *V. virginiensis**‡ (1 in 2004); *A. urticae*†‡; *N. polychloros*†‡ (3 in 2007); *N. antiopa*†‡; *A. io*†‡; *P. c-album*†‡; *B. selene*[1]†‡; *B. euphrosyne*†‡; *I. lathonia**‡ (1 in 2013); *A. adippe*†; *A. aglaja*[1]†‡; *A. paphia*†‡; *E. aurinia*[1]*‡; *M. cinxia*†‡; *P. aegeria*[1]†‡; *L. megera*[1]†‡; *M. galathea*†‡; *H. semele*[1]†‡; *P. tithonus*[2]†‡; *M. jurtina*[2]†‡; *A. hyperantus*[1]†‡; *C. pamphilus*[1]†‡; *D. plexippus*†‡, *D. chrysippus*‡. BMS (13 transects) 1988–present (2), 1989–present (2), 1992–1995, 1993–present, 1995–2007, 1996–present, 1997–2009, 1998–present, 1998–2013, 2001–2007 (2); WCBS (4 transects) 2009 (2), 2009–2010 and 2013–2015 (1) and 2014–2015 (1).

No records for Long, South Binness, Baker's, North Binness and Whale [SU60], or Gull [SZ49].

References

Annell (2016), Collins, R. in Masters (2000), Fearnehough (1937), Hardy (1995), Huggins (1972), Knill-Jones (1992, 1996, 2010), Redfern (1994), Russell (2014), Small and Taverner (1994).

Dorset

BROWNSEA [SZ08]: *T. sylvestris*†‡; *O. sylvanus*†‡; *E. tages*[1]†; *P. malvae*†; *C. croceus*†‡; *G. rhamni*[1]†‡; *P. brassicae*†‡; *P. rapae*†‡; *P. napi*[2]†‡; *A. cardamines*[2]†‡; *C. rubi*†‡ (1 in 2000); *F. quercus*†‡; *L. phlaeas*[1]†‡; *P. argus*[1]†; *A. agestis*‡ (1 in 1998); *P. icarus*[1]†‡; *C. argiolus*[2]†‡; *L. camilla*†‡; *V. atalanta*†‡; *V. cardui*†‡; *A. urticae*†‡; *N. polychloros*‡ (3 in 2007); A. *io*†‡; *P. c-album*†‡; *B. selene*[1]†; *A. aglaja*[1]†‡; *A. paphia*†‡; *P. aegeria*[1]†‡; *L. megera*[1]†‡; *M. galathea*†‡; *H. semele*[1]†‡; *P. tithonus*[2]†‡; *M. jurtina*[2]†‡; *A. hyperantus*[1]†‡; *C. pamphilus*[1]†‡.

GREEN [SZ08]: *C. croceus*‡; *G. rhamni*[1]‡; *P. rapae*‡; *P. napi*[2]‡; *A. cardamines*[2]‡; *F. quercus*‡; *L. phlaeas*[1]‡; *A. agestis*‡; *P. icarus*[1]‡; *C. argiolus*[2]‡; *V. atalanta*‡; *P. c-album*‡; *P. aegeria*[1]‡; *H. semele*[1]‡; *P. tithonus*[2]‡; *M. jurtina*[2]‡.

PORTLAND [SY67] ("fossil island"): *T. sylvestris*‡; *T. lineola*‡; *T. acteon*‡; *O. sylvanus*‡; *E. tages*[1]‡; *P. malvae*‡; *C. croceus*‡; *G. rhamni*[1]‡; *P. brassicae*‡; *P. rapae*‡; *P. napi*[2]‡; *A. cardamines*[2]‡; *L. phlaeas*[1]‡; *C. minimus*‡; *P. argus*[1]‡; *A. agestis*‡; *P. icarus*[1]‡; *P. coridon*‡; *P. bellargus*‡; *C. argiolus*[2]‡; *V. atalanta*‡; *V. cardui*‡; *A. urticae*‡; *N. polychloros*‡; *N. antiopa*‡; *A., io*‡; *P. c-album*‡; *A. aglaja*[1]‡; *A. paphia*‡; *P. aegeria*[1]‡; *L. megera*[1]‡; *M. galathea*‡; *H. semele*[1]‡; *P. tithonus*[2]‡; *M. jurtina*[2]‡; *A. hyperantus*[1]‡; *C. pamphilus*[1]‡; *D. plexippus*‡.

No records for Furzey, Drove and Green [SZ08], Pergin's [SY99] or Long/Round and Gigger's [SY98].

References

Cade (2008), George (2016), Shreeves (2016), Thomas and Webb (1984).

South Devon and south Cornwall

DRAKE'S [SX45]: *P. rapae*‡; *P. icarus*[1]‡; *V. atalanta*‡; *P. aegeria*[1]‡.

LOOE or ST GEORGE'S [SX25]: *O. sylvanus*†‡; *C. croceus*‡; *G. rhamni*[1]‡; *P. brassicae*†‡; *P. rapae*†‡; *P. napi*[2]‡; *L. phlaeas*[1]‡; *C. argiolus*[2]‡; *V. atalanta*†‡; *V. cardui*‡; *A. urticae*†‡; *A. io*‡; *P. c-album*‡; *A. paphia*‡; *P. aegeria*[1]†‡; *L. megera*[1]‡; *M. galathea*‡; *P. tithonus*[2]‡; *M. jurtina*[2]†‡; *A. hyperantus*[1]†‡. BMS 2013–present.

ST MICHAEL'S MOUNT [SW52]: *C. croceus*‡; *P. brassicae*†‡; *P. rapae*‡; *P. napi*[2]‡; *L. phlaeas*[1]‡; *P. icarus*[1]†‡; *C. argiolus*[2]†; *V. atalanta*†‡; *V. cardui*†‡; *A. urticae*‡; *P. aegeria*[1]‡; *L. megera*[1]‡; *P. tithonus*[2]‡; *M. jurtina*[2]‡.

No records for Ore Stone and Thatcher Rock [SX96], Burgh [SX64], Great Mew Stone [SX54], Gull Rock (Verryan) [SW93], Asparagus, Gull Rock (Kynance Cove) and Mullion [SW61] and St. Clement's [SW42].

References

Frost and Madge (1991), Goodere (2016), Spalding (1995), Spalding and Tremewan (1998), Summers (1976a), Tremewan (1995).

Isles of Scilly

RAT [SV91]: *V. atalanta*‡.

MENAWETHAN [SV91]: *P. icarus*[1]*; *M. jurtina*[4] (D)†.

GREAT ARTHUR [SV91]: *M. jurtina*[4]*.

GREAT INNISVOULS [SV91]: *L. phlaeas*[1]†; *P. icarus*[1]*; *M. jurtina*[4] (D)†.

GREAT GANILLY [SV91]: *P. rapae*†; *L. phlaeas*[1]†; *P. icarus*[1]†; *V. atalanta*†; *V. cardui*†; *A. io*†; *M. jurtina*[4]†.

ST MARTIN'S [SV91]: *C. croceus*†‡; *G. rhamni*[1]†‡ (1 in 2013); *P. brassicae*†‡; *P. rapae*†‡; *P. napi*[(2)]†; *L. phlaeas*[1]†‡; *L. boeticus*‡; *P. icarus*[1]†‡; *C. argiolus*[2]†‡; *V. atalanta*†‡; *V. cardui*†‡; *A. urticae*†‡; *A. io*†‡; *P. c-album*†‡; *P. aegeria*[3]†‡; *L. megera*[1]†; *M. jurtina*[4]†‡; *C. pamphilus*[1]†; *A. hyperantus*[1]†‡; *D. plexippus*‡. BMS 2012–present.

TEAN [SV91]: *P. rapae*†; *P. icarus*[3]†; *V. atalanta*†; *V. cardui**; *P. aegeria*[3]†; *M. jurtina*[4]†.

WHITE [SV91]: *C. croceus*‡; *L. phlaeas*[1]‡; *P. icarus*[1]‡; *A. urticae*‡; *P. aegeria*[3]†; *M. jurtina*[4]*‡.

ROUND [SV91]: *C. croceus**; *P. rapae**; *V. atalanta**; *V. cardui**.

ST MARY'S [SV81]: *C. hyale*†; *C. croceus*†‡; *G. rhamni*[1]†; *P. brassicae*†‡; *P. rapae*†‡; *P. napi*[(2)]†‡; *P. daplidice*†; *A. cardamines*[2]†; *L. phlaeas*[1]†‡; *L. boeticus*‡; *P. icarus*[1]†‡; *C. argiolus*[2]†‡; *V. atalanta*†‡; *V. cardui*†‡; *V. virginiensis*‡; *A. urticae*†‡; *N. polychloros*‡; *N. antiopa*‡ (1 in 2006); *A. io*†‡; *P. c-album*†‡; *I. lathonia*‡; *P. aegeria*[3]†‡; *L. megera*[1]†‡; *P. tithonus*[2]†‡; *M. jurtina*[4]†‡; *C. pamphilus*[1]†; *A. hyperantus*[1]‡; *D. plexippus*†‡.

SAMSON [SV81] *C. croceus*‡, *P. brassicae*‡; *P. rapae*†‡; *L. phlaeas*[1]‡; *P. icarus*[1]†‡; *V. atalanta**‡; *V. cardui**; *A. urticae*†‡; *A. io*‡; *P. tithonus*[2]‡; *M. jurtina*[4]†‡.

MERRICK [SV81]: *A. urticae*‡.

TRESCO [SV81]: *C. hyale**; *C. croceus*†‡; *G. rhamni*[1]*; *P. brassicae*†‡; *P. rapae*†‡; *P. napi*[(2)]†‡; *L. phlaeas*[1]†‡; *L. boeticus*‡; *P. icarus*[1]†‡; *C. argiolus*[2]†‡; *V. atalanta*†‡; *V. cardui*†‡; *A. urticae*†‡; *N. polychloros**; *A. io*†‡; *P. c-album*‡; *P. aegeria*[3]†‡; *P. tithonus*[2]†‡; *M. jurtina*[4]†‡; *A. hyperantus*[1]‡; *D. plexippus*†‡.

BRYHER [SV81]: *C. croceus*†; *G. rhamni*[1]*; *P. brassicae*†‡; *P. rapae*†‡; *P. napi*[(2)]*; *L. phlaeas*[1]†‡; *P. icarus*[1]†‡; *C. argiolus*[2]†; *V. atalanta*†‡; *V. cardui*†‡; *A. urticae*†‡; *A. io*†‡; *P. c-album*‡; *A. paphia*‡ (1 in 2008); *P. aegeria*[3]†‡; *L. megera*[1]‡; *M. jurtina*[4]†‡, *D. plexippus*‡.

ST HELEN'S [SV81]: *C. croceus**; *A. io*†; *M. jurtina*[4]*.

GUGH [SV80]: *C. croceus*‡; *P. rapae*‡; *P. napi*[(2)]†; *L. phlaeas*[1]‡; *P. icarus*[1]‡; *V. atalanta*‡; *V. cardui*‡; *A. urticae*‡; *A. io*‡; *P. aegeria*[3]†‡; *M. jurtina*[4]†‡; *D. plexippus*†‡.

ST AGNES [SV80]: *C. croceus*†‡; *G. rhamni*[1]†; *P. brassicae*†‡; *P. rapae*†‡; *P. napi*[(2)]†‡; *L. phlaeas*[1]†‡; *L. boeticus*‡; *P. icarus*[1]†‡; *C. argiolus*[2]†‡; *V. atalanta*†‡; *V. cardui*†‡; *V. virginiensis*‡; *A. urticae*†‡; *N. antiopa*‡; *A. io*†‡; *P. c-album*†‡, *P. aegeria*[3]†‡; *L. megera*[1]†; *P. tithonus*[2]‡; *M. jurtina*[4]†‡; *D. plexippus*†‡.

No records for Toll's, Ragged, Little Ganinick, Little Innisvouls, Nornour and Great Ganinick [SV91], Great Minalto, Mincarlo, Iliswilgig, Maiden Bower, Scilly Rock, Gweal, Northwethel and Men-a-vaur [SV81], Rosevean, Gorregan, Gilstone, Rosevear, Melledgan, Bishop Rock and Annet [SV80].

References

Adkin, R. (1892, 1894), Adkin, R.W. (1897, 1911), Agassiz (1982), Beavis (1976, 1995, 2004), Blair (1925), Bloomfield (1994), Clutterbuck (1940), Crewe (1877a, b), Dale (1890), Dannreuther (1933, 1944), Demuth (1974), Dobson (1981), Dowdeswell (1981), Dowdeswell and Ford (1948), Fassnidge (1933), Foggit (1981), Ford (1964), Frohawk (1935), Fryer (1926), Goodere (2016), Hale and Hicks (1995), Heckford (1978), Howarth (1971a), Huggins (1958, 1959), King (1938), Mere (1959, 1960), Moon (1994), Nicholson (1933), Norgate (1880), Richardson (1963), Richardson and Mere (1958), Simpson (1974), Smith (1933), Spalding (2017), Summers (1975a, 1976a, 1976b, 1977a), Sutton (1994), de Worms (1939).

North Cornwall and north Devon

LUNDY [SS14]: *T. sylvestris*†; *C. croceus*†‡; *G. rhamni*[1]*; *P. brassicae*†‡; *P. rapae*†‡; *P. napi*[2]†‡; *A. cardamines*[2]*‡; *C. rubi*†; *L. phlaeas*[1]†‡; *A. agestis*†; *P. icarus*[1]†; *C. argiolus*[2]†; *V. atalanta*†‡; *V. cardui*†‡; *A. urticae*†‡; *N. polychloros**; *A. io*†‡; *P. c-album*†; *B. selene*[1]*; *B. euphrosyne**; *A. aglaja*[1]†; *P. aegeria*[1]†‡ (1 in 1997); *L. megera*[1]†; *M. galathea*‡ (1 in 2015); *H. semele*[1]†‡ (1 in 2004); *P. tithonus*[2]†‡; *M. jurtina*[4]†‡; *A. hyperantus*[1]†‡; *C. pamphilus*[1]†‡; *D. plexippus*‡.

No records for the Brisons [SW33], Godrevy [SW54], Gulland Rock [SW87], Newland and the Mouls [SW98] and Little Shutter Rock, Rat and Mouse around Lundy [SS14].

References

Bristow (2016), Bristow, Mitchell and Bolton (1993), Bristowe (1939b), Carpenter (1943), Dymond (1974, 1975), Jones (1968), Longstaff (1907), Martineau (1894), Meakin (1995), Smith (1958), Tutt (1894), Walmsley and Warlsley (1977), Wollaston (1845).

Bristol Channel

STERT [ST24]: *P. rapae*‡; *L. phlaeas*[1]‡; *V. atalanta*‡; *M. jurtina*[2]‡.

STEEP HOLM [ST26]: *C. croceus**‡; *G. rhamni*[1]‡; *P. brassicae*†‡; *P. rapae*†‡; *P. napi*[1]*‡; *L. phlaeas*[1]*‡; *A. agestis*‡; *P. icarus*[1]*‡; *C. argiolus*[2]*‡; *V. atalanta**‡; *V. cardui**‡; *A. urticae*†‡; *A. io**‡; *P. c-album*‡; *A. aglaja*[1]*; *P. aegeria*[1]†‡; *L. megera*[1]†‡; *P. tithonus*[2]‡; *M. jurtina*[2]*‡; *C. pamphilus*[1]*; *A. hyperantus*[1]‡.

DENNY [ST48]: *P. brassicae*‡; *P. rapae*‡; *V. cardui*‡.

FLAT HOLM [ST26]: *O. sylvanus*‡; *C. croceus*‡; *G. rhamni*[1]‡; *P. brassicae*‡; *P. rapae*‡; *P. napi*[1]‡; *A. cardamines*[2]; *L. phlaeas*[1]‡; *P. icarus*[1]‡; *V. atalanta*‡;

V. cardui†‡; *A. urticae*†‡; *A. io*‡; *P. c-album*‡; *P. aegeria*[1]‡ (1 in 1996); *L. megera*[1]‡ (1 in 1996); *P. tithonus*[2]‡; *M. jurtina*[2]‡; *A. hyperantus*[1]‡. BMS 2004–2011.

No records for Sully [ST16].

References

Anthoney (2016), Ayling (2016), Griffiths and Bartlett (1914), Matthews (1932), Parsons (1978).

South Wales

BURRY HOLMS [SS49]: *C. croceus*‡; *V. cardui*‡; *M. galathea*‡; *H. semele*[1]‡; *M. jurtina*[2]‡.

WORM'S HEAD [SS38]: *C. croceus**; *P. brassicae*‡*; *P. icarus*[1]†; *V. atalanta**; *V. cardui**; *A. urticae**; *A. io**; *L. megera*[1]*; *H. semele*[1]†; *M. jurtina*[2]†; *C. pamphilus*[1]*.

CALDEY [SS19]: *O. sylvanus*†‡; *C. croceus**; *P. brassicae*†‡;; *P. rapae*†‡; *P. napi*[2]‡; *L. phlaeas*[1]†‡; *C. minimus*†‡; *P. icarus*[1]†‡; *V. atalanta*†‡; *V. cardui*‡; *A. urticae*†‡; *A. io*†‡; *P. c-album*‡; *P. aegeria*[1]†‡; *L. megera*[1]†‡; *H. semele*[1]*; *P. tithonus*[2]†‡; *M. jurtina*[2]†‡; *A. hyperantus*[1]†‡; *C. pamphilus*[1]†.

SKOKHOLM [SM70]: *T. sylvestris*‡; *O. sylvanus*†; *P. malvae* (?)†; *C. hyale*†; *C. croceus*†‡; *G. rhamni*[1]†; *P. brassicae*†‡; *P. rapae*†‡; *P. napi*[2]†‡; *A. cardamines*[2]†; *L. phlaeas*[1]†‡; *C. minimus*?†; *A. agestis*†; *P. icarus*[1]†‡; *P. bellargus*?†; *C. argiolus*[2]†; *V. atalanta*†‡; *V. cardui*†‡; *A. urticae*†‡; *A. io*†‡; *P. c-album*†; *B. selene*[1]†; *A. adippe*?†; *A. aglaja*[1]†‡; *A. paphia*†; *P. aegeria*[1]†; *L. megera*[1]†; *M. galathea*†; *H. semele*[1]†‡; *P. tithonus*[2]†‡; *M. jurtina*[2]†‡; *A. hyperantus*[1]†; *C. pamphilus*[1]†‡. BMS (2 transects) 1977–2004 and 2005–2007.

MEWSTONE [SM70]: *V. atalanta*‡; *V. cardui*‡.

SKOMER [SM70]: *T. sylvestris*†‡; *O. sylvanus*†‡; *C. croceus*†‡; *P. brassicae*†‡; *P. rapae*†‡; *P. napi*[2]†‡; *A. cardamines*[2]†; *L. phlaeas*[1]†‡; *P. icarus*[1]†‡; *C. argiolus*[2]‡; *V. atalanta*†‡; *V. cardui*†‡; *A. urticae*†‡; *A. io*†‡; *P. c-album**; *B. selene*[1]*; *A. aglaja*[1]†‡; *P. aegeria*[1]‡; *L. megera*[1]†‡; *H. semele*[1]†‡; *P. tithonus*[2]†‡; *M. jurtina*[2]†‡; *A. hyperantus*[1]†‡; *C. pamphilus*[1]†‡; *D. plexippus*†. BMS 1977–present.

GRASSHOLM [SM50]: *P. brassicae*†‡; *P. rapae**; *V. atalanta**; *V. cardui*†‡; *A. urticae**; *H. semele*[1]‡.

GARLAND STONE [SM71]: *V. atalanta*‡; *V. cardui*‡.

RAMSEY [SM72]: *T. sylvestris*‡; *O. sylvanus*†‡; *C. croceus*‡; *G., rhamni*[1]‡; *P. brassicae*†‡; *P. rapae*†‡; *P. napi*[2]†‡; *A. cardamines*[2]‡; *L. phlaeas*[1]†‡; *P. icarus*[1]†‡; *C. argiolus*[2]‡; *V. atalanta*†‡; *V. cardui*†‡; *A. urticae*†‡; *A. io*†‡; *A. aglaja*[1]†‡; *P. aegeria*[1]‡; *L. megera*[1]†‡; *H. semele*[1]†‡; *P. tithonus*[2]†‡; *M. jurtina*[2]†‡; *A. hyperantus*[1]‡; *C. pamphilus*[1]‡. BMS 2007–present.

CARREG RHOSON [SM62]: *P. brassicae*‡; *P. rapae*‡; *P. napi*[2]‡; *A. cardamines*[2]‡; *V. atalanta*‡; *V. cardui*‡; *A. io*‡; *P. c-album*‡; *P. aegeria*[1]‡; *L. megera*[1]‡; *P. tithonus*[2]‡; *M. jurtina*[2]‡.

NORTH BISHOP [SM62]: *P. rapae*‡; *P. napi*[2]‡; *A. cardamines*[2]‡; *V. atalanta*‡; *A. io*‡; *P. c-album*‡; *P. aegeria*[1]‡; *P. tithonus*[2]‡; *M. jurtina*[2]‡; *A. hyperantus*[1]‡.

No records for St. Catherine's [SS19], St. Margaret's [SN10], Sheep and Thorn [SM80], Gateholm and Midland [SM70], the Smalls [SM40], Ynys Bery, Ynys Eilun and Ynys Cantwr [SM72], Bishops (e.g., Bishop Rock South, apart from North Bishop) and Clerks [SM62], Ynys Lochtyn [SN35] and Cardigan [SN15].

References

Betts and Price (1995), Bradley (1954), Bullock (1996), Conder (1949), Evans (2015), Gander (2016), Hallet (1930), Heath (1975, 1976), Heron (1956), Redhead (2016), Slade (2016), Smith (1960).

North Wales

BARDSEY [SH12]: *C. croceus*†‡; *P. brassicae*†‡; *P. rapae*†‡; *P. napi*[(2)]†‡; *A. cardamines*[2]†; *L. phlaeas*[1]†‡; *P. icarus*[1]†‡; *V. atalanta*†‡; *V. cardui*†; *A. urticae*†‡; *A. io*†‡; *A. aglaja*[1]‡; *A. paphia**; *L. megera*[1]†; *H. semele*[1]†‡; *P. tithonus*[2]†‡; *M. jurtina*[2]†‡; *C. pamphilus*[1]†; *D. plexippus*‡.

ANGLESEY [SH46]: *T. sylvestris*†‡; *O. sylvanus*†‡; *E. tages*[1]†‡; *P. malvae*†; *C. croceus*†‡; *G. rhamni*[1]†‡; *P. brassicae*†‡; *P. rapae*†‡; *P. napi*[(2)]†‡; *A. cardamines*[2]†‡; *C. rubi*†; *F. quercus*†‡; *S. w-album*†; *L. phlaeas*[1]†‡; *P. argus*[1]†‡ (1 in 1999); *A. agestis*†‡; *P. icarus*[1]†‡; *C. argiolus*[2]†‡; *V. atalanta*†‡; *V. cardui*†‡; *A. urticae*†‡; *N. antiopa*†; *A. io*†‡; *P. c-album*†‡; *B. selene*[1]†‡; *B. euphrosyne*†; *A. adippe*†; *A. aglaja*[1]†‡; *A. paphia**; *E. aurinia*[1]†‡; *P. aegeria*[1]†‡; *L. megera*[1]†‡; *M. galathea*‡ (1 in 2013); *H. semele*[1]†‡; *P. tithonus*[2]†‡; *M. jurtina*[2]†‡; *A. hyperantus*[1]†‡; *C. pamphilus*[1]†‡ *D. plexippus*‡. BMS (7 transects) 1993–1994, 1999–2002 (2), 1978–1994, 1995–present, 1993–2001 and 1994–1998; WCBS (4 transects) 2013 (1), 2015 (1), 2013 (1), 2007 and 2009–2014 (1).

LLANDDWYN [SH36]: *T. sylvestris*‡; *O. sylvanus*‡; *E. tages*[1]‡; *C. croceus*‡; *P. brassicae*‡; *P. rapae*‡; *P. napi*[(2)]‡; *A. cardamines*[2]‡; *L. phlaeas*[1]‡; *P. icarus*[1]‡; *V. atalanta*‡; *V. cardui*‡; *A. urticae*‡; *A. io*‡; *P. c-album*‡; *B. selene*[1]‡; *A. aglaja*[1]‡; *P. aegeria*[1]‡; *H. semele*[1]‡; *P. tithonus*[2]‡; *M. jurtina*[2]‡; *A. hyperantus*[1]‡; *C. pamphilus*[1]‡; *D. plexippus*‡.

HOLY [SH27]: *T. sylvestris*‡; *O. sylvanus*†‡; *C. croceus*†‡; *G. rhamni*[1]‡; *P. brassicae*†‡; *P. rapae*†‡; *P. napi*[(2)]†‡; *A. cardamines*[2]†‡; *C. rubi*‡; *L. phlaeas*[1]†‡; *P. argus*[1]†‡; *P. icarus*[1]†‡; *C. argiolus*[2]‡; *V. atalanta*†‡; *V. cardui*†‡; *A. urticae*†‡; *A. io*†‡; *P. c-album*‡; *B. selene*[1]†‡; *A. aglaja*[1]†‡; *E. aurinia*[1]†; *P. aegeria*[1]‡; *L. megera*[1]†‡; *H. semele*[1]†‡; *P. tithonus*[2]†‡; *M. jurtina*[2]†‡; *A. hyperantus*[1]†‡; *C. pamphilus*[1]†‡. BMS (3 transects) 1979–1983, 1979–present, 2007–2009.

SOUTH STACK [SH28]: *P. brassicae*‡; *P. rapae*‡; *P. napi*[(2)]‡; *L. phlaeas*[1]‡; *P. icarus*[1]‡; *V. atalanta*‡; *A. urticae*‡; *A. io*‡; *P. c-album*‡; *L. megera*[1]‡; *H. semele*[1]‡; *P. tithonus*[2]‡; *M. jurtina*[2]‡.

PUFFIN [SH68]: *C. croceus**; *P. brassicae**; *P. rapae**; *P. napi*[(2)]*; *V. atalanta**; *V. cardui*†; *A. urticae**; *A. io**; *L. megera*[1]*; *H. semele*[1]*; *M. jurtina*[2]*.

ST TUDWAL'S WEST [SH32]: *D. plexippus*‡.

No records from Ynys Gwylan Fawr and Ynys Gwylan Bach [SH12], St. Tudwal's East (SH32), North Stack [SH28], the Skerries [SH29], Mouse Islands [SH49], Ynys Moelfre [SH58] and Ynys Dulas [SH59].

References

Askew (1974), Carter, Smith and Spence (1960), Darlington (1954), Dennis (1974), Graham (2016), Hancock (1978), Hardy (1995), McCormick (1994), Morgan (1969, 1988), Nye (1957), Sandison (2004), White (1892–93).

Irish Sea and north-west England

LITTLE HILBRE [SJ18]: *T. sylvestris*‡; *P. brassicae*‡; *P. rapae*‡; *P. napi*[2]‡; *V. atalanta*‡; *V. cardui*‡; *A. urticae*‡; *A. io*‡; *H. semele*[1]‡; *P. tithonus*[2]‡; *M. jurtina*[2]‡.

HILBRE [SJ18]: *T. sylvestris*†‡; *O. sylvanus*†‡; *C. croceus*†‡; *G. rhamni*[1]‡; *P. brassicae*†‡; *P. rapae*†‡; *P. napi*[2]†‡; *A. cardamines*‡; *L. phlaeas*[1]†‡; *P. icarus*[1]†; *C. argiolus*[2]†‡; *V. atalanta*†‡; *V. cardui*†‡; *A. urticae*†‡; *A. io*†‡; *P. c-album*†‡; *P. aegeria*[1]†‡; *L. megera*[1]†‡; *H. semele*[1]†‡; *P. tithonus*[2]†‡; *M. jurtina*[2]†‡; *C. pamphilus*[1]†‡.

PIEL [SD26]: *P. brassicae*‡; *P. rapae*‡; *P. napi*[2]‡; *L. phlaeas*[1]‡; *V. atalanta*‡; *V. cardui*‡; *A. urticae*‡; *H. semele*[1]‡.

FOULNEY [SD26]: *C. croceus*‡; *P. napi*[2]‡; *P. icarus*[1]‡; *V. cardui*‡; *P. aegeria*[1]‡.

ROA [SD26]: *P. brassicae*‡.

WALNEY [SD26]: *T. sylvestris*‡ (2 in 2014); *O. sylvanus*‡; *E. tages*[1]‡(?); *C. croceus*‡; *P. brassicae*‡; *P. rapae*‡; *P. napi*[2]‡; *A. cardamines*[2]‡; *L. phlaeas*[1]‡; *P. icarus*[1]‡; *C. argiolus*[2]‡; *V. atalanta*‡; *V. cardui*‡; *A. urticae*‡; *A. io*‡; *P. c-album*‡; *B. selene*‡(?); *A. aglaja*[1]‡; *P. aegeria*[1]‡; *L. megera*[1]‡; *H. semele*[1]‡; *P. tithonus*[2]‡; *M. jurtina*[3]‡; *C. pamphilus*[1]‡ (1 in 2007). BMS (2 transects) 2007–2010 and 2014–present; WCBS 2009.

MAN [SC16]: *O. sylvanus**; *C. croceus*†‡; *G. rhamni*[(2)](+D)†‡; *P. brassicae*†‡; *P. rapae*†‡; *P. napi*[(2)]†‡; *A. cardamines*[3]†‡; *L. phlaeas*[1]†‡; *C. minimus*(?)†; *P. icarus*[1]†‡; *C. argiolus*[2]†‡; *V. atalanta*†‡; *V. cardui*†‡; *A. urticae*†‡; *N. antiopa*†; *A. io*†‡; *P. c-album*†‡; *B. selene*[1]*; *B. euphrosyne**; *A. aglaja*[1]†‡; *E. aurinia*[1]*(?); *P. aegeria*[1]†‡; *L. megera*[1]†‡; *E. aethiops*[1]*(?); *H. semele*[1]†‡; *P. tithonus*[2]*; *M. jurtina*[3]†‡; *A. hyperantus*[1]*; *C. pamphilus*[1]†‡; *C. tullia*[(2)]*; *D. plexippus*†‡ (1999 and 2001). BMS 2002–2006; WCBS 2012–2015.

CHICKEN ROCK [SC16]: *V. cardui*†.

CALF OF MAN [SC16]: *C. croceus*†‡; *P. brassicae*†‡; *P. rapae*†‡; *P. napi*[(2)]†‡; *A. cardamines*[3]†; *L. phlaeas*[1]†‡; *P. icarus*[1]†‡; *C. argiolus*[2]‡; *V. atalanta*†‡; *V. cardui*†‡; *A. urticae*†‡; *A. io*†‡; *A. aglaja*[1]‡; *L. megera*[1]†‡; *H. semele*[1]†‡; *M. jurtina*[3]†‡; *C. pamphilus*[1]†‡; *D. plexippus*‡ (1 in 2001). BMS 1995–2006.

ST MICHAEL'S [SC26]: *P. brassicae*†; *V. atalanta*‡.

No records for Little Eye [SJ18], nor for the Burroo, the Stack, Kitterland [SC16] and St. Patrick's [SC28] in the vicinity of Man.

References

Adkin (1924), Asher (1996), Blackler (1940), Chalmers-Hunt (1970), Cowin (1955), Craggs (1982), Crow (1971), Curtis (2016), Dannreuther (1948a), Dean (1990), Dennis *et al.* (2010), Hardy *et al.* (2017), Hedges, A.V. (1947), Hedges, G. (2016), Hedges, J. (1980, 1981), Jeffcoate (2016), Majerus (1979, 1980), Rogers (1944), Seegar (1995, 1996), Shaw (1996, 2016), Thomson (1970), Tomlinson (2016), Turner (1924).

South-west Scotland and Firth of Clyde

ROUGH [NX85]: *B. selene*[1]‡.

HESTAN [NX85]: *A. io*‡; *H. semele*[1]‡; *M. jurtina*[(5)]‡.

LITTLE ROSS [NX64]: *V. cardui*‡.

MURRAY'S (Islands of Fleet) [NX55]: *P. napi*[4]‡; *L. phlaeas*[3]‡.

BARLOCCO (Islands of Fleet) [NX54]: *P. napi*[4]‡; *L. phlaeas*[3]‡; *P. icarus*[1]‡; *A. urticae*‡; *L. megera*[(2)]‡; *H. semele*[1]‡; *M. jurtina*[(5)]‡; *A. hyperantus*[(2)]‡.

ARDWALL (Islands of Fleet) [NX54]: *A. urticae*‡; *A. io*‡; *L. megera*[(2)]‡; *H. semele*[1]‡; *M. jurtina*[(5)]‡; *C. pamphilus*[1]‡.

AILSA CRAIG [NX09]: *E. tages*[1](?)†; *P. brassicae*†‡; *P. rapae*†‡; *P. napi*[4]†‡; *C. rubi*†; *L. phlaeas*[(3)]†‡; *A. artaxerxes*[1]* (recorded as *A. agestis*); *P. icarus*[1]†‡; *V. atalanta*†‡; *V. cardui*†; *A. urticae*†‡; *A. io**‡; *B. selene*[1](?)†; *A. aglaja*[1]*; *E. aethiops*[2]†; *H. semele*[1]†; *M. jurtina*[(5)]*; *A. hyperantus*[(2)]‡; *C. pamphilus*[1]*; *C. tullia*[(2)](?)*.

PLADDA [NS01]: *P. brassicae*†; *P. rapae*†; *P. napi*[4]†; *V. atalanta*†; *A. urticae*†‡; *H. semele*[1]†; *M. jurtina*[5]†; *C. pamphilus*[1]†.

LADY ISLE [NS22]: *P. brassicae*†; *P. rapae*†; *P. napi*[4]†; *A. cardamines*[2]‡; *L. phlaeas*[(3)]†; *P. icarus*[1]†; *V. atalanta*†; *V. cardui*†; *A. urticae*†; *A. io*†; *H. semele*[1]†; *M. jurtina*[(5)]†; *C. pamphilus*[1]†.

HOLY [NS02]: *P. brassicae*†‡; *P. rapae*†; *P. napi*[4]†‡; *A. cardamines*[2]‡; *L. phlaeas*[3]†‡; *P. icarus*[1]†‡; *V. atalanta*†‡; *V. cardui*‡; *A. urticae*†‡; *A. io*†; *B. selene*[1]†‡; *A. aglaja*[1]†‡; *E. aethiops*[2]†; *H. semele*[1]†‡; *M. jurtina*[5]†‡; *A. hyperantus*[(2)]‡; *C. pamphilus*[1]†‡; *C. tullia*[1]*.

HORSE [NS24]: *P. brassicae*†; *P. rapae*†; *P. napi*[4]†; *P. icarus*[1]†; *V. atalanta*†; *A. urticae*†; *A. io*†; *H. semele*[1]†; *M. jurtina*[(5)] †.

LITTLE CUMBRAE [NS15]: *P. brassicae*†; *P. rapae*†; *P. napi*[4]†; *L. phlaeas*[1]†; *P. icarus*[1]†; *V. atalanta*†; *V. cardui*†; *A. urticae*†‡; *A. io*†‡; *B. selene*[1]†; *A. aglaja*[1]†; *E. aethiops*[2]†; *H. semele*[1]†; *M. jurtina*[(5)]†; *C. pamphilus*[1]†; *C. tullia*[1]†.

GREAT CUMBRAE [NS15]: *C. croceus*‡; *P. brassicae*†‡; *P. rapae*†; *P. napi*[4]†‡; *A. cardamines*[2]‡; *C. rubi*†‡; *L. phlaeas*[1]†‡; *P. icarus*[1]†‡; *V. atalanta*†‡; *V. cardui*†‡; *A. urticae*†‡; *A. io*†‡; *B. selene*[1]†; *A. aglaja*[1]†; *E. aethiops*[2]†; *H. semele*[1]†‡; *M. jurtina*[(5)]†‡; *C. pamphilus*[1]†‡.

INCHMARNOCK [NS05]: *P. brassicae*†‡; *P. rapae*†‡; *P. napi*[4]†‡; *L. phlaeas*[3]†; *P. icarus*[1]†‡; *V. atalanta*†; *V. cardui*‡; *A. urticae*†‡; *A. io*†‡; *B. selene*[1]†; *A. aglaja*[1]†; *H. semele*[1]†; *M. jurtina*[5]†; *C. pamphilus*[1]†.

BUTE [NS05]: *O. sylvanus**(?); *C. croceus*†‡; *P. brassicae*†‡; *P. rapae*†‡; *P. napi*[4]†‡; *A. cardamines*[2]‡; *C. rubi*†‡ (1 in 2006); *F. quercus*‡ (1 in 2008); *L. phlaeas*[3]†‡; *P. icarus*[1]†‡; *V. atalanta*†‡; *V. cardui*†‡; *A. urticae*†‡; *A. io*†‡; *B. selene*[1]†‡; *A. aglaja*[1]†‡; *E. aurinia*[3]‡; *P. aegeria*[2]‡; *L. megera*[(2)]†; *E. aethiops*[2]†‡; *H. semele*[1]†‡; *M. jurtina*[5]†‡; *A. hyperantus*[(2)]‡; *C. pamphilus*[1]†‡; *C. tullia*[1]‡. WCBS 2011–2014.

BURNT ISLANDS [NS07]: *P. brassicae*†; *P. rapae*†; *P. napi*[4]†; *A. urticae*†; *H. semele*[1]†; *M. jurtina*[5]†; *C. pamphilus*[1]†.

SANDA [NR70]: *P. brassicae*†; *P. rapae*†‡; *P. napi*[4]†‡; *L. phlaeas*[3]†; *P. icarus*[1]†‡; *V. atalanta*†‡; *V. cardui*‡; *A. urticae*†‡; *A. io*†‡; *E. aethiops*[2]†; *H. semele*[1]†‡; *M. jurtina*[5]†‡; *C. pamphilus*[1]†‡.

GLUNIMORE [NR70]: *P. rapae*†; *A. urticae*†.

SHEEP [NR70]: *P. rapae*†; *P. napi*[4]†; *A. urticae*†; *H. semele*[1]†; *M. jurtina*[5]†.

DAVAAR [NR71]: *P. brassicae*†; *P. rapae*†; *P. napi*[4]†‡; *L. phlaeas*[3]†; *P. icarus*[1]†; *V. atalanta*†‡; *A. urticae*†; *A. io*†; *H. semele*[1]†; *M. jurtina*[5]†; *C. pamphilus*[1]†; *C. tullia*[1]†.

ARRAN [NR92]: *C. hyale*‡ (1 in 2007); *C. croceus*†‡; *P. brassicae*†‡; *P. rapae*†‡; *P. napi*[4]†‡; *A. cardamines*[2]‡; *C. rubi*†‡; *F. quercus*†‡; *L. phlaeas*[3]†‡; *C. minimus**; *P. icarus*[1]†‡; *V. atalanta*†‡; *V. cardui*†‡; *A. urticae*†‡; *A. io*†‡; *B. selene*[1]†‡; *B. euphrosyne**‡ (1 in 2012); *A. aglaja*[1]†‡; *P. aegeria*[2]†‡; *L. megera*[(2)]†; *E. aethiops*[2]†‡; *E. ligea**(?); *H. semele*[1]†‡; *M. jurtina*[5]†‡; *A. hyperantus*[(2)]*‡; *C. pamphilus*[1]†‡; *C. tullia*[1]*‡. BMS 1995–present; WCBS (4 transects) 2011 and 2012 (1), 2011 only (2), 2011–2014 (1).

SGAT MOR [NR96]: *P. rapae*†; *A. urticae*†.

GLAS EILEAN [NR98]: *P. brassicae*†; *P. rapae*†; *P. napi*[4]†; *A. urticae*†; *M. jurtina*[(5)] †.

No records for Inner Eilean and Outer Eilean [NS15], Gigalum [NR64]

References

Black (2016), Cooter (1978), Ellis (1981), Evans (1922), Fox (2016), Gibson (1976, 1982a, b, c, d, 1990, 1992, 1995), Haggart (1933), Hancock (1995), Lupton (1880), National Biodiversity Network (2016), Nelson (1980), Richardson (1960), Stewart (1925, 1933, 1939), Sutcliffe (1995), Thomson (1980), Watson (1893), Weir (1882b), White (1872), Wild (1922).

Inner Hebrides: Strathclyde

CARA [NR64]: *P. rapae*†; *P. napi*[4]†; *L. phlaeas*[3]†‡; *P. icarus*[1]†; *V. atalanta**; *V. cardui*†; *A. urticae*†; *A. io**; *H. semele*[1]†; *M. jurtina*[5]†; *C. pamphilus*[1]†‡.

EILEAN IMERSAY [NR44]: *P. napi*[4]‡; *P. icarus*[1]‡.

GIGHA [NR64]: *C. croceus*‡; *P. brassicae*†‡; *P. rapae*†‡; *P. napi*[4]†‡; *A. cardamines*[2]†‡; *C. rubi*†‡; *L. phlaeas*[3]†‡; *P. icarus*[1]†‡; *V. atalanta**‡; *V. cardui**‡; *A. urticae*†‡; *A. io*†‡; *P. aegeria*[2]‡; *E. aethiops*[2]‡; *H. semele*[3]‡; *M. jurtina*[5]†‡; *A. hyperantus*[(2)]‡; *C. pamphilus*[1]*‡. WCBS 2007.

EILEAN AN DROIGHINN [NR44]: *P. icarus*[1]‡; *M. jurtina*[5]‡.
ORSAY [NR15]: *P. icarus*[1]‡; *A. aglaja*[1]‡.
EILEAN CAM [NR15]: *E. aurinia*[3]‡.
EILEAN DUBH (1) (Loch Indaal) [NR25]: *C. pamphilus*[1]‡.
ISLAY [NR36]: *C. palaemon*‡ (1 in 2010); *T. sylvestris*‡ (1 in 2003); *C. croceus**‡; *P. brassicae*†‡; *P. rapae*†‡; *P. napi*[4]†‡; *A. cardamines*[2]†‡; *C. rubi*†‡; *L. phlaeas*[3]†‡; *P. argus*[1](?)†; *A. artaxerxes*[1](?)†; *P. icarus*[1]†‡; *V. atalanta*†‡; *V. cardui*†‡; *A. urticae*†‡; *A. io*†‡; *B. selene*[(2)]‡; *B. euphrosyne*‡; *A. adippe** (?*A. aglaja* misidentified?); *A. aglaja*[1]†‡; *E. aurinia*[3]†‡; *P. aegeria*[2]†‡; *L. megera*[(2)]†‡ (1 in 2005); *E. aethiops*[2]‡; *H. semele*[3]†‡; *M. jurtina*[5]†‡; *A. hyperantus*[2]†‡; *C. pamphilus*[1]†‡; *C. tullia*[1]†‡. BMS (3 transects) 1995–present (1), 2006–present (2).
EILEAN AM FRAOCH [NR46]: *M. jurtina*[5]‡.
EILEAN MÓR [NR67]: *M. jurtina*[(5)]‡.
NAVE [NR27]: *P. brassicae*‡; *A. aglaja*[1]‡.
DANNA [NR67]: *P. napi*[4]‡; *L. phlaeas*[3]‡; *P. icarus*[1]‡; *V. atalanta*‡; *V. cardui*‡; *A. urticae*‡; *A. io*‡; *E. aurinia*[3]‡; *P. aegeria*[2]‡; *E. aethiops*[2]‡; *H. semele*[(3)]‡; *M. jurtina*[(5)]‡; *A. hyperantus*[(2)]‡; *C. pamphilus*[1]‡.
JURA [NR56]: *C. palaemon*‡ (1 in 2010); *C. croceus*‡; *P. brassicae*†‡; *P. rapae*†‡; *P. napi*[4]†‡; *A. cardamines*[2]‡; *C. rubi*†‡; *L. phlaeas*[3]†‡; *P. icarus*[1]†‡; *V. atalanta*†‡; *V. cardui*†‡; *A. urticae*†‡; *A. io*†‡; *B. selene*[(2)]†‡; *A. aglaja*[1]†‡; *E. aurinia*[3]†‡; *P. aegeria*[2]†; *L. megera*[(2)]†‡ (1 in 2005); *E. aethiops*[2]‡; *H. semele*[(3)]†‡; *M. jurtina*[5]†‡; *A. hyperantus*[(2)]†‡, *C. pamphilus*[1]†‡; *C. tullia*[1]†‡. WCBS 2011.
EILEAN A'BHORRA [NR68]: *A. hyperantus*[(2)]‡.
ORONSAY [NR38]: *G. rhamni*[1]‡; *P. brassicae*‡; *P. rapae*‡; *P. napi*[4]‡†; *A. cardamines*[2]‡; *C., rubi*‡; *L. phlaeas*[(3)]†‡; *P. icarus*[1]‡; *V. atalanta*†‡; *V. cardui*‡; *A. urticae*‡; *A. io*‡; *B. selene*[(2)]‡; *A. aglaja*[1]‡; *E. aurinia*[3]‡; *H. semele*[(3)]‡; *M. jurtina*[5]‡; *A. hyperantus*[(2)]‡; *C. pamphilus*[1]‡.
EILEAN NAN RON [NR38]: *P. icarus*[1]‡; *A. urticae*‡; *A. io*‡; *E. aurinia*[3]‡; *H. semele*[(3)]‡; *C. pamphilus*[1]‡.
EILEAN GHAOIDEAMAL [NR38]: *P. napi*[4]‡; *L. phlaeas*[3]‡.
EILEAN MHUCAIG (STRAND ISLANDS) [NR39]: *P. napi*[4]‡; *C. rubi*‡; *V. cardui*‡; *A. urticae*‡; *A. io*‡; *C. pamphilus*[1]‡.
COLONSAY [NR38]: *C. croceus*‡; *G. rhamni*[1]‡; *P. brassicae*†‡; *P. rapae*†‡; *P. napi*[4]†‡; *A. cardamines*[2]‡; *C. rubi*†‡; *F. quercus*†‡; *L. phlaeas*[3]†‡; *P. icarus*[1]†‡; *V. atalanta*†‡; *V. cardui*†‡; *A. urticae*†‡; *A. io*†‡; *B. selene*[(2)]‡; *A. aglaja*[1]†‡; *E. aurinia*[3]†‡; *P. aegeria*[2]†‡; *E. aethiops*[2]‡ (1 in 1999); *H. semele*[(3)]†‡; *M. jurtina*[5]†‡; *A. hyperantus*[(2)]†‡; *C. pamphilus*[1]†‡; *C. tullia*[1]*‡.
SCARBA [NM60]: *P. napi*[4]‡; C. *rubi*‡; *P. icarus*[1]†‡; *A. io*‡; *A. aglaja*[1]‡; *E. aurinia*[3]‡ *M. jurtina*[5]‡; *C. pamphilus*[1]‡.
EILEACH AN NAOIMH (GARVELLACHS) [NM60]: *P. napi*[4]‡; *P. icarus*[1]‡; *V. atalanta*‡; *V. cardui*‡; *A. aglaja*[1]‡; *M. jurtina*[5]‡; *A. hyperantus*[(2)]‡; *C. pamphilus*[1]‡.
EILEAN BUIDHE [NM70]: *P. icarus*[1]‡; *M. jurtina*[5]‡.

SHUNA [NM70]: *P. napi*[4]‡; *C. rubi*‡; *L. phlaeas*[3]‡; *A. urticae*‡.

LUNGA [NM70]: *P. brassicae*†; *P. napi*[4]†‡; *C. rubi*†; *L. phlaeas*[3]†; *P. icarus*[1]†‡; *V. atalanta*‡; *B. selene*[(2)]†; *A. aglaja*[1]†‡; *E. aethiops*[2]†; *H. semele*[(3)]†; *M. jurtina*[5]†‡; *C. pamphilus*[1]†.

SOA [NM21]: *P. napi*[4]†; *P. icarus*[(2)]‡; *H. semele*[(3)]*.

GARBH EILEACH (GARVELLACHS) [NM61]: *P. napi*[4]†; *L. phlaeas*[3]†; *P. icarus*[1]†; *B. selene*[(2)]‡; *A. aglaja*[1]†; *E. aethiops*[2]‡; *M. jurtina*[5]†; *A. hyperantus*[(2)]†.

LUING [NM71]: *P. brassicae*‡; *P. rapae*‡; *P. napi*[4]†‡; *A. cardamines*[2]‡; *L. phlaeas*[3]‡; P. *icarus*[1]†‡; *V. atalanta*‡; *V. cardui*‡; *A. urticae*‡; *A. io*‡; *B. selene*[(2)]‡; *A. aglaja*[1]‡; *E. aurinia*[3]‡; *P. aegeria*[2]‡; *H. semele*[(3)]‡; *M. jurtina*[5]‡; *A. hyperantus*[(2)]‡; *C. pamphilus*[1]‡.

FRAOCH EILEAN [NM71]: *P. napi*[4]‡; *L. phlaeas*[3]‡; *A. urticae*‡; *A. io*‡; *E. aethiops*[2]‡; *H. semele*[(3)]‡; *A. hyperantus*[(2)]‡.

ORMSA [NM71]: *C. croceus*‡.

TORSA [NM71]: *P. napi*[4]‡; *P. icarus*[1]‡; *E. aurinia*[3]‡; *M. jurtina*[5]‡; *A. hyperantus*[(2)]‡; *C. pamphilus*[1]‡.

SEIL [NM71]: *C. croceus*‡; *P. brassicae*†; *P. rapae*†‡; *P. napi*[4]†‡; *A. cardamines*[2]‡; *C. rubi*‡; *L. phlaeas*[3]†‡; *P. icarus*[1]†‡; *V. atalanta*†‡; *V. cardui*†‡; *A. urticae*†‡; *A. io*†‡; *B. selene*[(2)]‡; *B. euphrosyne*†; *A. aglaja*[1]†; *E. aurinia*[3]‡; *P. aegeria*[2]†‡; *E. aethiops*[2]†; *H. semele*[3]†‡; *M. jurtina*[5]†‡; *A. hyperantus*[2]†‡; *C. pamphilus*[1]†‡.

EASDALE [NM71]: *P. brassicae*†; P. *rapae*†; *P. napi*[4]†‡; *A. cardamines*[2]‡; *L. phlaeas*[3]†; *P. icarus*[1]†‡; *V. atalanta*†; *V. cardui*‡; *A. urticae*†‡; *A. io*‡; *A. aglaja*[1]†; *E. aethiops*[2]†; *H. semele*[(3)]†‡; *M. jurtina*[5]†‡; *C. pamphilus*[1]†.

INSH [NM71]: *A. io*‡; *A. aglaja*[1]‡; *H. semele*[(3)]‡; *M. jurtina*[5]‡.

IONA [NM22]: *P. brassicae*†‡; *P. rapae*†‡; *P. napi*[4]†‡; *A. cardamines*[2]‡; *C. rubi*†; *P. icarus*[2]†‡; *V. atalanta*†‡; *V. cardui*‡; *A. urticae*†‡; *A. io*†‡; *B. selene*[(2)]‡; *A. aglaja*[(2)]‡; *H. semele*[3]†‡; *M. jurtina*[5]†‡; *C. pamphilus*[1]†‡.

KERRERA [NM82]: *P. brassicae*†‡; *P. rapae*†‡; *P. napi*[4]†‡; *A. cardamines*[2]‡; *C. rubi*†; *C. minimus**; *P. icarus*[1]†‡; *V. atalanta*†‡; *V. cardui*†‡; *A. urticae*†‡; *A. io*†‡; *B. selene*[(2)]†; *A. aglaja*[1]†‡; *E. aethiops*[2]†‡; *H. semele*[(3)]†‡; *M. jurtina*[5]†‡; *A. hyperantus*[(2)]‡; *C. pamphilus*[1]†‡; *C. tullia*[1]†.

DUTCHMAN'S CAP [NM23]: *P. napi*[4]‡; *V. atalanta*‡.

STAFFA [NM33]: *P. brassicae*‡; *P. napi*[4]‡; *P. icarus*[(2)]†‡; *V. cardui*‡; *A. urticae*†; *A. aglaja*[(2)]‡; *H. semele*[(3)]†; *M. jurtina*[5]†‡; *C. pamphilus*[1]†; *C. tullia*[1]†.

ULVA [NM43]: *P. brassicae*†; *P. napi*[4]†‡; *A. cardamines*[2]‡; *C. rubi*†‡; *L. phlaeas*[3]†; *P. icarus*[(2)]†‡; *V. atalanta**‡; *V. cardui*‡; *A. urticae*†‡; *A. io**‡; *B. selene*[(2)]†‡; *A. aglaja*[(2)]*‡; *E. aurinia*[3]*‡; *P. aegeria*[2]†‡; *E. aethiops*[2]†‡; *H. semele*[(3)]*; *M. jurtina*[5]†‡; *C. pamphilus*[1]*‡; *C. tullia*[1]‡.

MULL [NM53]: *O. sylvanus*†; *C. croceus*‡; *P. brassicae*†‡; *P. rapae*†‡; *P. napi*[4]†‡; *A. cardamines*[2]‡; *C. rubi*†‡; *L. phlaeas*[3]†‡; *P. icarus*[(2)]†‡; *V. atalanta*†‡; *V. cardui*†‡; *A. urticae*†‡; *A. io*†‡; *B. selene*[(2)]†‡; *B. euphrosyne*†† (1 in 2013); *A. aglaja*[2]††; *E. aurinia*[3]††; *P. aegeria*[2]†‡; *L. megera*[(2)]‡ (1 in 2004); *E. aethiops*[2]†‡; *H. semele*[3]†‡; *M. jurtina*[5]†‡; *A. hyperantus*[(2)]‡; *C. pamphilus*[1]†‡; *C. tullia*[1]†‡.

EILEAN MUSDILE [NM73]: *E. aurinia*[3]‡.
BERNERA IS. (nr. Lismore) [NM73]: *V. cardui*‡.
LISMORE [NM73]: *P. brassicae*‡; *P. rapae*‡; *P. napi*[4]†‡; *A. cardamines*[2]‡; *L. phlaeas*[3]†; *P. icarus*[1]†‡; *V. atalanta*†‡; *V. cardui*‡; *A. urticae*†‡; *A. io*†‡; *B. selene*[(2)]†‡; *A. aglaja*[(2)]‡; *E. aurinia*[3]†‡; *P. aegeria*[2]‡; *E. aethiops*[2]‡; *H. semele*[(3)]†‡; *M. jurtina*[5]†‡; *C. pamphilus*[1]†‡; *C. tullia*[1]†‡. BMS 2006–2007.
TIREE [NL94]: *P. brassicae*†‡; *P. rapae*†‡; *P. napi*[4]†‡; *A. cardamines*[2]‡; *P. icarus*[2]†‡; *V. atalanta*†‡; *V. cardui*†‡; *A. urticae*†‡; *A. io*†‡; *E. aurinia*[3]*‡; *H. semele*[(3)]†‡; *M. jurtina*[5]†‡; *C. pamphilus*[1]*‡; *C. tullia*[1]*.
LUNGA (TRESHNISH ISLES) [NM24]: *P. icarus*†; *V. cardui*†.
SGEIR A'CHAISTEIL [NM24]: *P. napi*[4]‡.
GOMETRA [NM34]: *P. rapae*‡; *P. napi*[4]‡; *P. icarus*[(2)]‡; *V. atalanta*‡; *A. urticae*‡; *B. selene*[(2)]‡; *M. jurtina*[5]‡; *C. pamphilus*[1]‡.
GUNNA [NM15]: *P. brassicae**; *P. rapae**; *P. napi*[4]*; *P. icarus*[2]*; *V. atalanta**; *V. cardui**; *A. urticae**; *E. aurinia*[3]*; *H. semele*[3]†; *M. jurtina*[5]*; *C. pamphilus*[1]*.
COLL [NM15]: *C. hyale**; *C. croceus*†; *P. brassicae*†‡; *P. rapae*†; *P. napi*[4]†‡; *C. rubi*†‡; *P. icarus*[2]†‡; *V. atalanta*†‡; *V. cardui*†‡; *A. urticae*†‡; *A. io*†‡; *B. selene*[2]†; *A. aglaja*[(2)]†‡; *H. semele*[3]†‡; *M. jurtina*[5]†‡; *C. pamphilus*[1]*‡; *C. tullia*[1]†‡.
CALVE [NM55]: *P. napi*[4]‡; *A. io*‡; *C. pamphilus*[1]‡.

No records for Texa [NR34], Eilean a'Chuirn and Eilean Craobhach [NR44], Brosdale [NR46], Eilean nan Gabhar, Eilean nan Coinein and Eilean nan Bhride (Small Isles) [NR56], Corr [NR67], Dubh Eilean [NR38], Eilean Macaskin, Garbh Reisa and Reisa an t'Struith [NR79], Dubh Artach [NM10], Eilean nan Gabhar, Reisa Mhic Phaide, Rubha Aird Luing, Eilean Arsa, Feola Meadhonach, Eilean Iosal and Fiola na Droma [NM70], Eilean Righ and Eilean Mhic Chrion [NM80], Erraid, Eilean nam Muc and Eilean Dubh (2) [NM21], Eilean a'Chalmain and Eilean Dubh (3) [NM31], Eilean Dubh Mor, Eilean Dubh Beg and A'Chuli [NM61], Rubha Fiola, Eilean Loisgte, Eilean Mhic Chiara and Belnahua [NM71], Bac Beag [NM23], Erisgeir and Little Colonsay [NM33], Inch Kenneth and Eorsa [NM43], Eilean nan Gamnha [NM83], Fladda [NM24], Eilean Dioghlum [NM34] and Eilean Mor (N. of Coll) [NM26].

References

Abercrombie (1953), Baxter and Rintoul (1948), Bowles and Fox (2007), Bryan (1987), Dowdeswell (1936), Dunn (1965), Graiff and Graiff (1995), Hancock (1923), Heslop Harrison (1938a, 1940c, e, f, g, 1941b, c, d, 1942e, 1943d, 1946d, e, 1947c, d, e, f, 1948a, b, 1949b, c, d, 1950d, 1956d), Hulme (1998), Laidlaw (1932), Miller and Owen (1952), Muir (1954), National Biodiversity Network (2016), Ogilvie (2016), Rintoul and Baxter (1937, 1950), Scottish Naturalist (1919), Shreeve (1992), Skeates (2016), Summers (1977b), Thomson (1970, 1980), Vine-Hall (1969), Waterston (1906), White (1872), Wilks (1941, 1945a,b), Young and Harper (1986).

Inner Hebrides: Highland

CARNA [NM65]: *V. atalanta*‡.

MUCK [NM47]: *P. brassicae*†; *P. napi*[4]†‡; *P. icarus*[2]†‡; *V. atalanta**‡; *V. cardui*‡; *A. urticae*†‡; *A. io**‡; *A. aglaja*[(2)]†‡; *P. aegeria*[2]†; *H. semele*[(3)]†‡; *M. jurtina*[5]†‡; *C. pamphilus*[2]†‡; *C. tullia*[1]‡.

SHONA [NM67]: *A. aglaja*[1]‡; *H. semele*[(3)]‡.

EILEAN NAN EACH [NM38]: *P. napi*[4]*; *A. urticae*†.

EIGG [NM48]: *C. croceus**‡; *G. rhamni*[1]‡ (1 in 2002); P. *brassicae**‡; *P. rapae**; *P. napi*[4]*‡; *A. cardamines*[2]‡; *C. rubi*†‡; *P. icarus*[2]*‡; *V. atalanta*†‡; *V. cardui*‡; *A. urticae**‡; *A. io**‡; *B. selene*[2]‡; *A. aglaja*[2]†‡; *E. aurinia*[3]‡; *P. aegeria*[2]†‡; *H. semele*[3]†‡; *M. jurtina*[5]†‡; *A. hyperantus*[(2)]*; *C. pamphilus*[2]*‡; *C. tullia*[1]†‡. BMS 1988–2014.

EILEAN IGHE [NM68]: *P. icarus*[1]‡; *M. jurtina*[5]‡.

LUINGA BEAG [NM68]: *P. napi*[4]‡; *P. icarus*[1]‡; *M. jurtina*[5]‡.

HYSKEIR (OIGH-SGEIR) [NM19]: *P. napi*[4]*; *P. icarus*[2]*; *V. atalanta**; *A. urticae**; *A. io**; *M. jurtina*[5]*.

RHUM [NM39]: *P. brassicae*†; *P. rapae*†; *P. napi*[4]†‡; *A. cardamines*[2]‡; *C. rubi*†‡; *L. phlaeas*[(3)]†; *P. icarus*[2]†‡; *V. atalanta*†‡; *V. cardui*†‡; *A. urticae*†‡; *A. io*†‡; *B. selene*[2]†‡; *B. euphrosyne**; *A. aglaja*[2]†‡; *E. aurinia*[3]*; *P. aegeria*[2]†‡; *E. aethiops*[2]*; *H. semele*[3]†‡; *M. jurtina*[5]†‡; *C. pamphilus*[2]†‡; *C. tullia*[1]†‡.

SANDAY [NG20]: *P. brassicae*†; *P. napi*[4]†‡; *C. rubi*†‡; *P. icarus*[2]†‡; *V. atalanta*†‡; *V. cardui*†‡; *A. urticae**; *B. selene*[(2)]†; *A. aglaja*[(2)]†; *P. aegeria*[2]†; *H. semele*[3]†‡; *M. jurtina*[5]†‡; *C. pamphilus*[(2)]†‡.

CANNA [NG20]: *C. croceus*†; *P. brassicae*†‡; *P. rapae*†‡; *P. napi*[4]†‡; *C. rubi*†‡; *P. icarus*[2]†‡; *V. atalanta*†‡; *V. cardui*†‡; *A. urticae*†‡; *A. io*†‡; *B. selene*[(2)]†‡; *A. aglaja*[2]†‡; *P. aegeria*[2]†; *H. semele*[3]†‡; *M. jurtina*[5]†‡; *C. pamphilus*[2]†‡; *C. tullia*[1]*.

SOAY [NG41]: *P. brassicae**; *P. napi*[4]†; *P. icarus*[2]*‡; *V. atalanta*‡; *V. cardui*‡; *A. urticae**; *A. io**; *B. selene*[2]*‡; *A. aglaja*[2]*‡; *H. semele*[3]*; *M. jurtina*[5]*‡; *C. pamphilus*[2]*‡; *C. tullia*[1]*‡.

PABAY [NG62]: *P. brassicae**; *P. napi*[4]*‡; *P. icarus*[2]*; *A. urticae**‡; *A. io*‡; *M. jurtina*[5]*; *C. pamphilus*[2]*‡; *C. tullia*[1]*.

EILEAN BAN [NG72]: *P. napi*[4]‡; *P. icarus*[2]‡; *V. atalanta*‡; *V. cardui*‡; *A. urticae*‡; *A. io*‡; *E. aethiops*[2]‡; *M. jurtina*[5]‡.

WIAY [NG23]: *E. aethiops*[2]†.

SCALPAY [NG63]: *P. brassicae**; *P. rapae**; *P. napi*[4]†‡; *A. cardamines*[2]‡; *C. rubi**; *P. icarus*[2]†; *V. atalanta**; *V. cardui**; *A. urticae**; *B. selene*[2]†‡; *B. euphrosyne**; *A. aglaja*[2]*‡; *P. aegeria*[2]‡; *E. aethiops*[2]†; *H. semele*[3]*; *M. jurtina*[5]†‡; *C. pamphilus*[2]†‡; *C. tullia*[1]†.

LONGAY [NG63]: *P. brassicae**; *P. napi*[4]*; *C. rubi**; *P. icarus*[2]*; *V. atalanta**; *A. urticae*†; *B. selene*[(2)]*; *A. aglaja*[(2)]*; *E. aethiops*[2]*; *H. semele*[(3)]*; *C. pamphilus*[2]*; *C. tullia*[1]*.

CROWLIN ISLES [NG63] (EILEAN MOR): *A. urticae*†, *M. jurtina*[5]‡.

SKYE [NG34]: *C. croceus*‡; *P. brassicae*†‡; *P. rapae*†‡; *P. napi*[4]†‡; *A. cardamines*[2]†‡; *C. rubi*†‡; *L. phlaeas*[3]‡; *P. icarus*[2]†‡; *V. atalanta*†‡; *V. cardui*†‡; *A. urticae*†‡; *N. antiopa**; *A. io*†‡; *B. selene*[2]†‡;

*B. euphrosyne**; *A. aglaja*[2]†‡; *P. aegeria*[2]†‡; *E. aethiops*[2]†‡; *H. semele*[(3)]†‡; *M. jurtina*[5]†‡; *A. hyperantus*[(2)]*; *C. pamphilus*[(2)]†‡; *C. tullia*[1]†‡; *D. plexippus**. WCBS (4 transects) 2009 (1), 2015 (1), 2007 and 2008 (2).

RAASAY [NG54]: *P. brassicae*†; *P. rapae*†; *P. napi*[4]†‡; *A. cardamines*[2]‡; *C. rubi**‡; *P. icarus*[2]†‡; *V. atalanta**‡; *V. cardui**‡; *A. urticae*†‡; *A. io**‡; *B. selene*[2]*‡; *B. euphrosyne**; *A. aglaja*[2]*‡; *P. aegeria*[2]†‡; *E. aethiops*[2]*‡; *H. semele*[3]†‡; *M. jurtina*[5]†‡; *C. pamphilus*[2]†‡; *C. tullia*[1]†‡.

FLADDAY [NG55]: *P. brassicae**; *P. napi*[4]*; *P. icarus*[(2)]*; *A. urticae**; *M. jurtina*[5]*; *C. pamphilus*[2]*; *C. tullia*[1]*.

SOUTH RONA [NG65]: *P. brassicae*†; *P. napi*[4]†; *P. icarus*[2]†; *V. atalanta*†; *V. cardui*†; *A. urticae*†; *B. selene*[(2)]*; *A. aglaja*[2]†; *E. aethiops*[2]*; *H. semele*[3]*; *M. jurtina*[5]†; *C. pamphilus*[2]†; *C. tullia*[1]*.

FLODIGARRY [NG47]: *P. icarus*[2]‡; *V. cardui*‡.

No records for Oronsay [NM55], Riska [NM67], Eilean Chatastail [NM48], Luinga Mhor [NM68], Ornsay [NG71], Eilean a'Mhal [NG72], Tarner and Harlosh [NG23], Crowlin Isles (Eilean Meadhonach and Eilean Beag) [NG63], Eilean Mor (near Dunvegan), Gairbh Eilein and Eilean Dubh [NG24], Mingay and Isay [NG25], Eilean Tigh [NG65], Eilean Creagach and Eilean Iosal [NG26], South Ascrib and Eilean Garave [NG36], Staffin [NG46] and Eilean Trodday [NG47].

References

Bird (1980), Bradley and Fletcher (1958), Campbell (1946, 1947a,b, 1948a,b, 1949a,b, 1950, 1951a, b, c, d, 1952, 1953, 1954, 1955a, b, 1958, 1967, 1969a, b, c, 1970, 1972, 1975a, b, 1978, 1980, 1984), Dale (1884d), Ellis (1896), Evans and Grimshaw (1916), Forbes (1876), Garrad (1973), Heslop Harrison (1937a, b, c, 1938a, b, c, 1939, 1941f, 1942d, e, f, g, 1943d, e, 1944, 1945a, b, d, 1946b, d, f, 1947a, b, g, 1948a, b, c, 1949b, c, d, 1950a, d, 1955b, 1956d, 1957, 1958b), Heslop Harrison and Morton (1952), Hull and Heslop Harrison (1937–1938), Hulme (1995), Kevan (1941), Kirkcaldy (1899), National Biodiversity Network (2016), Nonweiler (1949), Peacock, Smith and Davidson (1934–1935), Rutherford (1995), Shreeve (1994), Thomson (1970, 1980), White (1872), Wormell (1982, 1983).

Outer Hebrides

SHIANT ISLANDS (GARBH EILEAN and EILEAN AN TAIGHE) [NG49]: *P. napi*[4]*‡; *P. icarus*[2,3]*‡; *V. atalanta*‡; *V. cardui*‡; *M. jurtina*[5]*.

BERNERAY [NL58]: *P. brassicae**; *P. icarus*[2]*; *A. aglaja*[(2)]*; *H. semele*[3]*; *M. jurtina*[5]*‡; *C. tullia*[1]†.

MINGULAY [NL58]: *P. icarus*[2]*‡; *A. aglaja*[(2)]*‡; *H. semele*[3]*; *M. jurtina*[5]*‡; *C. pamphilus*[(2)]*; *C. tullia*[1]*.

PABBAY (south of Barra) [NL68]: *P. icarus*[3]*; *A. aglaja*[2]*; *H. semele*[3]*; *M. jurtina*[5]‡.

SANDRAY [NL69]: *P. icarus*[2]*‡; *V. cardui**; *A. aglaja*[2]*‡; *H. semele*[3]*; *M. jurtina*[5]‡.

FLODDAY (south of Barra) [NL69]: *P. icarus*[3]*; *A. aglaja*[3]*; *H. semele*[3]*.

MULDOANICH [NL69]: *P. icarus*[3]*; *A. aglaja*[1]*; *M. jurtina*[5]*.
UINESSAN [NL69]: *A. urticae**; *A. aglaja*[2]*; *H. semele*[3]*.
VATERSAY [NL69]: *P. brassicae**; *P. napi*[4]‡; *P. icarus*[2]†‡; *V. cardui*‡; *A. aglaja*[2]*; *H. semele*[3]*; *M. jurtina*[5]†.
BARRA [NF60]: *C. croceus**; *P. brassicae*†‡; *P. rapae*†; *P. napi*[4]†‡; *A. cardamines*[2](?)†; *P. icarus*[2]†‡; *V. atalanta*†‡; *V. cardui*†‡; *A. urticae*†‡; *A. io**; *A. aglaja*[2]†‡; *H. semele*[3]†‡; *M. jurtina*[5]†‡; *C. tullia*[1]*‡.
GIGHAY [NF70]: *P. napi*[4]†; *P. icarus*[2]†; *A. urticae*†; *A. aglaja*[2]†; *M. jurtina*[5]†.
HELLISAY [NF70]: *A. aglaja*[2]*.
OROSAY [NF70]: *A. io*‡.
FUDAY [NF70]: *P. napi*[4]‡; *P. icarus*[2]*‡; *A. urticae*‡; *A. aglaja*[2]*‡; *H. semele*[3]*‡; *M. jurtina*[5]‡.
ERISKAY [NF71]: *C. croceus**; *P. brassicae*†; *P. napi*[4]*‡; *P. icarus*[2]†‡; *V. atalanta**‡; *V. cardui*†‡; *A. urticae*†‡; *A. io*‡; *A. aglaja*[2]*‡; *H. semele*[3]*‡; *M. jurtina*[5]*‡; *C. pamphilus*[2]*‡.
FIARAY [NF7l]: *P. icarus*[3]*.
SOUTH UIST [NF73]: *P. brassicae*†‡; *P. rapae*†‡; *P. napi*[4]†‡; *A. cardamines*[2]†; *P. icarus*[2,3]†‡; *V. atalanta*†‡; *V. cardui*†‡; *A. urticae*†‡; *A. io*†‡; *A. aglaja*[2]†‡; *P. aegeria*[2]‡; *H. semele*[3]†‡; *M. jurtina*[5]‡, *A. hyperantus*[(2)]‡; *C. pamphilus*[2,3]†‡; *C. tullia*[1]†‡. BMS 2009–2011.
WIAY [NF84]: *C. pamphilus*[(2)] †.
EILEAN NA CILLE [NF84]: *P. napi*[4]‡; *P. icarus*[2]‡; *V. atalanta*‡; *A. urticae*‡; *A. io*‡; *M. jurtina*[5]‡.
BENBECULA [NF85]: *P. brassicae*†‡; *P. rapae*†; *P. napi*[4]†‡; *P. icarus*[2]†‡; *V. atalanta*†‡; *V. cardui*†‡; *A. urticae*†‡; *A. aglaja*[2]†‡; *M. jurtina*[5]†‡; *C. pamphilus*[(2)]†; *C. tullia*[1]†‡; *D. plexippus**.
CALAVAY [NF85]: *P. icarus*[2]†.
RONAY [NF85]: *P. icarus*[2]*; *M. jurtina*[5]*.
GRIMSAY (south of North Uist) [NF85]: *P. napi*[4]‡; *P. icarus*[2]‡; *V. atalanta*‡; *V. cardui*‡; *A. urticae*‡; *M. jurtina*[5]*‡; *C. tullia*‡.
MONACH ISLANDS (CEANN EAR AND CEANN IAR) [NF66]: *P. napi*[4]†‡; *P. icarus*[2]†‡; *V. cardui*‡; *A. urticae*†‡; *M. jurtina*[5]*.
BALESHARE [NF75]: *P. napi*[4]‡; *P. icarus*[2]*‡; *V. atalanta*‡; *V. cardui*‡, *A. urticae**‡; *M. jurtina*[5]*‡.
NORTH UIST [NF86]: *P. brassicae*†‡; *P. rapae*†‡; *P. napi*[4]†‡; *L. boeticus*‡ (1 in 2015); *P. icarus*[2]†‡; *V. atalanta*†‡; *V. cardui*†‡; *A. urticae*†‡; *A. io*†; *A. aglaja*[(2)]‡; *M. jurtina*[5]†‡; *C. pamphilus*[2]†‡; *C. tullia*[1]†‡.
PABBAY (north of North Uist) [NF88]: *P. icarus*[3]*; A. *aglaja*[2]*; *H. semele*[3]*; *M. jurtina*[5]‡.
BERNERAY (north of North Uist) [NF98]: *P. brassicae**‡; *P. rapae*†; *P. napi*[4]‡; *P. icarus*[2]*‡; *V. atalanta*‡; *V. cardui*‡; *A. urticae*‡; *A. io*†; *M. jurtina*[5]†‡; *C. tullia*[1]*.
KILLEGRAY [NF98]: *P. icarus*[2,3]*‡; *V. cardui**; *M. jurtina*[5]*‡.
ENSAY [NF98]: *P. napi*[4]*; *P. icarus*[2]*; *V. atalanta*†; *V. cardui*†; *A. urticae**; *M. jurtina*[5]*.
SHILLAY [NF89]: *P. icarus*[2]*; *M. jurtina*[5]*.
ST KILDA (HIRTA, SOAY and BORERAY) [NF09]: *C. croceus*†; *V. atalanta**‡; *V. cardui*†‡; *A. urticae*†; *C. pamphilus*[(2)]*.

SCALPAY (near East Loch Tarbert) [NG29]: *P. napi*[4]‡; *P. icarus*[2]‡.
SCOTASAY [NG19]: *P. icarus*[2]*; *M. jurtina*[5]*.
SCARP [NA91]: *P. brassicae**; *P. icarus*[2]*‡; *V. cardui*‡; *M. jurtina*[5]*.
MEALASTA [NA92]: *P. napi*[4]‡; *M. jurtina*[5]‡.
TARANSAY [NB00]: *P. icarus*[2]*; *A. urticae**; *V. cardui*‡; *M. jurtina*[5]†‡; *C. tullia*[1]*.
EILEAN CHALIUM CHILLE [NB32]: *A. urticae*‡.
LEWIS and HARRIS [NG08]: *C. croceus**‡; *P. brassicae*†‡; *P. rapae*†‡; *P. napi*[4]†‡; *A. cardamines*[2]‡; *P. icarus*[2]†‡; *V. atalanta*†‡; *V. cardui*†‡; *A. urticae*†‡; *A. io*†‡; *P. c-album*†; *P. aegeria*[2]‡; *H. semele*[3]†; *M. jurtina*[5]†‡; *C. pamphilus*[(2)]†‡; *C. tullia*[1]†‡.
FLODAY [NB13]: *P. icarus*[2]‡; *M. jurtina*[5]‡.
GREAT BERNERA [NB13]: *P. napi*[4]‡; *P. icarus*[2]*‡; *V. atalanta**; *A. urticae*‡; *M. jurtina*[5]‡; *A. hyperantus*[(2)]‡; *C. pamphilus*[(2)]‡; *C. tullia*[1]‡.
LITTLE BERNERA [NB14]: *P. icarus*[2]*‡; *A. urticae*‡.
NORTH RONA [HW83]: *V. cardui*‡; *A. urticae**.

No records for Eilean Mhuire (Shiant Islands) [NG49], Roisinis and Lingay (south of Barra) [NL68], Flodday (east of Barra), Garbh Lingay, Langay Fhada, Lamalum and Stack [NF70], Lingay (north of Barra) [NF71], Calvay and Gasaigh [NF81], Stuley [NF82], Ornish, Shillay Mor and Shillay Beg [NF83], Liursaigh Dubh, Gasaidh, Fodragaigh, Triallabreac, Kieravagh Islands (East and West) and Grimsay (south-east of Benbecula) [NF84], Gulalan [NF74], Garbh Eilean Mor and Eilean Flodaigh [NF85], Kirkibost [NF76], Vallay [NF77], Oronsay and Lingay (north of North Uist) [NF87], Boreray [NF88], Hermetray Bhacasaigh, Sromaigh, Taghaigh and Sursaigh [NF97], Lingay, Groay and Gilsay [NG07], Gasker [NA81], Kearstay [NA91], Flannans (Eilean Mor and Eilean Tighe) [NA74], Soay Beg and Soay Mor [NB00], Eilean Iubhard [NB30], Eilean Mor Laxay [NB32], Eilean Rosaidh, Eilean Thoraidh, Eilean Orasaigh, Tabhaigh Mhor and Tannaraidh [NB42], Eilean Orasaigh (Loch Roag) [NB23], Vuia Beg, Eilean Kearstay, Vuia Mor, Ceabaigh, Vacsay, Eilean Bhacasaigh and Pabaidh Mor [NB13], and Rockall [MC01].

References

Beirne (1943b), Blackie (1951), Campbell (1936, 1938, 1946, 1948a, b, 1949a), Carrington (1881), Corbet (1945), Dale (1884a, b, c, 1889), Dannreuther (1935, 1939), Duffield (2017), Evans (1912b), Forrest, Waterston and Watson (1936), Fremlin (1900), Grimshaw (1920), Hackett (1994, 1995), Heslop Harrison (1938a, c, 1940b, d, h, 1941a, b, e, 1942b, c, d, e, f, 1943e, 1944, 1945a, c, 1946a, d, f, g, 1947d, e, f, g, 1948a, b, 1949a, b, 1950a, b, c, d, 1953, 1954a, 1955a, 1956a, b, c, d, 1958b), Heslop Harrison and Morton (1953), Lack (1932), Leech (1951), MacGillivray (1842), [McArthur] (1901), Scottish Naturalist (1888), Shaw (1951), Shreeve (1995-2003), South (1888a, b), Thomson (1970, 1980), Trail (1888), Trevor (1994), Walker (1932), Waterston (1906), Weir (1881a), White, B. (1882), White, F.B.W. (1872).

North-west and north coast of Scotland

LONGA [NG77]: *M. jurtina*[5]†.
EILEAN FURADH MOR [NG79]: *P. icarus*[(2)]†; *H. semele*[(3)] †.
ISLE MARTIN [NH09]: *A. cardamines*[2]‡.
SUMMER ISLANDS (not distinguished) [NB90]: *E. aethiops*[2]†; *H. semele*[(3)]†; *M. jurtina*[5]†.
TANERA BEG [NB90]: *P. brassicae*†; *P. napi*[4]†; *P. icarus*[(2)]†‡; *E. aethiops*[2]†; *M. jurtina*[5]†.
TANERA MOR [NB90]: *P. brassicae*†‡; *P. napi*[4]†‡; *P. icarus*[(2)]‡, *V. atalanta*‡; *V. cardui*‡; *A. io*‡; *P. aegeria*[2]‡; *E. aethiops*[2]†; *H. semele*[(3)]†; *M. jurtina*[5]†‡; *C. tullia*[1]‡.
PRIEST [NB90]: *P. napi*[4]‡; *P. icarus*[(2)]‡; *V. atalanta*‡; *V. cardui*‡; *A. urticae*†; *H. semele*[(3)]‡; *M. jurtina*[5]‡.
GLAS-LEAC MOR [NB90]: *P. icarus*[(2)]‡; *V. cardui*‡; *M. jurtina*[5]‡.
ISLE RISTOL [NB91]: *P. napi*[4]‡; *P. icarus*[(2)]‡; *A. urticae*‡; *A. io*‡; *M. jurtina*[5]‡.
CARN NAN SGEIR [NC00]: *V. cardui*†.
SOYEA [NC02]: *P. icarus*[(2)]‡; *V. cardui*‡; *M. jurtina*[5]‡.
OLDANY [NC03]: *P. icarus*[(2)]‡; *V. atalanta*‡; *A. urticae*‡; *B. selene*[(2)]‡; *H. semele*[(3)]‡; *M. jurtina*[5]‡.
HANDA [NC14]: *P. brassicae*†‡; *P. rapae*†; *P. napi*[4]†‡; *P. icarus*[(2)]†‡; *V. atalanta*†‡; *V. cardui*†‡; *A. urticae*†‡; *A. io*†; *A. aglaja*[2]‡; *E. aethiops*[2]†‡; *M. jurtina*[5]†‡; *C. pamphilus*[(2)]†; *C. tullia*[1]†‡.
RABBIT ISLANDS (WEST ISLAND) [NC66]: *P. icarus*[(2)]‡; *A. aglaja*[2]‡; *M. jurtina*[5]‡.
NEAVE [NC66]: *P. napi*[4]‡; *P. icarus*‡; *A. urticae*‡; *A. aglaja*[2]‡; *M. jurtina*[5]‡.
EILEAN NAN RON [NC66]: *P. brassicae*‡; *P. napi*[4]‡; *P. icarus*[(2)]‡; *M. jurtina*[5]‡.
EILEAN HOAN [NC46]: *V. atalanta*‡; *V. cardui*‡; *A. urticae*‡.

No records for Ewe [NG88], Gruinard [NG99], Isle Martin [NH09], Bottle Island, Carn Iar, Carn Deas, Eilean Dubh (5) (Summer Isles), Glas-leac Beg and Eilean Fada Mor [NB90], Eilean Mullagrach [NB91], Horse and Meall nan Gabhar [NC00], Eilean Mor (Enard Bay) [NC01], Calbha Beag, Calbha Mor, Meall Mor and Meall Beg [NC13], Eilean a'Mhadaidh [NC14], Eilean Ard, Eilean an Roin Mor, Eilean an Roin Beag [NC15], Eilean an Eireannaich [NC25], Eilean Choraidh [NC45], Rabbit Islands (east island) [NC66] and Stroma [ND37].

References

Barbour (2016), Heslop Harrison (1940a), Thomson (1980), White (1872).

Orkney Islands

STACK [HX51]: *V. atalanta*‡; *V. cardui*‡.
SULE SKERRY [HX62]: *V. atalanta*‡; *V. cardui*‡.
SWONA [ND38]: *P. brassicae*‡; *P. napi*[4]‡; *L. phlaeas*[3]‡; *P. icarus*[2]‡; *V. atalanta*‡; *V. cardui*†; *A. urticae*†.
SOUTH WALLS [ND39]: *C. croceus*‡; *P. brassicae*‡; *P. napi*[4]‡; *P. icarus*[2]‡; *V. atalanta*‡; *V. cardui*‡; *A. urticae*‡; *A. io*‡; *M. jurtina*[5]‡; *C. tullia*[1]‡.

SOUTH RONALDSAY [ND48]: *P. brassicae*†‡; *P. rapae*†‡ (1 in 2002); *P. napi*[4]†‡; *L. phlaeas*[3]‡; *P. icarus*[2]†‡; *V. atalanta*†‡; *V. cardui*†‡; *A. urticae*†‡; *A. io*‡; *B. selene*[(2)]‡ (1 in 2006); *A. aglaja*[2]†; *M. jurtina*[5]†‡; *C. tullia*[1]†‡.

HOY [ND29]: *C. croceus*‡; *P. brassicae*†‡; *P. rapae*†; *P. napi*[4]†‡; *L. phlaeas*[3]*‡; *P. icarus*[2]†‡; *V. atalanta*†‡; *V. cardui*†‡; *A. urticae*†‡; *N. antiopa**‡ (1 in 2004); *A. io*†‡; *A. aglaja*[2]†; *M. jurtina*[5]†‡; *C. tullia*[1]†‡. WCBS (2 transects) 2013 (both).

SWITHA [ND39]: *P. napi*[4]‡; *V. atalanta*‡.

FLOTTA [ND39]: *P. napi*[4]‡; *P. icarus*[2]‡; *V. atalanta*‡; *V. cardui*‡; *A. urticae*‡; *M. jurtina*[5]‡; *C. tullia*[1]‡.

FARA [ND39]: *P. napi*[4]‡.

HUNDA [ND49]: *P. napi*[4]‡; *V. atalanta*‡; *V. cardui*‡; *A. urticae*‡; *M. jurtina*[5]‡.

BURRAY [ND49]: *P. brassicae*†‡; *P. napi*[4]†‡; *P. icarus*[2]†‡; *V. atalanta*‡; *V. cardui*†; *A. urticae*†‡; *A. io*‡; *A. aglaja*[2]†‡; *M. jurtina*[5]†‡.

GLIMPS HOLM [ND49]: *P. napi*[4]‡; *P. icarus*[2]‡; *V. atalanta*‡; *V. cardui*‡; *A. urticae*‡.

LAMB HOLM [HY40]: *P. brassicae*‡; *P. icarus*[2]‡; *V. cardui*‡; *A. urticae*‡; *M. jurtina*[5]‡.

CORN HOLM [HY60]: *V. cardui*‡.

COPINSAY [HY60]: *V. atalanta*‡; *V. cardui*‡.

GRAEMSAY [HY20]: *P. icarus*[2]‡; *V. atalanta*‡; *V. cardui*‡; *A. urticae*†; *M. jurtina*[5]‡.

MAINLAND ORKNEY [HY31]: *C. croceus**‡; *P. brassicae*†‡; *P. rapae*†‡ (1 in 2004); *P. napi*[4]†‡; *L. phlaeas*[3]‡; *P. icarus*[2]†‡; *V. atalanta*†‡; *V. cardui*†‡; *A. urticae*†‡; *A. io*†‡; *A. aglaja*[2]†‡ (1 in 2003); *M. jurtina*[5]†‡; *C. tullia*[1]†‡. WCBS (2 transects) 2006, 2010 and 2011 (1) and 2009 (1).

SHAPINSAY [HY41]: *P. brassicae*†‡; *P. napi*[4]‡; *P. icarus*[2]†‡; *V. atalanta*†‡; *V. cardui*‡; *A. urticae*‡; *M. jurtina*[5]‡.

AUSKERRY [HY61]: *P. brassicae*‡; *V. atalanta*‡; *V. cardui*‡; *A. io*‡.

BROUGH OF BIRSAY [HY22]: *P. brassicae*‡.

GAIRSAY [HY42]: *P. brassicae*‡; *P. napi*[4]‡; *P. icarus*[2]‡; *V. atalanta*‡; *V. cardui*‡; *A. urticae*‡; *M. jurtina*[5]‡.

WYRE [HY42]: *P. napi*[4]‡; *A. urticae*‡; *M. jurtina*[5]‡.

EGILSAY [HY42]: *P. napi*[4]‡; *P. icarus*[2]‡; *V. atalanta*‡; *V. cardui*‡; *A. urticae*‡.

STRONSAY [HY62]: *P. brassicae*‡; *P. napi*[4]‡; *V. atalanta*‡; *V. cardui*‡; *A. urticae*‡; *N. polychloros*‡? (1 in 2006).

LINGA HOLM [HY62]: *V. atalanta*‡.

ROUSAY [HY43]: *P. brassicae*†‡; *P. napi*[4]‡; *P. icarus*[2]†‡; *V. atalanta*‡; *V. cardui*†‡; *A. urticae*†‡; *M. jurtina*[5]‡.

EDAY [HY53]: *P. brassicae*‡; *P. napi*[4]‡; *P. icarus*[2]‡; *V. atalanta*‡; *V. cardui*‡; *A. urticae*‡; *M. jurtina*[5]‡.

FARAY [HY53]: *V. atalanta*‡; *V. cardui*‡.

HOLM OF FARAY [HY53]: *V. cardui*‡.

SANDAY [HY64]: *P. brassicae*†‡; *P. rapae*†; *P. napi*[4]†‡; *P. icarus*[2]†‡; *V. atalanta*†‡; *V. cardui*‡; *A. urticae**; *M. jurtina*[5]‡.
WESTRAY [HY44]: *P. brassicae*†‡; *P. napi*[4]‡; *V. atalanta*‡; *V. cardui*†‡; *A. urticae*†; *A. io*‡.
PAPA WESTRAY [HY45]: *P. brassicae*‡; *A. cardamines*[2]‡ (1 in 1999); *V. atalanta*‡; *V. cardui*‡; *A. urticae*‡.
HOLM OF PAPA [HY55]: *V. cardui*‡.
NORTH RONALDSAY [HY75]: *P. brassicae*†‡; *P. rapae*†; *P. napi*[4]†‡; *P. icarus*[2]‡; *V. atalanta*†‡; *V. cardui*†‡; *A. urticae*†‡; *N. antiopa*‡ (1 in 2002); *A. io*‡; *M. jurtina*[5]†.

No records for Sula Sgeir [HW63], Pentland Skerries [ND47], Rysa Little and Cava [ND39], Holm of Houton [HY30], Holm of Grimbister and Damsay [HY31], Helliar Holm [HY41], Eynhallow [HY32], Sweyn Holm [HY42], Muckle Green Holm [HY52], Papa Stronsay [HY62], Holm of Scockness [HY43], Calf of Eday [HY53] and Holm of Huip [HY63].

References

Archer (2003), Asher (1996), Bland (1977), Bristowe (1931a), Cheesman (1898), Scottish Naturalist (1888), Gauld (2016), Gibson (1974), Gregson (1885), Hanbury (1895), Heslop Harrison (1942a), Hodgson (1949a, b, 1951a, b, 1952a, b), Howard (1975), Lorimer (1981, 1983, 1988), Marwick (1931), Rockingham (1939), Shapiro (1970a), South (1888b), Stewart (1950), Tait (1878), Thomson (1970, 1980), Traill (1869, 1888), Weir (1882a), White, B. (1882), White, F.B.W. (1872), de Worms (1969, 1970, 1972).

Shetland Islands

FAIR ISLE [HZ27]: *P. machaon*[2]†; *C. croceus*†; *P. brassicae*†‡; *P. rapae*†; *P. icarus*[2]†; *V. atalanta*†‡; *V. cardui*†‡; *A. urticae*†‡; *A. io*†‡; *M. jurtina*[5]†; *C. pamphilus*[(1)]*.
FOULA [HT93]: *P. brassicae*†‡; *P. rapae*†; *V. atalanta*†‡; *V. cardui*†‡; *A. urticae*‡; *N. antiopa*†; *A. io*†‡.
ST NINIAN'S ISLE [HU32]: *V. cardui*‡; *A. urticae*‡.
MOUSA [HU42]: *P. brassicae*‡; *V. atalanta*†‡; *V. cardui*‡; *A. io*‡.
EAST BURRA [HU33]: *P. brassicae*‡.
WEST BURRA [HU33]: *P. brassicae*‡; *V. atalanta*†‡; *V. cardui*‡..
TRONDRA [HU33]: *P. brassicae*‡; *V. atalanta*‡; *V. cardui*‡; *A. io*‡.
MAINLAND SHETLAND [HU34]: *P. machaon*[2]†; *C. croceus**‡; *P. brassicae*†‡; *P. rapae**; *P. icarus*[2]†; *V. atalanta*†‡; *V. cardui*†‡; *A. urticae*†‡; *N. antiopa**‡(1 in 1996, 1 in 1999); *A. io*†‡; *C. tullia*[1]*(?); *D. plexippus**.
BRESSAY [HU54]: *C. croceus*‡; *P. brassicae*†‡; *P. napi*[4]†; *V. atalanta*†‡; *V. cardui*†‡; *N. antiopa*†; *A. io*†‡. WCBS 2013.
NOSS [HU54]: *P. brassicae*†‡; *V. atalanta*†‡; *V. cardui*†‡; *A. urticae*†‡; *A. io*†‡.
PAPA STOUR [HU16]: *P. brassicae*†; *V. atalanta*†‡; *V. cardui*†; *A. io*‡.

MUCKLE ROE [HU36]: *P. brassicae*‡; *V. atalanta*‡; *V. cardui*‡.

WHALSAY [HU56]: *P. brassicae*†‡; *V. atalanta*†‡; *V. cardui*†‡; *A. io*†‡.

HOUSAY (OUT SKERRIES) [HU67]: *P. brassicae*†‡; *V. atalanta*†‡; *V. cardui*‡; *A. urticae*†; *A. io*†‡.

YELL [HU49]: *P. brassicae*†‡; *P. rapae*†; *V. atalanta*†‡; *V. cardui*†‡; *A. urticae*†; *A. io*†‡.

FETLAR [HU69]: *P. brassicae*†‡; *V. atalanta*†‡; *V. cardui*†‡; *A. urticae*†‡; *A. io*†‡.

UYEA (south of Unst) [HU69]: *P. brassicae*†.

UNST [HU59]: *P. brassicae*†‡; *A. artaxerxes*[1]*(?); *V. atalanta*†‡; *V. cardui*†‡; *A. urticae*†‡; *N. polychloros*‡ (1 in 2002); *A. io*†‡.

No records for Little Havra and South Havra [HU32], Oxna, Papa and Linga (north of Burra) [HU33], Vaila [HU24], Hildasay and North Havra [HU34], South Isle of Gletness [HU45], Vementry [HU26], Papa Little and Linga [HU36], West Linga [HU56], Linga (near Samphrey), Fish Holm, Samphrey, Orfasay, Little Roe and Bigga [HU47], Lunna Holm [HU57], Grunay and Bruray (Out Skerries) [HU67], Lamba [HU38], Uynarey and Brother Isle [HU48], Uyea (north-west of Mainland Shetland) [HU39], Hascosay, Urie Lingey and Linga (north-east of Yell) [HU59], Haaf Gruney [HU69], Huney and Balta [HP60] and Muckle Flugga [HP61].

References

Anonymous (1993), Asher (1996), [Baldwin] (1995), Beirne (1943b, 1945), Briggs (1884), Bristowe (1931a), Carrrington (1880, 1881, 1883), Clarke (1897), Goater (1969, 1973), Grimshaw (1906), Hanbury (1895), Hardy (1956), Hare (1963), Harper (1974), Harvey (1990, 2016), Harvey, Riddiford and Riddiford (1992), Hughes (1935), Kennar (1912), King (1901), Kinnear (1976), Oldham (1931), Pennington (1993, 1994, 1996a, b, c), Pennington and Riddiford (1993), Ralston (1959, 1960), Robertson (1980), Rushton (1971), Scottish Naturalist (1888), Smith (1984), South (1888b, 1893), Thomson (1980), Traill (1888), Walker (1932), Waterston (1911, 1913), Weir (1880, 1881b, 1884), White, B. (1882), White, F.B.W. (1872), de Worms (1939, 1959).

East Scotland and Firth of Forth

INCH GARVIE [NT17]: *P. rapae*‡ (2002); *P. napi*[4]‡ (2005); *V. atalanta*‡ (2006).

CRAMOND [NT17]: *C. croceus*†; *P. brassicae*†; *P. rapae*†; *P. napi*[4]†; *L. phlaeas*[(1)]†; *P. icarus*[1]†; *V. atalanta*†; *V. cardui*†; *A. urticae*†; *A. io*†; *M. jurtina*[2]†; *C. pamphilus*[1]†.

INCHCOLM [NT18]: *P. brassicae*†‡; *P. rapae*†‡; *P. napi*[4]†‡; *A. cardamines*[2]‡; *L. phlaeas*[(1)]†‡; *P. icarus*[1]†; *V. atalanta*†‡; *V. cardui*†‡; *A. urticae*†‡; *A. io*†‡; *M. jurtina*[2]†; *A. hyperantus*[(2)]‡ (2003); *C. pamphilus*[1]†.

INCHMICKERY [NT28]: *P. rapae*†; *L. phlaeas*[(1)]†; *V. atalanta*†; *A. urticae*†‡.

INCHKEITH [NT28]: *P. brassicae*‡; *P. rapae*†‡; *P. napi*[4]†‡; *L. phlaeas*[(2)]†; *P. icarus*[1]†; *V. atalanta*†‡; *V. cardui*†‡; *A. urticae*†‡.

FIDRA [NT58]: *P. brassicae*†; *P. rapae*†; *P. napi*[4]†; *V. atalanta*†; *A. urticae*†.

BASS ROCK [NT68]: *P. brassicae**; *P. rapae*†; *P. napi*[4]†; *L. phlaeas*[(1)]†; *P. icarus*[1]†; *V. atalanta*†; *V. cardui*†; *A. urticae**.

MAY [NT69]: *C. croceus*†; *P. brassicae*†‡; *P. rapae*†‡; *P. napi*[4]†‡; *A. cardamines*[2]‡; *L. phlaeas*[(1)]†‡; *P. icarus*[1]†; *V. atalanta*†‡; *V. cardui*†‡; *A. urticae*†‡; *A. io*†‡; *P. c-album*‡; *M. jurtina*[2]†‡; *A. hyperantus*[(2)]‡; *C. pamphilus*[1]*.

No records for Bell Rock [NO72] and Craigleith [NT58].

References

Baxter (1950), Davidson (2016), Edwards (2016), Eggeling (1957, 1985), Evans (1912a, 1918), Grimshaw (1908), Morris (1995), Rintoul and Baxter (1920), Savill (1996), Smout (1995), Smout and Kinnear (1993), Stringer (1995), Thomson (1980), White (1872).

D. Islands off Ireland

RATHLIN [D14]: *C. croceus*†; *P. brassicae*†‡; *P. rapae*†‡; *P. napi*[3]†‡; *A. cardamines*[3]†‡; *L. phlaeas*[2]†‡; *P. icarus*[(2)]†‡; *V. atalanta*†‡; *V. cardui*†; *A. urticae*†‡; *A. io*‡; *A. aglaja*[1]†; *A. paphia*‡; *P. aegeria*[1]†; *L. megera*[1]†; *H. semele*[(5)]†‡; *M. jurtina*[3]†‡; *A. hyperantus*[1]†‡; *C. pamphilus*[1]†.

MUCK [D40]: *P. icarus*[(2)]‡; *V. atalanta*‡; *M. jurtina*[3]‡; *A. hyperantus*[1]‡.

COPELAND (GREAT COPELAND/BIG COPELAND) [J58]: *P. icarus*[(2)]†; *V. atalanta*‡; *V. cardui*‡; *A. urticae*‡; *M. jurtina*[3]†.

JOHN'S (near COPELAND†; also called CROSS or LIGHTHOUSE) [J58]: *C. croceus*†; *P. brassicae*†; *P. rapae*†; *P. napi*[3]†; *A. cardamines*[3]†; *L. phlaeas*[2]†; *V. atalanta*†; *V. cardui*†; *A. urticae*†; *A. io*†‡; *P. aegeria*[1]†; *L. megera*[1]†; *M. jurtina*[3]†; *A. hyperantus*[1]†.

LAMBAY [O35]: *G. rhamni*[2]*‡; *P. brassicae*†‡; *P. rapae*†‡; *P. napi*[3]†‡; *A. cardamines*[3]†; *L. phlaeas*[2]†; *P. icarus*[(1)]†‡; *V. atalanta*†‡; *V. cardui*†‡; *A. urticae*†‡; *A. io*†; *A. aglaja*[1]*; *P. aegeria*[1]†‡; *L. megera*[1]†; *H. semele*[(5)]*; *M. jurtina*[3]†‡; *A. hyperantus*[1]†‡.

IRELAND'S EYE [O24]: *P. brassicae*‡; *P. rapae*‡; *P. napi*[3]‡; *A. cardamines*[3]‡; *L. phlaeas*[2]‡; *C. minimus*‡; *P. icarus*[(2)]‡; *V. atalanta*‡; *V. cardui*‡; *A. urticae*‡; *A. io*‡; *L. megera*[1]‡; *H. semele*[(5)]‡; *M. jurtina*[3]‡; *A. hyperantus*[1]‡.

NORTH BULL (BULL) [O23]: *P. brassicae*‡; *P. rapae*‡; *P. napi*[3]‡; *P. icarus*[(2)]‡; *V. atalanta*‡; *V. cardui*‡; *A. urticae*‡; *A. io*‡; *E. aurinia*[2]‡; *H. semele*[(5)]‡; *M. jurtina*[3]‡; *A. hyperantus*[1]‡; *C. pamphilus*[1]‡.

DALKEY [O22]: *P. rapae*‡; *V. atalanta*‡; *A. urticae*‡; *A. hyperantus*[1]‡.

KEERAGH [S80]: *V. atalanta*‡; *V. cardui*‡; *A. urticae*‡.

LITTLE SALTEE [X99]: *P. brassicae*†‡; *P. rapae*†; *P. napi*[3]†; *L. phlaeas*[2]†; *P. icarus*[(2)]†; *V. atalanta*†; *V. cardui*†; *A. urticae*†‡; *A. io*‡; *M. jurtina*[3]†.

GREAT SALTEE [X99]: *L. juvernica* (*L. sinapis* [*sic*])†‡; *C. croceus*†‡; *P. brassicae*†‡; *P. rapae*†‡; *P. napi*[3]†‡; *A. cardamines*[3]†; *L. phlaeas*[2]†‡; *C. minimus*†; *P. icarus*[(2)]†‡; *V. atalanta*†‡; *V. cardui*†‡; *A. urticae*†‡;

*N. antiopa**; *A. io*†‡; *A. aglaja*[1]†; *P. aegeria*[1]†‡; *L. megera*[1]†‡; *P. tithonus*[2]†; *M. jurtina*[3]†‡; *A. hyperantus*[1]†‡, *D. plexippus*‡.

GREAT ISLAND [W86]: *A. urticae*‡; *A. io*‡.

LITTLE ISLAND [W77]: *V. atalanta*‡; *V. cardui*‡.

SHERKIN [W02]: *C. croceus*†‡; *P. brassicae*†‡; *P. rapae*†‡; *P. napi*[3]†‡; *A. cardamines*[3]†‡; *C. rubi*†‡; *L. phlaeas*[2]†‡; *P. icarus*[(2)]†‡; *C. argiolus*[2]†‡; *V. atalanta*†‡; *V. cardui*†‡; *A. urticae*†‡; *A. io*†‡; *A. aglaja*[1]†‡; *A. paphia*†; *E. aurinia*[2]†‡; *P. aegeria*[1]†‡; *L. megera*[1]†‡; *H. semele*[(5)]†‡; *P. tithonus*[2]†‡; *M. jurtina*[3]†‡; *A. hyperantus*[1]†‡; *C. pamphilus*[1]†‡.

CLEAR (CAPE CLEAR) [V92]: *C. hyale*†; *C. croceus*†‡; *G. rhamni*[2]†‡; *P. brassicae*†‡; *P. rapae*†‡; *P. napi*[3]†‡; *A. cardamines*[3]†‡; *C. rubi*†‡; *L. phlaeas*[2]†‡; *C. minimus**‡; *P. icarus*[(2)]†‡; *C. argiolus*[2]‡; *V. atalanta*†‡; *V. cardui*†‡; *A. urticae*†‡; *A. io*†‡; *A. aglaja*[1]†‡; *E. aurinia*[2]†; *P. aegeria*[1]†‡; *L. megera*[1]†‡; *H. semele*[(5)]†‡; *P. tithonus*[2]†‡; *M. jurtina*[3]†‡; *A. hyperantus*[1]†‡; *C. pamphilus*[1]†‡; *D. plexippus*†.

WHIDDY [V94]: *P. brassicae*†‡; *P. rapae*†; *P. napi*[3]†; *A. cardamines*[3]†; *L. phlaeas*[2]†; *P. icarus*[(2)]†; *C. argiolus*[2]†; *V. atalanta*†‡; *A. urticae*†; *A. io*†; *P. aegeria*[1]†‡; *M. jurtina*[3]†.

GARINISH (1) (GARNISH OR ILNACULLIN) [V95]: *P. brassicae*†; *P. rapae*†; *P. napi*[3]‡; *P. icarus*[2]‡; *C. argiolus*[2]*‡; *V. cardui*†; *A. urticae*‡; *A. io*‡; *A. paphia*‡; *P. aegeria*[1]‡; *H. semele*[(5)]‡; *P. tithonus*[2]‡; *M. jurtina*[3]‡.

BEAR (BERE) [V74]: *A. cardamines*[3]‡; *C. rubi**; *V. cardui*‡; *A. urticae*†‡; *A. io**‡; *A. aglaja*[1]†; *H. semele*[(5)] †.

DURSEY [V44]: *C. croceus*‡; *P. brassicae*‡; *P. rapae*‡; *P. napi*[3]†‡; *A. cardamines*[3]‡; *C. rubi*‡; *L. phlaeas*[2]*‡; *P. icarus*[(2)]* ‡; *C. argiolus*[2]‡; *V. atalanta*‡; *V. cardui*‡; *A. urticae*‡; *A. io*‡; *A. aglaja*[1]‡; *A. paphia*‡ (1 in 1996); *P. aegeria*[1]‡; *L. megera*[1]‡; *H. semele*[(5)]*‡; *M. jurtina*[3]‡; *A. hyperantus*[1]‡, *D. plexippus*‡.

SCARRIF [V45]: *L. phlaeas*[2]†; *P. icarus*[(2)]†; *V. atalanta*†; *V. cardui*†; *A. urticae*†; *L. megera*[1]†; *M. jurtina*[3]†.

GREAT SKELLIG (SKELLIG MICHAEL) [V26]: *V. atalanta*†; *V. cardui*‡; *A. urticae*‡.

PUFFIN [V36]: *P. brassicae*†; *P. rapae*†; *A. urticae*†; *A. io*†; *M. jurtina*[3]†.

VALENTIA (VALENCIA) [V37]: *C. croceus**; *P. brassicae*‡; *P. rapae*†; *P. napi*[3]†; *A. cardamines*[3]†; *C. rubi**; *L. phlaeas*[2]†; *A. urticae*†; *A. aglaja*[1]†; *E. aurinia*[2]*; *P. aegeria*[1]†; *L. megera*[1]†; *H. semele*[(5)]*; *M. jurtina*[3]†; *A. hyperantus*[1]‡.

INISHVICKILLANE [V29]: *P. brassicae*†; *P. napi*[3]†; *P. icarus*[(2)]†; *V. atalanta*†; *A. urticae*†; *M. jurtina*[3]†.

GREAT BLASKET [V29]: *P. brassicae*†; *P. rapae*†; *P. napi*[3]†‡; *L. phlaeas*[2]†; *V. atalanta*†‡; *V. cardui*†; *A. urticae*†‡; *A. io*†; *A. aglaja*[1]‡; *P. aegeria*[1]†; *M. jurtina*[3]†; *A. hyperantus*[1]†; *C. pamphilus*[1]†.

MUTTON [Q97]: *P. brassicae*†; *P. napi*[3]†; *M. jurtina*[3].

INISHEER (INISHERE) [L90]: *E. tages*[2]‡; *P. brassicae*‡; *P. rapae*‡; *P.napi*[3]‡; *A. cardamines*[3]‡; *L. phlaeas*[2]‡; *C. minimus*‡; *P. icarus*[(2)]‡; *V. atalanta*‡;

V. cardui‡; *A. urticae**‡; *A. aglaja*[1]‡; *A. paphia*‡; *P. aegeria*[1]‡; *L. megera*[1]‡; *H. semele*[(4)]‡; *M. jurtina*[3]*‡ *A. hyperantus*[1]‡; *C. pamphilus*[1]‡.

INISHMAAN [L90]: *E. tages*[2]‡; *C. croceus*‡; *G. rhamni*[2]‡; *P. brassicae*†‡; *P. rapae*‡; *P. napi*[3]‡; *A. cardamines*[3]‡; *L. phlaeas*[2]‡; *C. minimus*‡; *P. icarus*[(2)]‡; *V. atalanta*‡; *V. cardui*‡; *A. urticae*†‡; *A. io*‡; *B. euphrosyne*‡; *A. aglaja*[1]‡; *P. aegeria*[1]‡; *L. megera*[1]‡; *M. jurtina*[3]†‡; *C. pamphilus*[1]‡.

INISHMORE [L80]: *E. tages*[2]†‡; *C. croceus*†; *G. rhamni*[2]†‡; *P. brassicae*†‡; *P. rapae*†‡; *P. napi*[3]†‡; *A. cardamines*[3]‡, *L. phlaeas*[2]†; *C. minimus*†‡; *P. icarus*[(2)]†‡; *V. atalanta*†‡; *V. cardui*†‡; *A. urticae*†‡; *A. io*†‡; *B. euphrosyne*‡; *A. aglaja*[1]†; *A. paphia*†; *E. aurinia*[2]‡; *P. aegeria*[1]†‡; *L. megera*[1]†‡; *H. semele*[(4)]†; *M. jurtina*[3]†‡; *A. hyperantus*[1]†; *C. pamphilus*[1]†‡.

GORUMNA [L82]: *P. brassicae*†; *P. rapae*†; *P. napi*[3]†; *A. cardamines*[3]†; *L. phlaeas*[2]†; *P. icarus*[(2)]†; *P. aegeria*[1]†; *L. megera*[1]†; *H. semele*[(5)]†‡; *M. jurtina*[3]†; *A. hyperantus*[1]†.

OMEY [L55]: *P. machaon*[2]‡ (2006); *L. phlaeas*[2]‡; *V. atalanta*‡; *V. cardui**; *A. urticae*‡; *E. aurinia*[2]; *M. jurtina*[3]‡.

INISHARK [L46]: *P. napi*[3]†; *M. jurtina*[3]†; *A. hyperantus*[1]†; *C. pamphilus*[1]†.

INISHBOFIN (1) [L56]: *C. croceus*‡; *G. rhamni*[2]‡; *P. brassicae*†‡; *P. rapae*†‡; *P. napi*[3]†‡; *L. phlaeas*[2]†; *P. icarus*[(2)]†‡; *V. atalanta*‡; *V. cardui*†‡; *A. urticae*†‡; *A. io*‡; *P. aegeria*[1]†; *L. megera*[1]†‡; *M. jurtina*[3]†‡; *A. hyperantus*[1]†; *C. pamphilus*[1]†‡.

INISHTURK [L67]: *P. rapae*‡; *P. napi*[3]†‡; *V. atalanta*‡; *V. cardui*‡; *A. urticae*†‡; *A. io*‡; *M. jurtina*[3]‡; *A. hyperantus*[1]†; *C. pamphilus*[1]†.

CAHER [L67]: *V. cardui*‡; *A. io*‡; *M. jurtina*[3]‡.

CLARE [L68]: *P. brassicae*†‡; *P. rapae*†; *P. napi*[3]†‡; *C. rubi*†; *L. phlaeas*[2]†; *C. minimus*†; *P. icarus*[(2)]†‡; *V. atalanta*‡; *V. cardui**; *A. urticae*†; *A. io*†; *P. aegeria*[1]*; *L. megera*[1]†‡; *H. semele*[(5)]*; *M. jurtina*[3]†‡; *A. hyperantus*[1]†; *C. pamphilus*[1]†.

ACHILLBEG [L79]: *C. croceus*‡.

INISHTURKBEG [L99]: *P. brassicae*‡; *V. atalanta*‡; *V. cardui*‡; *A. io*‡; *P. aegeria*[1]‡; *M. jurtina*[3]‡.

ACHILL [F60]: *C. croceus*‡; *P. brassicae*†‡; *P. rapae*†; *P. napi*[3]†‡; *A. cardamines*[3]‡; *C. rubi**‡; *L. phlaeas*[2]†‡; *C. minimus**; *P. icarus*[(2)]†‡; *C. argiolus*[2]*; *V. atalanta*†‡; *V. cardui*†‡; *A. urticae*†‡; *A. io*†‡; *A. aglaja*[1]†; *E. aurinia*[2]†; *P. aegeria*[1]‡; *L. megera*[1]†‡; *H. semele*[(5)]‡; *M. jurtina*[3]†‡; *A. hyperantus*[1]†; *C. pamphilus*[1]†‡; *C. tullia*[1]*.

BARTRAGH [G23]: *L. phlaeas*[2]‡; *C. minimus*‡; *P. icarus*[(2)]‡; *A. aglaja*[1]‡; *M. jurtina*[3]‡; *C. pamphilus*[1]‡; *C. tullia*[1]‡.

DERNISH (1) [G65]: *E. tages*[2]‡; *A. cardamines*[3]‡; *P. aegeria*[1]‡.

INISHMURRAY [G55]: *P. brassicae*‡; *P. rapae*‡; *P. napi*[3]‡; *L. phlaeas*[2]‡; *P. icarus*[(2)]‡; *V. atalanta*‡; *V. cardui*‡; *A. urticae*‡; *A. io*‡; *P. aegeria*[1]‡; *L. megera*[1]‡; *M. jurtina*[3]‡; *A. hyperantus*[1]‡.

INCH [C32]: *P. brassicae*‡; *P. rapae*‡; *P. napi*[3]‡; *A. urticae*‡; *A. io*‡; *P. aegeria*[1]‡.

INISHTRAHULL [C36]: *P. brassicae*†; *P. rapae*†; *P. napi*[3]†‡; *A. cardamines*[3]‡; *L. phlaeas*[2]†; *V. atalanta*†‡; *V. cardui*†‡; *A. urticae*†‡; *A. io*‡; *A. paphia*†; *P. aegeria*[1]†; *H. semele*[(5)]†; *M. jurtina*[3]†.

UMFIN [B72]: *P. icarus*[(2)]‡.

ARAN (ARRANMORE) [B72]: *P. brassicae*†‡; *P. napi*[3]‡; *A. cardamines*[3]‡; *C. rubi*‡; *L. phlaeas*[2]†‡; *P. icarus*[(2)]‡; *V. atalanta*†‡; *V. cardui*‡; *A. urticae*†‡; *A. io*‡; *A. aglaja*[1]‡; *E. aurinia*[2]†‡; *P. aegeria*[1]‡; *L. megera*[1]‡; *H. semele*[(5)]†; *M. jurtina*[3]†‡; *A. hyperantus*[1]‡; *C. pamphilus*[1]‡†.

OWEY [B72]: *P. icarus*[(2)]‡; *M. jurtina*[3]‡.

EIGHTER [B71]: *P. icarus*[(2)]‡; *A. urticae*‡; *E. aurinia*[2]‡; *L. megera*[1]‡; *H. semele*[(5)]‡; *M. jurtina*[3]‡.

INISHFREE UPPER [B71]: *P. brassicae*‡; *L. phlaeas*[2]‡; *P. icarus*[(2)]‡; *E. aurinia*[2]‡; *P. aegeria*[1]‡; *C. pamphilus*[1]‡.

CRUIT [B72]: *C. croceus*‡; *G. rhamni*[2]‡; *P. brassicae*‡; *P. rapae*‡; *P. napi*[3]‡; *A. cardamines*[3]‡; *L. phlaeas*[2]‡; *C. minimus*†‡; *P. icarus*[(2)]†‡; *V. atalanta*‡; *V. cardui*‡; *A. urticae*†‡; *A. io*†‡; *A. aglaja*[1]†‡; *E. aurinia*[2]†‡; *P. aegeria*[1]‡; *L. megera*[1]‡; *H. semele*[(5)]†‡; *M. jurtina*[3]†‡; *A. hyperantus*[1]‡; *C. pamphilus*[1]†‡.

GOLA [B72]: *P. napi*[3]‡; *P. icarus*[(2)]‡; *A. urticae*‡; *A. io*‡; *A. aglaja*[1]‡; *H. semele*[(5)]‡; *M. jurtina*[3]‡; *A. hyperantus*[1]‡; *C. pamphilus*[1]‡.

INISHMEANE [B72]: *P. icarus*[(2)]‡; *M. jurtina*[3]‡; *A. hyperantus*[1]‡; *C. pamphilus*[1]‡.

INISIRRER (INISHSIRRER) [B73]: *P. icarus*[(2)]‡; *A. aglaja*[1]‡; *A. paphia*‡; *E. aurinia*[2]‡; *M. jurtina*[3]‡; *A. hyperantus*[1]‡; *C. pamphilus*[1]‡.

TORY [B84]: *C. croceus*‡; *P. brassicae**‡; *P. rapae**‡; *P. napi*[3]*‡; *C. rubi*‡; *L. phlaeas*[2]*‡; *P. icarus*[(2)]*‡; *V. atalanta**‡; *V. cardui*‡; *A. urticae**‡; *A. io**‡; *L. megera*[1]‡; *H. semele*[(5)]*‡; *M. jurtina*[3]*‡; *A. hyperantus*[1]‡.

Records for butterflies on islands in Strangford Lough are listed in Appendix 12.

No records for the Maidens or Hulin Rocks [D41], Skerry Islands [C84], Cannon Rock [J65], Burial [J66], Mew [J68], Rockabill [O36], Colt and St. Patrick's [O26], Shenick's [O25], Tuskar Rock [T20], Begerin [T02], Bannow [S80], Spike [W86], Fota (Foaty) [W77], Rocky and Haulbowline [W76], Inchydoney [W43], Rabbit [W23], Ringarogy, Spanish and Heir (Hare or Inishodriscol) [W02], Inishbeg and Mannin [W03], Skeam East, Skeam West, Calf Islands (East), Calf Islands (Middle), Castle, Calf Islands (West) and Long [V92], Horse (1) [V93], Chapel [V94], Carbery and Furze [V83], Fastnet Rock [V81], Roancarrigmore [V74], Dinish (1) [V64], Inishfarnard (V65), Deenish (1) [V45], Abbey [V55], Little Skellig (Skellig Rock Little) [V26], Sherky and Garinish (2) [V66], Rossdohan, Rossmore and Ormond's [V76], Dinish (2), Cappanacush, Dunkerron West, Greenane and Dunkerron East [V86], Beginish (1), Church and Lamb [V47], Tearaght [V19], Inishabro and Beginish (2) [V29], Inishtooskert (Blasket Islands) [Q20], Fenit [Q71], Gurrig [Q52], Illaunboe, Ilauntannig, Reenafardarrig, Illaunimmil and Inishtooshkert (Magharee Islands or the Seven Hogs) [Q62], Carrig [Q94], Tarbert [R04], Scattery and Inishbig (Hog) [Q95], Foynes,

Aughinish (1), Inishcorker, Inishtubbrid, Canon and Inishmacowney [R25], Inishloe and Dernish (2) [R35], Green [R45], Coney (1), Feenish, Inishmacnaghtan, Inishmore (Deer), Deenish (2) and Horse (2) [R36], Aughinish (2) [M21], Eddy, Tawin and Tawin East [M31], Brannock Island Middle, Brannock Island South, Brannock Island East and Rock [L71], Finish, Masson (Mason), Mweenish and Inishmuskerry [L72], Birmore, Crappagh, Lettermullen (Lettermullan), Inisherk, Furnace, Dinish (3), Illauneeragh, Inishbarra and Lettermore [L82], Inchamakinna and Rossroe [L92], Croaghnakeela [L63], St. Macdara's, Rusheennachola, Freaghillaun, Inishlackan and Illaungorm North [L73], Inishlusk, Inishtravin, Ilauneeragh West and Illaunmore [L83], Annaghvaan, Clynagh, Roeilaun and Inisheltia [L93], Inishdawros [L64], Inishnee [L74], Illaunamid, Illaunurra and Horse (3) [L54], Turbot (Talbot or Inish Toirbirt), Inishturk (Inishturk South), Cruagh, High and Friar [L55], Crump [L66], Davillaun, Inishgort (1) and Inishlyon [L56], Annagh Islands (West), Annagh Islands (Middle), Annagh Islands (East), Inisheeny, Dorinish (Dorinish More and Dorinish Beg), Inishraher, Crovinish, Illanataggart, Inishgort (2), Inishlyre, Collanmore (Collan More), Island More, Knockycahilaun and Clynish [L98], Inishcooa, Inishilra, Moynish More, Inisherkin, Inishgowla (1), Inishnacross, Inishtubbrid, Inishkeel and Inishbobunnan [L89], Inishbee, Derrinish, Inishcottle, Inishnakillew, Inishoo, Inishgowla (2), Inishdaf, Rosbarnagh, Inishturlin, Inishquirk, Rosturk and Rosmore [L99], Annagh and Inishbiggle [F70], Duvillaun More and Duvillaun Beg [F51], Black Rock [F51], Inishkea South and Inishkea North [F52], Barranagh [F62], Inishglora, Inishkeeragh (1) and Eagle [F63], Kid [F74], Illanmaster [F94], Inishmulclohy (Coney (2)) and Oyster [G63], Conor's [G65], Inishpat and Rooney's [G97], Aughnish [C22], Rotten [G77], Rathlin O'Birne [G47], Inishkeel [B70], Inishkeeragh (2) [B61], Inishal, Rutland (Inishmacadum) and Inishcoo [B71], Inishfree Lower and Inishinny [B72], Inishbofin (2) and Inishdooey [B83] and Island Roy [C13].

References

Aldwell (2010), Baring (1915, 1918), Baynes (1957, 1964, 1973), Dannreuther (1950), Dublin Naturalists' Field Club (2010 to 2015), Dunn (1969) [Hogan and Haliday] (1855), Howarth (1971b), Huggins (1928, 1956), Kane (1907, 1912), Lansbury (1961), de Lattin (1952), Lavery (1989), Lavery and Cronin (1996), Mackenzie (1996), Nash (2016), Redway (1981), Rees and Sutton (1960), Rippey (1988 to 2016), Wedd (2004).

Appendix 4. Sources of Personal Observations on Island Records

Many records accumulated in Appendix 3 have been sent to us directly (occasionally indirectly) from private sources. In this appendix we have combined the lists of private sources for the 1996 book (Dennis and Shreeve, 1996) and this new title. Our grateful thanks to all those named below, and the multitude of individuals not possible to list here but who are recorded in the regional databanks for the Butterflies for the New Millennium Atlas Databank (see Asher *et al.*, 2011), and who have kindly contributed to the data on butterflies on British and Irish offshore islands. Without their generosity this work would not have been possible. Our knowledge of butterflies on the Irish offshore islands is largely owing to many detailed communications from Ian Rippey, to whom we are especially indebted, as we are to all those who are responsible for the records.

Date of letters, emails and locations covered are given in parentheses in the entries below. Details of current regional recorders can be found on the Butterfly Conservation website at https://butterfly-conservation.org/2390/recording-contacts.html

Aldwell, R. (2010; www.butterflyireland.com/2010_donegal_butterflies_june_10.htm).

Annell, R. (13.v.2016: data file for Hayling Island 1996–2015; 17.v.2016: data file for Isle of Wight 1996–2015).

Anthoney, M. (14.v.2016: data file for Denny 2000 and 2003).

Archer, M. (30.xi.2003: *V. cardui* on North Rona; *V. cardui* and *V. atalanta*, Sule Skerry).

Asher, J. Butterfly Net, *Millennium Atlas Project* (3.v.1996: Fair Isle, Shetland Islands, Orkney Islands and Isle of Man; *Nymphalis antiopa* and *Danaus plexippus* records for 1995).

Austin, R.A. (9.vi.2016: eleven data files for Guernsey 1990–1996, 2006–2011, 2013–2015).

Ayling, D. (25.v.2016: one data file for Steep Holm 1988–2008 and one for Stert 2007).

Barbour, D. (26.v.2016: eight data files for north-west Scotland 1995–2014).

Beavis, I. (2.vi.2004: Isles of Scilly).

Betts, M. and Price, S. (2.vi.1995, 16.vi.1995: Skokholm) including Macrolepidoptera lists in the Skokholm bird observatory reports of Downhill, I.R. (1961), Scott, D.A. (1968), Neale, A. (1970) and Lawman, J. (1977).

Black, J. (12.v.2016: advice of nil records from Ayrshire islands).

Bloomfield, M. (13.v.1994: Tresco).

Brazil, A. (9.vi.2016: data file for Scolt Head 1996–2015; 10.vi.2016: data file for Blakeney for 2015).

Brereton, T. (6.xii.2016: data file for Jersey 2004–2010).

Bristow, R. (18.v.2016: one data file for Lundy 1987–2015 and one for Drake's Island 1999).

Bullock, I. (20.xi.1996: Grassholm. *V. cardui* on 25.vi.1996).

Clough, S. (4.viii.2017: advice of nil records for Whitton Island).

Clynes, W.W. (12.iii.1996: Brownsman and Staple in Farnes).

Crow, P.N. (1971: *C. minimus* near Douglas, Man).

Curtis, G. (26.vi.2016: summary of Manx records).

Curtis, R. (10.i.2014: Isles of Scilly; *L. boeticus* on St Agnes).

Davidson, D. (14.v.2016: data file for Fife Islands 1996–2015 and Isle of May report).

Davison, J. (1.viii.2017: advice of nil records from Read's Island).

Dublin Naturalists' Field Club (website www.butterflyireland.com, viewed 2016: many observations for Irish islands 2010–2015).

Duffield, S. (4.vi.2016: three data files for Outer Hebrides 1997–2012 and 2014; 5.ix.2017: advice regarding *A. cardamines* and *V. cardui* in Hebrides).

Dunn, T.C. (12.x.1986: Lindisfarne).

Easterbrook, M. (9.vi.2016: data files for Eastchurch, Elmley, Minster Marshes, Minster on Sea, Shellness, Warden Point (all on Sheppey) 2010–2015; 16.vi.2016: data files for same places 2003–2009).

Edwards, G. (12.v.2016: advice of nil records from Angus).

Evans, S. (25.vii.2015: *V. atalanta*, *P. c-album*, *P. aegeria*, *M. jurtina* on Caldey Island).

Fox, M. (13.v.2016, 8.vi.2016: advice regarding Clyde islands).

Gander, L. (18.v.2016: data for Ynys Lochtyn (nil return) and Cardigan Island).

Gauld, S. (19.vi.2016: data file for Orkney 1996–2015).

George, R.A.B. (14.v.2016: data file for Dorset islands 1996–2015).

Gibson, J.A. (21.iv.1995, 28.vi.1995: Bute, Inchmarnock, Great Cumbrae, Little Cumbrae, Glas Eilean, Ailsa Craig, Horse, Lady, Holy, Pladda, Sanda, Davaar, Minard, Sgat Mor, Sheep, Glunimore, Burnt Islands).

Goodere, R. (6.vi.2016: one data file for Cornish islands 1965–2015 and one for Scillies 1996–2015).

Graham, A. (13.v.2016: data file for Anglesey and North Wales islands 1997–2015).

Graiff, R.D. and D. (29.ix.1995: Colonsay and Oronsay).

Hackett, P. (23.vi.1994, 2.ix.1994. 29.viii.1995: Outer Hebrides).

Harding, P.T. (ITE, BRC Database at Monks Wood Experimental Station, Abbots Ripton, Huntingdon, to 16.iii.1988).

Hardy, P.B. (24.v.1995: Wight and Holy Island).

Harvey, P. (20.vi.2016: data file for Shetland 1996–2015).

Hedges, G. (17.v.2016: data file for Walney and neighbouring islands 1996–2015).

Hodges, M. via McElwee, S. (24.i.1996: Inner Farne Islands).

Hulme, D.C. (5.v.1995: Handa, Tanera Mór, Tanera Beg, Eilean Fadadh Mór, Priest, Cam nan Sgeir, Mull, Staffa, Skye and Crowlin Isles).

Hull, M. (18.iii.1996: Herm).

Jeffcoate, G. (29.vi.2016: annual recording reports for Isle of Man 2011–2013).

Jeffcoate, S. (2.vii.2007: note on Calf of Man).

Lavery, T.A. and Cronin, K., Irish Lepidoptera Records Database (15.iv.1996: Lesser Saltee, Whiddy, Great Skellig, Great Blasket, Inishvickillane, Mutton, Inishark, Inishbofin, Inishturk).

Lockley, R.M. *Letters from Skokholm* including reference to Dent (1947, p. 218) in Betts, M. (16.vi.1995: Skokholm).

Long, R.C. (5.vi.2016: information on geography of smaller Channel islands).

Long, R.C. and Austin, R.A. (23.x.1986: Channel Islands).

Mackenzie, N. (16.vi.1996: Garinish, Ireland).

McCormick, M. via Hardy, P.B. (6.ix.1994: Bardsey).

McElwee, S. compiling records from Baxter, B., Littlewood, N. and Patterson, S. (24.i.1996: Outer Farne Islands).

McGowan, D. (7.ix.2017: data files from Jersey, 2004–2014).

Meakin, M. (22.viii.1995: Lundy).

Moon, J.R. (16.v.1994: Isles of Scilly; 22.viii.1994: Alderney).

Morgan, M.J. (19.x.1988: Puffin Island).

Morris, R. (16.xii.1995: Inchcolm, Fidra, Inchkeith).

Nash, D. (5.vi.2016: data file for Irish islands 1996–2015).

National Biodiversity Network (https://nbn.org.uk, viewed 2016: data for Inner Hebrides and Clyde islands).

Ogilvie, M. (5.vi.2016: three data files for Islay, Jura and Colonsay 1996–2015).

Owen, D.F. (25.viii.1995: Steep Holm).

Parker, R. (3.iv.2015: summary of records for Orfordness and Havergate 2002–2013).

Parrack, J.D. (1994: Farne Islands).

Peet, T.N.D. (15.ii.1987: Guernsey).

Pennington, M. (ix.1994: Shetland Islands).

Perkins, M. (13.vi.2016: four data files: for Coquet 1997–2003, Farne Island 1999–2015, Lindisfarne 1996–2015, St Mary's 1997–2015).

Redfern, A.J. (17.v.1994: Isle of Wight).
Redhead, D. (14.vi.2016: one data file for Skomer 1996–2007 and one for other Pembrokeshire islands 1996–2015).
Rippey, I. (22.xii.1988, 23.v.1989, 28.vi.1989, 7.vii.1989, 18.viii.1994, 15.i.2000; 14.iv.2000, 27.iv.2000, 17.v.2000, 3.vii.2000, 1.x.2000, 9.x.2000, 6.viii.2001, 21.xi.2001, 22.iv.2002, 2.v.2002, 9.vi.2002, 10.vii.2002, 13.vii.2002, 26.viii.2002, 25.iv.2003, 26.iv.2003, 3.ix.2003, 12.v.2004, 20.vii.2004, 18.viii.2004, 14.vii.2005, 7.xii.2005, 10.xii.2005, 15.i.2006, 25.iv.2006, 11.vii.2006, 26.vi.2007, 16.viii.2007, 1.iv.2008, 4.i.2009, 22.vi.2009, 12.viii.2010, 17.viii.2010, 24.viii.2010, 21.ii.2011, 20.iii.2011, 10.v.2011, 1.i.2012, 14.ii.2012, 20.ix.2012, 12.v.2016, 13.v.2016, 14.v.2016, 15.v.2016: all Irish offshore islands).
Robertson, I.S. (28.x.1980: Fair Isle. Copy from Heath, J.).
Rutherford, R.I. (26.ix.1995: *P. napi* on Luing and Sell, *V. atalanta* on Eigg).
Sandison, W. (8.xi.2004: Llanddwyn island, Anglesey).
Savill, R.E. (23.iii.1996: Bass Rock and Cramond).
Seegar, V. (20.iii.1995, 9.vi.1996: including records from Hilbre Bird Observatory); with Sawaik, A. (9.vi.1996: *C. rubi* on Mull).
Shaw, B.T. (15.ii.1996: Hilbre from Hilbre Bird Observatory reports 1984–1994; 19.ii.1996; Isles of Scilly; 12.v.2016; data file for Hilbre Island 2007–2014).
Shreeve, T.G. (personal communications to RLHD: 1992, Easdale and Seil; 1994, Rhum; 1995–2003, Grimsay; Baleshare; Berneray, north of North Uist; 1999, Great Bernera).
Shreeves, W. (12.v.2016: advice regarding contact for Dorset islands).
Skeates, A. (13.v.2016: advice regarding contact for Inner Hebrides).
Slade, D. (31.v.2016: data file for Burry Holms 1997–2012).
Small, I. and Taverner, J. (15.vii.1994: Hayling Island).
Smith, R. (15.v.2013, 17.v.2013, 1.iv.2015: two data files for Essex islands 1998–2012 and 2013–2014).
Smout, A.M., Fife Regional Council (1.xi.1995: Inchcolm and Inchkeith).
Spalding, A. (29.iii.1995: St Michael's Mount; 2017: data file for Scillies).
Stringer, G. and Wight, A. with Farne Islands' wardens (2.ix.1994, 10.ix.1994, 27.x.1994, 1.xi.1994: Farne Islands).
Stringer, G. from RSPB Vane Farm Nature Centre (8.xi.1995: Fidra and Inchmickery in the Firth of Forth).
Sutcliffe, R. (18.v.1995: Ailsa Craig from Rodway, G. and Hancock, E.G.).
Sutton, R. (6.v.1994, 12.v.1994: Isles of Scilly).
Tanner, C.M. (14.viii.1995: Lindisfarne and Longstone in Farnes).
Tomlinson, P. (30.vi.2016: data file for Isle of Man 1890–2012).
Tremewan, W.G. (24.viii.1995: *P. icarus* on St Michael's Mount); with Spalding, A. (24.vi.1996: Looe Island).
Vicary, G. (24.x.1994: Channel Islands).
Walton, J. (19.v.1995: Brownsman, Farnes).
Wedd, D. (19.v.2016: summary of Alderney species).

Appendix 5. Regional Atlases of British and Irish Butterflies

Anonymous (2014) *Butterflies of North-east England*. Natural History Society of Northumbria, Newcastle.

Arnold, V.W., Baker, C.R.B., Manning, D.V. and Woiwod, I.P. (1997) *The Butterflies and Moths of Bedfordshire*. Bedfordshire Natural History Society, Bedford.

Asher, J. (1994) *The Butterflies of Berkshire, Buckinghamshire and Oxfordshire*. Pisces Publications (Nature Conservation Bureau Ltd), Newbury. Records from 1987 to 1993.

Blencowe, M. and Hulme, N. (2017) *The Butterflies of Sussex – a Twenty-first Century Atlas*. Pisces Publications (Nature Conservation Bureau Ltd), Newbury.

Bristow, C.M., Mitchell, S.H. and Bolton, D.E. (1993) *Devon Butterflies*. Devon Books, Tiverton. Records from 'before 1960' to 1992.

Corke, D. (1997) *The Butterflies of Essex*. Lopinga Books, Saffron Walden.

Duncan, I., Seal, P., Tilt, J., Wasley, R. and Williams, M. (2016) *Butterflies of the West Midlands – Birmingham and the Black Country, Herefordshire, Shropshire, Staffordshire and Worcestershire*. Pisces Publications (Nature Conservation Bureau Ltd), Newbury.

Dunn, T.C. and Parrack, J.D. (1992) *The Moths and Butterflies of Northumberland and Durham*. Northern Naturalists' Union, Stockton. Records from before 1900 to 1991.

Field, R., Perrin, V., Bacon, L. and Greatorex-Davies, N. (2006) *The Butterflies of Cambridgeshire*. Butterfly Conservation Cambridgeshire and Essex Branch, Cambridge. Records from 1990 to 2005.

Frost, H.M. (ed.) (2005) *The Butterflies of Yorkshire*. Butterfly Conservation Yorkshire, Wareham. Records from 1995 to 2003.

Futter, K., Sutcliffe, R., Welham, D., Welham, A., Rostron, A.J., MacKay, J., Gregory, N., McCleary, J., Tait, T.N., Black, J. and Kirkland, P. (2006) *Butterflies of South West Scotland: An Atlas of their Distribution*. Butterfly Conservation Glasgow and South West Scotland branch, Glasgow.

Garland, S.P. (1981) *Butterflies of the Sheffield Area*. Sorby Natural History Society and Sheffield City Museums, Sheffield. Records from 1960 to 1980.

Goddard, D. and Wyles, A. (2012) *Butterflies of Northamptonshire*. Butterfly Conservation Bedfordshire and Northamptonshire branch.

Hall, M.R. (1991) *An Atlas of Norfolk Butterflies 1984–1988*. BBCS Ltd, Loughborough.

Hardy, P.B. (1998) *Butterflies of Greater Manchester.* PGL Enterprises, Sale. Records from 1980 to 1997.

Harrison, F. and Sterling, M.J. (1985) *Butterflies and Moths of Derbyshire, Part 1.* Derbyshire Entomological Society, Derby. Species accounts, no distribution maps. (Parts 2 and 3 cover moths.)

Lorimer, R.L. (1983) *The Lepidoptera of the Orkney Islands.* E.W. Classey, Faringdon.

Mendel, H. and Piotrowski, S.H. (1986) *The Butterflies of Suffolk – an Atlas and History.* Suffolk Naturalists' Society, Ipswich.

Meredith, G. (2000) *The Butterflies of Gloucestershire.* http://www.gloucestershire-butterflies.org.uk/bflyglos/bflyglos.html

Ni Lamhna, E. (1980) *Distribution Atlas of Butterflies in Ireland: European Invertebrate Survey.* Foras Forbatha, Dublin.

Oates, M., Taverner, J. and Green, D. (2000) *The Butterflies of Hampshire.* Pisces Publications (Nature Conservation Bureau Ltd), Newbury.

Philp, E.G. (1993) *The Butterflies of Kent.* Transactions of the Kent Field Club, Sittingbourne. Records from 1981 to 1990.

Plant, C.W. (1987) *The Butterflies of the London Area.* Natural History Museum, London.

Riley, A.M. (1991) *A Natural History of the Butterflies and Moths of Shropshire.* Swan Hill Press, Shrewsbury. Records from 1970 to 1989, includes moths (maps for butterflies only).

Sawford, B. (1987) *The Butterflies of Hertfordshire.* Castlemead Publications, Ware.

Shaw, B.T. (1998) *The Butterflies of Cheshire.* National Museums and Galleries on Merseyside, Liverpool.

Smout, A.M. and Kinnear, P. (1993) *The Butterflies of Fife.* Fife Nature, Glenrothes.

Thomas, J.A. and Webb, N. (1984) *Butterflies of Dorset.* Dorset Natural History and Archaeological Society, Dorchester. Records from pre-1900 to 1984.

Thomson, G. (1980) *The Butterflies of Scotland, A Natural History.* Croom Helm, London. All records to 1979.

Wacher, J., Worth, J. and Spalding, A. (2003) *A Cornwall Butterfly Atlas.* Pisces Publications (Nature Conservation Bureau Ltd), Newbury.

Warmington, K. and Vickery, M. (2003) *Warwickshire's Butterflies, Their Habitats and Where to Find Them.* Butterfly Conservation Warwickshire branch.

Whitehead R.W. (c. 1986) *Butterflies of Merseyside.* BBCS Merseyside branch, Craigfechan. Species accounts; no maps.

Appendix 6. Migration Records and Measures for Butterfly Species Resident on the British and Irish Islands

It should be mentioned at the outset in introducing the movement (transfer) records below that a distinction exists between dispersal and migration in butterfly ecology (Van Dyck and Baguette, 2005; Dingle and Drake, 2007) (see Chapter 3).

- The term **dispersal** is applied to movements (which may be linear and/or non-linear transfers) between habitat patches associated with daily activities, the sum of such movements observed from a population appearing to be random and non-directional (Dennis, 1992) but which alternatively display population spatial bias (Baker, 1978, 1984). These are the movements that connect local populations (**metapopulations**) (see Fig. 1.3).
- On the other hand, the term **migration** is generally used to describe predictable long-distance seasonal movements made by large numbers of individuals in approximately the same direction and at approximately the same time (Dennis, 1992); these movements have been shown to be reversed during different seasons (viz. spring, late summer/autumn) (Baker, 1978, 1984).

Thus, although we refer to **migration ability** and **migration capacity** in this book – migration applied in the sense of Baker (1984) as inter-habitat movements – only the last two categories in the table below (i.e. RLDM, OOM) would be strictly regarded as true migration events. However, without additional information the preceding seven categories (VA to SR) could be the product of migration or dispersal, as distinguished above. In this work, as reference to islands is continually made using the terms **immigration** and **emigration**, the word migration equates more with the sense of migration coined by Baker (1984) as a more general term to describe all movements outside habitats unrestricted as to distance and direction, numbers or timing.

Despite the fact that **mobility** (movement) is one of the key life history attributes, knowledge of the ability of butterflies to move beyond the strict confines of habitats (dispersal) is still poorly known; it is dependent on individual transfer ability referred to here as migration ability (dependent on an individual's **bauplan**) and, more broadly but less easily determined, on transfer capacity, here termed migration capacity (driven by opportunities for transfers). Much work has been carried out on movement *within* vegetation units (using **mark–release–recapture** methods) acting as surrogates for habitats (Warren, 1992a) but, apart from long-distance migrants that undergo seasonal reversals in migration (true migrants in the sense of Dingle and Drake, 2007; e.g. *Pieris brassicae*, *Pieris rapae*, *Colias croceus*, *Vanessa cardui*; Williams, 1965; Baker, 1978; Stefanescu *et al.*, 2013; Chapman *et al.*, 2015), no systematic data collection has been carried out on the ex-habitat movements of species since the work of Baker (1984) and Cook *et al.* (2001) using very different approaches.

The study by Cook *et al.* (2001) should help readers to understand the issues in **matrix** transfers. It focused on two zones in northern England, each contrasting in biotopes (urban and woodland), divided into habitat and matrix on the occurrence of each butterfly species' larval host plant in hectare units (100 m squares). It found that the vagrancy indices do not simply reflect individual abundances of species and this suggests an innate component to transfer ability (dispersal) beyond the influence of **source mass** (potential numbers of vagrants or migrants) (see Fig. 6.6). The degree of vagrancy agreed closely for the two data sets and correlated highly with the movement indices (Shreeve and Dennis, 1996) reproduced and enhanced in the table below (variables *MI-1*, *MI-2*). The **vagrancy index** (*Vi*) is:

$$Vi = (bc - ad)/[(a + b)(b + d)]$$

where *a* is the number of squares with both the host plant and butterfly species; *b* the number of squares with the butterfly species in the absence of the host plant; *c* the number of squares with the host plant in the absence of the butterfly species; and *d* the number of squares in which both the butterfly species and host plant are absent. See Cook *et al.* (2001) for the derivation of this measure and its enhancement.

The list below was devised in the earlier study (Dennis and Shreeve, 1996, p. 47) to add some detail to the earlier observations on matrix transfers from studies of population structure (for examples, see Warren, 1992). It is a simple categorization based on evidence for transfers across specific barriers (e.g. terrestrial matrix outside habitats, water bodies) and the ability to undertake mass movements. Since first constructing this table in 1996, some additional detail has been added through the kind help of contributors; each addition is precious to understanding differences in transfers (dispersal and migration) across species. The vagrancy index above suggested that some species were being underestimated for movements across the landscape. An indication of obvious gaps within this simple categorization of movement data is the apparent absence of vagrants (VA, ex-habitat records) for some species. But every species should have records of vagrants, as ex-habitat movements are critical for producing new colonies when conditions in old ones become unsuitable through vegetation succession (viz. a population that has no dispersing individuals rapidly becomes extinct) (Dennis, 1977). One feature should be clear: not only do different species respond differently to barriers such as water and roads (Jacobson *et al.*, 2016) but so do individuals of the same species (e.g. *Pieris napi* and *Anthocharis cardamines*) (Dennis, 1982a, b, 1986, 2010).

The issue of comparative movements of species is one area in which quantitative observations are desperately needed, so that the suitable dimensions of metapopulations for species conservation can be gauged. The island data in this work have provided valuable insights into the ability of species to cross over water, thereby to colonize new land; readers should no longer be surprised if any British butterfly species is found at sea or has crossed to an offshore island. Evidence points to repeated colonization of offshore islands by many species (Dennis *et al.*, 2010).

Mobility classes from populations studies

M_1: mobility classes 1 (from Thomas, 1984); C, closed populations; O, open populations.

M_2: mobility classes 2 (from Pollard and Yates, 1993); S, sedentary; I, intermediate; W, wide ranging.

Mobility records for specific criteria ranked in order of effect (obstacle overcome)

1 **VA:** Vagrants recorded ex-habitat.
2 **GR:** Suburban garden records.
3 **UR:** Records from the central business districts of large towns, cities and conurbations (e.g. inner London, Manchester, Glasgow, Oxford, Birmingham and Sheffield centres).
4 **RE:** Documented range expansions.
5 **MM:** Documented mass movements with numbers moving ex-habitat in the same direction across open country or water, often with other species.
6 **SR:** At-sea records from light vessels or documented records of sea crossings.
7 **OWM:** Evidence of crossing to an island over water (sea crossings > 10 km).
8 **RLDM:** Frequent long-distance reversed mass migrations.
9 **OOM:** Over-ocean movement (e.g. Atlantic Ocean).

The data on movements and migration (**VA-OOM**) are recorded as presence or absence of records (coded as: 0 no record, 1 ≥ one record) irrespective of the number of records. Doubtful records are excluded and no scaling is given to the number of records in any category since none of these data have been systematically collected.

Mobility indices

MI-1: Movement index. Sum of binary states for seven variables, **VA** to **OOM** (scores 0 to 9).

MI-2: Movement index. Sum of ranks for binary states for seven variables, **VA** to **OOM** (scores 0 to 45). There are good arguments for adding further weighting for long-distance over-sea migrations and perhaps for ordering the ranks differently. Note differences in notation from Dennis and Shreeve (1996).

Table Appendix 6.1. Measures of movement and mobility in butterflies.

Species	M1	M2	VA	GR	UR	RE	MM	SR	OWM	RLDM	OOM	*MI-1*	*MI-2*
C. palaemon	C	S	0	0	0	0	0	0	0	0	0	0	0
T. sylvestris	C	S	1	1	1	1	0	1	0	0	0	5	16
T. lineola	C	S	1	1	1	1	0	0	0	0	0	4	10
T. acteon	C	S	0	0	0	1	0	0	0	0	0	1	4
H. comma	C	I	1	1	0	1	0	0	0	0	0	3	7
O. sylvanus	C	S	1	1	1	1	0	1	0	0	0	5	16
E. tages	C	S	1	1	0	0	0	0	0	0	0	2	3
P. malvae	C	S	1	1	0	0	0	0	0	0	0	2	3
P. machaon	C	I	1	1	0	0	1	1	1	0	0	5	21
L. sinapis/ juvernica	C	I	1	0	0	1	0	0	0	0	0	2	5
C. croceus	O	W	1	1	1	1	1	1	1	1	0	8	36
G. rhamni	O	W	1	1	1	1	1	1	1	0	0	7	28
P. brassicae	O	W	1	1	1	1	1	1	1	1	0	8	36
P. rapae	O	W	1	1	1	1	1	1	1	1	0	8	36
P. napi	O	I	1	1	1	1	1	1	1	0	0	7	28
A. cardamines	O	I	1	1	1	1	0	1	0	0	0	5	16
C. rubi	C	S	1	1	1	0	0	0	1	0	0	4	13
T. betulae	C	I	0	1	1	1	0	0	0	0	0	3	9
F. quercus	C	S	1	1	1	0	1	0	0	0	0	4	11
S. w-album	C	S	1	1	1	0	0	0	0	0	0	3	6
S. pruni	C	S	0	0	0	0	0	0	0	0	0	0	0
L. phlaeas	C	I	1	1	1	1	1	1	0	0	0	6	21
L. dispar	C	S	0	0	0	0	0	0	0	0	0	0	0
C. minimus	C	S	1	1	1	1	0	0	0	0	0	4	10
P. argus	C	S	1	1	0	0	0	0	0	0	0	2	3
A. agestis	C	I	1	1	1	1	0	0	0	0	0	4	10
A. artaxerxes	C	S	1	1	0	0	0	0	0	0	0	2	3
P. icarus	C	I	1	1	1	1	1	1	1	0	0	7	28
P. coridon	C	S	1	1	0	0	0	0	1	0	0	3	10
P. bellargus	C	S	0	0	0	1	0	0	0	0	0	1	4
C. argiolus	O	W	1	1	1	1	1	1	0	0	0	6	21
P. arion	C	S	0	0	0	0	0	0	0	0	0	0	0
H. lucina	C	S	0	0	0	0	0	0	0	0	0	0	0
L. camilla	C	I	1	1	1	1	0	1	0	0	0	5	16
A. iris	C	I	1	1	0	1	0	0	0	0	0	3	7
V. atalanta	O	W	1	1	1	1	1	1	1	1	0	8	36
V. cardui	O	W	1	1	1	1	1	1	1	1	1	9	45
A. urticae	O	W	1	1	1	1	1	1	1	1	0	8	36
A. io	O	W	1	1	1	1	1	1	1	1	0	8	36
N. polychloros	O	W	1	1	1	0	1	1	1	0	0	6	24
P. c-album	O	I	1	1	1	1	1	1	0	0	0	6	21
B. selene	C	S	1	1	0	0	0	1	0	0	0	3	9
B. euphrosyne	C	S	1	1	0	0	0	0	0	0	0	2	3
A. adippe	C	I	1	1	1	0	0	0	0	0	0	3	6
A. aglaja	C	I	1	1	1	1	0	1	0	0	0	5	16
A. paphia	C	I	1	1	1	1	1	1	0	0	0	6	21

Continued

Table Appendix 6.1. Continued.

Species	M1	M2	VA	GR	UR	RE	MM	SR	OWM	RLDM	OOM	*MI-1*	*MI-2*
E. aurinia	C	I	1	1	0	1	0	0	0	0	0	3	7
M. cinxia	C	I	0	0	0	0	0	0	0	0	0	0	0
M. athalia	C	S	0	0	0	0	0	0	0	0	0	0	0
P. aegeria	C	S	1	1	1	1	0	1	0	0	0	5	16
L. megera	C	I	1	1	1	1	0	1	0	0	0	5	16
E. epiphron	C	S	0	0	0	0	0	0	0	0	0	0	0
E. aethiops	C	S	1	1	0	0	0	0	0	0	0	2	3
M. galathea	C	S	1	1	1	1	0	0	0	0	0	4	10
H. semele	C	S	1	1	0	0	0	1	1	0	0	4	16
P. tithonus	C	S	1	1	1	1	0	0	0	0	0	4	10
M. jurtina	C	S	1	1	1	1	1	1	1	0	0	7	28
A. hyperantus	C	S	1	1	1	1	0	0	1	0	0	5	17
C. pamphilus	C	S	1	1	1	0	0	1	0	0	0	4	12
C. tullia	C	S	1	0	0	0	0	0	0	0	0	1	1

Entries of '1' indicate a positive observation from anywhere in the UK and Ireland. The data from 1996 (Dennis and Shreeve, 1996) have been upgraded from our records and from those kindly provided by the following regional recorders and others to whom we are most grateful: Stephen B. Smith, James Black (Ayrshire), David Hinde (Flintshire and Denbighshire), Andrew Wood (Herts and Middlesex), Rupert Adams (Cheshire and Wirral), David Hatton (Yorkshire VC64), Simon Metcalfe (Lothians), Harry E. Clarke (Surrey), Paul Harvey (Shetland), Alan Skeates (VC103), Robert Smith (Essex), William Shreeves (Dorset), Simon Metcalfe (Lothians), David Redhead (Pembrokeshire), David Nash (Ireland).

Appendix 7. Migration Capacity and Colonization Ability Measures for British Butterflies

The ability to migrate to and colonize new habitats (islands) depends on a large number of variables. The two indices, **migrfac** and **colonfac**, were developed to describe migration capacity and colonization ability in British butterflies. Both indices are here given as **standardized scores** and extracted from sets of variables using ordination (see Dapporto and Dennis, 2013).

Migration capacity (migrfac) is built from four variables. One of these is potentially a measure of migration ability (wingspan); the other three are measures of migration opportunity (viz. flight period; number of adult broods; movement index) (see Appendix 6). Colonization ability is built from ten variables: number of biotopes exploited; number of larval host plants used; host plant abundance in natural vegetation classification categories; niche breadth (mean of **Ellenberg values**, except salinity) for all larval host plants; larval host plant abundance in eight biotopes; larval host plant density; nectar plant specialism; number of nectar sources; symbiosis with ants; and number of utility resources (outlets) used by pupae and adults, excluding nectar plants. The full data set for variables entered into the indices is provided in Supplementary Appendix 3.2.

It should be obvious that these measures present a first attempt to quantify the abilities of species to migrate to and colonize new habitats and islands. In time, it should be possible fully to distinguish measures of **migration ability** (relating to physical and physiological attributes) from those of **migration capacity** (dependent on migration opportunities or triggers). It should also be possible to develop more realistic measures of each species' (and individual's) migration ability, migration capacity and colonization ability. For instance, as an index of migration ability, at the very least we need measures of variation in **aspect ratio** and **wing loading** among butterfly species; measures of wingspan are not an adequate surrogate (Sekar, 2012).

Table Appendix 7.1. Migrfac and colonfac values for species.

Species name	Species code	migrfac	colonfac
C. palaemon	Cpal	0.14601	0.42763
T. sylvestris	Tsyl	0.37614	0.74566
T. lineola	Tlin	0.27903	0.76606
T. acteon	Tact	0.15896	0.35480
H. comma	Hcom	0.25265	0.33192
O. sylvanus	Osyl	0.39619	0.76823
E. tages	Etag	0.32381	0.58112
P. malvae	Pmal	0.30412	0.73999
P. machaon	Pmac	0.70121	0.47347
L. sinapis	Lsin	0.41382	0.61445
C. croceus	Ccro	0.79927	0.60392
G. rhamni	Grha	0.60242	0.51857
P. brassicae	Pbra	0.89164	0.74324
P. rapae	Prap	0.84409	0.72153
P. napi	Pnap	0.84902	0.81830
A. cardamines	Acar	0.52478	0.79501
C. rubi	Crub	0.29346	0.77454
T. betulae	Tbet	0.15037	0.27768
F. quercus	Fque	0.28919	0.45790
S. w-album	Swal	0.21130	0.39189
S. pruni	Spru	0.08426	0.24623
L. phlaeas	Lphl	0.73426	0.70707
L. dispar	Ldis	0.18384	0.24482
C. minimus	Cmin	0.40133	0.24812
P. argus	Parg	0.27845	0.63502
A. agestis	Aage	0.45290	0.63170
A. artaxerxes	Aart	0.27014	0.33274
P. icarus	Pica	0.71457	0.73952
P. coridon	Pcor	0.33282	0.37928
P. bellargus	Pbel	0.43573	0.32069
C. argiolus	Carg	0.59238	0.75109
P. arion	Pari	0.27106	0.25477
H. lucina	Hluc	0.11338	0.45999
L. camilla	Lcam	0.49074	0.23774
A. iris	Airi	0.36113	0.42197
V. atalanta	Vata	0.82275	0.65985
V. cardui	Vcar	0.86697	0.83767
A. urticae	Aurt	0.88963	0.59685
A. io	Aio	0.72430	0.59910
N. polychloros	Npol	0.62146	0.60328
P. c-album	Pcal	0.76553	0.69418
B. selene	Bsel	0.44769	0.58360
B. euphrosyne	Beup	0.40437	0.54789
A. adippe	Aadi	0.45883	0.49566
A. aglaja	Aagl	0.52473	0.58682
A. paphia	Apap	0.57930	0.39775

Continued

Table Appendix 7.1. Continued.

Species name	Species code	migrfac	colonfac
E. aurinia	Eaur	0.29803	0.62815
M. cinxia	Mcin	0.26162	0.43882
M. athalia	Math	0.34436	0.61254
P. aegeria	Paeg	0.82333	0.83495
L. megera	Lmeg	0.72401	0.81842
E. epiphron	Eepi	0.19005	0.32765
E. aethiops	Eaet	0.36203	0.48154
M. galathea	Mgal	0.39746	0.71307
H. semele	Hsem	0.45610	0.74673
P. tithonus	Ptit	0.43941	0.82351
M. jurtina	Mjur	0.56049	0.81884
A. hyperantus	Ahyp	0.41164	0.74273
C. pamphilus	Cpam	0.70933	0.74051
C. tullia	Ctul	0.21629	0.53259

See Supplementary Appendix 3.2 for values of variables used to build both indices.
Each value is the mean of scores for each of the four (migrfac) and ten (colonfac) variables, each variable initially scaled from 0 to 1 by dividing the difference of each species' value from the minimum value for all species by the difference between the maximum and minimum values.

Appendix 8. A Brief Guide to Island Studies: Observations and Analysis

A. A Guide to Making Observations on Butterfly and Moth Species During Visits to Islands

These notes are intended only as a brief guide to how and what to observe and record during short visits to islands. Substantial changes in recording butterflies on British and Irish islands have occurred since the previous audit (Dennis and Shreeve, 1996). A number of the smaller islands now have **Butterfly Monitoring schemes** (BMS) (95 offshore transect sites, of which 52 were operating in 2015; islands > 1978: n = 24; 2015: n = 14) and the **Wider Countryside Butterfly Survey** (WBCS) (38 offshore sites, of which eight were operating in 2015; islands > 2006: n = 18; 2015, n = 5) recording butterflies on them (see Appendix 3) (http://www.ukbms.org/wcbs.aspx; T. Brereton, personal communication). The current notes are mainly, then, for islands lacking these recording schemes – islands often with few inhabitants or uninhabited. Much of recording is obvious and common sense, but nonetheless it may prove useful to have a reminder of what to do for those who intend to plan for their visit. Needless to say, all records are worthwhile, even those made casually without preparation. The issues addressed and the points made do not differ in substance from those associated with local recording, for example UK county or regional atlases. However, we would encourage efforts to determine the breeding status of species observed on islands, together with resources used by butterflies (e.g. substrates oviposited on; larval host plants; and adult food sources, typically nectar flowering plants) which can then usefully be added to data accumulated by one of us (PBH; see Dennis, 2010 for summaries).

Before visits

It is particularly useful to note species previously found on the island to be visited. The present list (see Appendix 3) used in conjunction with an authoritative text (Asher *et al.*, 2001; Thomas and Lewington, 2010) should indicate species most likely to be observed in the adult stage during a visit. Bearing in mind the primary influence of nearest sources on the species richness of an island's butterfly fauna, it is also useful to make a note of additional species recorded at the nearest sources (mainland and other islands) and within the 100 km square of the island (Emmet and Heath, 1989).

Prior to the visit, familiarize yourself with the features (i.e. wing patterns) and biology of species (broods, likely developmental stage at the planned time for the visit, etc.) likely to be seen. In particular, be aware of other butterfly or moth species that could be confused with butterflies that may be encountered. Also become familiar with biotopes (vegetation and substrate types) that species typically inhabit, their larval host plants and details that will help in finding and recognizing early stages of each species (see Dickson, 1992).

During visits

Equipment

At the very least take a detailed map (OS 1:25,000 scale), a notebook, identification texts for butterflies (Thomas, 1986; Thomas and Lewington, 2010) and moths (Waring and Townsend, 2009; Sterling and Parsons, 2012), and for flowering plants, grasses and sedges (Hubbard, 1968; Fitter *et al.*, 1978, 1984; Pankhurst and Allinson, 1985; Price, 2016). Of course much can be downloaded these days without using what are often heavy hard copies of sources but this account assumes that you do not have access to applications for mobiles and 'tablets' and cannot get internet access.

Take a pocket lens (×10 magnification) and a folding net for checking identifications. It is not necessary to 'thrash' at passing butterflies; when studying a variety of species in the field it was found that gentle approaches were more successful and created less alarm than aggressive tennis strokes (Dennis and Williams, 1987). Fine felt-tip permanent markers (various colours) are invaluable if more precise population estimates are to be made using mark–release–recapture methods (unique numbers for species are easily established by applying the 1-2-4-7/10-20-40-70 code to distinct areas on right and left hindwings respectively (Ehrlich and Davidson, 1960). Warren (1992a) is an excellent if dated guide and reference source to population studies in butterflies (see Henderson, 2003 for broader techniques in ecology). A digital camera and accessories are essential equipment and greatly assist in obtaining a valid record for a species, particularly if there are legal restrictions on catching and taking specimens (for the Butterfly Conservation UK policy on collecting, breeding and photography, see http://butterfly-conservation.org/files/bc-policy_collecting-breeding--photography-2010.pdf). The authors encourage respect for life and appeal to readers not to kill Lepidoptera unnecessarily. A pair of binoculars (low-power; wide field of view) is also important for identifying species in inaccessible locations and observing behaviour without causing disturbance. Most moths can also be identified against a good reference text (e.g. Skinner, 1984), but unfortunately some can only be determined to species level by dissection.

Conditions

Take note of the time at the start of your observations during your visit. Note down the weather conditions, in particular the duration of bright sunshine and cloud, and a record of wind speed. The Beaufort scale is useful for the latter. If weather conditions are being monitored on the island, obtain a record of the shade temperatures during the visit. Many digital watches now have inbuilt thermometers as well as compasses and, increasingly, GPS facilities for determining locations.

Cover

For small islands, try to visit as much of the island as possible. For large islands, visit as many different biotopes (habitats) as possible. If you have obtained a map at a scale of 1:25,000 or larger, you can plan your route; also search the web for photographs of the island and view it in Google Earth to familiarize yourself with the terrain. There is good reason for preparation, as visits to islands are potentially hazardous; on this, the earlier ventures of teams under Professor Heslop Harrison provide interesting, if disturbing, reading (see References). Therefore, we advise that the fieldwork guidelines of the Institute of Biology (1992) or of university institutions (e.g. University of Birmingham, UK: *Safety Policy: Rules and Guidance for the Safe Conduct of Fieldwork, Expeditions and Outdoor Activities USP/90/FW/18*) are followed.

Observations

If you are uncertain about an observation, take photographs as a record. It is important to have unusual identifications confirmed by colleagues.

An OS map (UK; Ireland) will allow you to give pinpoint references to locations on the island and enable you to plan your route and divide it into sections, much as for the transects carried out in the ITE Butterfly Monitoring Scheme (Pollard and Yates, 1993; http://www.ukbms.org/). Make notes on the biotopes (habitats) of sections and photograph them. Then, in years to come, new records can be related to any changes in habitat structure. Simple methods exist for habitat surveys (Handbook for Phase 1 habitat survey, 1990; now JNCC (2010), Handbook for Phase 1 habitat survey – a technique for environmental audit) and the subjective assessment of plant abundance; also, simple methods are available for describing habitat structure and composition such as the Braun-Blanquet and Domin techniques (Kershaw, 1964; Gilbertson *et al.*, 1985) and the national vegetation communities classification (Rodwell, 1991–1995). For those island visitors intent on more serious study, numerous texts on field biology techniques have been produced over the years (Henderson, 2003 is a useful place to start).

Count numbers of adults seen of both sexes separately if possible. If you make a note of your route, your visit becomes a transect and you can calculate the distance covered and the area over which observations have been made as well as the time spent in each biotope. The standard reference for transect work (Pollard and Yates, 1993) has not as yet been updated, though more sophisticated techniques are continuing to emerge for interpreting and processing the data (Isaac *et al.*, 2011a; Dennis, E.B., 2015; Dennis, E.B. *et al.*, 2016a, b).

Make observations on all aspects of butterfly behaviour, but especially those activities that allow determination of the breeding status of species, such as:

- teneral adults, (that is, adults that have recently eclosed and are still drying their wings);
- state of wing wear of individuals seen; a simple four-state scoring scale (e.g. 1, fresh; 2, no wear but not fresh; 3, significant scale loss and wing-edge tattering; 4, worn and tattered) can help to classify individuals seen;
- mating behaviour, such as territorial disputes (i.e. spiral interactions; horizontal chases), attempted courtships or copulation;
- egglaying – and note the substrate or plant and part of the plant on which the egg(s) is/are placed; and
- presence of eggs and larvae on plants and feeding damage on plants.

Survey work

For those intending to make more detailed observations on island butterfly populations, simple surveying techniques are described in Bennett and Humphries (1974) and Henderson (2003). Another useful source is Dennis (1992), especially Chapters 2–5 by Professor Tim G. Shreeve, Dr Keith Porter and Dr Martin S. Warren. On some islands, especially those that have nature reserves, extensive monitoring of butterfly populations is already being carried out, as noted above (e.g. Skokholm and Skomer). By far the best example of biotope cover and documentation of butterfly population changes from the use of BMS transects on a small island is that for Jersey (Long, 2015). Only using intensive long-term monitoring schemes is it possible to determine changes and turnover (colonizations, extinctions) in species status on islands.

Colvin and Reavey (1993, *A Directory for Entomologists*) includes a great deal of other useful information on insect studies.

After visits and publication of observations

All observations on islands (as for the whole of the UK, Ireland and Isle of Man) present valuable additions to our knowledge. Most importantly, they inform on changes to the faunas of islands. A number of nationwide journals readily publish observations of Lepidoptera:

- *Bulletin of the Amateur Entomologists' Society* (https://www.amentsoc.org/publications/bulletin/)
- *Entomologist's Record and Journal of Variation* (https://www.amentsoc.org/publications/entomologists-record/)
- *Atropos* (http://www.atropos.info/site/index.php/)
- *Entomologist's Gazette* (http://www.pemberleybooks.com/).

The most appropriate destination for island butterfly data is the computerized Butterfly Net, which is mapping records for the New Millennium Atlas Project. Data can be sent to regional branch recorders (see http://butterfly-conservation.org/ for links).

B. A Note on Analyses in this Text

The chapters in this book contain results from a range of analyses carried out on island data over the past 30 years. Analysis of island records has involved both the selection of islands and the selection of species (see Chapter 3, last section). This is not a simple matter, as the island records are not the product of a systematic survey based on an appropriate sampling design. The data for each island were not originally collected with any analysis in mind. Moreover, the quality of records for islands clearly varies in a number of important respects: in the number of visits made to islands, their timing and their geographical coverage. Observations have also varied in their thoroughness. Some recorders have searched for early stages but most have not made any attempt to validate observations as evidence of breeding on islands. Thus, selection of islands for analysis has, of necessity, had to be arbitrary and based on the occurrence of marker species (long-distance migrants) that are regarded as being an indication of adequate sampling of islands.

It is not possible to give a detailed account of procedures applied but a brief outline is given below of techniques highlighted in the accounts (terms in bold type are defined in the Glossary). The reader interested in the statistical techniques (including numerical taxonomy) used herein is advised to refer to one or more of the many texts and websites now available on statistical methods (Sneath and Sokal, 1973; Ludwig and Reynolds, 1988; Michener and Brunt, 2000; Legendre and Legendre, 2000; Dytham, 2003). Similarly, a wide range of statistical software packages is available for computing statistics and easily found using web searches; many are expensive to obtain (e.g. STATISTICA; SPPS; SYSTAT; MINITAB) but some are free and prove to be invaluable tools for many researchers worldwide without access to funding (i.e. PAST; http://folk.uio.no/ohammer/past/). Many of the operations carried out in this work have been done in STATISTICA (Statsoft Inc.) and by using PAST. Increasingly, researchers are moving to use 'R', which is an open source programming language and software environment for statistical computing and graphics that is supported by the R Foundation for Statistical Computing (https://cran.r-project.org/doc/contrib/Verzani-SimpleR.pdf; https://onlinecourses.science.psu.edu/statprogram/r; for texts on R see https://www.r-project.org/doc/bib/R-books.html).

Statistical methods have a number of basic objectives, such as: (i) description and assessing the probability of events; (ii) comparisons; and (iii) determining relationships and associations. For most butterfly biogeographical research, the focus is on object (**OTU**, **operational taxonomic unit** such as an island or species) × attribute (variables such as geographical locations, life history, resources, etc.) data. The statistical procedures used depend primarily on measurement scales (categorical, binary, ordinal or ranked, interval, interval–ratio), indicative of levels of precision of measurements. Thus, measures of central tendency and variation in a sample of data for an ordinal scale variable (**median**, quartiles) contain less information than those for a variable measured on an interval–ratio scale (**mean**, **standard deviation**, **variance**, **coefficient of variation**). However, processing data on higher scales makes a greater number of assumptions about the form (frequency distribution) of the data; ordinal scale statistics make no assumption other than equality of differences among ranks (they are frequency distribution-free), whereas interval–ratio scale statistics assume that the data have a **normal distribution**, with numbers for **variates** or classes of variates approximating a bell-shaped (Gaussian) curve. The advantage of ordinal statistics is that they make fewer assumptions of the data; the disadvantage is that there is a loss of power when ranking measurements: higher levels of measurement can be re-coded to lower scales but not vice versa. A number of techniques are available for modifying (transforming) the variates of data scored on interval–ratio scales so that appropriate statistics can be applied.

In this book, reference is made to results from comparative tests and measures of association and relationship appropriate to different scales of measurement. Thus, comparisons include the **chi-square test** and **Fisher exact test** (categorical data), the **Mann–Whitney** ***U*** **test** (ordinal data) and ***t*-test** and **ANOVA** (for interval–ratio data). Similarly correlations have applied **Spearman rank correlation** r_s (ordinal data) and **Pearson's correlation coefficient** ***r*** (interval–ratio data). Correlation coefficients (r, r_s Kendall's tau, etc.) typically range from −1 (perfect negative relationship) through 0 (no relationship) to +1 (perfect positive relationship).

Relationships between a dependent variable, such as numbers of species on islands, and potential predictors (e.g. latitude, island area, island isolation) have been sought using various forms of **regression analysis techniques**, of which there are many; for instance, multiple linear regression (which assumes normality of **residuals** from the relationship) for dependent variables measured on an interval–ratio scale, and **logit** (or

logistic) regression for those scored on a binary scale (1 and 0: present/ absent). Regression analysis facilitates a detailed examination of the relationships between variables. The standard model is:

$$Y = a + bX + e$$

where a is the intercept, the value for Y (dependent variable) when X (predictor variable) is zero; b are regression coefficients describing the relationship (slope of the regression line) between dependent and predictor variables; e is a measure of error. In multiple **mode** (more than one predictor) this becomes:

$$Y = a + bX_1 + bX_2 + bX_3 + bX_n + e$$

The higher the value for b, the steeper the slope of the regression line; with higher values for b, each increment for the variable Y is influenced by a smaller increment of variable X. For standardized data in which the variates (value for each variable) are given in standard deviation units, the slope is given by beta, which is identical to the correlation coefficient. An ***F* statistic** measures the significance of the relationship. Perhaps the most important aspect of regression analysis is the examination of residuals, the difference between observed values for variable Y and those predicted from values of variable X in the regression analysis. They are illustrated by the scatter of points about the least-squares regression line (for multiple variables, between the observed and predicted values of the regression analysis). Large residuals are indicative of a poor fit by the regression equation. The immediate objective has been to seek the statistical significance of comparisons and relationships; (***P* (probability)** < 0.05 or < 1 in 20 of occurring by chance alone is typically taken as a measure of statistical significance). Ultimately, the objectives of regression in this work have been to determine the biological or geographical significance of the findings and the probability of events (incidences of species on islands, especially for islands lacking records).

In regression analysis where a dependent variable is assessed against a number of 'explanatory' variables, the multivariable equivalent of a paired relationship (correlation) described by Pearson's r is R (nothing to do with the statistical programming language R!); a useful rule-of-thumb for interpreting r or R is to square it (r^2 and R^2); this is the **coefficient of determination**, the percentage variation in the dependent variable explained by the predictor variables. This measure refers only to 'explanatory' variables entered into the analysis: usually some procedure is applied that limits entry (i.e. stepwise forwards or backwards entry) to just those variables accounting for statistically significant additional variation.

In **logistic regression** (logit and logistic regression differ largely in the reports they provide: logit reports coefficients, while logistic reports odds ratios), probabilities (P) of the incidence of a species over islands are determined from parameter estimates using:

$$P = exp(Y')/(1+exp(Y'))$$

where $Y' = a + bX_1 + bX_2...bX_n$, with X_1 to X_n being predictor variables, a and b parameters of the regression equation.

More complex procedures have also been used to determine relationships among a number of **operational taxonomic units** (**OTU**s; species; islands) and variables simultaneously (multivariate as opposed to univariate techniques). Two forms herein used are clustering techniques and ordination techniques. **Clustering techniques** (e.g. nearest neighbour, complete or maximum linkage, unweighted pair group method with arithmetic mean UPGMA, etc.; see Sneath and Sokal, 1973), of which there are legion, typically identify groups and linkages, producing sets (natural clusters) of and links (edges) between OTUs or variables. The SAHN clusters produce a sequential agglomerative hierarchical non-overlapping **dendrogram**. In single linkage, fusion is determined by the two closest objects in clusters; whereas in complete linkage clustering, fusion is dictated by the two most distant objects in clusters. In UPGMA (unweighted pair group method with arithmetic mean) clustering, the distance between two clusters is calculated as the average distance between all pairs of objects in the two different clusters. They were used more widely in a previous study (Dennis and Shreeve, 1996). **Ordination techniques** (e.g. **principal components analysis** (PCA), factor analysis (FA), **discriminant function analysis**, **non-metric scaling** (NMMS)) produce plots in one (rarely), two or more dimensions (see Sneath and Sokal, 1973 for an introduction). These procedures are applied to a matrix (table) of OTUs and variables, the order of which (rows and columns) can be transposed for different tasks. Relationships among variables are based on correlation coefficients and those between OTUs on similarity or distance coefficients (**Jaccard coefficients**, percentage difference and many more).

PCA/FA are extremely useful as they have the objective of boiling down a larger number of correlated variables to a smaller set of uncorrelated variables; generally only those that account for some arbitrary amount (usually 5% or more) of the variation are then followed up. Non-metric scaling is particularly helpful as it can be used for ordering OTUs or variables. This produces a ‘map’ of points, representing islands or species, in two or more dimensions. The points are effectively moved about until their relative positions in the mapped space equate as well as possible with the ranked differences in their affinities. The programme runs through a number of iterations from a starting configuration (a Guttman–Lingoes initial configuration based on PCA). Put simply, this seeks to place units (e.g. islands, species) in independent axes, each of which maximally describes the variation among the units. A number of coefficients measure distortion against the original similarity or distance coefficients (OTUs) or correlations (variables) (e.g. **Kruskal’s stress**; Guttman’s alienation K). **Discriminant function analysis** has as its prime objective the maximization of predictions (correct probability of occurrence and absence) for pre-defined classes; when just two classes are involved (e.g. species on an island and absent from it) this procedure has much in common with logistic (logit) regression though the assumptions around the data of using it are stricter. It has use in establishing probabilities of incidences of species on islands for which there are no records.

The book also cites work (Dennis *et al.*, 2012a) that has used more complex analyses such as RLQ analysis and fourth-corner analysis, which allow the joint analysis of more than two data matrix tables (Dray and Legendre, 2008). One of the features of biogeography is the growing sophistication of analytic techniques but the reader should be aware that the results have to make biological sense and ultimately be tested and checked. Thus, a cautionary note on the predictions in this book: they are based on far less-than-perfect data (just records of species, not recorded evidence of breeding) and solutions (most analyses predict the absence of species where they are known to occur, as well as predicting where species may be found on islands as yet lacking records!). Ultimately, common sense is required in judging findings and the human brain can often be, in effect, a superior multivariate analytic tool for this purpose.

Appendix 9. Predictions for Resident Butterfly Species in the British Isles

An important question is: what species are expected to occur in mainland Britain? Two sets of data are available for making predictions: (i) the **Climatic Risk Atlas** (Settele *et al.*, 2008); and (ii) the **Contemporary Geography Study** (this book) (see Chapter 4). The full data for the present study are contained in Supplementary Appendix 4.1 and the corresponding analyses in Supplementary Appendix 4.2.

It is noteworthy that both data sets predict species that should occur on the British mainland (but are absent), but also predict the absence of species that actually occur on the British mainland. The two sets of data also make significantly different predictions.

A. Table Appendix 9.1. Comparison of current resident British mainland species with those predicted from the Climatic Risk Atlas (CRA) (Settele *et al.*, 2008). Nomenclature and listing as in Kudrna *et al.* (2015).

Actual British residents	CRA prediction 0	CRA prediction 1	Row totals
0	215 (178)	19 (56)	234
1	7 (44)	51 (14)	58
All Groups	222	70	292

0, absent; 1, present; numbers, frequencies of species observed; expected frequencies in brackets.

Pearson Chi-square: 162.45, $df = 1$, $P < 0.0001$

British resident species predicted *not* to occur in the CRA assessment ($n = 7$): *Hesperia comma; Thymelicus acteon; Papilio machaon; Satyrium pruni; Argynnis adippe; Melitaea athalia; Melitaea cinxia*

B. Table Appendix 9.2. Comparison of current resident British mainland species with those predicted from geographical variables in the Contemporary Geography Study (see Supplementary Appendix 4.1).

Actual British residents	Prediction (geography) 0	Prediction (geography) 1	Row totals
0	338 (304)	13 (47)	351
1	16 (50)	42 (8)	58
All Groups	354	55	409

0, absent; 1, present; numbers, frequencies of species observed; expected frequencies in brackets

Pearson Chi-square: 201.90, $df = 1$, $P < 0.0001$

(i) Continental species predicted to occur as residents in the British Isles ($n = 13$): *Pyrgus alveus; Aporia crataegi; Pontia daplidice; Cyaniris semiargus; Glaucopsyche alexis; Lycaena hippothoe; Lycaena virgaureae; Plebejus idas; Argynnis niobe; Brenthis ino; Issoria lathonia; Nymphalis antiopa; Lasiommata maera.*

(ii) British resident species predicted *not* to occur in the British Isles ($n = 16$): *Thymelicus acteon; Thymelicus sylvestris; Aricia agestis; Hamearis lucina; Phengaris arion; Polyommatus bellargus; Polyommatus coridon; Satyrium pruni; Thecla betulae; Apatura iris; Euphydryas aurinia; Limenitis camilla; Melitaea cinxia; Erebia aethiops; Erebia epiphron; Pyronia tithonus.*

C. Table Appendix 9.3. Comparison of predicted resident species for the British Isles: predictions based on geography in the Contemporary Geography Study (nine continental variables in one PCA vector) versus those based on the Climatic Risk Atlas (Settele *et al.*, 2008).

Prediction from geography	CRA prediction 0	CRA prediction 1	Row totals
0	206 (180)	31 (57)	237
1	16 (42)	39 (13)	55
All groups	222	70	292

0, absent; 1, present; numbers, frequencies of species observed; expected frequencies in brackets

Pearson Chi-square: 81.91, $df = 1$, $P < 0.0001$

The differences indicate that modelling advances are required to determine better the factors underlying the presence and absence of species in Britain.

Appendix 10. Factors Affecting Butterfly Species Richness on the Offshore British and Irish Islands

For the study of species richness on islands (Chapter 5), the following variables have been included:

S, island species richness;
FS1, species at mainland faunal source (within 50 km radius);
FS2, species at nearest larger (by area or number of species) faunal source (within 50 km);
LAT, island latitude;
LONG, island longitude;
A, island area (ha);
ELEV, island maximum elevation (m);
I1, closest distance of island to mainland source (km);
I2, closest distance of island to nearest larger source (km);
SAT, number of satellite islands (satellites 0-0, 1-1; 2-2 to 10; 3 -11+);
POPNO, human population total;
POPDENS, human population density (n/km^2).

A. Comparison of Statistics for Geographical Variables

Comparison of statistics for geographical variables for the islands studied around Britain and Ireland (1996 study versus additions for 2017 study). Data for Channel Islands excluded. Each study includes islands with the four migrants (*Pieris brassicae*, *P. rapae*, *Vanessa atalanta*, *V. cardui*).

Table Appendix 10.1. Original 1996 study: statistics for geographical variables (*n* = 68).

Island variable	Median	Minimum	Maximum	Lower quartile	Upper quartile
Latitude	55.65	49.90	60.6	52.00	57.00
Longitude	–5.54	–10.00	0.9	–6.35	–3.26
Area	1240.00	6.00	213,730.0	195.00	11,535.00
Elevation (max)	127.00	4.00	986.0	46.50	307.50
Isolation (I1)	8.58	0.02	230.0	2.08	38.75
Isolation (I2)	2.60	0.02	90.0	0.70	7.05
Satellites	1.00	0	3.0	0	2.00

Table Appendix 10.2. Additions for the 2017 study: statistics for geographical variables (*n* = 55 islands not in the 1966 study; the 2017 study in the book included a total of 123 islands).

Island variable	Median	Minimum	Maximum	Lower quartile	Upper quartile
Latitude	53.40	49.94	59.24	51.79	55.63
Longitude	–4.73	–10.54	1.56	–6.17	–1.44
Area	299.50	1.00	10,800.00	21.40	1060.00
Elevation (max)	44.00	2.00	462.00	6.00	89.00
Isolation (I1)	1.50	0.00	81.00	0.20	5.00
Isolation (I2)	0.63	0.00	21.50	0.13	2.00
Satellites	1.00	0	3.00	0	2.00

Table Appendix 10.3. Comparison of data for the original 68 islands versus that for the additional 55 islands (**Mann–Whitney *U* tests**).

Island variable	*U* value	*Z*	*P*
Latitude	1416.00	2.31	0.02
Longitude	1501.50	–1.87	0.06 NS
Area	1120.00	3.81	0.0001
Elevation (max.)	933.00	4.76	<0.001
Isolation (I1)	983.00	4.51	<0.001
Isolation (I2)	984.50	4.50	<0.001
Satellites	26460.00	–0.15	0.88 NS

B. Pearson Correlations

Table Appendix 10.4. Pearson correlations among species richness and geographical variables (Pearson correlations on transformed (normalized) data) (see above introduction for variable codes).

Variable	S	FS1	FS2	LAT	LONG	A	ELEV	I1	I2	SAT	POPNO	POPDENS
S	1.00	0.48	0.55	–0.46	*0.03*	0.33	*0.04*	–0.44	–0.22	–0.28	0.42	0.42
FS1	0.11	1.00	0.88	–0.83	0.41	–0.40	–0.59	–0.59	–0.47	–0.40	*–0.03*	0.33
FS2	0.13	0.82	1.00	–0.69	0.30	–0.36	–0.50	–0.75	–0.45	–0.57	*–0.05*	0.23
LAT	*–0.01*	–0.86	–0.74	1.00	*–0.03*	0.35	0.45	0.43	0.35	0.28	*0.07*	–0.32
LONG	0.21	0.10	*–0.04*	0.25	1.00	*–0.14*	–0.49	–0.25	–0.22	–0.11	*0.11*	*0.17*
A	0.56	–0.27	–0.19	0.26	*0.02*	1.00	0.60	0.26	0.27	0.31	0.79	0.26
ELEV	0.36	–0.40	–0.33	0.31	–0.21	0.58	1.00	0.43	0.46	0.20	*0.19*	–0.25
I1	*–0.02*	–0.56	–0.76	0.56	0.18	0.14	0.25	1.00	0.64	0.65	*0.05*	*–0.14*
I2	0.28	–0.15	–0.19	0.14	*–0.06*	0.26	0.32	0.27	1.00	0.30	*0.07*	*–0.18*
SAT	*–0.06*	–0.45	–0.59	0.41	0.08	0.16	0.16	0.75	0.13	1.00	0.24	*0.09*
POPNO	0.56	0.14	0.10	–0.03	0.25	0.76	0.30	0.01	0.15	0.06	1.00	0.78
POPDENS	0.36	0.27	0.19	–0.17	0.19	0.41	0.00	–0.12	–0.03	–0.03	0.84	1.00

Lower triangle: $N = 910$ except for population variables where $n = 454$

Upper triangle: $n = 123$ (islands with all four migrants *Pieris rapae*, *P. brassicae*, *Vanessa atalanta*, *V. cardui*), except for population variables where $n = 101$.

Correlations in italics are not significant. The correlations given for species richness (S) in the lower triangle (red) show, when compared with the top line, how they are affected by the lack of any data for most of the smaller islands.

C. Principal Components Analysis of Geographical Variables

Principal components analysis (varimax rotation) of nine geographical variables: faunal source mass (FS1, FS2), latitude (LAT), longitude (LONG), area (A), maximum elevation (ELEV), isolation (I1, I2), number of satellites (SAT), over 910 islands. See Fig. 5.3.

Table Appendix 10.5. Eigenvalues for the principal components.

	Eigenvalue	% Total variance	Cumulative eigenvalue	Cumulative %
1	3.98	44.17	3.98	44.17
2	1.53	17.01	5.51	61.18
3	1.03	11.42	6.53	72.60
4	0.94	10.40	7.47	83.00

Table Appendix 10.6. Factor **loadings** (varimax normalized) for variables.

Variable	Factor 1	Factor 2	Factor 3	Factor 4	Communality after 4 factors
FS1	**–0.8574**	–0.3385	0.1598	0.1602	0.901
FS2	**–0.9140**	–0.1651	0.0293	–0.0507	0.866
LAT	**0.8127**	0.3352	0.2183	–0.1859	0.855
LONG	0.0718	–0.0408	**0.9829**	–0.0206	0.973
A	0.0715	**0.8648**	0.1172	0.1308	0.784
ELEV	0.2214	**0.8044**	–0.2244	0.1772	0.778
I1	**0.8217**	–0.0359	0.1406	0.4011	0.857
I2	0.0706	0.2919	–0.0425	**0.8234**	0.770
SAT	**0.7338**	–0.1038	0.0586	0.3649	0.686
Expl.Var	3.5088	1.7481	1.1293	1.0836	
Rel. Var.	0.3899	0.1942	0.1255	0.1204	

All variables have been log transformed. Key loadings are shown in **bold blue**.

D. Comparison of Geographical Variables (All Islands)

Table Appendix 10.7. Comparison of geographical variables for British and Irish offshore islands (all islands) (Mann–Whitney *U* Test; Ireland $n = 280$ islands; Britain $n = 629$ islands).

	Rank sum Irish islands	Rank sum British islands	*U*	*Z*	*P*
Latitude (LAT)	82163.0	331432.0	42823.0	–12.38	0.0001
Longitude (LONG)	54916.0	358679.0	15576.0	–19.83	0.0001
Area (A)	127813.5	285781.5	87646.5	0.11	0.91 NS
Elevation (max) (ELEV)	106010.0	283393.0	72340.0	–2.42	0.016
Isolation (I1)	76112.5	337482.5	36772.5	–14.03	0.0001
Isolation (I2)	120997.5	292597.5	81657.5	–1.75	0.08 NS
Satellites (SAT)	98766.5	314828.5	59426.5	–7.83	0.0001

E. Summaries of Multiple Regression Analyses

Summaries of multiple regression analyses of species richness of British and Irish offshore islands against geographical variables.

(i) 1996 study islands

Islands studied in 1996 (islands that have all four migrants *Pieris rapae*, *P. brassicae*, *Vanessa cardui*, *V. atalanta*), except Channel islands; $n = 68$ islands.

Table Appendix 10.8. Regression summary for dependent variable: S (total number of species including migrants).

Variables	β	SE_β	b	SE_b	$t_{(63)}$	R	R^2	R^2 change	P
Intercept			12.95	3.037	4.26				<0.0001
FS2	0.46	0.116	0.41	0.105	3.96				0.0002
A	0.55	0.069	0.08	0.010	8.00	0.69	0.47	0.47	<0.0001
LAT	–0.39	0.091	–2.96	0.703	–4.21	0.82	0.68	0.20	<0.0001
I1	–0.21	0.091	–0.06	0.025	–2.27	0.86	0.73	0.06	0.026

Beta (β) and b are standardized and raw slope coefficients of the regression equations, each with their standard errors (SE); t is Student's t-test with degrees of freedom; R is the multiple correlation coefficient; R^2 is the coefficient of determination.

$F_{(4,63)} = 48.35$, $P < 0.0001$, standard error of estimate: 0.211.

Variables log transformed. **Standardized residuals**: Grassholm (–4.78), Puffin (–2.05), Fair Isle (2.09).

(ii) 2017 study islands

Islands studied in 2017 (islands that have all four migrants *Pieris rapae*, *P. brassicae*, *Vanessa cardui*, *V. atalanta*) except Channel islands; $n = 123$ islands, included 68 islands from the 1996 study.

Table Appendix 10.9. Regression summary for dependent variable: total number of species, including common migrants.

Variables	β	SE_β	b	SE_b	$t_{(119)}$	R	R^2	R^2 change	P
Intercept			8.755	2.548	3.44				0.0008
FS2	0.60	0.076	0.554	0.071	7.86	0.55	0.30	0.30	<0.0001
A	0.63	0.059	0.090	0.008	10.68	0.78	0.61	0.31	<0.0001
LAT	–0.26	0.075	–2.092	0.599	–3.49	0.81	0.65	0.04	0.0007

Beta (β) and b are standardized and raw slope coefficients of the regression equations, each with their standard errors (SE); t is Student's t-test with degrees of freedom; R is the multiple correlation coefficient; R^2 is the coefficient of determination.

$F_{(3,119)} = 73.63$, $P < 0.0001$, **standard error of estimate:** 0.235.

Variables log transformed. Standardized residuals: Staple (–2.107), Foulness (–2.046), Skokholm (2.273), Grassholm (–3.976), Piel (–2.574).

See Supplementary Appendix 6.1b for predicted species richness.

(iii) 2017 study islands

Islands studied in 2017 (islands that have two or three of the long-distance continental-wide migrants: *Colias crcoeus, Vanessa cardui, V. atalanta*) except Channel islands; $n = 191$ islands.

Model based on four factors built up on nine variables: faunal source mass (FS1, FS2), latitude (LAT), longitude (LONG), isolation (ISOL1, ISOL2), area (A), maximum elevation (ELEV), satellites (SAT). The factors are: GeogF1: FS1, FS2, LAT, ISOL1, SAT; GeogF2: AREA, ELEV; GeogF3: LONG (+15 to avoid negative numbers); GeogF4: ISOL2 (see part C above). All variables are log transformed for normality.

Table Appendix 10.10. Regression summary for dependent variable: total number of species including common migrants.

Variables	β	SE_β	b	SE_b	$t_{(188)}$	R	R^2	R^2 change	P
Intercept			2.089	0.047	44.90				<0.0001
GeogFact1	−0.679	0.049	−0.458	0.033	−13.89	0.63	0.40	0.40	<0.0001
GeogFact2	−0.399	0.049	−0.250	0.031	−8.16	0.75	0.56	0.16	<0.0001

Beta (β) and b are standardized and raw slope coefficients of the regression equations, each with their standard errors (SE); t is Student's t-test with degrees of freedom; R is the multiple correlation coefficient; R^2 is the coefficient of determination.

$F_{(2,188)} = 117.81$, $P < 0.0001$, standard error of estimate: 0.476.

Standardized residuals: West Wideopen (−2.070), Coquet (−3.778), Burry Holms (−2.256), Mewstone (−3.371), Garland Stone (−3.125), Eilean Hoan (−2.111), Sule Skerry (−2.518), Copinsay (−2.213), Great Copeland (−2.178), Kerragh (−2.400), Little Island, East Cork (−4.378), Great Skellig (−2.495).

See Supplementary Appendix 6.1b for predicted species richness.

Appendix 11. Summary of Predictions for Species Incidences on British and Irish Offshore Islands

Predictions for each butterfly species over all islands ($n = 910$ islands) are based on **logit regression analysis** (forwards, stepwise entry) parameters for geographical and source-influence variables determined for a subset of 191 islands that have records for two or more of three long-distance migrants that engage in seasonal reversal of movement (*Colias croceus*, *Vanessa cardui*, *V. atalanta*).

Predictions for two sets of logit regressions are compared.

The first set is based on five factors (**principal components**, varimax rotated, normalized) extracted from seven geographical variables:

F1: island isolation (I1) and satellites (SAT);

F2: island area (A) and maximum elevation (ELEV);

F3: longitude (LONG);

F4: island isolation (I2);

F5: latitude (LAT) (see Supplementary Appendices 6.1a and 6.1b).

The second set is based on one of two contributory source variables:

FS1: number of 10 km square records within 50 km of the nearest mainland source;

FS2: number of 10 km square records within 50 km of the nearest larger source, island or mainland.

For any species, the source selected is that producing the higher Z value from a **Mann–Whitney U test** for the 191 islands.

Tables Appendix 11.1 and 11.2 give a factor analysis for the seven geographical variables.

Table Appendix 11.1. Eigenvalues for geographical variables.

Factor extracted	Eigenvalue	% Total variance	Cumulative eigenvalue	Cumulative %
F1	2.619	37.42	2.619	37.42
F2	1.494	21.34	4.113	58.76
F3	0.974	13.92	5.088	72.68
F4	0.823	11.76	5.911	84.44
F5	0.540	7.72	6.451	92.16

Table Appendix 11.2. Factor **loadings** (varimax normalized) of variables.

	Factor 1	Factor 2	Factor 3	Factor 4	Factor 5	Communality 5 factors
LAT	0.3087	0.1760	0.1682	0.0261	**0.8891**	0.9458
LONG	0.0600	–0.0407	**0.9700**	–0.0187	0.1343	0.9645
A	0.0835	**0.9439**	0.1143	0.0686	0.0010	0.9157
ELEV	0.0720	**0.7697**	–0.2910	0.1942	0.3223	0.8240
I1	**0.8378**	0.0421	0.0807	0.1994	0.3582	0.8783
I2	0.1019	0.1599	–0.0231	**0.9759**	0.0343	0.9900
SAT	**0.9571**	0.1035	0.0122	–0.0102	0.0768	0.9329
Explained variance	1.7394	1.5541	1.0741	1.0357	1.0479	
Proportion of total variance	0.2485	0.2220	0.1534	0.1480	0.1497	

All variables but satellites are log transformed. Bold **loadings** indicate focus of each factor.

A. Results of Logit Regressions

(i) Table Appendix 11.3. Polarity of effect (sign) and significance for logit regressions describing species incidences on islands[1].

	Geography factors						Nearest source mass	
SPECIES	F1	F2	F3	F4	F5	Loss [$\chi_{(df)}$], P	Source variable[2]	Loss [$\chi_{(df)}$], P
C. palaemon								
T. sylvestris	– **		+ **		–****	79.21 (3), ****	+ ****	60.11 (1), ****
T. lineola			+ †		– ***	89.35 (2), ****	+ ****	78.24 (1), ****
T. acteon								
H. comma								
O. sylvanus	– **	+ †	+ *		–****	91.41 (4), ****	+ ****	75.26 (1), ****
E. tages					– *	8.69 (1), †	+ ****	18.49 (1), ***
P. malvae								
P. machaon								
L. sinapis sl								
C. croceus								
G. rhamni	– *				–****	71.57 (2), ****	+ ****	47.46 (1), ****
P. brassicae	– †	+ ****	+ **		– †	52.14 (4), ****	+ ns [0.07]	3.53 (1), ns [0.06]
P. rapae	–****	– ***	– *	–†	–****	94.32 (5), ****	+ ****	33.48 (1), ****
P. napi	–****	+ ****			– *	46.69 (3), ****	+ ****	40.20 (1), ****
A. cardamines	–****	+ ****			–****	62.80 (3), ****	+ ***	21.50 (1), ****
C. rubi	– *	+ ****			– *	57.53 (3), ****	+ ****	26.25 (1), ****

Continued

(i) Table Appendix 11.3. Continued.

	Geography factors						Nearest source mass	
SPECIES	F1	F2	F3	F4	F5	Loss [$\chi_{(df)}$], *P*	Source variable[2]	Loss [$\chi_{(df)}$], *P*
T. betulae								
F. quercus								
S. w-album								
S. pruni								
L. phlaeas	–****	+*			–****	113.62 (3), ****	+****	73.42 (1), ****
L. dispar								
C. minimus	–*	+*	–*		–**	38.72 (4), ****	+****	18.81 (1), ***
P. argus								
A. agestis	–†		+†		–*	85.39 (3), ****	+****	74.03 (1), ****
A. artaxerxes								
P. icarus	–****	+****			–**	54.35 (3), ****	+****	25.64 (1), ****
P. coridon								
P. bellargus								
C. argiolus	–**	+†			–****	95.12 (3), ****	+****	58.71 (1), ****
P. arion								
H. lucina								
L. camilla								
A. iris								
V. atalanta								
V. cardui								
A. urticae	–*	+****				29.75 (2), ****	+**	12.66 (1), **
A. io	–*	+****			–*	37.92 (3), ****	+*	11.37 (1), **
N. polychloros								
P. c-album	–*		+**		–****	104.14 (3), ****	+****	72.49 (1), ****
B. selene	–†	+****				45.40 (2), ****	+****	51.38 (1), ****
B. euphrosyne		+****				37.61 (1), ****	+*	6.79 (1), *
A. adippe								
A. aglaja	–**	+****	–†			72.57 (3), ****	+****	33.17 (1), ****
A. paphia	–*		–*		–**	31.43 (3), ****	+***	16.26 (1), ***
E. aurinia		+****	–*			36.61 (2), ****	+***	16.20 (1), ***
M. cinxia								
M. athalia								
P. aegeria	–***	+***			–****	102.19 (3), ****	+****	49.53 (1), ****
L. megera	–****	+†			–****	107.39 (3), ****	+****	70.15 (1), ****
E. epiphron								
E. aethiops	–***	+****			+***	64.20 (3), ****	+****	72.62 (1), ****
M. galathea	–*				–***	56.34 (2), ****	+****	44.91 (1), ****
H. semele	–****	+****	–†			56.23 (3), ****	+****	62.19 (1), ****
P. tithonus	–**		+*		–****	144.12 (3), ****	+****	129.32 (1), ****
M. jurtina	–****	+****			–****	75.53 (3), ****	+****	33.88 (1), *****
A. hyperantus	–****	+**		+†	–****	73.77 (4), ****	+****	39.66 (1), ****
C. pamphilus	–****	+****			–**	74.19 (3), ****	+****	65.93 (1), ****
C. tullia		+****	–†		+**	100.46 (3), ****	+****	45.51 (1), ****

The chi square (χ^2) with degrees of freedom gives the significance of the logit equation procedure (maximum likelihood quasi-Newton procedure). Variable influence: – negative, + positive; † $P < 0.05$, *$P < 0.01$,** $P < 0.001$,*** $P < 0.0001$, $p < 0.00001$. [1]Logit regression equations based on 191 islands with ≥2 migrants (*Colias croceus, Vanessa atalanta, V. cardui*). [2]Variable FS1 or FS2. See Appendix 1 for full names of species and Supplementary Appendix 6.1a for detailed results.

(ii) Table Appendix 11.4. Predictions for species incidences on offshore islands based on logit regression analysis: comparison of five geographical factors and two source variables.

SPECIES	Observed island incidences (all 910 islands)	Predicted island incidences (on geog. factors)[1]	Predicted island incidences (on source variables)[1]	% observed incidences correctly predicted (on geog. factors)[2]	Total % surveyed correctly predicted (on geog. factors)[3]	% observed incidences correctly predicted (on source variable)[2]	Total % surveyed correctly predicted (on source variable)[3]
C. palaemon	2	–	–	–	–	–	–
T. sylvestris	28	60	73	59.3	92.1	44.4	87.4
T. lineola	16	48	56	87.5	98.4	87.5	97.4
T. acteon	2	–	–	–	–	–	–
H. comma	1	–	–				
O. sylvanus	30	54	79	58.6	91.6	55.2	90.1
E. tages	11	0	49	0	94.8	0	94.2
P. malvae	6	–	–	–	–	–	–
P. machaon	7	–	–	–	–	–	–
L. sinapis sl	2	–	–	–	–	–	–
C. croceus	85	–	–	–	–	–	–
G. rhamni	38	80	76	57.9	86.9	42.1	85.3
P. brassicae	194	422	910*	94.2	82.2	100	81.2
P. rapae	163	381	741	89.2	83.2	85.4	76.4
P. napi	222	624	801	94.0	81.7	95.3	84.3
A. cardamines	86	127	293	68.0	75.9	46.2	63.9
C. rubi	49	23	75	31.7	81.7	17.1	77.0
T. betulae	1	–	–	–	–	–	–
F. quercus	9	–	–	–	–	–	–
S. w-album	3	–	–	–	–	–	–
S. pruni	0	–	–	–	–	–	–
L. phlaeas	131	414	488	86.7	80.6	77.3	76.4
L. dispar	0	–	–	–	–	–	–

Continued

(ii) Table Appendix 11.4. Continued.

SPECIES	Observed island incidences (all 910 islands)	Predicted island incidences (on geog. factors)[1]	Predicted island incidences (on source variables)[1]	% observed incidences correctly predicted (on geog. factors)[2]	Total % surveyed correctly predicted (on geog. factors)[3]	% observed incidences correctly predicted (on source variable)[2]	Total % surveyed correctly predicted (on source variable)[3]
C. minimus	19	14	18	16.7	90.1	22.2	91.6
P. argus	6	–	–	–	–	–	–
A. agestis	20	57	17	80.0	96.3	65.0	94.2
A. artaxerxes	3	–	–	–	–	–	–
P. icarus	235	562	817	92.4	79.6	93.8	79.1
P. coridon	3	–	–	–	–	–	–
P. bellargus	3	–	–	–	–	–	–
C. argiolus	38	66	92	66.7	90.1	41.7	82.7
P. arion	0	–	–	–	–	–	–
H. lucina	1	–	–	–	–	–	–
L. camilla	4	–	–	–	–	–	–
A. iris	1	–	–	–	–	–	–
V. atalanta	227	–	–	–	–	–	–
V. cardui	217	–	–	–	–	–	–
A. urticae	237	719	910	98.8	85.9	100.0	84.3
A. io	172	546	910	90.7	73.8	100.0	72.8
N. polychloros	10	–	–	–	–	–	–
P. c-album	44	81	91	75.6	92.7	58.4	86.4
B. selene	45	25	109	38.5	83.8	33.3	80.6
B. euphrosyne	14	7	1	28.6	93.2	0.0	92.6
A. adippe	4	–	–	–	–	–	–
A. aglaja	104	121	217	67.1	77.5	43.4	68.6
A. paphia	18	4	10	6.3	91.6	0.0	90.6
E. aurinia	37	11	39	11.1	84.3	7.41	84.8
M. cinxia	1	–	–	–	–	–	–

M. athalia	0	–	–	–	–	–	–
P. aegeria	101	225	429	77.9	81.2	65.1	72.3
L. megera	73	190	278	76.1	85.3	62.7	78.5
E. epiphron	0	–	–	–	–	–	–
E. aethiops	33	34	93	44.4	89.5	59.3	90.6
M. galathea	14	45	45	50.0	94.8	57.1	94.2
H. semele	123	277	357	66.3	70.2	73.0	75.9
P. tithonus	49	109	167	79.6	72.8	84.1	91.6
M. jurtina	257	572	815	92.7	84.8	93.3	81.2
A. hyperantus	80	100	247	71.2	82.2	42.2	71.2
C. pamphilus	145	211	453	77.8	77.5	77.8	75.4
C. tullia	52	44	112	58.5	85.9	34.2	81.7

Predictions made for 910 islands using two sets of logit regression based on (i) five geography factors, and (ii) the best predictor of two source variables (10 km square records within 50 km radius of source), nearest mainland source or nearest larger source; – numbers too few (or many) for logit regression or variable(s) fails significance. [1]Predicted incidences for 910 islands in Appendix 3; [2]% correct predictions on 191 islands that have at least two of three migrant species (*C. croceus, V. atalanta, V. cardui*); [3]% correct predictions for presence and absence of a species on 191 islands.

B. Results of Discriminant Function Analyses

(i) Table Appendix 11.5. Predictions for species incidences on offshore islands based on discriminant function analysis of five geographical factors and two source variables.

SPECIES	Significant variables	$F_{[df]}$ statistic (all $P < 0.0001$)	Predicted island incidences[1]	% observed incidences correctly predicted[2]	Total % surveyed correctly predicted[3]
C. palaemon	–	–	–	–	–
T. sylvestris	FS2, F2, F3, F5, F1	37.93 [5,185]	85	81.48	92.15
T. lineola	FS1, F3	254.45 [2,188]	58	93.75	97.91
T. acteon	–	–	–	–	–
H. comma	–	–	–		
O. sylvanus	FS2, F2, F5, F3, F1	50.24 [5,185]	94	89.66	93.19
E. tages	FS1	32.52 [1,189]	141	90.00	82.72
P. malvae	–	–	–	–	–
P. machaon	–	–	–	–	–
L. sinapis sl	–	–	–	–	–
C. croceus	–	–	–	–	–
G. rhamni	F5, FS1	54.80 [2,188]	195	73.68	83.25
P. brassicae	F2, FS1, F3	18.88 [3,187]	213	74.84	75.39
P. rapae	F5, F1, F2, F4, FS1	23.07 [5,185]	324	81.54	81.68
P. napi	FS2, F2, F5, FS1	21.14 [4,185]	500	79.19	78.95
A. cardamines	FS2, F2	39.67 [2,188]	198	73.08	72.77
C. rubi	F2, FS2	46.41 [2,188]	111	80.49	79.58
T. betulae	–	–	–	–	–
F. quercus	–	–	–	–	–
S. w-album	–	–	–	–	–
S. pruni	–	–	–	–	–
L. phlaeas	FS2, F5, F1, F2	45.28 [4,186]	387	82.86	82.20
L. dispar	–	–	–	–	–
C. minimus	FS1, F3, F2	20.48 [3,187]	141	55.56	85.34
P. argus	–	–	–	–	–
A. agestis	FS1, F2, F3, F1	80.42 [4,186]	62	80.00	95.81
A. artaxerxes	–	–	–	–	–
P. icarus	FS2, F2, F5, FS1	19.62 [4,186]	373	77.24	76.96
P. coridon	–	–	–	–	–
P. bellargus	–	–	–	–	–
C. argiolus	F5, FS1, F2	51.09 [3,187]	147	88.89	88.48
P. arion	–	–	–	–	–
H. lucina	–	–	–	–	–

Continued

(i) Table Appendix 11.5. Continued.

SPECIES	Significant variables	$F_{[df]}$ statistic (all $P < 0.0001$)	Predicted island incidences[1]	% observed incidences correctly predicted[2]	Total % surveyed correctly predicted[3]
L. camilla	–	–	–	–	–
A. iris	–	–	–	–	–
V. atalanta	–	–	–	–	–
V. cardui	–	–	–	–	–
A. urticae	F2, FS2, F4	15.03 [3,187]	281	73.29	74.35
A. io	F2, FS1	24.16 [2,188]	280	69.06	71.20
N. polychloros	–	–	–	–	–
P. c-album	FS1, F5, F3, F1	54.70 [4,186]	127	87.80	90.05
B. selene	FS2, F2	76.78 [2,188]	118	84.62	85.34
B. euphrosyne	F2, FS2, F4	20.28 [3,187]	45	92.86	86.39
A. adippe	–	–	–	–	–
A. aglaja	F2, FS2, F3	43.68 [3,187]	175	76.32	81.15
A. paphia	FS1, F2	15.90 [2,188]	137	62.50	84.29
E. aurinia	F2, FS2, F3	21.60 [3,187]	149	88.89	79.58
M. cinxia	–	–	–	–	–
M. athalia	–	–	–	–	–
P. aegeria	F5, F1, F2, FS2	35.93 [4,186]	283	84.88	83.25
L. megera	FS1, F2, F5, F1	39.258 [4,186]	269	83.58	84.82
E. epiphron	–	–	–	–	–
E. aethiops	FS2, F1, F2	78.55 [3,187]	34	88.89	93.19
M. galathea	FS1, F2	86.63 [2,188]	58	78.57	95.29
H. semele	FS2, F2, F1, F3	33.70 [4,186]	318	79.78	80.10
P. tithonus	FS1, F5, F3	140.55 [3,187]	186	93.18	92.15
M. jurtina	FS2, F5, F2, FS1, F1, F3	20.16 [6,184]	536	83.33	83.77
A. hyperantus	FS2, F2, F5, F1	25.63 [4,186]	255	86.36	81.68
C. pamphilus	FS2, F2, F5	56.13 [3,187]	229	76.77	81.15
C. tullia	F2, FS2, F4	65.72 [3,187]	102	95.12	89.01

Predictions for 910 islands are based on a traditional discriminant function analysis (forwards stepwise entry) based on a selection of five geography factors and two source variables for 191 islands having two or more long distance migrants (*C. croceus, V. atalanta, V. cardui*). Variables as described above. [1]Predicted incidences for 910 islands in Appendix 3; [2]% correct predictions for known incidences on 191 islands that have at least two of three migrant species; [3]% correct predictions for presences and absences of species on 191 islands. See Supplementary Appendix 6.1a part 2 for the discriminant analyses (basic) results, and Supplementary Appendix 6.1b sheet 3 for species' predictions for 910 islands.

C. Comparisons of Results from Logit and Discriminant Function Analysis Models

The logit and discriminant function analysis models applied in the above Tables 11.3, 11.4 and 11.5 have been compared using non-parametric tests (**Friedman ANOVA** and **Kendall coefficient of concordance**). Although the discriminant analysis does not predict the greatest numbers of islands (Fig. App. 11.1), the results are more reliable in that discriminant function analysis has produced a higher percentage of correct predictions for species known to occupy (or to have occupied) the islands (Fig. App. 11.2).

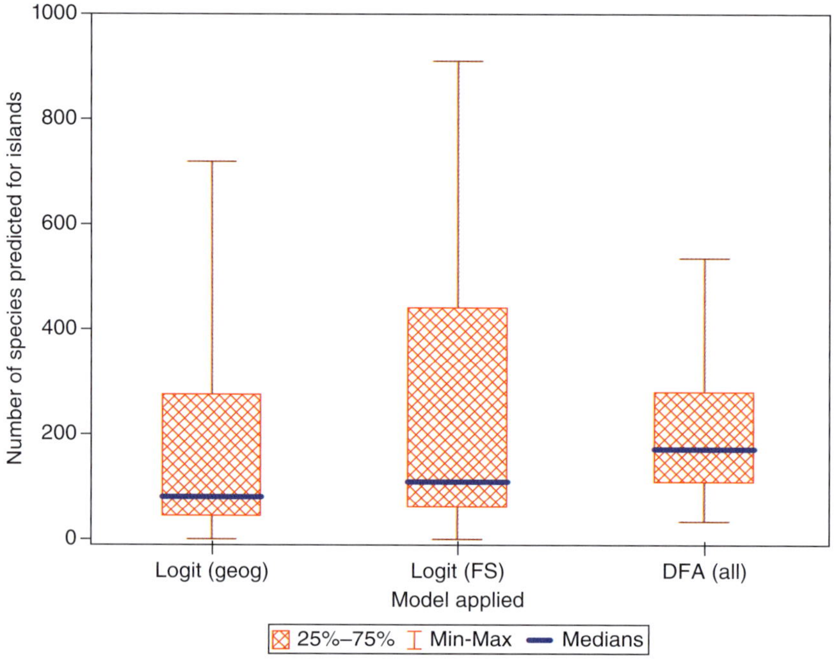

Fig. App. 11.1. Bar chart illustrating comparison of correct predictions for species recorded on 191 islands. Three models are compared: logit (based on geography factors), logit (based on one of two faunal source variables) and discriminant analysis (based on all factors and variables). In each model, variables are selected by stepwise forwards entry.

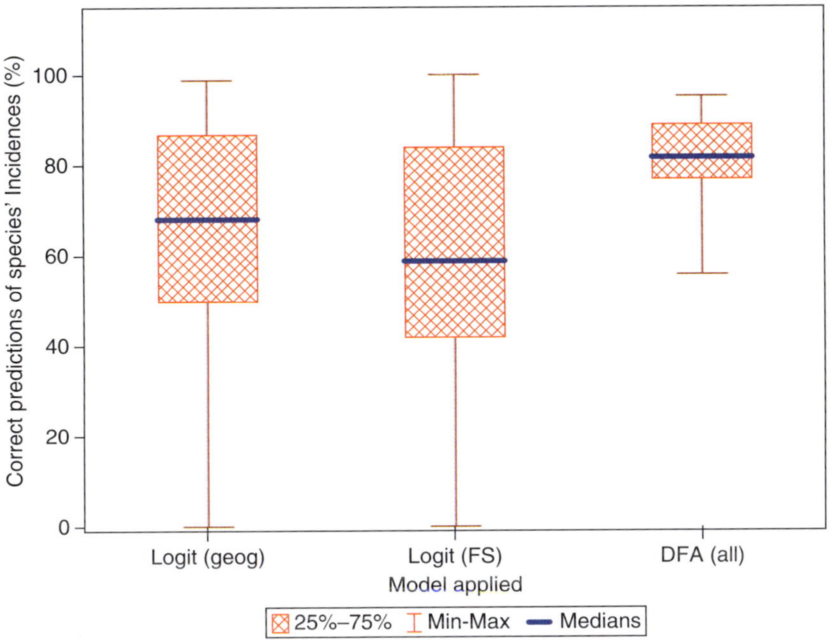

Fig. App. 11.2. Bar chart illustrating comparison of predictions for numbers of species recorded for 910 islands. The models are as in Fig. App. 11.1.

Table Appendix 11.6. Comparison of models for predictions of numbers of species (910 islands).

	Average rank	Sum of ranks	Mean	Standard deviation
Logit (geography factors)	1.344	43.00	182.22	212.37
Logit (faunal source variable)	2.359	75.50	279.94	300.94
DFA (all variables)	2.297	73.50	199.94	127.98

DFA, discriminant function analysis; Friedman ANOVA and Kendall coefficient of concordance; ANOVA χ^s (N = 32, df = 2) = 21.06, P = 0.00003; coefficient of concordance = 0.329; average rank r = 0.31.

Table Appendix 11.7. Comparison of models for correct predictions of species incidences across 191 islands.

	Average rank	Sum of ranks	Mean	Standard deviation
Logit (geography factors)	1.909	63.00	63.46	27.62
Logit (faunal source variable)	1.621	53.50	56.50	30.95
DFA (all variables)	2.470	81.50	81.46	8.87

DFA, discriminant function analysis; Friedman ANOVA and Kendall coefficient of concordance; ANOVA χ^2 (N = 33, df = 2) = 12.67, P = 0.0018; coefficient of concordance = 0.192; average rank r = 0.17.

Appendix 12. Butterflies Recorded on Strangford Lough Islands, Northern Ireland, During the Butterfly Conservation UK Big Butterfly Counts (July/August 2012–2017 Inclusive).[1]

Records of butterflies on the islands in Strangford Lough, Northern Ireland, are not included in Appendix 3. These are special islands, enclosed by land, each the remnant of a **drumlin** (glacial deposit) and composed of till (boulder clay). The records below emanate from recent surveys of the **Big Butterfly Count** (Butterfly Conservation, UK).

Abbreviations and Notes

Grid references in square brackets.

Species in round brackets are predicted to occur on islands from **logistic regression** analyses (standardized on two visits when number of visits are one of the predictor variables) based on two models: 1. Islands visited (v); 2. Islands having one or more butterfly records (r). Codes 3 and 4 indicate probabilities greater than the median and upper quartile, respectively, for islands known to have the species noted.

Species in curly brackets are those predicted to occur from **discriminant function analysis** based on a single randomly selected visit.

* Islands not visited.

** Fossil islands not included in analyses.

2016 and 2017 records (two unplaced in J5466): † confirmation of earlier records; ‡ new records.

Names of islands are those shown on the 1:25,000 OSNI Activity Map of Strangford Lough (www.osni.gov.uk; 2008). Islands within Strangford Narrows are excluded.

Islands

Chapel Island [J561512]: *P. brassicae*‡, *P. rapae*†, *P. napi*†, *P. icarus*, *L. phlaeas*, *A. urticae*†, *A. io*, *V. atalanta*†, *P. aegeria*‡, *M. jurtina*†, *A. hyperantus*†.

Jackdaw Island [J556510]: *P. rapae*†, *P. napi*†, *A. urticae*, *V. atalanta*†, *M. jurtina*†, *A. hyperantus*†.

Shark [J545504]: *P. brassicae*, *P. napi*†, *P. icarus*, *A. io*, *V. atalanta*†, *A. urticae*‡, *M. jurtina*†, *A. hyperantus*†.

Launches Little [J540504]: (*P. napi*, v3), *P. brassicae*‡, *A. urticae*‡, A. *io*‡, *V. atalanta*‡, *M. jurtina*†, *A. hyperantus*†, *P. aegeria*.

Launches Long [J537500]: *P. napi*‡, *A. urticae*‡, *M. jurtina*†, *A. hyperantus*†, *P. aegeria*†. (Two visits in 2016; not clear if one may have belonged to Launches Little.)

Gores [J519493]: *P. rapae*, *P. napi*, *A. urticae*, *A. io*, *V. cardui*, *M. jurtina*‡, *A. hyperantus*†. {*P. icarus*}

Hare [J506492]: *P. rapae*, *P. napi*‡, *P. icarus*†, *L. phlaeas*, *A. urticae*, *M. jurtina*†, *A. hyperantus*†, *P. aegeria*.

Gibbs [J509495]: *P. napi*‡, *P. icarus*‡, *A. urticae*‡, *M. jurtina*†, *A. hyperantus*†, *P. aegeria*‡. {*P. icarus*} (Two visits in 2016.)

Rat Island [J522500]: *P. brassicae*, *P. rapae*, *A. urticae*, *A. io*, *V. atalanta*, *M. jurtina*. {*A. hyperantus*} (Visited in 2016 but no butterflies seen.)

Salt Island [J530501]: *P. rapae*, *P. napi*†, *P. icarus*, *L. phlaeas*, *A. urticae*, *A. io*, *V. atalanta*, *M. jurtina*†, *A. hyperantus*†. (Two visits in 2016.)

Green Island, Killyleagh [J536511]: *P. napi*, *P. icarus*‡, *A. urticae*†, *V. atalanta*, *V. cardui*‡, *M. jurtina*†, *A. hyperantus*†. (Two visits in 2016.)

Dunnyneill Island (paired linked at low water) [J547538]: *P. napi*‡, *A. urticae*†, *A. hyperantus*. (South islet visited in 2016.)

Dodds [J530547]: *A. urticae*, *M. jurtina*‡, *A. hyperantus*‡.

Island Taggart [J533546]: *P. brassicae*†, *P. napi*†, *P. icarus*†, *A. urticae*, *M. jurtina*†, *A. hyperantus*†. (Two visits in 2016.)

Simmy [J532557]: nil. (*P. napi* v3; *P. icarus* v4, r4) {*P. icarus*}

Pawle [J544574]: *P. brassicae*, *P. napi*, *A. urticae*, *M. jurtina*, *A. hyperantus*.

Islandmore [J540581]: *P. napi*, *A. urticae*, *M. jurtina*, *A. hyperantus*.

Dunsy Island [J544590]: *P. icarus*, *A. io*‡, *M. jurtina*†, *A. hyperantus*.

Castle [J537593]: nil. (*P. napi* v4; *P. icarus* v4, r4; *M. jurtina* v4, r4; *A. hyperantus* v3, r3) {*P. icarus*; *A. hyperantus*}

Darragh [J538601]: *P. icarus*, *A. urticae*, *M. jurtina*, *A. hyperantus*.

Conly Island [J530608]: *P. napi*, *A. urticae*, *M. jurtina*, *A. hyperantus*. {*P. icarus*}

Roe Island [J543610]: *A. urticae*, *M. jurtina*, *A. hyperantus*.

Parton [J541605]: *P. napi*, *V. atalanta*‡, *A. hyperantus*.

Green [J547596]: *P. napi*, *A. urticae*, *M. jurtina*†, *A. hyperantus*.

Great Minnis [J550600]: *P. napi*†.

Little Minnis [J550603]: *P. napi*‡, *M. jurtina*, *A. hyperantus*.

Drummond [J551608]: *P. rapae*, *P. napi*‡, *V. atalanta*‡, *V. cardui*‡, *A. urticae*, *M. jurtina*, *A. hyperantus*.

Innísharoan Island [J543615]: *P. rapae*‡, *A. urticae*, *M. jurtina*†, *A. hyperantus*.

Inishanier Island [J548614]: *L. phlaeas*‡, *M. jurtina*†, *A. hyperantus*.
Craigaveagh Rock [J539613]: *A. io*‡, *M. jurtina*, *A. hyperantus*.
Trasnagh Island [J537620]: *P. napi*‡, *M. jurtina*†, *A. hyperantus*.
Hen Island [J523622]: nil. (*P. icarus* v3, r3)
Rainey [J527630]: *P. napi*‡, *P. aegeria*‡. (*P. napi* v4; *P. icarus* v3, r3; *M. jurtina* v3, r3; *A. hyperantus* v3, r3) {*A. hyperantus*}
Calf [J532631]: *V. atalanta*‡, *M. jurtina*‡ (*P. napi* v3)
Lythe Rock [J533633]: nil.
Mahee Island [J533638]: *P. rapae*, *P. napi*, *L. phlaeas*, *A. urticae*, *V. atalanta*, *M. jurtina*. {*P. icarus*}
Bird Island [J569616]: *A. urticae*†, *M. jurtina*‡.
Reagh – slipway [J526648]: *P. brassicae*‡, *P. icarus*‡, *A. hyperantus*‡ (*P. napi* v4; *P. icarus* v3, r3; *M. jurtina* v3, r3; *A. hyperantus* v3, r3). {*P. icarus*}
Boretree West [J539678]: nil.
Boretree East [J542680]: *P. brassicae*‡, *P. napi*‡, *A. urticae*.
Peggys Island [J551685]: *M. jurtina*.
Gull Rock [J541683]: nil.
North Boretree [J546683]: nil.
Turley Rock [J550680]: nil.
Reids [J550676]: nil.
Chanderies [J547674]: nil.
Hare Island [J552670]: nil.
Chapel Island [J554672]: *P. brassicae*‡, *P. napi*‡, *P. icarus*, *V. atalanta*‡, *V. cardui*‡, *M. jurtina*†.
Mid Island [J566672]: *P. brassicae*‡, *P. napi*‡, *P. icarus*†, *L. phlaeas*, *V. atalanta*‡, *A. urticae*, *P. aegeria*‡, *M. jurtina*†, *A. hyperantus*.
South Island [J566666]: *P. icarus*, *A. urticae*, *M. jurtina*, *A. hyperantus*.
Gabbock [J560656]: *P. napi*†.
Sheelah's [J572645]: nil.
South Sheelah's [J569641]: nil.
Long Skart [J567630]: *P. brassicae*‡, *V. atalanta*‡, *M. jurtina*‡.
Round Skart Island [J565626]: nil.
Horse Island [J597608]: nil. (*P. napi* v4; *P. icarus* v4, r4)
Gransha Point [J584593]: *P. icarus*.
Mid Island-Gransha [J592595]: *V. cardui*. {*P. icarus*}
Long Island [J583572]: *A. urticae*‡, *A. hyperantus*†.
Round Island [J577565]: *A. urticae*, *M. jurtina*‡.
Phersons Island [J596565]: *P. icarus*, *A. urticae*, *M. jurtina*, *A. hyperantus*, *P. aegeria*.
Ballyhenry Point [J574520]: *P. napi*‡, *M. jurtina*, *A. hyperantus*. (Record of 11 *G. rhamni* in 2016 suspect.)
Dunsy Rock* [J551587]: nil.
Greenisland Rock* [J544601]: nil.
Shamrock Island* [J527602]: nil. (*P. napi* v3; *P. icarus* v4, r4)
Bush Island* [J521601]: nil. (*P. icarus* v4, r4)
Braddock Island* [J528617]: nil. (*P. icarus* v4, r4)

Bird Island by Mahee* [J538648]: *P. napi*‡. (*P. icarus* v3, r3)
Gull Rock by Mahee* [J541642]: nil. (*P. icarus* v3, r3)
Watson's Island* [J508636]: nil. (*P. icarus* r3)
Wood Island* [J511641]: nil. (*P. napi* v4; *P. icarus* v3, r3; *M. jurtina* v3, r3; *A. hyperantus* v3, r3)
Long Island by Drum Hill* [J511646]: nil. (*P. napi* v3; *P. icarus* v4, r4)
Rolly Island* [J520650]: nil. (*P. napi* v4; *P. icarus* v4, r4; *M. jurtina* v3, r3; *A. hyperantus* v3, r3)
Cross Island* [J530647]: nil. (*P. napi* v3; *M. jurtina* v3, r3; *A. hyperantus* v3, r3)
Sketrick Island* [J527624]: nil. (*P. napi* v4; *P. icarus* v4, r4; *M. jurtina* v4, r4; *A. hyperantus* v3, r3)
Horse Island by Ringneill Br.* [J505664]: nil. (*P. icarus* v3, r3)
Rough Island* [J495688]: *P. brassicae*‡. (*P. napi* v3; *P. icarus* v3, r3)
Ogilby Island* [J505689]: nil.
Black Island* [J505685]: nil.
Salt Rock* [J543681]: nil.
Whaup Rock* [J545673]: nil.
Island by Greyabbey* [J578675]: nil. (*P. icarus* v4, r4)
Scady Rocks* [J589596]: nil. (*P. icarus* v3, r3)
Near Craiglee Rock* [J597583]: nil. (*P. icarus* v4, r4)
Sconce Rock* [J595582]: nil. (*P. icarus* v3, r3)
Island by Deer Park (1)* [J597561]: nil. (*P. icarus* v4, r4)
Island by Deer Park (2)* [J597560]: nil. (*P. icarus* v4, r4)
Bird Island by Hare* [J507494]: nil. (*P. icarus* v3, r3)
Shones Island* [J530495]: nil. (*P. icarus* v3, r3) (Visited in 2016, none seen.)
Island nr Comber Road* [J490713]: nil. (*P. icarus* v4, r4)
Pig Island* [J540700]: *P. brassicae*‡, *C. argiolus*‡, *V. atalanta*‡, *V. cardui*‡, *M. jurtina*‡, *P. aegeria*‡.
Black Rock* [J543559]: nil. (Visited in 2016, none seen.)
Castle** [J512486]: *P. rapae*, *P. napi*‡, *L. phlaeas*, *P. icarus*‡, *A. urticae*†, *A. io*, *M. jurtina*†, *A. hyperantus*†. (Two visits in 2016.)
Ballywallon Island** [J593562]: *P. napi*, *P. icarus*, *L. phlaeas*, *M. jurtina*, *A. hyperantus*, *P. aegeria*.

Note

[1] See Enright *et al.* (2017) for full details of analysis; the data in this table have been upgraded for 2016. In 2016, an additional 83 records were accumulated (http://www.bigbutterflycount.org/); 40 confirmed previous records on 21 islands, a further 43 were new records on 26 islands, four supporting predictions from 2012–2015 data. In 2017, an additional 43 records were made (the record of *M. galathea* for Mid island is discounted); 20 were new records and 23 confirmed previous records.

Appendix 13. Probabilities of Individual Butterfly Transfers to Islands and Locating a Resource on the Island

Recently a model has been produced by Professor Richard J.D. Tilley (Tilley and Dennis, 2017) to account for the probability of an individual or individuals transferring from a mainland source to an island, and then the probability of it or them finding a key resource. This model clearly shows that transfers to islands, and finding resources on them, are far more likely than previously supposed, particularly islands within 5 km of the mainland source shore. Below are some basic details from that work.

A. Island Location: Transference to an Island

Professor Tilley supposed that in order to encounter an island any butterfly flies away in a random direction from a coastal source in a more or less straight line regardless of the land or sea surface (in reality, different butterfly species will more or less readily fly over land than water and be more or less attracted by the coastal margin on leaving it). Most of these paths out from the coast will take the butterfly away from the shore and a small number will be in the direction of any island present. The modelled behaviour assumes that the butterfly does not change course (i.e. the range of perception of the butterfly is considered to be small). Within these constraints, Professor Tilley showed that if there is an island situated offshore at a distance d (m) from a departure point (Fig. App. 13.1) then the probability (p) of a butterfly hitting the island is:

$$p = \beta/\alpha$$

The probability of missing the island is then q:

$$q = 1 - p = 1 - (\beta/\alpha)$$

where the angles α and β can be measured in radians or degrees.

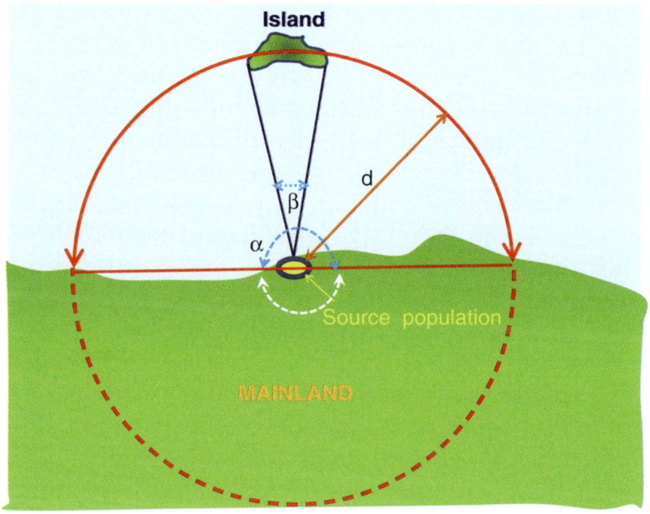

Fig. App. 13.1. The basic angles required to estimate the probability of a butterfly encountering an island. The probability will be affected by the bias for movement over land matrix rather than sea matrix.

For a second island situated further offshore, it was shown that the probability of the butterfly reaching it is given by the probability of it missing the first island multiplied by the simple probability of hitting the second as if it existed in isolation.

Of course, the probability of a butterfly (belonging to any species) reaching an island depends not just on one butterfly leaving one population, but on a number of them emigrating from multiple populations along the mainland coastline; this in practice is measured by **source mass**. But Professor Tilley has shown that, in general, the probability ($P_{r/n}$) that at least r butterflies will arrive at an island when a subset (n) of a combined coastal population of butterflies all take a random linear flight path (random in direction but linear from the point of exit from a habitat) is given by the series:

$$P_{r/n} = 1 - \{(n!/n!)\,(1-p)^n + [n!/1!(n-1)!]\,p(1-p)^{n-1} + [n!/2!(n-2)!]\,p^2(1-p)^{n-2} + [n!/3!(n-3)!]p^3(1-p)^{n-3} + \ldots + [n!/(r-1)!(n-(r-1))!]\,p^{r-1}(1-p)^{n-(r-1)}\}$$

where $n!$ (factorial n) = $n \times (n-1) \times (n-2) \times (n-3) \times \ldots 3 \times 2 \times 1$ and p is the probability of an isolated individual reaching the island via a random flight path.

Thus, the probability of a single butterfly attaining Flat Holm from Lavernock Point is 1.9% but if the number of butterflies emigrating from Lavernock is 400 then the probability that at least one individual will make the crossing is 100% (see map in Fig. 6.5). The probability of any species crossing to an island is enhanced, of course, by the existence of multiple populations for that species occurring along a source coastline.

The probability of transfers to an island is greatest when, in these conditions, an island is enclosed within a bay of the mainland land mass.

In reality, the prospect of any particular species transferring to an island depends much on its facility for colonization (**colonfac**) and migration capacity (**migrfac**), both of which affect its population structure and dynamics at mainland sources (Fig. App. 13.2) (see Appendix 7). Dispersal from populations (habitats) varies within and among species, both (i) independently of population size (an inherent capacity for vagrancy depending on **CSR life history strategies**) but, importantly, also (ii) as a function of population size. Regarding the latter, the numbers leaving a population (habitat) will tend to increase with population size purely as a consequence of the stochastic process of contact with the habitat boundary, but are also likely to increase with population density (density-dependent dispersal) in response to limited carrying capacity of the habitat (resource limitations, competition, predation, etc.). Studies of organisms find that although the *proportion* of individuals leaving habitats tends to increase

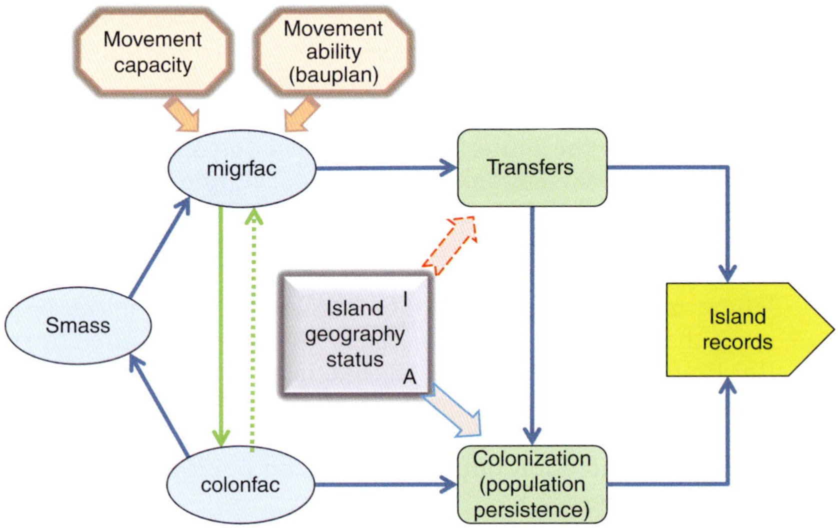

Fig. App. 13.2. The relationships between potential for movement (migrfac), colonization ability (colonfac) and source mass and the transfer to, and colonization of, islands by butterfly species. Migrfac is influenced by movement ability (bauplan; especially wing expanse, wing loading, aspect ratio, wing toughness, thorax: body mass) and movement capacity (flight period, voltinism, population size and density) (see Appendix 7). Source population is dependent on number of populations and population size, influenced by opportunities for population growth (flight period, voltinism, **intrinsic rate of population increase**). Records of species are made of those engaged in brief stays and of colonists. The greater the ability to colonize sites, the greater the source mass is likely to be; in turn this leads to the greater the number dispersing. Migrfac and colonfac therefore influence each other and also source mass, the last of which determines the number of migrants. The numbers of island arrivals and colonists are directly limited by island isolation (I) and area (A).

at very low and high density (fewest at moderate densities) (Travis *et al.*, 1999; Strevens and Bonsall, 2011), including butterflies (Shapiro, 1970b; Dennis and Bramley, 1985; Baguette *et al.*, 1998; Enfjäll and Leimar, 2005; but see Kuussaari *et al.*,1996 for inverse density-dependent migration in *Melitaea cinxia*), importantly, the actual *numbers* dispersing will increase in larger populations.

There is not space here to consider the complexities of the relationship between emigration and population size. In brief, the number of individuals leaving a population is intimately linked with population growth and the carrying capacity of the population. Population growth can be described by:

$$N_{t+1} = RN_t e^{r(1-RN_t/K)}$$

where N_t is the number of individuals *i...j...n* in generation *t*, *r* is the **intrinsic rate of population increase**, *R* is the proportion (lim 0,1) of individuals resident in the population, in which *1–R* emigrate, *K* is the carrying capacity of the habitat (Thomas and Hanski, 1997) and *e* is Euler's mathematical constant 2.71828 ($log_e(x) = 1$). If the number of individuals emigrating is, as is likely to be the case, a positive function of population size (density) and the intrinsic rate of population increase, and an inverse function of the carrying capacity of the habitat (negative), then – in the bounded system of a habitat – numbers emigrating may be affected by a negative feedback caused by the depression on population size and density by emigration itself. This leads to a logistic model of population growth (population reaches an asymptote) rather than an exponential model of increased growth. This follows from the simple relationship:

Intrinsic rate of growth (*r*) = (natality + immigration) – (mortality + emigration).

A key to understanding the potential for emigration from insect (butterfly) populations is their massive facility for population growth (intrinsic rate of increase), and the life history time lags between production and loss which can rapidly, if temporarily, allow population size to exceed carrying capacity (Schowalter, 2016).

Back to more practical ground, the extent to which emigration (as transfers) is homogeneous among species has been tested directly with reference to near offshore islands, for example, in relation to records of species for two islands in the Bristol Channel (Flat Holm and Steep Holm; see Fig. 6.5 and Plates 1.11 and 1.12). Analyses of the data for these islands and the adjoining mainland of Wales and England indicated that the probability of species transferring to these islands is affected by three closely correlated variables; in fact, the species were divided into two distinct groups (present and absent on the islands) on the first axis of a **principal components analysis** which describes the joint influence of migration capacity (migrfac), colonization ability (colonfac) and the number of records at mainland sources (**source mass** or *Smass*) (see Fig. 6.6). The **eigenvalues** and **loadings** of the principal components analysis (PCA) are provided below. Each variable has been standardized; the species values for migrfac and colonfac are found in Appendix 7. Source mass is a standardized and weighted measure of records within the 100 km square in which the islands occur. In unstandardized form:

$$Smass = (R70*0.25) + (R99*0.5) + (R10*10) + (R5*5.5) + (R1*1)$$

where $R70$ is the number of 10 km squares with only pre-1970 records (weighted 0.25), $R99$ is the number of squares with records no later than 1995–1999 (weighted 0.5), and $R10$, $R5$ and $R1$ are the number of squares with records for the period 2000–2004, weighted respectively for the number of records in squares (10 for > 10 records, 5 for 2–9 records, and 1 for 1 record, respectively).

The first two axes of the PCA account for > 90% of the variance (Table Appendix 13.1). The loadings demonstrate a large 'size' factor in which source mass is prominent, whereas the second factor distinguishes migrfac (positive) and colonfac (negative) (Table Appendix 13.2).

It is valuable to consider the relationship among the variables migrfac, colonfac and *Smass*, particularly as there is **positive feedback** (reinforcement of effect) between them (Fig. Appendix 13.2).

B. The Law of Random Search: Seeking Resources on an Island

Search theory was originally developed during the Second World War to optimize the chances of locating a concealed enemy 'target' (Koopman, 1980) – much the same problem a butterfly has of finding a resource on entering a new biotope. For a target known to occur within a specific area (A), a good approximation to the probability (p) of locating the target is governed by three assumptions:

Table Appendix 13.1. Eigenvalues.

Factor	Eigenvalue	% Total variance	Cumulative eigenvalue	Cumulative %
1	2.182	72.73	2.182	72.73
2	0.531	17.69	2.712	90.42
3	0.288	9.58	3.000	100.00

Table Appendix 13.2. Factor–variable correlations (factor loadings).

Variable	Factor 1	Factor 2	Factor 3
Migrfac	0.829	0.504	0.244
Colonfac	0.821	–0.526	0.221
Source mass	0.906	0.016	–0.423

1. The target can occur anywhere in A, and need not be static.
2. The searcher's path is random in A; that is, the path can twist and go back on itself.
3. On any portion of the path the observer always detects the target within the lateral range of $w/2$ (where w is width of perception) on either side of the path and never beyond (Fig. Appendix 13.3).

These assumptions lead to the *law of random search*: the probability p of encountering the target is given by:

$$p = 1 - e^{-(wL)/A}$$

where L (m) is the path length that the searcher travels, w (m) the effective width over which target recognition occurs and A (m^2) is the area to be searched. Clearly, the probability of a successful encounter increases as the path length and path width increase, and decreases as the area to be searched increases (Tilley and Dennis, 2017).

When there are a number of potential targets present, distributed at random over the search area, the law of random search equation needs to be slightly modified. In the case of n targets, the probability of locating one is:

$$p = 1 - e^{-(wL)n/A}$$

The area to be searched is effectively reduced by a factor n. As there is usually an equal number of females to males in a population, and many of these will already have been mated, the probabilities need to be multiplied by ½ for a male finding a receptive female (in fact < ½, if only because resident females are already likely to have been mated).

These equations have huge relevance for the success of butterflies entering islands locating crucial resources (females finding host plants; males and females finding mates; both sexes finding nutrition after a long journey), as well as different resources required in succession. Although many species are capable of locating distant resources, and use sensory and landscape cues for finding them, many butterflies can only find suitable host plants in their near vicinity (e.g. *Pieris brassicae*, *Vanessa atalanta*; Dennis,

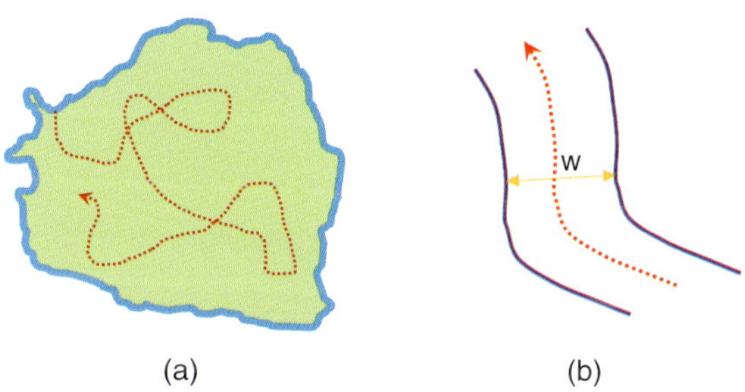

(a) (b)

Fig. App. 13.3. Butterfly searching for a resource in a habitat or Island. (a) Random search path taken by butterfly over an island. (b) Path width (w) is the width of the path over which a target can be recognized.

personal observation). A word of caution is appropriate here: finding a resource is one thing; persisting in a biotope with a resource is quite another. It is the quality of resources that matters; very few host plant zones are suitable for growth and development and large areas of host plant are vacant of butterflies expected to occur on them (see **resource-based habitat** in Glossary) (Dennis, 2010).

Appendix 14. Species Richness and Species Occurrence Predictions for the Isle of Man

This appendix on the Isle of Man provides details of the analyses referred to in Chapter 7 (Hardy *et al.*, 2017). See Appendix 3 for records of species and Supplementary Appendix 7.1 for the full data file for the Isle of Man.

A. Species Richness Predictions

Numbers of species have been estimated for the Isle of Man *post hoc* by comparing recent species numbers in identically sized, contiguous areas (fourteen 10 km squares) to that making up the island for each 100 km square (OS letter coding applied) over the whole of Britain and Ireland, including some larger islands that exceed the same area and have at least 14 contiguous 10 km squares (records for 1995–1999; 2005–2009) (Fox *et al.*, 2011). Where possible, the shape of the Isle of Man has been placed centrally in each 100 km square. Predictor variables are latitude, longitude (corrected for negative numbers by adding 12 degrees), altitude (m) and number ($\leq$ 14) of coastal squares in each sample. As residuals are approximately normally distributed, the analytical method used is simple multiple regression (stepwise forwards entry). Long-distance migrants from Europe (i.e. *Colias croceus*, *Vanessa atalanta*, *V. cardui*) were excluded from analyses.

Numbers of species for virtual Isle of Man units for 100 km squares are significantly related to latitude and longitude but not altitude or numbers of coastal squares, such that: $\boldsymbol{F}_{(2,57)} = 200.21$, $\boldsymbol{P} < 0.0001$, **SE of the estimate** 4.26, $\boldsymbol{R}^2 = 0.875$ (Table Appendix 14.1). The Isle of Man is predicted to have 25.79 species (95% confidence limits 24.69–26.90). With the removal of two outliers (100 km squares TA in Britain and S in Ireland)

Table Appendix 14.1. Multiple regression parameters for numbers of species in 100 km squares (fourteen 10 km sample squares per 100 km square as virtual Isles of Man) of Britain and Ireland on latitude, longitude, altitude and numbers of coastal squares (regression excluding: Isle of Man).

	β	β_{SE}	b	b_{SE}	$t_{(57)}$	P
Intercept			208.93	12.28	17.01	<0.0001
Latitude	−0.762	0.047	−3.61	0.22	−16.17	<0.0001
Longitude	0.456	0.047	1.67	0.17	9.67	<0.0001

Altitude and numbers of coastal squares not significant.

Table Appendix 14.2. Multiple regression parameters for numbers of species in 100 km squares (fourteen 10 km sample squares per 100 km square as virtual Isles of Man) of Britain and Ireland on latitude, longitude, altitude and numbers of coastal squares (regression excluding: Isle of Man and outliers 100 km grid squares TA in Britain and S in Ireland).

	β	β_{SE}	b	b_{SE}	$t_{(57)}$	P
Intercept			227.37	10.18	22.34	<0.0001
Latitude	−0.839	0.041	−4.02	0.198	−20.25	<0.0001
Longitude	0.507	0.041	1.89	0.154	12.25	<0.0001
Altitude	0.103	0.047	0.004	0.002	2.21	0.032

Numbers of coastal squares not significant.

the relationship is improved with the significant inclusion of altitude: $F_{(3,54)} = 235.72$, $P < 0.0001$, SE estimate: 3.27, $R^2 = 0.929$ (Table Appendix 14.2). The Isle of Man is predicted to have 26.01 species (95% confidence limits 25.13–26.89) – nine species more than currently occur and one more than the number recorded historically prior to 1982.

B. Species Occurrence Predictions

Having determined the number of species expected on the Isle of Man, attention was then turned to the presence and absence of each species. **Logit regression** of species incidence on the Isle of Man (presence/absence in the period 1995–1999, 2005–2009) was carried out using a **gravity model** based on **source mass** (10 km square records) and source to island distance summed for eight coastal fringe compass sectors centred on the Isle of Man (Hardy *et al.* 2017).

For this analysis, species were regarded as missing on the Isle of Man if they were not recorded on the island during the period 1995–2009, or if considered to be infrequent vagrant individuals (i.e. *Gonepteryx rhamni*). Long-distance migrants from Europe (i.e. *Colias croceus*, *Vanessa atalanta*, *V. cardui*) were excluded from consideration.

A single predictor variable is used to determine the incidence of species on the island: a summed gravity model algorithm from eight sectors (centred on the mid-point for the Isle of Man) of source records for the coastlines of Britain and Ireland. The basic (economic) gravity model, excluding parameters and exponents, describing the influence of (flow from) a mass in relationship to an island is:

$$G_i = M_j/D_{ij}$$

where G is the gravitational influence, M is population size (source mass: records) and D is the distance from the source to the island. The variable applied for the Isle of Man study is:

$$I = \Sigma_{1<j<k}((R_j/R_{jmax})/(D_j/D_{max}))$$

where I is the likelihood of occurrence, R_j is coastal records for each of $k = 8$ sectors each of 45 degrees, radiating from the Isle of Man to the British and Irish coasts, R_{jmax} is the maximum number of records possible per sector, D is the distance from a source sector to the Isle of Man, and D_{max} is the maximum median distance from all source sectors to the Isle of Man. No exponent for D is used as all the distances lie beyond the negative exponential fall (10 km) expected of most resident butterfly species movements. I is further scaled from 0 to 1. In the case of the Isle of Man study, the records are the number of filled 10 km squares for each species in the coastal source region of each sector (counting only two 10 km squares inland, i.e. squares actually along the coastline and just one row of squares behind (inland) from them) standardized by the available number of squares per sector (the number of 10 km squares differs for each source sector). D is the median distance from the mainland source sector to the Isle of Man standardized by the maximum distance isolating the island from the eight sectors (180 km). An initial inspection was first made of species in a space of I values for each sector in a principal components analysis; as these are highly correlated, logit regression was carried out on species incidences on the Isle of Man against the gravity model variable I summed over the eight sectors.

Parameters of the logit regression equation are: log likelihood −7.74, deviance 15.47 (38 *df*), gravity measure Wald statistic 9.81, $P = 0.002$; 90% correct classification.

Table Appendix 14.3 contains details of observed species for the Isle of Man and the surrounding mainland source margins around the Irish Sea. It also provides predictions and confidence bands for the predictions.

Table Appendix 14.3. Records and predictions for species on the Isle of Man.

Species	First record	Possible extinction date	Observed during past three decades (suspected resident)	Prediction value	Lower 95% confidence level	Upper 95% confidence level
T. sylvestris	N		0	0.03	0.00	0.22
O. sylvanus	1871	1950s	0	0.53	0.22	0.82
E. tages	1893	1900s	0	0.14	0.04	0.44
P. malvae	N		0	0.01	0.00	0.14
L. sinapis sl	N		0	0.04	0.01	0.26
G. rhamni	1941	†	0†	0.06	0.01	0.29
P. brassicae	1876		1	0.99	0.80	1.00
P. rapae	1876		1	1.00	0.80	1.00
P. napi	1871		1	1.00	0.84	1.00
A. cardamines	1871		1	0.96	0.62	1.00
C. rubi	N		0	0.05	0.01	0.28
F. quercus	N		0	0.04	0.01	0.25
S. w-album	N		0	0.02	0.00	0.18
L. phlaeas	1871		1	0.99	0.75	1.00
C. minimus	1971‡	1970s	0	0.03	0.00	0.22
P. argus	N		0	0.01	0.00	0.16
A. agestis	N		0	0.02	0.00	0.18
A. artaxerxes	N		0	0.02	0.00	0.20
P. icarus	1871		1	0.99	0.78	1.00
C. argiolus	1885		1	0.31	0.11	0.63
H. lucina	N		0	0.01	0.00	0.13
A. urticae	1856		1	1.00	0.84	1.00
A. io	1885		1	0.98	0.70	1.00
P. c-album	1995		1	0.13	0.03	0.42
B. selene	1863	1890s	0	0.12	0.03	0.39
B. euphrosyne	1863	1890s	0	0.02	0.00	0.19
A. adippe	*		0	0.01	0.00	0.15
A. aglaja	1871		1	0.58	0.25	0.85
A. paphia	N		0	0.03	0.00	0.23
E. aurinia	1893	1890s	0	0.03	0.00	0.22
P. aegeria	1903	1900s	1	0.86	0.45	0.98
L. megera	1871		1	0.99	0.72	1.00
E. epiphron	N		0	0.01	0.00	0.13
E. aethiops	1869	1870s	0	0.01	0.00	0.14
H. semele	1862		1	0.88	0.48	0.98
P. tithonus	1896	1950s	0	0.28	0.09	0.60
M. jurtina	1871		1	1.00	0.83	1.00
A. hyperantus	1885	1930s/40s	0	0.83	0.42	0.97
C. pamphilus	1871		1	0.97	0.66	1.00
C. tullia	1893	1900s	0	0.02	0.00	0.20

Analysis based on species at Irish Sea coastal sources. Species included in the model are found within eight mainland source sectors around the Irish Sea (migrants *C. croceus*, *V. atalanta* and *V. cardui* excluded from table). First record and possible date of extinction (decade following last known record)

Continued

Table Appendix 14.3. Continued.

from Chalmers-Hunt (1970). *Polygonia c-album* first Manx record dates to 2 August 1995 (Jerry Jones, at SC253846) and observed in eight 10 km squares by 2014 (Fox *et al.*, 2015). N, never reported as present; *mistaken record for *Argynnis aglaia*; †vagrants occurring at irregular intervals and poor resources for persistence; *Gonepteryx rhamni* has been released on several occasions and not recorded during several years (e.g. in 2011–2013 Annual Butterfly Reports; G. Jeffcoate, personal communication), but was likely resident in 1980 (Majerus, 1980); ‡single specimen in the collection of Peter Crow, obtained in July 1971 (confirmed by RLHD). *Aphantopus hyperantus* recorded for the Ballaugh Curragh at https://data.nbn.org.uk/Reports/Sites/GA000374292/Groups. *Apatura iris* was allegedly observed in a wood near Patrick between 1914 and 1918 (see comments of Chalmers-Hunt, 1970). Nomenclature after Kudrna *et al.* (2015).

Appendix 15. Modelling Evolutionary Changes on a Fossil Island: the Great Orme's Head, North Wales

Size of organisms on islands often differs from conspecifics on the mainland. As mentioned in Chapter 8, on the fossil island of the Great Orme's Head in North Wales, two unrelated butterfly species (the Grayling *Hipparchia semele* and the Silver-studded Blue *Plebejus argus*) are both smaller and emerge earlier than elsewhere in Britain, significantly so compared with adjacent populations on the Creuddyn Peninsula on which Llandudno occurs. Here, three simple process-response models are used to explore – more deeply – the concatenation of factors that may be responsible. The three figures in this Appendix illustrate potential proximal and distal factors argued to be influencing small size and early emergence in two butterfly species (so-called *Hipparchia semele thyone* Thompson and *Plebejus argus caernensis* Thompson) on the Great Orme's Head.

One factor considered to be of potential importance for evolution of dwarfism in these butterflies on the Orme is a contracted period of larval development time: early emergence but then rapid growth (Dennis, 1972a, b). Figure App. 15.1 illustrates the factors leading to an early growing season, thus early post-winter hatching from eggs (*P. argus*), early feeding (herbivory) and development of larvae on the Great Orme's Head. The seaboard position and the North Atlantic drift up St George's Channel maintain high temperatures (**specific heat capacity** effects) above the water surface, countering any tendency for frosts to occur along coastal margins at this latitude. The prominent position of the headland in the Irish Sea in a rain shadow of the Welsh mountains also results in low cloud cover (rainfall on the Orme is as low as in East Anglia) and thus high radiation receipt throughout winter and spring days. Radiation receipt is enhanced by the low western horizon, by the sun reflected off the sea, and the angle of receipt is maximized by the steep scarp slopes. One of the strange effects occasionally seen from high on the west side of the Orme

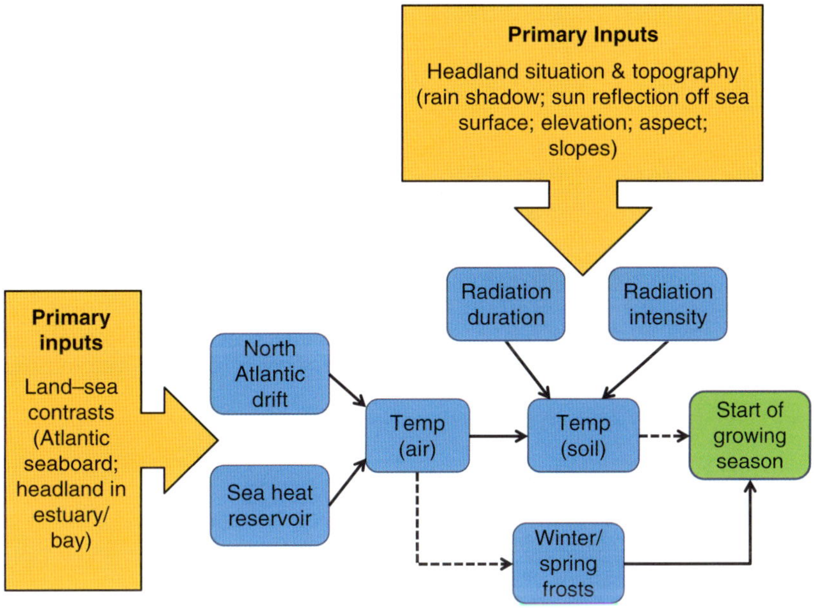

Fig. App. 15.1. Basic geographical factors influencing an early growing season for plants on the Great Orme's Head, particularly the sea surrounds warmed by the North Atlantic drift and winter high radiation levels facilitated by the unhindered west to south-west aspects of the Orme's slopes. Arrows: continuous, positive relationships; pecked, negative relationships.

in the evening is the apparent occurrence of two suns low in the sky over Anglesey, each equally radiant (one is the powerful reflectance off the sea). High radiation raises the soil surface temperatures, encouraging fast early growth of plants, including the host plants, and butterflies. The microhabitat of the dry bare soil surface in which eggs and early instars of both species are found is an extremely hot one.

However, the same factors (high radiation levels and soil surface temperatures) also propel conditions that lead to rapid senescence in plants. In Fig. App. 15.2 are illustrated the factors that result in rapid deterioration of host plant quality through plant senescence. The first factor is the headland geology. The Great Orme's Head is a geological island, a boat-shaped syncline of Carboniferous limestone strata unconformably set as an **outlier** on a lower Palaeozoic (Silurian) sedimentary base (see Plate 1.17). These rocks are highly permeable (pervious), rainwater finding its way through them down joints and bedding planes widened by solution (carbonation). Moreover, the limestone soils are immature (thin) and recent in origin (**rendzinas**), previous soil cover having been removed by glacial erosion during the Devensian Maximum Glaciation, the **regolith** having few insoluble residues and then these too washed away by deglacial flood meltwater. Despite the high **infiltration capacity** of the soils and regolith, the thin soil cover over steep slopes results in high down-slope wash during rain storms, causing further erosion which then reinforces (amplifies) (via Melton's 1958 positive feedback system) subsequent discharge down the steep slopes (scars) into the sea. In other words, water is lost very rapidly from the Orme following rainstorms (and then annual precipitation is low) either into the scree and rock or downslope, and retention of soil water is very low. Thus, plants desiccate rapidly and host quality quickly deteriorates. But this is not all. The Orme, as seen in Fig. App. 15.1, is in a very high-radiation environment. Furthermore, being exceptionally exposed to frequent prevailing westerly gales, it is in a very high wind-speed environment. Both high radiation levels and high wind speeds cause high evapotranspiration rates on a regime already lacking soil moisture. This results in low vegetation growth, further reduced by historic high levels of grazing, which erases shelter, increases exposure, thereby increasing evapotranspiration, another synergistic impact

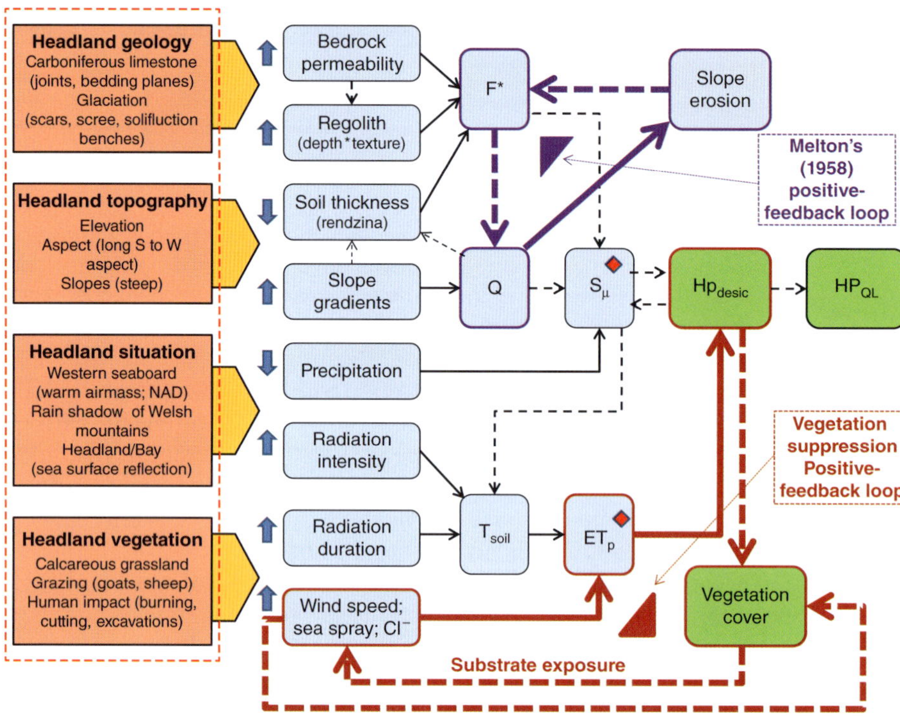

Fig. App. 15.2. Bare bones model of geographical factors making for low soil moisture levels (low precipitation, the high permeability of the substrates and rapid run-off from precipitous slopes) and high evapotranspiration rates (high wind speed, salt spray, long exposure to high radiation levels) on the Great Orme's Head, both affecting summer plant desiccation. Two critical positive feedback loops are illustrated, one having profound impact on plant desiccation and texture, the other (Melton's feedback) affecting geomorphic processes (regolith, soil thickness) on all steep slopes, as found on the Great Orme's Head. The red diamonds indicate critical factors in the system affecting plant quality. Arrows: continuous, positive relationships; pecked, negative relationships. Blue vertical, tendency of change, increase (up) and decrease (down).

(a **positive feedback** loop) in that vegetation is depressed. Sea spray (chloride ions) may also have a significant impact on host plant quality; Cl^- concentrations depend on proximity to the sea, storm frequency, wind direction and topography, and are likely to be high in the storm environment of the Orme's western slopes. Elsewhere, **chloride ion** deposition can reach high levels within coastal margins (e.g. 175 kg/ha), as much as 50 kg/ha 6 km inland (Hutton, 1976; Yaalon in Xu *et al.*, 2000), and may result in reduced plant growth and small, scorched and thicker leaves, all of which have a negative impact on plant quality.

These factors are likely to set the scene for rapid growth and early development and short development times of the two butterflies (Fig. App. 15.3). Host plants have an early start to the growing season but are soon short of water and perennially under stress. The butterflies, in early vulnerable stages of development, with immature mouthparts, are forced to feed up

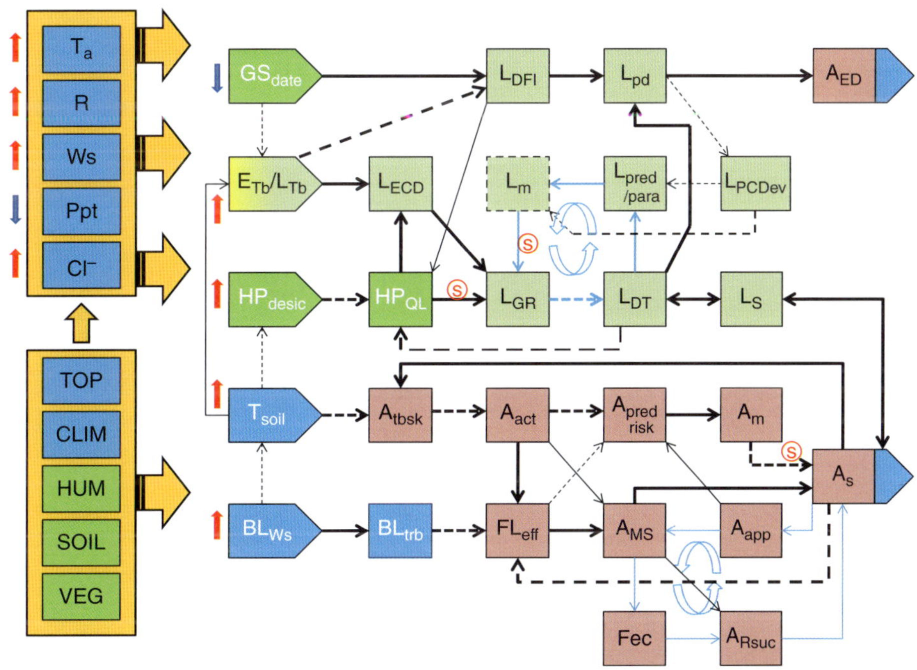

Fig. App. 15.3. Potential concatenation of geographical factors on butterfly wing size and emergence timing in *Hipparchia semele* and *Plebejus argus* on the Great Orme's Head. Boxes: yellow, eggs; pale green, larvae; pale red, adults; bright blue, climate and sea variables; bright green, land and plant variables. Vertical arrows, red up (increase), blue down (decrease), indicate expected polarity of change on the Great Orme's Head; wide orange arrows: basic influences without specific polarity (sign); other arrows: unbroken, positive correlations/effects; pecked lines, negative correlations/effects; thick lines, main paths; thin lines, other paths; blue arrows and circular arrows indicate two well known positive and negative feedback scenarios. Encircled 'S' indicates selection for reversal of trait character. Note the strong positive feedback between larval and adult size shown by a double-headed arrow.

Key to box symbols for Fig. App. 15.2 and 15.3:

Adult activity A_{act}
Adult conspicuousness A_{app}
Adult basking time to reach optimum body temperature A_{tbsk}
Adult emergence date A_{ED}
Adult mating success A_{MS}
Adult mortality A_m
Adult predation risk $A_{pred\ risk}$
Adult reproductive success $A_{rsuccess}$
Adult size A_S
Ambient temperature T_a
Ambient radiation R
Boundary layer turbulence BL_{trb}
Boundary layer wind speed BL_{Ws}
Climate factors CLIM
Chloride ion Cl^-
Edaphic factors SOIL
Egg temperature E_{Tb}
Evapotranspiration (potential) ET_p
Fecundity (measured on adults) Fec
Flight efficiency (ability to manoeuvre) FL_{eff}
Growing season (start of) GS_{date}
Host plant desiccation HP_{desic}
Host plant growth and quality HP_{QL}
Human factors HUM
Infiltration capacity (soil; includes seepage above water table) F*
Larva: body temperatures L_{Tb}
Larva: date of feeding initiation L_{DFI}
Larval development time L_{DT}
Larva: efficiency of conversion of digested food L_{ECD}
Larval growth rate L_{GR}
Larval mortality L_m
Larval predation/parasitization $L_{pred/para}$
Larva: facility for (pupal) cell development (in soil) L_{PCDev}
Larva: pupation date L_{pd}
Larval size L_S
North Atlantic drift NAD
Precipitation Ppt
Radiation R
Soil water (moisture) S_μ
Surface (precipitation) runoff from slopes Q
Temperature (screen/ambient) T_a
Topographical factors (e.g. altitude, aspect, slope) TOP
Vegetation factors VEG
Wind speed Ws

quickly, before plants become too tough. The fastest rates will be on south and west scars and screes, which sites are strongholds for the key host plants of both species. The top part of the model (Fig. App. 15.3) attempts to illustrate these issues for early developmental stages: the basic factors driving the system are boxed on the left and the main paths of impact are shown by prominent arrows. The fast growth rate leads not just to early emergence and rapid maturity, but also (as typical of butterflies undergoing fast growth rates) to smaller size. At the same time, in the high-energy (wind speeds) environment of the Orme, small adult winged insects are also favoured over larger ones, which take longer to heat up to become active and are less manoeuvrable. The high soil temperatures and high radiation levels will favour small butterflies that heat up rapidly, engage mates more readily and avoid predators more easily. Moreover, small adults will also tend – as part of a positive feedback cycle – to produce smaller larvae. There is, then, likely to be an advantage to being small as a larva and as an adult and the two are causally linked. The main counter influence will be a tendency for larger individuals to have higher potential fecundity; but, if larger butterflies take longer to heat up when basking, and are more vulnerable to predation in the cooler turbulent airstream of the Orme, it is likely that **realized fecundity** of larger butterflies is less than that of smaller butterflies.

The key to understanding the situation in which adults function is the sharp lapse rate in conditions (sharp drop in ambient temperatures, rapid increase in wind speed) from the ground into the boundary layer above it (Bennett *et al.*, 2015); the sheltered, dry, highly radiated soil thermal conditions are ideal for rapid heating of small butterflies and the high wind speeds and wind chill a few feet above ground favour individuals with short wings. The short growing season of the Orme selects for faster growing rates and increased selectivity of host plants in early instars that facilitate faster growth rates in caterpillars, early emergences of adult butterflies and small size, which then also favours adults. However, both species have very long flight periods and the Grayling becomes significantly larger during the season (Dennis *et al.*, 2018). The likely basis for this is the variety of contrasting slopes and aspects, as well as substrates, over the Orme, which select for different growth and size patterns on hotter and cooler slopes, reflected in species richness patterns (see Fig. 4.3). Variable seasonal conditions and continued immigration from later emerging populations with larger wing expanses (i.e. from Conwy Mountain) provide a constant supply of larger later-emerging butterflies.

The situation is fascinating as, although a number of moth species have similarly been affected (H.N. Michaelis, personal communication) no other butterfly species on the headland is significantly small or early in emergence (e.g. *Maniola jurtina*; Shreeve *et al.*, 1995). An essential context for evolution has long been held to be isolation of an evolving population, but if selection pressures are high enough, then evolutionary changes will occur in **sympatric** and **parapatric** populations (Ford, 1964; Via, 2009). In *Hipparchia semele thyone* and *Plebejus argus caernensis*, we see the beginnings of events that lead to speciation; but such natural experiments, here triggered by warmer, drier climates in the Pre Boreal and Boreal (see Fig. 8.2), are by no means always successful and are unlikely to be here, if only because of the short duration of interglacial stages compared with glacial stages.

Although processes of differentiation now operate sympatrically and parapatrically among populations of the Great Orme's Head Grayling (i.e. distinct population sub-units occur in contact and overlap; Dennis, 1977), there are reasons for suspecting that in the early Holocene isolation may well have contributed to the evolution to these two butterfly 'races', arising from vicariance or remote colonization (**peripatry**). During the Holocene, the Orme has been isolated by woodland at its base extending out over the now eroded glacial till (as seen in fossil tree trunks on the beaches), but at times it was also an island, owing to a succession of marine transgressions (relatively high sea levels) in the Irish Sea basin during the later Holocene, especially between 3ka and 2ka BP (see Tooley, 1978, p. 130 for details). It is intriguing to speculate on the changing geography of the Orme; the tombolo on which Llandudno now stands comprises a considerable accumulation of windblown sand from the Conwy estuary during historical times, and the geological map (BGS Llandudno Sheet 94, 1:50,000 series) reveals underlying marine alluvium and silt, evidence of inundation by the sea. In a borehole taken by M.J.C. Nutt (Foraky Ltd for IGS, courtesy of Scott Renshaw, BGS) at the Oval, Llandudno (SH 77548191) in June 1971, marine deposits extended 1.17 m below current ordnance datum (mean high water mark, MHWM); some 29.54 m of unconsolidated deposits were disclosed to the solid Ordovician [*sic*], passing through earlier clays and peats to the Devensian till, and then down through earlier marine and glacial facies predating the Devensian.

Other butterfly species, lacking signs of early emergence and dwarfism on the Orme, evidently were not restricted to the short herb biotopes characteristic of the scars and screes of the fossil island, whereas the Grayling and Silver-studded Blue would not have been able to inhabit the thicker turf and scrub-covered northern slopes and summit area.

The concepts in this brief section illustrate the complex issues that could be involved in explaining ecological and evolutionary changes in populations on islands. The essential work has not even been started on *Hipparchia semele thyone* and the first author has long been keen to see this research under way by the next generation of butterfly biologists. One clue in the concatenation of factors is that the same author has never seen such small specimens of *Plebejus argus* on the Orme as in 2017, a year with low winter and spring rainfall (see www.metoffice.gov.uk/climate/uk/summaries).

Glossary of Terms

Terms in italics refer to statistical issues

adaptive radiation: The evolutionary diversification of a species or single ancestral lineage into various forms that are each adaptively specialized to a specific environmental niche. Adaptive radiation generally proceeds most rapidly in environments where there are numerous unoccupied niches or where competition for resources is minimal.

affinities (mean similarities): The average of a number of similarity coefficients between one geographical unit and all others in a specified region.

allochthonous: Not indigenous; population (species) originating or formed in a place other than where found.

allopatry: Species occurring in geographically separate units; **allopatric speciation** involves some form of geographical barrier.

allozymes: Also called **alloenzymes**. Variant forms of an enzyme that are coded by different alleles at the same locus (see **SNPs**). Allozymes are opposed to **isozymes**, which are enzymes that perform the same function but which are coded by genes located at different loci.

anacladogenesis: Anagenesis (*q.v.*) but in which the progenitor species survives with little change alongside (the same island or archipelago) the derived species.

anagenesis: A mechanism in which one species evolves into another (the progenitor becoming extinct) by evolutionary changes within a lineage (phyletic transformation or gradualism).

ANOVA: Analysis of variance; a parametric test for comparing means of a variable across multiple population samples.

Anthropocene: Term proposed by Creutzen for the present geological epoch (from the time of the Industrial Revolution onwards), during which humanity has begun to have a significant impact on the environment. Some regard it as having begun about 1950, defined by the creation of > 200 new Earth minerals (Hazen *et al.*, 2017) and radioactive elements being dispersed across the planet by nuclear bomb tests, as this provides a distinct horizon marker in sedimentary deposits.

aptery: The state of an animal completely lacking any kind of wings.

aspect ratio: Wing length divided by wing area ($4L^2/A$, where L is wing length and A is the total area of both pairs of forewings and hindwings when linked) (Betts and Wootton, 1988).

association coefficients: Measures of similarity or (inversely) dissimilarity over areal units or regions, usually ranging from 0 (no relationship) to 1 (perfect relationship). A large number of such coefficients are available and differ in their calculation and interpretation (cf. Jaccard's S_J (*q.v.* ignores joint absences whereas percentage dissimilarity includes them).

asthenosphere: The highly viscous, easily deformed region of the upper mantle of the Earth. It lies at depths between approximately 80 km and 200 km below the surface crust (lithosphere).

autochthonous (and **autochthonous speciation**): Indigenous; population (species) originating or formed in the place found.

balanced polymorphism: An equilibrium mix of homozygotes and heterozygotes in a population maintained by natural selection against both homozygotes. This occurs typically owing to the heterozygotes having greater fitness than either homozygote (e.g. haemoglobin variants in a malarial environment).

bauplan: The generalized structural body plan that characterizes a group of organisms.

beta coefficient: A measure of how strongly each predictor variable influences the dependent variable in a regression analysis. Beta (β) is measured in units of standard deviation.

Big Butterfly Count (UK): A nationwide survey of butterflies instigated and administered by Butterfly Conservation, UK. Counts of butterflies are taken over short periods in mid-summer by the general public. Participants can survey as many 'spots' in the landscape as they like, with each being observed for 15 min duration. The method has potential limitations, including restricted sampling time, sampling in inadequate conditions and poor identification of species. See website for Butterfly Conservation.

biota: The animal and plant life of a particular region, biotope, or geological period.

biotope: An area of uniform environmental conditions (substrates, vegetation) providing a living place for a specific assemblage of plants and animals. Biotope is almost synonymous with the traditional use of the term habitat. It translates as an area where life exists.

bottlenecks: Relates to a population (size or genetics); a bottleneck (or genetic bottleneck) is a sharp reduction in the size of a population due to environmental events (such as earthquakes, floods, fires, disease or droughts) affecting the resource base of a habitat (see **resource-based habitat**).

BP: Before present 1950. Note: radiocarbon [^{14}C] years BP is not the same as years calibrated (calendar) years BP (cal BP) (see Reimer, 2012).

brachyptery: The state of an animal with short or reduced wings.

Butterfly Monitoring Scheme (BMS): A transect scheme for assessing butterfly populations developed by Dr Ernest Pollard (Centre for Ecology and Hydrology) (Pollard and Yates, 1993). The United Kingdom Butterfly Monitoring Scheme (UKBMS, launched in 2006) is a merger of the long-running Butterfly Monitoring Scheme (BMS) (established by Centre for Ecology and Hydrology (CEH) in 1976) and Butterfly Conservation's coordination of 'independent' transects (established in 1998). Since 2009, the UKBMS has incorporated the Wider Countryside Butterfly Survey (*q.v.*).

CBD: Central business district of a large town or city.

chi square **($\chi 2_{df}$)** ***test***: A comparative test for nominal (categorical) data. Differences are summarized by the test statistic and the so-called **degrees of freedom** (*df*), the latter of which measures the size of the test, specifically the number of rows-1 times columns-1 in a contingency table. The probability value gives the likelihood of the test statistic occurring by chance.

chloride ion: Anion (negatively charged ion) Cl^- forms when chlorine (a halogen element) gains an electron or when a compound such as hydrogen chloride is dissolved in water or other polar solvents. Chloride salts such as sodium chloride are often very soluble in water and a significant component of sea spray (Xu *et al.*, 2000).

choros model: The term describes dimensional space, the total effect of a region's or island's area and habitat diversity on species richness. Thus, choros (K) is the product of area and habitat (biotope) diversity ($K = H^*A$), where H is the number of habitats, A is the total area. Species richness is then $S = cK^z$ and in effect a corollary of the Arrhenius model $S = cA^z$ (see Triantis *et al.*, 2003).

clade: A group of organisms that consists of a common ancestor and all its lineal descendants, and represents a single branch on the tree of life.

cladistics: Biological classification; organisms are classed based on shared derived characteristics traced through a group's potential ancestors (Wiley, 1981).

cladogenesis: An evolutionary splitting event where a parent species splits into two distinct species, forming a clade (*q.v.*).

cladogram: A branching diagram (dendrogram) showing the cladistic relationship between a number of species.

climatic climax (also **climax community** or **climatic climax community**): A biological community of plants, animals and fungi which, through the process of ecological succession and the development of vegetation in an area over time, has reached a steady state in relation to climate; this is a historic term as a vegetation steady state may be determined by other factors (geology, soil, human activity).

Climatic Risk Atlas (CRA): The risk to butterfly species and their populations from climate change, both natural and human generated (see Settele *et al.*, 2008).

cline: A gradual change in an inherited characteristic over the geographical range of a species; such gradients are often associated with an environmental gradient or transition such as altitude or climate.

cluster analysis (clustering algorithms: single linkage, maximum linkage, UPGMA): Methods for grouping a set of objects in such a way that objects in the same group (called a cluster) are more similar (in some sense or another) to each other than to those in other groups (clusters). See Appendix 8.

coefficient of determination (R^2 or r^2): Pearson's correlation coefficient squared, gives the amount of variation (percentage when multiplied by 100) in one variable accounted for by one ($\boldsymbol{r^2}$) or more ($\boldsymbol{R^2}$) other variable(s) with which it is correlated.

coefficient of variation (CV): A standardized measure of variation in which the standard deviation is divided by the mean for a set (sample) of observations.

colonfac: An index of colonization ability in British butterflies (see Dennis *et al.*, 2012a; Dapporto and Dennis, 2013).

colonization: The act of an animal (or plant) establishing itself in an area (island).

communalities: The sum of the squared factor loadings for all factors for a given original variable, that is, the variance in that variable accounted for by all the factors.

Contemporary Geography Study (CGS): An analysis of the presence or absence of a species in a region based on its presence/absence in adjoining regions (see Dennis *et al.*, 2002; Dennis and Shreeve, 2003).

cryptic refugia: Hidden refuges for species as during the past glacial maximum advance.

cryptic species: One of two or more morphologically indistinguishable biological groups that are incapable of interbreeding or at least considered incapable of panmixis (random mating within a breeding population).

cryptoturnover: Missed colonization and extinction events caused by intermittent sampling, thus owing to missed sequences of records (Schoener, 2010).

CSR life history strategies: CSR analysis categorizes plants (e.g. butterfly larval host plants) into ruderals (**R**), stress tolerators (**S**) and competitors (**C**) (see Dennis *et al.*, 2004; Grime *et al.*, 1988).

cuesta: A landscape feature that has a steep cliff or escarpment on one side and a gentle dip or back slope on the other accordant with the dip of the strata.

Dansgaard-Oeschger (D-O) oscillations: Rapid millennial-scale climate fluctuations that occurred 25 times during the last glacial period (Devensian into Late Glacial) evident in ice cores and marine records (see Genty *et al.*, 2003).

dendrogram: A tree diagram, especially one showing taxonomic relationships.

depauperate islands: Islands with a fauna lacking a great many species found in similar biotopes elsewhere (this lack of species is especially common on islands).

depleted faunas: Faunas of an area (island) lacking certain species expected to occur there.

Devensian: Last glacial stage in the British Isles dating 115,000–10,000 ka ^{14}C BP (equivalent to Weichselian on the continent).

discriminant function analysis (DFA): A statistical procedure used to predict a categorical dependent variable (called a grouping variable) by one or more continuous or binary independent variables (called predictor variables). Unlike cluster analysis (*q.v.*), the groups are known *a priori*, but DFA can be used to place individuals of unknown provenance with a specified degree of confidence.

disharmonic faunas: Faunas biased in their representation of higher taxa (genera and above) compared with nearby mainland sources; characteristic of island biotas (*q.v.*).

dispersal: The movement of organisms away from their point (habitat) of origin.

Doggerland: An area now beneath the southern North Sea once connecting Britain to continental Europe during the Devensian and early Holocene. It included various biotopes occupied by Mesolithic man. It was flooded by rising sea levels around 8500–8200 years BP; its final submergence was likely caused by the tsunami resulting from the Storegga Slide off Norway.

drumlin: Elliptical mounds formed of till by a glacial advance having a generally steeper 'stoss' end than the 'lee'.

dummy variable: Also known as an indicator variable, design variable, Boolean indicator, categorical variable, binary variable, or qualitative variable. A 'proxy variable' (a numeric stand-in for quantitative facts), one that takes the value 0 or 1 in regression analysis to indicate the absence or presence of some categorical effect that may be expected to shift the outcome.

ecological release: The expansion of geographical range, range of biotopes and/or resource usage by an organism on changes to the community structure in its own habitat or when arriving in a potential new habitat. The release traditionally refers to a change in competition from other organisms (Paine, 1966) and is indicative of the difference between the fundamental and realized niche (*q.v.*) of a species. Ecological release also occurs when a species expands its niche within its own habitat or into a new habitat (Cox and Ricklefs, 1977). The phenomenon describes a state for many species upon initial colonization of islands, where communities may have lower species diversity, or where the existing predators/ parasites are not yet capable of exploiting the newly arriving species.

ecological scale: The spatial scale of resource distributions in a species' habitat or matrix.

ecotone: A transitional zone between two ecological communities, as between woodland and meadow. An ecotone has its own characteristics (an edge biotope) as well as sharing certain characteristics of the two communities.

ectotherm: An organism that regulates its body temperature largely by exchanging heat with its surrounding environment ('cold blooded').

eigenvalue: In simple terms, the eigenvalue for a given component/factor measures the variance accounted for by that factor in all the variables being processed.

Ellenberg values: Indicator scores for abiotic conditions (e.g. light, acidity, salinity, soil water) in the environment (Ellenberg, 1988; Hill, M.O. *et al.*, 1999).

emigration: The movement of an individual out from a natal habitat. In island biogeography, it is more typically used to describe the movement of an individual (species) from a mainland source and other (island) source.

endemism: The restriction of species to a land area (these may be autochthonous (*q.v.*) or allochthonous (*q.v.*) in origin). In a study of butterflies on European islands (Dennis *et al.*, 2008), endemicity was measured in two ways: from a European (European endemics, Ee) and individual island (island endemics, Ei) vantage.

endogenetic processes: Processes that originate below the Earth's surface (i.e. by faulting, earthquakes and volcanic activity).

equilibrium theory of island biogeography: A theory based on the supposed balance between opposing forces, i.e. rates of immigration and extinction on islands.

eustacy: Changes in sea level caused by the changing volume of water in the sea or configuration of the ocean basins.

evolutionary scale: Gradual change in the characteristics of a population of animals or plants typically over geological time or at least over successive generations; i.e. speciation processes.

exogenetic processes: Processes operating at or near the surface of the Earth (e.g. weathering, erosion, deposition by marine, fluvial, glacial agents).

extent element: Species belonging to the faunal element with the widest range across the whole of northern Europe, many extending into southern Europe (see Dennis *et al.*, 1991).

extinction: In this book, most typically the loss of a species from an island. Extinction is often qualified as population, regional, species, etc.

extinction debt: Anticipated loss of species from an area (island) following changes to its habitat or resource base, caused by change in quality or quantity of resources.

extreme environmental events (EEEs): High-magnitude events of low frequency, typically described in standard deviation units (> 95%, > 99% of the distribution of occurrences for a normal frequency distribution) generated by weather and climate, but also geo-agents (volcanicity, tsunami, etc.) and human agencies.

eyot (also **ait**): Small island in a river channel, formed by fluvial deposition. Eyots tend to shift in shape, size and position with seasonal erosion and deposition.

faunal element: A spatial (geographical) unit of a faunal group (a collection of species that share similar distributions) distinct from other collections of species of the same taxon (for more technical description see Dennis and Schmitt, 2009, p. 251).

faunal region: A unique geographical area characterized by a set of taxa distinct from that of other geographical areas (Dennis and Schmitt, 2009).

Fisher Exact Test: A test of statistical significance (named after Ronald Fisher) used in the analysis of contingency tables (categorical data), typically employed when sample sizes are smaller than feasible for the chi-square test (*q.v.*), though it can be used for all sample sizes.

Fleuve Manche: The fossil anastomosing river system (confluence of great rivers from Britain, France, Netherlands) occupying the English Channel during each glacial stage and the immediate aftermath prior to inundation by eustatic sea level rise.

fluctuating asymmetry: Small random deviations from perfect symmetry in bilaterally paired structures (e.g. wings); asymmetry is thought to act as a measure of stress in development, the assumption being that perfect symmetry is indicative of minimal stress (Windig *et al.*, 2000).

fluvio-glacial landforms: Depositional and erosion landforms created by glacial meltwater.

fossil island: a piece of land along the coast that has been an island in recent (Holocene) times.

founder event: Initiation of a population or colony (usually by dispersal) by a few members of an original population; founder colonization is associated with a substantial loss of genetic variation.

Friedman ANOVA: A non-parametric alternative for multiple comparisons to the parametric analysis of variance (ANOVA (*q.v.*)).

F statistic **(F_{df})**: *F* tests are used to determine the statistical significance of group means in analysis of variance (ANOVA (*q.v.*)) and the significance of parameters in regression analysis. In regression, the *F* value is the ratio of the mean regression sum of squares divided by the mean error sum of squares. The size of the test is dependent on the number of observations measured by degrees of freedom (*df*).

Gausse's exclusion principle: The competitive exclusion principle which states that two species competing for the same limiting resource cannot coexist (have equivalent population parameters) (Metzger *et al.*, 2009).

generalists (generalism): Species that use a wide variety of resources. Note: a butterfly species may be a generalist for one resource (nectar) but a specialist for another/others (larval host plant resource) (see Dennis, 2010).

general linear model (GLM): The term general linear model usually refers to conventional linear regression models for a continuous response variable given continuous and/or categorical predictors. It includes multiple linear regression, as well as ANOVA (*q.v.*). The term **generalized linear model (GLIM** or GLM) refers to a larger class of models popularized by McCullagh and Nelder (1982; 2nd edition 1989).

genetic drift: The change in the frequency of a gene variant (allele) in a population due to random sampling of organisms (also known as the Sewall Wright effect).

genome: An organism's complete set of DNA, including all of its genes.

gigantism: The inclination for some isolated populations to undergo significant increases in body size relative to populations on a mainland; reduced competition is considered one cause.

glacial landforms: Landforms created by the action of glaciers directly (erosion and deposition) as opposed to fluvio-glacial landforms produced by meltwater.

gravity model: Measure of influence of source populations on a target location. Basically, $G = \Sigma P_j/d_{ij}$, where G is gravitational influence, P is population size of sources, d is distance from sources, i is the target location and j the influential source(s). In the Isle of Man study (this volume), the model was built on eight sectors of coastal sources, the records for 10 km squares were standardized on the number of squares available in each sector (two 10 km coastal squares inland), and the distances standardized on the maximum distance recorded.

habitat: In traditional usage, the site (biotope, vegetation type) where an animal lives. See **resource-based habitat** for comparison (Dennis, 2010).

Heinrich events: Events lasting < 1000 years that often coincide with indications of global climate fluctuations, representing episodic discharges of icebergs from northern hemisphere ice sheets; a number occurred during the Devensian (*q.v.*). Melting of the icebergs released fresh water and debris in the North Atlantic, the cold and fresh water altering the density-driven, thermohaline circulation patterns of the ocean.

heterotroph: An organism unable to manufacture its own food; it obtains its food and energy by consuming organic substances, usually plant or animal matter. All animals, protozoans, fungi, and most bacteria are heterotrophs.

heterozygote: An individual having two different alleles of a particular gene.

Holocene: The latest epoch (apart from the Anthropocene (*q.v.*)), an interglacial, commencing at c. 11.7–11.5 ka cal BP.

immigration: The arrival of an individual (propagule) at a destination, which may or may not be a potential habitat (resource-based habitat). This term is typically used in reference to the arrival of an individual (species) at an island in island biogeography. In ecology generally, immigration may be to any terrestrial site (a potential habitat) occupied or unoccupied by that species.

infiltration capacity: The maximum rate at which water can enter soil; when this is exceeded, surface runoff results.

intrinsic rate of population increase: The rate at which a population increases in size (exponential) if there are no density-dependent forces (i.e. population carrying capacity from resource availability, competition and predation) regulating the population.

introgressive hybridization: The movement of a gene (gene flow) from one species into the gene pool of another by the repeated backcrossing of an interspecific hybrid with one of its parent species.

island vagrants: Visitors (immigrants) to islands but not necessarily able to breed on it, owing to lack of opportunity or suitable resources or conditions.

isobase: A line drawn on a map or chart connecting places of equal or constant land upheaval above sea level at any one point in time.

isostacy: Changes in sea level owing to relative adjustments in the land surface caused by the imposition or removal of mass (water, ice).

Jaccard's coefficient S_j: A similarity coefficient. $S_j = a/(a+b+c)$, where a is the number of jointly occurring species in two populations, b is the number of species present in population 2 but absent in population 1, and c is the number of species present in population 1 but absent from population 2. Joint absences (d) are ignored.

ka: Thousand years.

Kendall coefficient of concordance: This coefficient expresses the simultaneous association (relatedness) between k sets of rankings (i.e. cases; correlated samples). In effect, it is the average of all Spearman's r_s (*q.v.*) between cases. Kendall concordances vary from 0 to +1. Values close to zero represent lack of agreement in the rankings of the variables (e.g. objects) among cases, while values close to 1 represent perfect agreement in the rankings.

keratophyre: A volcanic rock of intermediate composition (in silica content). It is similar to trachyte but has a richer sodium component than the plagioclase minerals found in trachyte (Schermerhorn, 1973).

kettle hole: A hollow created when buried blocks of glacier ice melt out; these are eventually subjected to vegetation succession of bog communities.

Krummholz: Stunted, deformed vegetation (in subarctic and subalpine tree line landscapes, coastal fringe) shaped by continual exposure to strong, cold winds.

Kruskal stress: Given by $\varphi = \Sigma(d_{ij}-\delta_{ij})^2$, where d_{ij} are the observed distances (or dissimilarities) and δ_{ij} are the reproduced distances in a non-metric multidimensional scaling (NMMS) (*q.v.*) ordination.

Late Glacial: Period between the end of the Devensian Glaciation (deglaciation) and the start of the Holocene.

law of mass action: The rate of an elementary reaction (a reaction that proceeds through only one transition state) is proportional to the product of the concentrations of the participating molecules; i.e. the rate of any given chemical reaction is proportional to the product of the activities (or concentrations) of the reactants.

life history characteristics: Traits that affect the life table of an organism and can be imagined as various investments in growth, reproduction and survivorship.

life history strategy: Different modes of living for distributing resources among the potential activities (growth, development, maintenance, reproduction) in the process of maximizing fitness or the number of offspring the organism leaves in the next generation. A continuum prevails in life history strategies between what are referred to as ***r*-selected species** (maximum productivity) and ***K*-selected species** (maximizing survival).

lifetime resource track: The geographical path taken by an individual from birth to death; usually in reference to flight path from adult eclosion to death.

loadings (principal components analysis): Correlations between a component in principal components analysis (or factors in factor analysis) and the original variables.

Loch Lomond Readvance: The re-development of ice sheets and corrie glaciers in Scotland, the Lake District and Wales during the Younger Dryas between c. 12.9 and 11.7 ka cal BP.

loess: Wind-blown sand deposits (loosely compacted yellowish-grey deposit), usually reprocessed fluvio-glacial deposits.

logistic (logit) regression: A regression model in which the dependent variable is categorical, typically binary (present or absent) as in a species' incidence on islands.

macroevolution: Evolution on a scale at or above the level of species (contrasts with microevolution, which refers to smaller changes such as allele frequencies within a population or populations).

Mann–Whitney U test: A test used to compare differences between the scores of two independent groups when the dependent variable is treated as adhering to an ordinal (ranked) scale (equivalent of *t*-test on normally distributed continuous data). For multiple groups a Kruskal-Wallis ANOVA (*q.v.*) can be applied.

Mantel Test: A statistical test of the correlation between two association matrices.

marginality: Limits on the persistence of populations placed on them by resource scarcity and extreme conditions; where these fail to cause local population extinction, individuals respond (adapt) in a variety of ways to the changes in conditions (Shreeve *et al.*, 1996).

mark–release–recapture (MRR): Also given as capture–mark–release–recapture. A technique that allows individuals in a population to be identified on recapture (Ehrlich and Davidson, 1960) for the determination of population parameters (changes in population numbers, life span, residence, movement distances, etc.).

mass effects (ecology): The outcome of any processes that depends on the numbers and density of individuals in a population; related to the chemical **law of mass action**, in that increased density increases the probability of individuals coming into contact and/or integrating/leaving a site.

matrix: The zone outside the habitat. In resource-based habitat context, the matrix may well comprise resources for a species but not the full complement of resources and conditions sufficient to maintain a local population. There is also internal matrix (resourceless trivial space within the bounds of a species' resource-based habitat). The ultimate matrix in island biogeography for butterflies is the sea (water bodies).

mean: The arithmetic mean or average of a set of variates; the mean is the sum of measurements (e.g. of butterfly wing expanse) divided by the number of individuals (usually a sample is taken, which has an associated measure of error).

median: A simple measure of central tendency. The median is the middle value of a set of ranked values. If there is an odd number of observations, the median is the middle value. If there is an even number of observations, the median is the average of the two middle values.

Mesolithic: The middle part of the Stone Age, c. 10–3 ka BCE, characterized by adaptation to a hunting, collecting and fishing economy between the Palaeolithic and Neolithic.

metapopulation: A group of interacting local populations of the same species separated by empty space (matrix).

microevolution: Evolutionary change within a species or population(s) over a short period of time.

microgranite: A medium-grained intrusive igneous rock containing small randomly interlocking crystals. For the complex-sounding arfvedsonite–aegirine–microgranite on Ailsa Craig, see Harrison *et al.* (1987).

migration: Movement of an individual across matrix (non-habitat); typically sea in island biogeography.

migration ability: The anatomical, physical and physiological traits that affect the distance a butterfly individual can fly, e.g. wing aspect ratio, wing span, wing loading, thoracic musculature, wing shape, wing tensile strength, body mass disposition (thorax mass:total body mass), etc. (Betts and Wootton, 1988). Migration ability is potentially a measurable variable among individuals in any population for any species; a sample of individuals will have a characteristic frequency distribution for migration ability.

migration capacity: All the factors, including opportunities (i.e. flight periods, mass of numbers, population density, numbers of broods, etc.) as well as bauplan (*q.v.*) (migration ability) that influence the numbers moving out from habitats as well as the distances a butterfly species can fly. It is a measure currently limited to a species or at best a population. Although any one individual in a population will have a probability of leaving a habitat and flying a certain distance, logistic difficulties deter measures of migration capacity for individuals.

migrfac: A measure of a species' migration capacity that depends not just on migration ability or flight endurance, but also on opportunity for individuals (i.e. flight period, voltinism, population numbers) (see **Mass effects**) to transfer increasing distances.

mobility: The ability of insects to move between locations.

mode: A measure of central tendency. It is the observation that occurs most frequently (if no number is repeated, then there is no mode for the list) or the class of values (i.e. in a histogram) that occurs most frequently.

mtDNA: Mitochondrial DNA is located in mitochondria, cellular organelles within eukaryotic cells that convert chemical energy from food into a form that cells can use, adenosine triphosphate (ATP).

multiple regression analysis: An extended version of simple linear regression, used to predict the value of a variable based on the values of two or more other variables. The variable we want to predict is called the dependent variable and those used to predict it are referred to as independent or predictor variables. Different procedures may be used. A common one (forwards stepwise entry) is to enter the predictor variables one by one in order of their importance, rejecting those that have no effect.

multivariate analysis (statistical techniques): The statistical process of simultaneously analysing multiple independent (or predictor) variables with multiple dependent (outcome or criterion) variables using matrix algebra (most multivariate analyses involve the determination of correlations among both dependent variables and predictor variables).

multivoltine: Having multiple broods each year.

N and ***n***: Number of variates (individual cases), with N the complete total and n a sample.

nanism: Dwarf forms of species (on islands) found elsewhere.

near-island effect (NIE): Extinctions on near islands are countered by a rescue effect facilitated by islands of even a small size presenting relatively large targets.

negative feedback: Changes in a chain of effects that dampen or buffer the initial input and stabilize a system (see **positive feedback**).

Neolithic: Period in the development of human technology (New Stone Age; farming, domestication of animals) beginning c. 10.2 ka BC in some parts of the Middle East, later in other parts of the world.

nestedness: State of faunas (species assemblages) in which biotas (*q.v.*) of sites with lower numbers of species tend to be subsets of the biotas at richer sites.

niche: The role (function and/or position) of a species within an ecological community (i.e. producer, consumer, predator, etc.); relational position of an organism or a population in a particular ecosystem. Hutchinson (1957) defined it as the conceptual multidimensional space (n-dimensional hypervolume) of resources available to and used by a species within a community.

non-metric [multidimensional] scaling (NMMS): An ordination technique in which units are moved in 1, 2,...k spaces until the distances between them equate as closely as possible with similarities (distances) describing their relationships. In the case where geographical distances are input, the plot in two dimensions should produce a plot close to their relative geographical positions. See Appendix 8.

non-stationarity: A non-stationary time series occurs when either or both the mean and variance in a condition changes; when change is systematic (directional), this produces a trend.

normal distribution: A frequency distribution (an arrangement of a data set) in which most values cluster in the middle of the range (the mean, median and mode coincide) and the rest taper off symmetrically towards either end. Values can be described as being so many standard deviations above or below the mean, called **z scores** (*q.v.*).

operational taxonomic unit (OTU): The 'thing' or focus of study (e.g. species, taxa, areas, regions).

ordination techniques: A variety of methods that attempt to expose the relationships between OTUs (*q.v.*).

outlier (in statistics): See **standardized residual**.

outlier (in geology): An area of younger rock completely surrounded by older rocks (opposite of inlier in geology usage).

P (probability): The typical format for the significance of a test or relationship, $P < 0.05$ (1 in 20) being taken as a statistically significant result in much biological work. **NS** is used to indicate a non-significant relationship. Note: as the number of cases (islands) in a study is increased, *P* will also decrease (the probability of a statistic occurring purely by chance increases) even if a statistical parameter (e.g. **Jaccard's** S_J; **Pearson's correlation**) (*q.v.*) also decreases. In the case of correlations, the key feature is to consider the **coefficient of determination** (*q.v.*) (r^2 or R^2) for pairs of variables.

parapatry: The occupying of populations of geographical areas that are partially overlapping or have a partial barrier between them. In parapatric speciation, reproductive isolation between populations of a species develops while the affected populations continue to exchange genes. Whether such populations are continuous or discontinuous, typically mating occurs non-randomly and gene flow is unequal.

partial regression analysis: A form of regression that allows the unique contribution of a variable or variables to be determined by first accounting the contribution from other relevant potential predictors.

Pearson's correlation r: A measure of association between two interval–ratio scale variables made over a number of cases; it ranges between +1 and −1 inclusive, where +1 is a perfect positive linear correlation, 0 indicates no linear correlation, and −1 is a perfect negative linear correlation; r^2 is the coefficient of determination (*q.v.*), the amount of variation in one variable accounted for by another variable.

peninsula effect: The tendency for species richness to decrease along a gradient from the axis to the most distal point off a peninsula or headland (e.g. Italy; Great Orme's Head, North Wales).

peripatry and **peripatric speciation**: A variant of allopatric (*q.v.*) speciation, in which evolution (speciation) occurs in an isolated peripheral population formed by colonization or vicariance (*q.v.*).

permafrost: Ground, including rock or soil, at or below the freezing point of water (0°C, 32°F) for 2 or more years. Severe permafrost climates, distinguished by ice wedges, have mean annual air temperatures of −6° to −8°C or colder, as these form when substrate temperatures fall below −17°C.

phagy: A variable term in this work describing the number of plants exploited by a butterfly larva; more typically as a suffix, '-phagy', indicating 'feeding on'.

phenology: The study of periodicity in plant and animal life cycle events (broods, seasonal stage occurrence); how these are influenced by seasonal and other variations in climate, as well as habitat factors.

phenotype: An organism's observable traits, including morphology, development, biochemical or physiological properties, and behaviour.

phylogenetic analysis: The study of the evolutionary history and relationships among individuals or groups of organisms (e.g. species, or populations). Phylogenetic analysis has advanced into phylogeography, linking geographical distributions of genetic lineages. It is a method of mapping the DNA sequence variants to their geography to show the present-day distribution of genetic lineages (Avise *et al.*, 2016).

plagioclimax: A relatively stable plant community that has arisen as a result of persistent human intervention in the natural succession of communities (e.g. heather moor management).

positive feedback: A process occurring in a feedback loop in which the effects of a change on a system include an increase in the magnitude of the initial perturbation; e.g. X produces more of Y, which in turn produces more of X. These changes tend to move a system away from its equilibrium state and make it more unstable.

***principal components analysis* (*PCA*)**: An ordination technique in which new variables are extracted from original variables based on the covariances among them; the solution is mathematically unique, the new variables or components are orthogonal (unrelated) to one another and can be interpreted by reference to the component loadings (*q.v.*). Factor analysis is an extension of this technique with different algorithms used for defining the path of vectors through the cloud of variables; the axes are typically rotated to focus on groups of variables (i.e. maximize loadings on axes) as in a varimax rotation (*q.v.*) in which the axes are still orthogonal to one another, but there are also oblique (correlated) solutions.

pro-glacial lake: A lake formed by the damming action of a moraine or ice dam during the retreat of a melting glacier, or by meltwater trapped against an ice sheet.

pseudosinks: Habitat patches able to support a population – but whose carrying capacity is significantly lower than apparent – with the support of immigration (Watkinson and Sutherland, 1995); cf. **sink** (*q.v.*).

pseudoturnover: Inability of observers to score rare species each year, owing for instance to limited censuses of islands (Schoener, 2010).

Quaternary: The most recent of the three periods of the Cenozoic Era following the Neogene Period; it spans from 2.59 ± 0.005 million years ago to the present and is divided into two basic epochs: the **Pleistocene** (2.59 million years BP to 11.7 ka BP) and the **Holocene** (> 11.7 ka BP).

R^2*, *r^2: See ***Coefficient of determination*** and ***Pearson's correlation r***.

raised bog: A peat bog in which growth of peat layers comprising the vegetation of sphagnum communities is most rapid at the centre, giving it a domed shape; growth initially occurs in hollows of standing surface water but later depends on rainwater. The centre is nutrient depleted but the surrounds can be areas of nutrient accumulation with richer plant communities (Rodwell, 1991–1995).

***rank-abundance curve* (*Whittaker plot*)**; A diagram used in ecology to display relative species abundance, a component of biodiversity. The

Y-axis depicts relative abundance (log scale usually) and X-axis abundance rank, with the most abundant species ranked 1.

rarity: The quality of having limited ranges and distributions. When categorized, usually taken as 5% of a fauna with the smallest ranges. A variety of approaches can be taken (Gaston, 1994); in a study of butterflies on European islands, rarity was measured quantitatively as the inverse of species distribution cover (Dennis *et al.*, 2008).

realized fecundity: The actual number of offspring produced in a lifetime.

refuge, **refugium** (plural: **refugia**): A location that supports an isolated or relict population of a once more widespread species. This isolation (causing allopatry (*q.v.*)) can be due to a range of factors that restrict populations (i.e. climatic changes, urbanisation, deforestation, etc.).

regolith: A layer of loose, heterogeneous superficial material covering solid rock, the product of weathering and mass-wasting.

regression analysis techniques: Statistical methods for determining the relationships among variables. They includes a wide variety of techniques for modelling and analysing several variables, when the focus is on the relationship between a dependent variable and one or more independent (or predictor) variables.

relative turnover rate (*T*): For an island's species, $T_{t1,t2} = (n_e + n_i)/(N_{t1} + N_{t2})$, where t_0, t_1, t_2...t_m are time intervals (years), n_e is the number of species extinctions on an island, n_i the number of novel colonizations on an island, and N the total number of species at different times.

relaxation fauna: The equilibrium number of species (a balance maintained by colonization and extinction of species) on land bridge islands following their separation from the mainland.

relicts: Organisms (or ecosystems) that were abundant over a large area but now survive over a much smaller area.

rendzina: A humus-rich shallow soil type, with a dark layer over pale weather bedrock, usually formed over calcareous sediments.

rescue effect: The rescue effect (see **sink**), the maintenance of a species' population, is greater for islands close to mainland sources, owing to a greater number of immigrants.

residual: The difference between the observed value of the dependent variable (y) in regression analysis (*q.v.*) and the predicted value ($\hat{y}$). Each data point has one residual. Thus a residual = observed value minus predicted value.

resource-based habitat (RBH): The collection (set) of ground-based resources and conditions sufficient to maintain a population or cohort of a species (Dennis *et al.*, 2014, p. 6).

routine movements: See **trivial flights**.

ruderal strategies: The strategies of plant or animal species that prosper in situations of high-intensity disturbance and low-intensity stress.

search theory: Probabilistic methods for determining moving and stationary targets (Hohzaki, 2009).

seral stage: A stage in ecological (vegetation) succession from bare ground to a climatic climax forest cover.

single nucleotide polymorphism (SNP): Variation in a single nucleotide (organic molecules that serve as the monomers, or subunits, of the nucleic acids DNA (deoxyribonucleic acid) and RNA (ribonucleic acid)), that occurs at a specific position in the genome; each variant is referred to as an allele and will occur at a particular frequency in a species.

sink: An unsuitable habitat patch in which death rates are greater than birth rates, resulting in a population decline toward extinction unless enough individuals transfer to it from a source patch to maintain persistence of the population, referred to a rescue effect (*q.v.*) (Pulliam, 1988).

small-island effect (SIE): A pattern that species richness on islands has – below a certain threshold area – of varying independently of island area (Lomolino, 2000; Triantis *et al.*, 2006); see Dennis (2001) and Burns *et al.* (2009) for statistical artefacts.

Sokal and Michener's S_{MM}: A similarity coefficient between two OTUs (e.g. species for islands) described by $(a+d)/(a+b+c+d)$ where a, b, c and d describe joint presence and/or absence in a contingency table (see ***Jaccard's coefficient***). In its complementary form, $1-S_{SM}$, it is equal to the squared Euclidean distance, i.e. $d = \sqrt{1-S_{SM}}$.

source: In ecology: a habitat patch with a local population in which birth rates exceed death rates and from which immigrants transfer to other patches (Pulliam, 1988). In biogeography: an island or mainland acting as a source of species for smaller and/or more isolated islands.

source mass: The number and size of populations of species at mainland (or other sources) for islands.

spatial autocorrelation: Correlation across geographical space. Variability in the measurements of a variable correlates with geographical distance of the units on which they are based; analysis of spatial data usually requires removing the artefact of spatial autocorrelation (see: https://en.wikipedia.org/wiki/Spatial_analysis).

Spearman's r_s *or* ρ: A non-parametric measure of rank correlation (statistical dependence between the ranking of two variables); as with Pearson's correlation (*q.v.*) for interval–ratio data it varies from +1, through 0 to –1.

specialist (specialism): A species characteristically associated (adapted) with a narrow range of environmental conditions and resources (e.g. monophagy for larval host plants; cf. generalists (*q.v.*) and generalism). Most organisms grade between extremes of generalism and specialism.

species: A species (of butterfly) is often defined as a group (set) of individuals that actually or potentially interbreed in nature; this follows Ernst Mayr's (1942, 1963) biological species definition (species are groups of interbreeding natural populations that are reproductively isolated from other such groups). Care in using this definition needs to be taken as examples are provided in this study of so-called species that exchange genes (introgression) at spatial boundaries. See Descimon and Mallet (2009) for full discussion of the many alternative definitions that have been developed for different purposes.

species turnover: Species replacement. Changes in species composition on islands resulting from some species becoming extinct and others immigrating. See **relative turnover rate**.

specific heat capacity: The amount of heat per unit mass required to raise the temperature by one degree Celsius. For example, a metre-thick depth of sea water cooled by 0.1°C releases sufficient heat to raise the temperature of a c. 30 m layer of air by 10°C (Barry and Chorley, 1976).

standard deviation: A measure of deviation about the mean (symbol σ). The sample standard deviation is the root of the sum of deviations of variates squared from the mean divided by the number of sampled individuals minus one: $= \sqrt{[\Sigma(x_i - \bar{x})^2/(n-1)]}$. See http://stattrek.com/statistics/formulas.aspx.

standard error of the estimate: A measure of the dispersion (or variability) in the predicted scores in a regression analysis. SE estimate = $\sqrt{\Sigma[(Y-Y')^2]/n}$, where Y is an actual score in the sample and Y' is a predicted score.

standard error of the mean: Usually given the notation $SE_{\bar{x}}$ it is usefully a measure of the bounds within which the population mean is likely to occur. It is calculated as: $SE = \sigma/\sqrt{n}$ where σ is the sample standard deviation and n is the size of the sample. It is the standard deviation of the sampling distribution of the mean; i.e. a measure comparable with the standard deviation of a sample mean but which measures the average dispersion of the means of samples of a given size about the population mean. The population mean lies with 95% probability 1.96 *SE*s either side of population mean estimated by repeated sampling.

standardized residual (regression analysis): A measure of the size of residuals from a regression line when standardized (divided) by the standard deviation of residuals. Residuals that are outliers are > 2 standard deviations from the regression line (< 5% probability of occurring in a normally distributed dependent variable).

standardized scores: The values of a continuous data set approaching a normal distribution expressed in standard deviation units. Standardized score $z = (x-\bar{x})/\sigma$, where x is a variate, $\bar{x}$ is the mean and σ is the standard deviation.

stationarity: A stationary time series has unchanging statistical properties; the mean, variance, autocorrelation, etc. are all constant (do not differ significantly) over time.

Storegga tsunami: Devastating waves caused by underwater landslides at the edge of Norway's continental shelf north-west of the Møre coast. This collapse involved an estimated 290 km (180 miles) length of coastal shelf, with a total volume of 3500 km^3 (840 cubic miles) of debris.

stress tolerators: Species adapted to (thrive in) environmentally stressful conditions (salt-tolerant, drought-tolerant, etc.).

Student's t-test: A parametric statistical hypothesis developed by W.S. Gosset (pseudonym 'Student') to determine if the means of two sets of data are significantly different from each other (see Appendix 8).

subspecies (ssp): A category in biological classification ranking below a species. A subspecies designates a population of a particular geographical region considered (but often not on the attributes describing it) genetically distinguishable from other such populations of the same species but capable of interbreeding successfully with them where their ranges overlap. No clear distinction actually exists between a subspecies and a so-called 'race'; the latter if anything is of lower taxonomic rank and informal in usage. Bearing in mind the possible plastic determination of physical differences (e.g. distinguishing some subspecies of British butterflies), the subspecies concept is often regarded as defunct (Braby *et al.*, 2012).

sympatry: The occurrence of species in the same geographical area (overlapping distributions or ranges, and coming into contact). Sympatric speciation is the process through which new species evolve from a single ancestral species while inhabiting the same geographical region

target effect: The larger an island and the closer it is to a mainland, the greater the angle subtended by the island to a source shore and the greater the probability of transfers from source species to it.

taxon cycle: Sequential phases of expansion and contraction of the ranges of species, associated generally with shifts in ecological distribution; typically a four-stage process is envisaged whereby an expansion phase of an immigrant is followed by phenotypic differentiation and specialization, then extinction with residual populations surviving on only large high islands (Jønsson *et al.*, 2014).

terrane: A fault-bounded area or region with a distinctive stratigraphy, structure and geological history, typically a fragment of crustal material formed on, or broken off from, one tectonic plate and attached to the crust lying on another plate.

transient communities: Communities of species showing rapid turnover in species content.

trivial flights (**movements**): Short movements by adults that typically occur within the habitat in the course of daily activity, movements between resource units and individuals of the same and other species; such movements also occur outside a habitat. The movements are only trivial in distance terms and certainly not as regards butterfly biology; thus, Van Dyck and Baguette (2005) refer to them as **routine movements**.

trophic: A term that relates to feeding and obtaining nutrition. Trophic levels are hierarchical levels in an ecosystem, consisting of organisms sharing the same function in the food chain and the same nutritional relationship to the primary sources of energy.

turnover: See **species turnover**.

vagrancy index: A measure of a species' ability to move outside habitat units (Cook *et al.*, 2001).

vagrants: Individuals of a species found outside habitats (as understood from a resource-based habitat viewpoint).

variance: A measure of deviation in variates about the mean related to the standard deviation such that, for a sample, it is the sum of the squared deviations from the mean divided by the number of individuals minus one. Sample variance = $\Sigma(x_i - \bar{x})^2/(n - 1)$ where x is a variate and $\bar{x}$ is the mean. See http://stattrek.com/statistics/formulas.aspx.

variate: A single observed value of a random variable, considered to be one of many possible realizations.

varves: Pairs of thin layers of clay and silt of contrasting colour and texture which represent the seasonal deposits (summer and winter) of a single year's sediment layer in a lake. Varves are used to determine the chronology of glacial sediments.

vicariance: The separation of a population or larger unit (species) into disjunct groups as the result of a physical barrier.

voltinism: Number of broods each year.

Wider Countryside Butterfly Survey (WCBS): Scheme established in 2009 to improve data on the population status of butterflies across the countryside as a whole. The WCBS method is based on the British Trust for Ornithology (BTO) Breeding Bird Survey (BBS), counting along two parallel transects 1 km long subdivided into ten sections, located within randomly selected 1 km squares. The differences are that butterflies are counted in a more restricted area than for birds and at different times of the day. Only two to four visits annually are required, compared with 26 visits on the traditional UK Butterfly Monitoring Scheme (UKBMS) 'Pollard walk' transects (see http://www.ukbms.org/wcbs.aspx).

Wilcoxon Matched Pairs Test: The Wilcoxon signed-ranks test is a non-parametric equivalent of the paired *t*-test (see **Student's *t*-test**).

Windermere interstadial: A relatively warm period during the Late Glacial (at the end of the last (Devensian) glaciation in Britain) dating to c. 13 k–11 ka ^{14}C BP. It includes the Bølling, Older Dryas and Allerød chronozones of Scandinavia distinguished in pollen curves by a sharp rise followed by a marked fall in birch and juniper.

wing loading: Measure of flight efficiency where wing load = body mass/wing area.

Younger Dryas: A geological cold period from c. 12.9 ka to c. 11.7 ka BP, named after an indicator genus, the alpine-tundra wildflower *Dryas octopetala*. It incorporates the Loch Lomond Readvance in Britain. The earlier Upper Dryas in Scandinavia is not recognized in Britain.

Z score: A standard score that indicates how many standard deviations an element (variate) is from the mean. It can be calculated from: $z = (x - \mu)/\sigma$ where z is the *z*-score, x is the value of the element, μ is the population mean, and σ is the standard deviation.

Bibliography

Abercrombie, R.G. (1953) *Callophrys rubi* in Iona. *Entomologist* 86, 197.

Ackery, P.R. (1984) Systematic and faunistic studies on butterflies, pp. 9–21. In: Vane-Wright, R.I. and Ackery, R.R. (eds) *The Biology of Butterflies*. Academic Press, London.

Addison, K., Edge, M.J. and Watkins, R. (1990) *The Quaternary of North Wales: Field Guide*. Quaternary Research Association, Coventry.

Adkin, R. (1892) [South London Ent. and NHS exhibition – Lepidoptera from Isles of Scilly]. *Entomologist* 25, 98.

Adkin, R. (1894) On a collection of Lepidoptera from the Scilly Isles. *Entomologist* 27, 9–12.

Adkin, R. (1924) Divergence of character [*Aglais urticae* on Man]. *Entomologist's Record and Journal of Variation* 37, 11.

Adkin, R.W. (1897) *Colias edusa* in the Isles of Scilly. *Entomologist* 30, 269.

Adkin, R.W. (1911) *Colias edusa* and *Sphinx convolvuli* at Scilly. *Entomologist* 44, 324.

Agassiz, D. (1982) *A Revised List of the Lepidoptera of the Isles of Scilly*. Isles of Scilly Museum Association, Tresco.

Albre, J., Gers, C. and Legal, L. (2008) Molecular phylogeny of the *Erebia tyndarus* (Lepidoptera, Rhopalocera, Nymphalidae, Satyrinae) species group combining CoxII and ND5 mitochondrial genes: a case study of a recent radiation. *Molecular Phylogenetics and Evolution* 47, 196–210.

Allan, P.B.M. (1949) Dispersal of *Lysandra coridon* Poda. (Lep., Lycaenidae). *Entomologist* 82, 10.

Allan, P.B.M. (1956) The middle copper. *Entomologist's Record and Journal of Variation* 68, 68–73.

Alley, R.B., Meese, D.A., Shuman, C.A., Gow, A.J., Taylor, K.C. *et al.* (1993) Abrupt increase in Greenland ice accumulation at the end of the Younger Dryas event. *Nature* 362, 527–529.

Altizer, S., Hobson, K.A., Davis, A.K., De Roode, J.C. and Wassenaar, L.I. (2015) Do healthy monarchs migrate farther? Tracking natal origins of parasitized vs. uninfected monarch butterflies overwintering in Mexico. *PLoS ONE* 10, e0141371. doi: 10.1371/journal.pone.0141371.

Ammann, B. (ed.) (2000) Biotic responses to rapid climatic changes around the Younger Dryas. *Palaeogeography, Palaeoclimatology, Palaeoecology* 159, 313–347.

Andersen, S.T. (1978) Local and regional vegetational development in eastern Denmark in the Holocene. *Danmark Geologiske Undersogelse Årbog* 1976, 5–27.

Angel, H. (1975) *Photographing Nature: Insects.* Fountain Press, London.

Anonymous (1872–73) Butterflies at sea [editorial]. *Entomologist's Monthly Magazine* 9, 246.

Anonymous (1992) [photograph of *Junonia lemonias*]. *Butterfly Conservation News* 50, 40.

Anonymous (1993) Update [on Lepidoptera]. *Shetland Entomological Group Newsletter* 4 (Dec. 1993), 8–10.

Anonymous (2017) *List of Lights, Radio Aids and Fog Signals. British Isles, English Channel and North Sea.* Pub. 114. National Geospatial-Intelligence Agency, Springfield, Virginia.

Armstrong, R.A. and Bradwell, T. (2010) The use of lichen growth rings in lichenometry: some preliminary findings. *Geografiska Annaler: Series A, Physical Geography* 92, 141–147.

Arnold, H.R. (1993) *Atlas of Mammals in Britain.* ITE research publication no. 6. HMSO, London.

Asher, J., Warren, M., Fox, R., Harding, P., Jeffcoate, G. and Jeffcoate, S. (2001) *The Millennium Atlas of Butterflies in Britain and Ireland.* Oxford University Press, Oxford.

Askew, R.R. (1974) Insects from Bardsey Island. *Entomologist's Gazette* 25, 45–51.

Atkinson, T.C., Briffa, K.R. and Coope, G.R. (1987) Seasonal temperatures in Britain during the past 22,000 years reconstructed using beetle remains. *Nature* 325, 587–592.

Atmar, W. and Patterson, B.D. (1993) The measure of order and disorder in the distribution of species in fragmented habitat. *Oecologia* 96, 373–382.

Atmar, W. and Patterson, B.D. (1995) *The Nestedness Temperature Calculator, a visual BASIC program, including 294 presence/absence matrices.* AICS Research Inc., University Park, New Mexico and the Field Museum, Chicago, Ilinois.

Avise, J.C., Bowen, B.W. and Ayala, F.J. (2016) In the light of evolution. X: Comparative phylogeography. *Proceedings of the National Academy of Sciences* 113, 7957–7961.

Baguette, M., Vansteenwegen, C., Convi, I. and Néve, G. (1998) Sex-biased density-dependent migration in a metapopulation of the butterfly *Proclossiuna eunomia. Acta Oecologica* 19, 17–24.

Bai, Y., Ma, L.B., Xu, S.-Q. and Wang, G.-H. (2015) A geometric morphometric study of the wing shapes of *Pieris rapae* (Lepidoptera: Pieridae) from the Qinling Mountains and adjacent regions: an environmental and distance-based consideration. *Florida Entomologist* 98, 162–169.

Bailey, N.T.J. (1995) *Statistical Methods in Biology.* Cambridge University Press, Cambridge.

Baker, G. (1900) *Lampides boeticus* bred in Guernsey. *Entomologist's Monthly Magazine* 36, 240.

Baker, R.R. (1969) The evolution of the migratory habit in butterflies. *Journal of Animal Ecology* 38, 703–746.

Baker, R.R. (1978) *The Evolutionary Ecology of Animal Migration.* Hodder and Stoughton, London.

Baker, R.R. (1984) The dilemma: when and how to go or stay. In: Vane-Wright, R.I. and Ackery, P.R. (eds) *The Biology of Butterflies.* Academic Press, London, pp. 279–296.

Baldwin, T. (1995) Shetland swallowtail. *Butterfly Conservation News* 60, 16.

Balfour-Browne, F. (1958) The origin of our British Swallowtail and our Large Copper butterflies. *Entomologist's Record and Journal of Variation* 70, 33–36.

Ball, S.G. and Morris, R.K.A. (2000) *Provisional atlas of British hoverflies (Diptera, Syrphidae).* NERC: Centre for Ecology and Hydrology, Huntingdon.

Barber, D.C., Dyke, A., Hillaire-Marcel, C., Jennings, A.E., Andrews, J.T. *et al.* (1999) Forcing of the cold event of 8,200 years ago by catastrophic drainage of Laurentide lakes. *Nature* 400, 344–348.

Barber, K.E. (1993) Peatlands as scientific archives of past biodiversity. *Biodiversity and Conservation* 2, 474–489.

Baring, C. (1915) Notes on the fauna and flora of Lambay. *Irish Naturalist* 24, 71.

Baring, C. (1918) Lepidoptera of Lambay. *Irish Naturalist* 27, 63.

Barry, R.G. and Chorley, R.J. (1976) *Atmosphere, weather and climate.* Methuen and Co. Ltd, London.

Barton, N.H. (1989) Founder effect speciation. In: Otte, D. and Endler, J.A. (eds) *Speciation and its Consequences.* Sinauer, Sunderland, Massachusetts, pp. 229–256.

Baselga, A. (2012) The relationship between species replacement, dissimilarity derived from nestedness, and nestedness. *Global Ecology and Biogeography* 21, 1223–1232.

Baxter, E.V. and Rintoul, L.J. (1948) Peacock butterfly in Argyll and Mull. *Scottish Naturalist* 1948, 212.

Baxter, E.V. and Rintoul, L.J. (1950) Peacock and painted lady butterflies [on Isle of May]. *Scottish Naturalist* 1950, 50.

Baynes, E.S.A. (1957) The Lepidoptera of Tory island, county Donegal. *Entomologist* 90, 310–317.

Baynes, E.S.A. (1964) *A Revised Catalogue of Irish Macrolepidoptera.* E.W. Classey, Hampton.

Baynes, E.S.A. (1973) Butterflies and moths. In: Sharrock, J.T.R. (ed.) *The Natural History of Cape Clear Island.* Poyser, Berkhamsted, pp. 1–207.

Beavis, I.C. (1976) Butterflies of the Isles of Scilly during August 1975. *Entomologist's Record and Journal of Variation* 88, 194–195.

Beavis, I.C. (2004) Resident and regular migrant butterflies on the Isles of Scilly. *Entomologist's Record and Journal of Variation* 116, 97–102.

Becker, B. and Kramer, B. (1991) Dendrochronology and radiocarbon calibration of the early Holocene. In: Barton, N., Roberts, A.J. and Roe, D.A. (eds) *The Late Glacial in North-west Europe: Human Adaptation and Environmental Change at the End of the Pleistocene.* Council for British Archaeology, London, pp. 22–24.

Beirne, B.P. (1943a) The distribution and origin of the British Lepidoptera. *Proceedings of the Royal Irish Academy* 49(B), 27–59.

Beirne, B.P. (1943b) The relationship and origins of the Lepidoptera of the Outer Hebrides, Shetlands, Faroes and Iceland. *Proceedings of the Royal Irish Academy* 49(B), 91–101.

Beirne, B.P. (1945) The Lepidoptera of Shetland. *Entomologist's Record and Journal of Variation* 57, 37–40.

Beirne, B.P. (1947) The origin and history of the British Macrolepidoptera. *Transactions of the Royal Entomological Society of London* 98, 275–372.

Beirne, B.P. (1955) Natural fluctuations in the abundance of British Lepidoptera. *Entomologist's Gazette* 6, 21–52.

Bell, E., Boyle, D., Floyd, K., Garner-Richards, P., Swann, B. *et al.* (2011) The ground-based eradication of Norway rats (*Rattus norvegicus*) from the Isle of Canna, Inner Hebrides, Scotland. In: Veitch, C.R., Clout, M.N. and Towns, D.R. (eds) *Island Invasives: Eradication and Management.* IUCN, Gland, Switzerland, pp. 269–274.

Bennett, D.P. and Humphries, D.A. (1974) *Introduction to Field Biology.* Edward Arnold, London.

Bennett, K.D. (1989) A provisional map of forest types for the British Isles 5000 years ago. *Journal of Quaternary Science* 4, 141–144.

Bennett, K.D., Tzedakis, P.C. and Willis, K.J. (1991) Quaternary refugia of north European trees. *Journal of Biogeography* 18, 103–115.

Bennett, N.L., Severns, P.M., Parmesan, C. and Singer, M.C. (2015) Geographic mosaics of phenology, host preference, adult size and microhabitat choice predict butterfly resilience to climate warming. *Oikos* 124, 41–53.

Bennie, J., Huntley, B., Wiltshire, A., Hill, M.O. and Baxter, R. (2008) Slope, aspect and climate: spatially explicit and implicit models of topographic microclimate in chalk grassland. *Ecological Modelling* 216, 47–59.

Berry, R.J. (2009) *The Natural History of Islands*. HarperCollins, London.

Betts, C.R. and Wootton, R.J. (1988) Wing shape and flight behaviour in butterflies (Lepidoptera: Papilionoidea and Hesperioidea): a preliminary analysis. *Journal of Experimental Biology* 138, 271–288.

Bink, F.A. (1992) *Ecologische Atlas van de dagvlinders van Noordwest-Europa*. Schuyt and Co., Haarlem.

Birchall, E. (1866) Catalogue of the Lepidoptera of Ireland. *Proceedings of the Natural History Society of Dublin* 5, 57–85.

Bird, R.J. (1980) Casual observations of insects in the Isle of Skye, July 1979. *Bulletin of the Amateur Entomologists' Society* 39, 22–24.

Birkett, N.L. (1995) Is the Silver-studded Blue extinct in north-west England? *Butterfly Conservation News* 59, 24–25.

Birks, H.J.B. (1986) Late-Quaternary biotic changes in terrestrial and lacustrine environments, with particular reference to north-west Europe. In: Berglund, B.E. (ed.) *Handbook of Holocene Palaeoecology and Palaeohydrology*. Wiley, Chichester, pp. 375–397.

Blackie, J.E.H. (1951) Hebridean satyrids. *Entomologist* 84, 264.

Blackler, H. (1940) Migration of large white butterflies observed at Hilbre Point, Hoylake, Cheshire. *Entomologist* 73, 210.

Blair, K.G. (1925) Lepidoptera of the Scilly Isles. *Entomologist* 58, 3–10, 38.

Bland, K.P. (1977) Lepidoptera from South Ronaldsay, Orkney. *Entomologist's Record and Journal of Variation* 89, 70–72.

Blyth, S.F.P. (1901) *Pieris daplidice* in Jersey. *Entomologist* 34, 291.

Bolotov, I.N. (2006) Diurnal butterflies (Lepidoptera, Rhopalocera) of the Solovetskie Islands (Northwestern Russia, the White Sea). *Entomological Review* 86, 516–523.

Bolotov, I.N. and Shutova, E.V. (2006) Patterns of formation of island fauna of butterflies (Lepidoptera, Diurna) at the northern forest boundary in the region of Pleistocene Continental glaciations (by the example of White Sea Islands). *Biology Bulletin of the Russian Academy of Sciences* 33, 260–268.

Boulton, G.S., Peacock, J.D. and Sutherland, D.G. (1991) Quaternary. In: Craig, G.Y. (ed.) *Geology of Scotland* (3rd edn). The Geological Society, London.

Bourillet, J.-F., Reynaud, J.Y., Baltzer, A. and Zaragosi, S. (2003) The 'Fleuve Manche': the submarine sedimentary features from the outer shelf to the deep-sea fans. *Journal of Quaternary Science* 18, 261–282.

Bourillet, J.-F., Zaragosi, S. and Mulder, T. (2006) The French Atlantic margin and deep-sea submarine systems. *Geo-Marine Letters* 26, 311–315.

Bowen, D.Q., Rose, J., McCabe, A.M. and Sutherland, D.G. (1986) Quaternary glaciations in England, Ireland, Scotland, and Wales. *Quaternary Science Reviews* 5, 299–340.

Bowen, D.Q., Phillips, F.M., McCabe, A.M., Knutz, P.C. and Sykes, G.A. (2002) New data for the last glacial maximum in Great Britain and Ireland. *Quaternary Science Reviews* 21, 89–101.

Bowles, N. and Fox, R. (2007) Wildlife reports: butterflies. *British Wildlife* 19, 57.

Bowles, N. and Fox, R. (2014) Wildlife reports: butterflies. *British Wildlife* 26, 52–54.

Box, G.E.P. (1976) Science and Statistics. *Journal of the American Statistical Association* 71, 791–799.

Brabt, M.F., Farias Quipildor, G.E., Vane-Wright, R.I. and Lohman, D.J. (2015) Morphological and molecular evidence supports recognition of *Danaus petilia* (Stoll 1790) (Lepidoptera: Nymphalidae) as a species distinct from *D. chrysippus* (Linnaeus, 1758). *Systematics and Biodiversity* 13, 386–402.

Braby, M.F., Eastwood, R. and Murray, N. (2012) The subspecies concept in butterflies: has its application in taxonomy and conservation biology outlived its usefulness? *Biological Journal of the Linnean Society* 106, 699–716.

Bradley, J.D. (1954) Lepidoptera records from Skokholm. *Entomologist* 87, 146–147.

Bradley, J.D. and Fletcher, D.S. (1958) Lepidoptera records from the Isle of Rhum. *Entomologist* 91, 126–131.

Brakefield, P.M. (1984) The ecological genetics of quantitative characters of *Maniola jurtina* and other butterflies. In: Vane-Wright, R.I. and Ackery, R. R. (eds) *The Biology of Butterflies*. Academic Press, London, pp. 167–190.

Brakefield, P.M. and Shreeve, T.G. (1992) Diversity in populations. In: Dennis, R.L.H. (ed.) *The Ecology of Butterflies in Britain*. Oxford University Press, Oxford, pp. 178–196.

Bretherton, R.F. (1968) The butterflies in Britain in relation to those on the adjacent parts of the continent. *Proceedings of the British Entomological and Natural History Society* 1, 7–15.

Briggs, C.A. (1884) A week's collecting on Unst. *Entomologist* 17, 197–201.

Bristow, C.R., Mitchell, S.H. and Bolton, D.E. (1993) *Devon Butterflies* [includes Lundy]. Devon Books, Tiverton.

Bristow, R. (1994) *Satyrium w-album* (Knoch) (Lep.: Lycaenidae) in a moth trap. *Entomologist's Record and Journal of Variation* 106, 73.

Bristowe, W.S. (1931a) Notes on a migration of the 'Red Admiral' (*Pyrameis atalanta*). *Entomologist's Monthly Magazine* 67, 205–206.

Bristowe, W.S. (1931b) A small collection of insects from Lundy Island. *Entomologist's Monthly Magazine* 67, 232.

Brock, J.P. (1990) Origins and phylogeny of butterflies. In: Kudrna, O. (ed.) *Butterflies of Europe: 8. Introduction to Lepidopterology*. Aula-Verlag, Wiesbaden, pp. 209–233.

Broecker, W.S. (1994) Massive iceberg discharges as triggers for global climate change. *Nature* 372, 421–424.

Bronk Ramsey, C., Staff, R.A., Bryant, C.L., Brock, F., Kitigawa, H. *et al.* (2012) A complete terrestrial radiocarbon record for 11.2 to 52.8 kyr BP. *Science* 338, 370–374.

Brooks, A.J., Bradley, S.L., Edwards, R.J. and Goodwyn, N. (2011) The palaeogeography of Northwest Europe during the last 20,000 years. *Journal of Maps* 2011, 573–587.

Brooks, S.J., Matthews, I.P., Birks, H.J. and Birks, H.J.B. (2012) High resolution Late-glacial and early-Holocene summer air temperature records from Scotland inferred from chironomid assemblages. *Quaternary Science Reviews* 41, 67–82.

Brown, D.C.G. (2013) *Erebia aethiops* at Grassington. *Atropos* 50, 90 [introduction].

Brown, J.H. and Kodric-Brown, A. (1977) Turnover rates in insular biogeography: effects of immigration on extinction. *Ecology* 58, 145–449.

Brown, R.H. (1887) Moths settling on water. *Entomologist* 20, 283.

Bryan, M.D. (1987) Island interludes – Guernsey and the Isle of Mull in 1986. *Entomologist's Record and Journal of Variation* 99, 125–128.

Bryant, E.A. and Haslett, S.K. (2007) Catastrophic wave erosion, Bristol Channel, UK: Impact of tsunami? *Journal of Geology* 115, 253–269.

Bubová, T., Kulma, M., Vrabec, V. and Nowicki, P. (2016) Adult longevity and its relationship with conservation status in European butterflies. *Journal of Insect Conservation* 20, 1021–1032.

Buckstone, A.A.W. (1938) Migration of *Polygonia c-album*. *Entomologist* 71, 110.

Bulman, C.R., Wilson, R.J., Holt, A.R., Bravo, L.G., Early, R.I. *et al.* (2007) Minimum viable metapopulation size, extinction debt, and the conservation of a declining species. *Ecological Applications* 17, 1460–1473.

Burek, C.V. (1978) Quaternary deposits on the Carboniferous limestone of Derbyshire. PhD thesis. University of Leicester. Available at: http://hdl.handle.net/2381/7935 (accessed 4 March 2018).

Burg, N.A., Pradhan, A., Gonzalez, R.M., Morban. E.Z., Zhen EW, Sakchoowong, W. and Lohman, D.J. (2014) Inferring the provenance of an alien species with DNA barcodes: The Neotropical butterfly Dryas iulia in Thailand. *PLoS ONE* 9(8): e104076. doi:10.1371/journal.pone.0104076

Burn, C. and Zhang, Y. (2009) Permafrost and climate change at Herschel Island (Qikiqtaruq), Yukon Territory, Canada. *Journal of Geophysical Research* 114, F02001. doi: 10.1029/2008JF001087.

Burns, K.C., McHardy, R.P. and Pledger, S. (2009) The small-island effect: fact or artefact? *Ecography* 32, 269–276.

Bush, M.B. and Whittaker, R.J. (1991) Non-equilibration in island theory of Krakatau. *Journal of Biogeography* 18, 341–356.

Cade, M. (2008) Portland, Dorset (Coastal Report). *Atropos* 33, 51–58

Campbell, J.L. (1936) Immigrant Lepidoptera in the Outer Hebrides, 1936 (*Colias croceus* on Barra). *Entomologist* 69, 265–266.

Campbell, J.L. (1938) The Macrolepidoptera of the parish of Barra. *Scottish Naturalist* 1938, 153–163.

Campbell, J.L. (1946) Catalogue of a collection of Macro-Lepidoptera made in the Hebrides between 1936 and the present date. *Entomologist* 79, 49–53.

Campbell, J.L. (1947a) Hebridean notes. *Entomologist* 80, 116–117.

Campbell, J.L. (1947b) Migrant Lepidoptera on the Isle of Canna during August 1947. *Entomologist* 80, 269.

Campbell, J.L. (1948a) Immigrant Lepidoptera in the Hebrides. *Scottish Naturalist* 1948, 47–48.

Campbell, J.L. (1948b) On the spread of the peacock butterfly in the Highlands and Islands. *Scottish Naturalist* 1948, 126.

Campbell, J.L. (1949a) Migrants in the Hebrides. *Entomologist* 82, 189.

Campbell, J.L. (1949b) Release of marked *Vanessa cardui. Entomologist* 82, 209.

Campbell, J.L. (1950) Lepidoptera taken at Heisker lighthouse in 1949. *Scottish Naturalist* 1950, 122.

Campbell, J.L. (1951a) An experiment in marking migratory butterflies. *Entomologist* 84, 1–6.

Campbell, J.L. (1951b) *Callophrys rubi* on the Isle of Canna. *Entomologist* 84, 159.

Campbell, J.L. (1951c) *Pararge aegeria* on the Isle of Canna. *Entomologist* 84, 237.

Campbell, J.L. (1951d) *Vanessa atalanta* [Canna], *Entomologist* 84, 237.

Campbell, J.L. (1952) Notes on Canna butterflies. *Entomologist* 85, 263.

Campbell, J.L. (1953) *Pieris brassicae* larvae in November and December; a further note. *Entomologist* 86, 51.

Campbell, J.L. (1954) The Macrolepidoptera of the Isle of Canna. *Scottish Naturalist* 1954, 101–121.

Campbell, J.L. (1955a) *Vanessa cardui* on the Isle of Canna. *Entomologist* 88, 231.

Campbell, J.L. (1955b) *Vanessa atalanta* on the Isle of Canna. *Entomologist* 88, 236.

Campbell, J.L. (1958) Isle of Canna report for 1958. *Entomologist's Record and Journal of Variation* 70, 282–283.

Campbell, J.L. (1967) Isle of Canna, 1966. *Entomologist's Record and Journal of Variation* 79, 97–99.

Campbell, J.L. (1969a) Isle of Canna report for 1968. *Entomologist's Record and Journal of Variation* 81, 67–70.

Campbell, J.L. (1969b) Rise and decline of *Vanessa io* in the Small Isles (Inner Hebrides). *Entomologist's Record and Journal of Variation* 81, 117–118.

Campbell, J.L. (1969c) Isle of Canna report for 1969. *Entomologist's Record and Journal of Variation* 81, 323–325.

Campbell, J.L. (1970) Macrolepidopera Cannae. The butterflies and moths of Canna. *Entomologist's Record and Journal of Variation* 82, 211–214, 235–242, 292–299.

Campbell, J.L. (1972) Isle of Canna notes for 1970 and 1971. *Entomologist's Record and Journal of Variation* 84, 196–198.

Campbell, J.L. (1975a) Isle of Canna report, 1972–1974. *Entomologist's Record and Journal of Variation* 87, 10–12.

Campbell, J.L. (1975b) On the rumoured presence of the large blue butterfly (*Maculinea arion*) in the Hebrides. *Entomologist's Record and Journal of Variation* 87, 161–166.

Campbell, J.L. (1978) *Pieris rapae* on the Isle of Canna. *Entomologist's Record and Journal of Variation* 89, 279.

Campbell, J.L. (1980) Isle of Canna: invasion of painted ladies and red admirals. *Entomologist's Record and Journal of Variation* 92, 256–257.

Campbell, J.L. (1984) Isle of Canna, 1984: the peacock returns. *Entomologist's Record and Journal of Variation* 96, 187.

Campbell, R.C. (1989) *Statistics for Biologists*. Cambridge University Press, Cambridge.

Campbell-Taylor, J.E. (1931a) *Melitaea aurinia* Rott. *Entomologist* 64, 188–189.

Campbell-Taylor, J.E. (1931b) The flying power of *Melitaea aurinia*. *Entomologist* 64, 278.

Carlquist, S. (1974) *Island Biology*. Columbia University Press, New York.

Carlson, A.E. (2013) The Younger Dryas Event. In: Elias S.A. (ed.) *The Encyclopedia of Quaternary Science*, Vol. 3. Elsevier, Amsterdam, pp. 126–134.

Carpenter, G.D.H. (1943) New records of insects and woodlice from Lundy Island. *Entomologist's Monthly Magazine* 87, 121–123.

Carrington, J.T. (1880) The Lepidoptera of the Shetland Isles. *Entomologist* 13, 238.

Carrington, J.T. (1881) [Lepidoptera of Shetland Isles and Hebrides.] *Entomologist* 14, 303.

Carrington, J.T. (1883) Lepidoptera on Unst. *Entomologist* 16, 236–237.

Carter, D.J. (1967) The Lepidoptera, Trichoptera and Odonata of Cape Clear Island. *Cape Clear Bird Observatory Report* 8, 45–46.

Carter, H.H., Smith, V.C. and Spence, T.F. (1960) Puffin Island 1959. *Proceedings of the Birmingham Natural History and Philosophical Society* 19, 37–53.

Cartwright, R.A. and Cartwright, D.B. (1976) *The Holy Island of Lindisfarne and the Farne Islands*. David and Charles, Newton Abbot.

Castle-Russell, S.G. (1932) Flying habits of *Melitaea aurinia*. *Entomologist* 65, 42–43.

Cesaroni, D., Lucarelli, M., Allori, P., Russo, F. and Sbordoni, V. (1994) Patterns of evolution and multidimensional systematics in Graylings (Lepidoptera, *Hipparchia*). *Biological Journal of the Linnean Society* 52, 101–119.

Chalmers-Hunt, J.M. (1970) The butterflies and moths of the Isle of Man. *Transactions of the Society for British Entomology* 19, 1–171.

Chapman, J.W., Reynolds, D.R. and Wilson, K. (2015) Long-range seasonal migration in insects: mechanisms, evolutionary drivers and ecological consequences. *Ecology Letters* 18, 287–302.

Charman, D.J. (2010) Centennial climate variability in the British Isles during the mid–late Holocene. *Quaternary Science Reviews* 29, 1539–1554.

Cheesman, E.M. (1898) Lepidoptera captures in the Orkney Isles. *Entomologist's Record and Journal of Variation* 10, 204–206.

Chernov, Y.I. and Tatarinov, A.G. (2006) Butterflies (Lepidoptera, Rhopalocera) in the Arctic fauna. *Entomological Review* 86, 760–786.

Chiverrell, R.C. and Thomas, G.S.P. (2010) Extent and timing of the Last Glacial Maximum (LGM) in Britain and Ireland: a review. *Journal of Quaternary Science* 25, 535–549.

Chorley, R.J. (1959) The shape of drumlins. *Journal of Glaciology* 3, 339–344.

Chorley, R.J. and Kennedy, B.A. (1971) *Physical Geography. A Systems Approach*. Prentice-Hall International Inc., London.

Clapham, A.R., Tutin, T.G. and Moore, D.M. (1989) *Flora of the British Isles*. Cambridge University Press, Cambridge.

Clark, C.D., Hughes, A.L.C., Greenwood, S.L., Jordan, C. and Sejrup, H.P. (2010) Pattern and timing of retreat of the last British-Irish Ice Sheet. *Quaternary Science Reviews* 44, 112–146.

Clarke, W.E. (1897) *Vanessa antiopa* in Shetland. *Annals of Scottish Natural History* 1897, 48.

Clutterbuck, C.G. (1940) Lepidoptera in the Isles of Scilly in August 1939. *Entomologist* 73, 129–131.

Coles, J. and Hall, D. (1998) *Changing Landscapes: The Ancient Fenland*. Cambridge County Council, Cambridge.

Collier, N., Mackay, D.A. and Benkendorff, K. (2008) Is relative abundance a good indicator of population size? Evidence from fragmented populations of a specialist butterfly (Lepidoptera: Lycaenidae). *Population Ecology* 50, 17–23.

Collier, R.V. (1986) The conservation of the chequered skipper in Britain. *Focus on Nature Conservation* 16, p. 16.

Colvin, M. and Reavey, D. (1993) *A directory for entomologists*. Pamphlet No. 14. Amateur Entomologists' Society, Feltham.

Conant, S. (1988) Geographic variation in the Laysan Finch (*Telespyza cantans*). *Evolutionary Ecology* 2, 270–282.

Conder, P.J. (1949) Observations on a migration of *Pieris brassicae* L. at Skokholm Island, Pembrokeshire, in August 1947. *Proceedings of the Royal Entomological Society of London Series A General Entomology* 24, 35–38.

Coney, G.B. (1904) Lepidoptera in Jersey, 1903. *Entomologist* 37, 127–131.

Conover, D.O. and Present, T.M.C. (1990) Countergradient variation in growth rate: compensation for length of the growing season among Atlantic silversides from different latitudes. *Oecologia* 83, 316–324.

Conservation Committee for Butterfly Conservation (1995) Lepidoptera restoration: Butterfly Conservation's policy, code of practice and guidelines for action. *Butterfly Conservation News* 60, 20–21.

Cook, L.M. (2008) Species richness in Madeira land snails and its causes. *Journal of Biogeography* 35, 647–653.

Cook, L.M. and Saccheri, I.J. (2013) The peppered moth and industrial melanism; evolution of a natural selection case study. *Heredity* 110, 207–212.

Cook, L.M., Dennis, R.L.H. and Mani, G.S. (1999) Melanic morph frequency in the peppered moth in the Manchester area. *Proceedings of the Royal Society of London Biological Sciences* 266, 293–297.

Cook, L.M., Dennis, R.L.H. and Hardy, P.B. (2001) Butterfly-hostplant fidelity, vagrancy and measuring mobility from distribution maps. *Ecography* 24, 497–504.

Cook, L.M., Dennis, R.L.H. and Dockery, M. (2004) Fitness of insularia morphs of the peppered moth *Biston betularia*. *Biological Journal of the Linnaean Society* 82, 359–366.

Cook, L.M., Dockery, M. and Dennis, R.L.H. (2011) Melanic peppered moths in Manchester: the end of an era. *Entomologist's Gazette* 62, 91–99.

Coope, G.R. (1979) Late Cenozoic fossil Coleoptera: evolution, biogeography and ecology. *Annual Review of Ecology and Systematics* 10, 247–267.

Coope, G.R. (1987) The response of Late Quaternary insect communities to sudden climatic changes. In: Gee, J.H.R. and Giller, P.S. (eds.) *Organization of Communities: Past and Present*. Blackwell Scientific Publishing, Oxford, pp. 421–438.

Coope, G.R., Lemdahl, G., Lowe, J.J. and Walkling, A. (1998) Temperature gradients in northern Europe during the last glacial-Holocene transition (14–9 ^{14}C kyr BP) interpreted from Coleopteran assemblages. *Journal of Quaternary Science* 13, 419–433.

Cooter, J. (1978) Some insects from Ailsa Craig. *Entomologist's Monthly Magazine* 113, 168.

Corbet, A.S. (1945) The Lepidoptera of St. Kilda. *Entomologist* 78, 166–168.

Coulson, S.J. (2015) The alien terrestrial invertebrate fauna of the High Arctic archipelago of Svalbard: potential implications for the native flora and fauna. *Polar Research* 34, 1–6.

Courtney, S.P. (1980) Studies on the biology of the butterflies *Anthocharis cardamines* L and *Pieris napi* L. in relation to speciation in the Pierinae. PhD thesis, University of Durham, UK.

Courtney, S.P. (1988) Oviposition on peripheral hosts by dispersing *Pieris napi* (L.) (Pieridae). *Journal of Research on the Lepidoptera* 26, 58–63.

Courtney, S.P., Hill, C.P. and Westerman, A. (1982) Pollen carried for long periods by butterflies. *Oikos* 38, 260–263.

Cowin, W.S. (1955) Our missing butterflies [Man]. *Peregrine* 2, 6–8.

Cox, G.W. and Ricklefs, R.E. (1977) Species diversity and ecological release in Caribbean land bird faunas. *Oikos*, 28, 113–122.

Craggs, J.D. (1982) *Hilbre: the Cheshire island: its History and Natural History*. Liverpool University Press, Liverpool.

Crewe, H.H. (1877a) Entomology at Tresco and the Scilly Isles. *Entomologist* 10, 295–297.

Crewe, H.H. (1877b) Notes on the Lepidoptera of the Scilly Isles. *Entomologist's Monthly Magazine* 14, 148–150.

Curtis, R.J., Botham, M.S., Brereton, T.M. and Isaac, N.J.B. (2015a) The rise and demise of the Glanville fritillary on the Isle of Wight. *Journal of Insect Conservation* 19, 305–311.

Curtis, R.J., Brereton, T.M., Dennis, R.L.H., Carbone, C. and Isaac, N.J.B. (2015b) Butterfly abundance is determined by resource availability and mediated by species traits. *Journal of Applied Ecology* 52, 1676–1684.

Curtis, W.P. (1931) The flying powers of *Melitaea aurinia*. *Entomologist* 64, 211–212.

Curtis, W.P. (1932) The flying powers of *Melitaea aurinia*. *Entomologist* 65, 73–80.

Dale, C.W. (1884a) Capture of insects in the Hebrides and in St. Kilda. *Scottish Naturalist* 1884, 284.

Dale, C.W. (1884b) Captures on South Uist and St. Kilda. *Entomologist's Monthly Magazine* 20, 213–214.

Dale, C.W. (1884c) Additions to the entomology of the Isle of Harris. *Entomologist's Monthly Magazine* 20, 256.

Dale, C.W. (1884d) Captures in the Isle of Skye. *Entomologist's Monthly Magazine* 20, 257.

Dale, C.W. (1889) The insect fauna of St. Kilda. *Entomologist* 22, 12–13.

Dale, C.W. (1890) Insects in the Scilly Isles. *Entomologist's Monthly Magazine* 26, 238.

Dannreuther, T. (1933) Migration records, 1932 (*Pieris rapae* on Round Island). *Entomologist* 66, 268.

Dannreuther, T. (1935) Migration records, 1934 [*Vanessa cardui* on Sandray]. *Entomologist* 68, 209.

Dannreuther, T. (1936) Migration records, 1936. *Entomologist* 69, 225–230.

Dannreuther, T. (1939a) Migration record for 1938 [*Aglais urticae* on North Rona]. *Entomologist* 72, 119–122.

Dannreuther, T. (1939b) Migration records, 1939. *Entomologist* 72, 273–283.

Dannreuther, T. (1944) Migration records, 1943 [*Colias croceus* on Round Island]. *Entomologist* 77, 112.

Dannreuther, T. (1948a) Migration records, 1947 [*Vanessa cardui* on Chicken Rock, Man]. *Entomologist* 81, 73–83.

Dannreuther, T. (1948b) Insect immigration: Scotland an end point. *Scottish Naturalist* 1948, 74–81.

Dannreuther, T. (1950) Migration records, 1950 [Great Saltee]. *Entomologist* 84, 85–90.

Dannreuther, T. (1951) Insect immigration in 1950. *Scottish Naturalist* 1951, 122–127.

Dapporto, L. and Dennis, R.L.H. (2008) Species' richness, rarity and endemicity on Italian offshore islands: complementary signals from island-focused and species-focused analyses. *Journal of Biogeography* 35, 664–674.

Dapporto, L. and Dennis, R.L.H. (2013) The generalist-specialist continuum: testing predictions for distribution and trends in British butterflies. *Biological Conservation* 157, 229–236.

Dapporto, L., Bruschini, C., Baracchi, D., Cini, A., Gayubo, S. *et al.* (2009) Phylogeography and counter-intuitive inferences in island biogeography: evidence from morphometric markers in the mobile butterfly *Maniola jurtina* (Linnaeus) (Lepidoptera, Nymphalidae). *Biological Journal of the Linnean Society* 98, 677–692.

Dapporto, L., Schmitt, T., Vila, R., Scalercio, S., Biermann, H. *et al.* (2011) Phylogenetic island disequilibrium: evidence for ongoing long-term population dynamics in two Mediterranean butterflies. *Journal of Biogeography* 38, 854–867.

Dapporto, L., Ramazzotti, M., Fattorini, S., Vila, R., Talavera, G. and Dennis, R.L.H. (2013) Recluster: an unbiased clustering procedure for beta-diversity turnover. *Ecography* 36, 1070–1075.

Dapporto, L., Fattorini, S., Vodă, R., Dincă, V. and Vila, R. (2014) Biogeography of western Mediterranean butterflies: combining turnover and nestedness components of faunal dissimilarity. *Journal of Biogeography* 41, 1639–1650.

Dapporto, D., Ciolli, G., Dennis, R.L.H., Fox, R. and Shreeve, T.G. (2015) A new procedure for extrapolating turnover regionalization at mid–small spatial scales applied to British butterflies. *Methods in Ecology and Evolution* 6, 1287–1297.

Dapporto, L., Cini, A., Menchetti, M., Vodăl, R., Bonelli, S. *et al.* (2017) Rise and fall of island butterfly diversity: understanding genetic differentiation and extinction in a highly diverse archipelago. *Diversity and Distributions* 23, 1169–1181. doi: 10.1111/ddi.12610.

Darlington, A. (1954) Ecological notes on some insects of Bardsey during August 1954. *Report of Bardsey Bird and Field Observatory* 2, 35–38.

Darlington, P.J. (1957) *Zoogeography: the Geographical Distribution of Animals.* John Wiley and Sons, New York.

Davis, B.A.S., Brewer, S., Stevenson, A.C., Guiot, J. *et al.* (2003) The temperature of Europe during the Holocene reconstructed from pollen data. *Quaternary Science Reviews* 22, 1701–1716.

Dawkins, R. (2009) *The Greatest Show on Earth: the Evidence for Evolution.* Free Press, New York.

de Jong, F.M.W., de Snoo, G.R. and van de Zande, J.C. (2008) Estimated nationwide effects of pesticide spray drift on terrestrial habitats in the Netherlands. *Journal of Environmental Management* 86, 721–730.

De Keyser, R., Breuker, C., Hails, R., Dennis, R.L.H. and Shreeve, T. (2015) Why small is beautiful: wing colour is free from thermoregulatory constraint in a small *Polyommatus* butterfly. *PLoS ONE* 10(4), e0122623.

De Palma, A., Dennis, R.L.H., Brereton, T., Leather, S.R. and Oliver, T.H. (2016) Large reorganizations in butterfly communities during an extreme climatic event. *Ecography* 40, 577–585.

Dean, T. (1990) *The Natural History of Walney Island.* Faust Publications, Burnley, UK.

Delaney, C. (2003) The Last Glacial Stage (the Devensian) in northwest England. *North-West Geography* 3, 27–37.

Dempster, J.P., King, M.L. and Lakhani, K.H. (1976) The status of the swallowtail butterfly in Britain. *Ecological Entomology* 1, 71–84.

Dempster, J.P., Lakhani, K.H. and Coward, P.A. (1986) The use of chemical composition as a population marker in insects: a study of the brimstone butterfly. *Ecological Entomology* 11, 51–56.

Demuth, R.P. (1974) Late autumn in the Scilly Isles. *Entomologist's Record and Journal of Variation* 86, 72–73.

Dengler, J. (2010) Robust methods for detecting a small island effect. *Diversity and Distributions* 16, 256–266.

Dennis, E.B. (2015) Development of statistical methods for monitoring insect abundance. PhD thesis, University of Kent, UK.

Dennis, E.B., Morgan, B.J.T., Freeman, S.N., Brereton, T. and Roy, D.B. (2016a) A generalized abundance index for seasonal invertebrates. *Biometrics* 72, 1305–1314. doi: 10.1111/biom.12506.

Dennis, E.B., Morgan, B.J.T., Brereton, T., Freeman, S.N. and Roy, D.B. (2016b) Dynamic models for longitudinal butterfly data. *Journal of Agricultural, Biological, and Environmental Statistics* 21, 1–21.

Dennis, R.L.H. (1972a) *Plebejus argus* (L.) *caernensis* Thompson. A stenoecious geotype. *Entomologist's Record and Journal of Variation* 84, 132–140.

Dennis, R.L.H. (1972b) *Eumenis semele* (L.) *thyone* Thompson (Lep., Satyridae). A micro-geographical race. *Entomologist's Record and Journal of Variation* 84, 1–11, 38–44.

Dennis, R.L.H. (1974) *Lysandra coridon* (Poda) and *L. bellargus* (Rott.) in North Wales? *Entomologist's Record and Journal of Variation* 86, 24.

Dennis, R.L.H. (1977) *The British Butterflies. Their Origin and Establishment.* E.W. Classey, Faringdon, UK.

Dennis, R.L.H. (1982a) Patrolling behaviour in orange tip butterflies within the Bollin valley in north Cheshire, and a comparison with other pierids. *Vasculum* 67, 17–25.

Dennis, R.L.H. (1982b) Observations on habitats and dispersal made from oviposition markers in north Cheshire *Anthocharis cardamines* (L.) (Lep., Pieridae). *Entomologist's Gazette* 33, 151–159.

Dennis, R.L.H. (1982c) *Erebia aethiops* (Esper) (Satyridae) on Arnside Knott. Confidential Report for the Nature Conservancy Council, Peterborough.

Dennis, R.L.H. (1985a) Small plants attract attention! Choice of egg-laying sites in the green-veined white butterfly (*Artogeia napi*) (L.) (Lep: Pieridae). *Bulletin of the Amateur Entomologists' Society* 44, 77–82.

Dennis, R.L.H. (1985b) British butterfly distributions. Space-time predictability. In: Heath, J. (ed.) *Proceedings of the 3rd Congress of Lepidopterology, Cambridge* 1982, 55–62.

Dennis, R.L.H. (1986) Motorways and cross-movements. An insect's 'mental map' of the M56 in Cheshire. *Bulletin of the Amateur Entomologists' Society* 45, 228–243.

Dennis, R.L.H. (ed.) (1992) *The Ecology of Butterflies in Britain.* Oxford University Press, Oxford.

Dennis, R.L.H. (1993) *Butterflies and Climate Change.* Manchester University Press, Manchester, UK.

Dennis, R.L.H. (1997a) An inflated conservation load for European butterflies: increases in rarity and endemism accompany increases in species richness. *Journal of Insect Conservation* 1, 43–62.

Dennis, R.L.H. (1997b) A record of *Hipparchia semele* (L.) (Lep.: Satyridae) for Grassholm Island, Wales. *Entomologist's Record and Journal of Variation* 109, 88.

Dennis, R.L.H. (2001) Progressive bias in species status is symptomatic of fine-grained mapping units subject to repeated sampling. *Biodiversity and Conservation* 10, 483–494.

Dennis, R.L.H. (2004) Butterfly habitats, broad scale biotope affiliations and structural exploitation of vegetation at finer scales: the matrix revisited. *Ecological Entomology* 29, 744–752.

Dennis, R.L.H. (2010) *A Resource-based Habitat View for Conservation. Butterflies in the British Landscape.* Wiley-Blackwell, Oxford.

Dennis, R.L.H. (2015a) A decade of the resource-based habitat paradigm: the semantics of habitat loss. In: Elias, S. (ed.) *Reference Module on Earth Systems and Environmental Sciences.* Elsevier SclTech Connect, Waltham, Massachusetts, pp. 1–12. Available at: http://www.sciencedirect.com/science/referenceworks/9780124095489 (accessed 4 March 2018).

Dennis, R.L.H. (2015b) Butterflies on a dragon's head; butterflies in a dragon's head. In: Dyer, L. and Forister, M. (eds) *The Lives of Lepidopterists.* Springer International, Cham, Switzerland, pp. 213–230.

Dennis, R.L.H. and Bardell, P. (1996) The impact of extreme weather events on local populations of *Hipparchia semele* (L.) (Nymphalidae) and *Plebejus argus* (L.) (Lycaenidae): hindsight, inference and lost opportunities. *Entomologist's Gazette* 47, 211–225.

Dennis, R.L.H. and Bramley, M.J. (1985) The influence of man and climate on dispersion patterns within a population of adult *Lasiommata megera* (L.) (Satyridae) at Brereton Heath, Cheshire (UK). *Nota Lepidopterologica* 8, 309–324.

Dennis, R.L.H. and Eales, H.T. (1997) Patch occupancy in *Coenonympha tullia* (Lepidoptera: Satyridae): habitat quality matters as much as patch size and isolation. *Journal of Insect Conservation* 1, 167–176.

Dennis, R.L.H. and Eales, H.T. (1999) Probability of site occupancy in the large heath butterfly *Coenonympha tullia* determined from geographical and ecological data. *Biological Conservation* 87, 295–301.

Dennis, R.L.H. and Schmitt, T. (2009) Faunal structures, palaeogeography and historical inference. Chapter E2 in: Settele, J., van Dyck, H., Konvicka, M. and Shreeve, T.G. (eds) *Ecology of Butterflies in Europe.* Cambridge University Press, Cambridge, pp. 250–280.

Dennis, R.L.H. and Shreeve, T.G. (1989) Butterfly morphology variation in the British Isles: the influence of climate, behavioural posture and the hostplant-habitat. *Biological Journal of the Linnean Society* 38, 323–348.

Dennis, R.L.H. and Shreeve, T.G. (1991) Climatic change and the British butterfly fauna: Opportunities and constraints. *Biological Conservation* 55, 1–16.

Dennis, R.L.H. and Shreeve, T.G. (1996) *Butterflies on British and Irish Offshore Islands: Ecology and Biogeography.* Gem Publishing Company, Wallingford, UK.

Dennis, R.L.H. and Shreeve, T.G. (1997) Diversity of butterfly species on British islands: ecological influences underlying the roles of area, isolation and faunal source. *Biological Journal of the Linnaean Society* 60, 257–275.

Dennis, R.L.H. and Shreeve, T.G. (2003) Gains and losses of French butterflies: tests of predictions, under-recording and regional extinction from data in a new atlas. *Biological Conservation* 110, 131–139.

Dennis, R.L.H. and Sparks, T.H. (2006) When is a habitat not a habitat? Dramatic resource use changes under differing weather conditions for the butterfly *Plebejus argus. Biological Conservation* 129, 291–301.

Dennis, R.L.H. and Sparks, T.H. (2007) Climate signals are reflected in an 89 year series of British Lepidoptera records. *European Journal of Entomology* 104, 763–767.

Dennis, R.L.H. and Thomas, C.D. (2000) Bias in butterfly distribution maps: the influence of hot spots and recorder's home range. *Journal of Insect Conservation* 4, 73–77.

Dennis, R.L.H. and Williams, W.R. (1986) Butterfly 'diversity': regressing and a little latitude. *Antenna* 10, 108–112.

Dennis, R.L.H. and Williams, W.R. (1987) Mate-location behaviour in the butterfly *Ochlodes venata* (Br. and Grey) (Hesperiidae). Flexible strategies and spatial components. *Journal of the Lepidopterists' Society* 41, 45–64.

Dennis, R.L.H., Porter, K. and Williams, W.R. (1984) Ocellation in *Coenonympha tullia* (Mueller) (Lep., Satyridae). I. Structures in correlation matrices. *Nota Lepidopterologica* 7, 199–219.

Dennis, R.L.H., Porter, K. and Williams, W.R. (1986) Ocellation in *Coenonympha tullia* (Mueller) (Lep., Satyridae). II. Population differentiation and clinal variation in the context of climatically-induced anti-predator defence strategies. *Entomologist's Gazette* 37, 133–172.

Dennis, R.L.H., Williams, W.R. and Shreeve, T.G. (1991) A multivariate approach to the determination of faunal structures among European butterfly species (Lepidoptera, Rhopalocera). *Zoological Journal of the Linnean Society* 101, 1–49.

Dennis, R.L.H., Shreeve, T.G. and Williams, W.R. (1995a) Affinity gradients among European butterflies: evidence for an historical component to species distributions. *Entomologist's Gazette* 46, 141–153.

Dennis, R.L.H., Shreeve, T.G. and Williams, W.R. (1995b) Taxonomic differentiation in species richness gradients among European butterflies (Papilionidea, Hesperiodea): contribution of macroevolutionary dynamics. *Ecography* 18, 27–40.

Dennis, R.L.H., Shreeve, T.G. and Sparks, T.H. (1998a) The effects of island area, isolation and source population size on the presence of the grayling butterfly *Hipparchia semele* (L.) (Lepidoptera: Satyrinae) on British and Irish offshore islands. *Biodiversity and Conservation* 7, 765–776.

Dennis, R.L.H., Sparks, T.H., and Shreeve, T.G. (1998b) Geographical factors influencing the probability of *Hipparchia semele* (L.) (Lepidoptera: Satyrinae) occurring on British and Irish offshore islands. *Global Ecology and Biogeography Letters* 7, 205–214.

Dennis, R.L.H., Williams, W.R. and Shreeve, T.G. (1998c) Faunal structures among European butterflies: evolutionary implications of bias for geography, endemism and taxonomic affiliations. *Ecography* 21, 181–203.

Dennis, R.L.H., Sparks, T.H. and Hardy, P.B. (1999) Bias in butterfly distribution maps: the effects of sampling effort. *Journal of Insect Conservation* 3, 33–42.

Dennis, R.L.H., Donato, B., Sparks, T.H. and Pollard, E. (2000a) Ecological correlates of island incidence and geographical range among British butterflies. *Biodiversity and Conservation* 9, 343–359.

Dennis, R.L.H. Shreeve, T.G., Olivier, A. and Coutsis, J.G. (2000b) Contemporary geography dominates butterfly diversity gradients within the Aegean archipelago (Lepidoptera: Papilionoidea, Hesperioidea). *Journal of Biogeography* 27, 1365–1384.

Dennis, R.L.H., Shreeve, T.G., Sparks, T.H. and L'Honoré, J.E. (2002) A comparison of geographical and neighbourhood models for improving atlas databases. The case of the French butterfly atlas. *Biological Conservation* 108, 143–159.

Dennis, R.L.H., Shreeve, T.G. and Van Dyck, H. (2003) Towards a functional resource-based concept for habitat: a butterfly biology viewpoint. *Oikos* 102, 417–426.

Dennis, R.L.H., Hodgson, J.G., Grenyer, R., Shreeve, T.G. and Roy, D.B. (2004) Host plants and butterfly biology. Do host plant strategies drive butterfly status? *Ecological Entomology* 29, 12–26.

Dennis, R.L.H., Shreeve, T.G., Arnold, H.R. and Roy, D.B. (2005) Does diet breadth control herbivorous insect distribution size? Life history and resource outlets for specialist butterflies. *Journal of Insect Conservation* 9, 187–200.

Dennis, R.L.H., Shreeve, T.G. and Van Dyck, H. (2006a) Habitats and resources: the need for a resource-based definition to conserve butterflies. Extinction Risk Issue. *Biodiversity and Conservation* 15, 1943–1966.

Dennis, R.L.H., Shreeve, T.G., Isaac, N.J.B., Roy, D.B., Hardy, P.B. *et al.* (2006b) The effects of visual apparency on bias in butterfly recording and monitoring. *Biological Conservation* 128, 486–492.

Dennis, R.L.H., Dapporto, L., Shreeve, T.G., John, E., Coutsis, J.G. *et al.* (2008) Butterflies of European islands: the implications of the geography and ecology of rarity and endemicity for conservation. *Journal of Insect Conservation* 12, 205–236.

Dennis, R.L.H., Dapporto, L., Sparks, T.H., Williams, S.R., Greatorex-Davies, J.N. *et al.* (2010) Turnover and trends in butterfly communities on two British tidal islands: stochastic influences and deterministic factors. *Journal of Biogeography* 37, 2291–2304.

Dennis, R.L.H., Dapporto, L., Fattorini, S. and Cook, L.M. (2011a) The generalism–specialism debate: the part played by generalists in the life and death of species. *Biological Journal of the Linnean Society* 104, 725–737.

Dennis, R.L.H., Dennis, M.P., Hardy, P.B., and Kinder, P.M. (2011b) Range extension in butterflies: dispersal capacity, colonisation potential and geographical outliers in the distribution of *Aphantopus hyperantus* (Linnaeus, 1758) (Lepidoptera: Nymphalidae, Satyrinae). *Entomologist's Gazette* 62, 83–87.

Dennis, R.L.H., Hardy, P.B. and Dapporto, L. (2012a) Nestedness in island faunas: novel insights into island biogeography through butterfly community profiles of colonization ability and migration capacity. *Journal of Biogeography* 39, 1412–1426.

Dennis, R.L.H., Hodgson, J.G., Hardy, P.B. and Dapporto, L. (2012b) Strategies for size and growth in butterflies: counterintuitive trends and unique solutions to achieving maturity. *Journal of Natural History* 46, 2415–2437.

Dennis, R.L.H., Dapporto, L., Dover, J.W. and Shreeve, T.G. (2013) Corridors and barriers in biodiversity conservation: new perspectives from the resource-based habitat viewpoint for butterflies. *Biodiversity and Conservation* 22, 2709–2734.

Dennis, R.L.H., Dapporto, L. and Dover, J.W. (2014) Ten years of the resource-based habitat paradigm: the biotope-habitat issue and implications for conserving butterfly diversity. *Journal of Insect Biodiversity* 2, 1–32. Available at: http://www.insectbiodiversity.org/index.php/jib/article/view/55/pdf_20 (accessed 4 March 2018).

Dennis, R.L.H., Hardy, P.B., Middlebrook, I. and Botham, J.S. (2018) The Great Orme's Head grayling butterfly *Hipparchia semele* (Linnaeus, 1758) (Lepidoptera: Satyrinae): a unique population? *Journal of Insect Conservation* (in press).

Descimon, H. and Mallet, J. (2009) Bad species. In: Settele, J., Shreeve, T. G., Konvička, M. and Van Dyck, H. (eds) *Ecology of Butterflies in Europe.* Cambridge University Press, Cambridge, pp. 219–249.

Devoy, R.J.N. (1995) Deglaciation, earth crustal behaviour and sea-level changes in the determination of insularity: a perspective from Ireland. In: Preece, R.C. (ed.) *Island Britain: a Quaternary Perspective.* Geological Society Special Publication 96. Geological Society, London, pp. 181–208.

Dickson, R. (1992) *A Lepidopterist's Handbook.* Amateur Entomologists' Society, Feltham, UK.

Dincă, V., Lukhtanov, V.A., Talavera, G. and Vila, R. (2011a) Unexpected layers of cryptic diversity in wood white *Leptidea* butterflies. *Nature Communications* 2, article 324, 8 pp. doi: 10.1038/ncomms1329.

Dincă, V., Dapporto, L. and Vila, R. (2011b) A combined genetic-morphometric analysis unravels the complex biogeographical history of *Polyommatus icarus* and *Polyommatus celina* common blue butterflies. *Molecular Ecology* 20, 3921–3935.

Dincă, V., Wiklund, C., Lukhtanov, V.A., Kodandaramaiah, U., Norén, K. *et al.* (2013) Reproductive isolation and patterns of genetic differentiation in a cryptic butterfly species complex. *Journal of Evolutionary Biology* 26, 2095–2106.

Dingle, H. and Drake, V.A. (2007) What Is migration? *BioScience* 57, 113–121.

Dobson, K.S. (1981) *Pararge aegeria insula* Howarth (Lep., Satyridae) in 1980. *Entomologist's Gazette* 32, 113.

Dowdeswell, W.H. (1936) Lepidoptera of Cara Island. *Entomologist* 69, 49–53.

Dowdeswell, W.H. (1981) *The Life of the Meadow Brown.* Heinemann, London.

Dowdeswell, W.H. and Ford E.B. (1948) Butterfly migration noted in the Isles of Scilly in 1947. *Entomologist* 81, 141.

Dray, S. and Legendre, P. (2008) Testing the species traits–environment relationships: the fourth-corner problem revisited. *Ecology* 89, 3400–3412.

Duffey, E. (1968) Ecological studies on the large copper (Haw.) *Lycaena dispar* (Haw.) *batavus* (Obth.) at Woodwalton Fen National Nature Reserve, Huntingdonshire. *Journal of Applied Ecology* 5, 69–96.

Duffey, E. (1977) The reestablishment of the large copper butterfly *Lycaena dispar batava* Obth. on Woodwalton Fen National Nature Reserve, Huntingdonshire, England 1969–73. *Biological Conservation* 12, 143–158.

Duncan, K. (1991) A comparison of the temperature records of Edinburgh and central England. *Weather* 46, 169–173.

Dunn, T.C. (1965) Lepidoptera seen on the Isle of Colonsay 1964. *Entomologist* 98, 243–247.

Dunn, T.C. (1969) Collecting in the Aran Islands. *Entomologist's Gazette* 20, 271–278.

Dunn, T.C. and Parrack, J.D. (1986) The moths and butterflies of Northumberland and Durham. Part I: Macrolepidoptera. *The Vasculum*, Supplement No. 2. Northern Naturalists' Union, Chester-le-Street, Co. Durham, UK.

Dymond, J.N. (1974) Butterflies in 1972. *Report of Lundy Field Society* 23, 38.

Dymond, J.N. (1975) Butterflies and moths. *Report of Lundy Field Society* 24, 24–27.

Dytham, C. (2003) *Choosing and Using Statistics: a Biologist's Guide.* Blackwell Publishing, Oxford.

Eales, H.T. (1995) A revision of the status of the large heath butterfly *(Coenonympha tullia)* in Northumberland. Unpublished report for Butterfly Conservation, Wareham.

Eaton, G. (2011a) Attack and defence: observations of behaviour in the grayling *H. semele* (L.). [wing opening by females] *Atropos* 43, 50–53.

Eaton, G. (2011b) Egg laying grayling (*Hipparchia semele*} in N.W. Surrey. *Atropos* 44, 55.

Eaton, G. (2011c) Tale of three grayling [female wings open during courtship]. *Atropos* 47, 55–57.

Eggeling, W.J. (1957) A list of the butterflies and moths recorded from the Isle of May. *Scottish Naturalist* 1957, 75–83.

Eggeling, W.J. (1985) *The Isle of May. A Scottish Nature Reserve.* Larien Press, Perthshire, UK.

Ehrlich, P.R. and Davidson, S.E. (1960) Techniques for capture-recapture studies of Lepidoptera populations. *Journal of the Lepidopterists' Society* 14, 227–229.

Elias, S. (2001) Coleoptera and Trichoptera. In: Smol, J.P., Birks, H.J.B. and Last, W.M. (eds) *Tracking Environmental Change Using Lake Sediments.* Zoological Indicators 4. Kluwer, Dordrecht, pp. 67–80.

Ellenberg, H. (1988) *Vegetation Ecology of Central Europe.* Cambridge University Press, Cambridge.

Ellis, H.A. (1994) The status of the wall brown butterfly, *Lasiommata megera*, in Northumberland, 1965–91, in relation to local weather. *Transactions of the Natural History Society of Northumbria* 56, 135–152.

Ellis, P.M. (1896) *Vanessa antiopa* at Skye. *Entomologist* 29. 316.

Ellis, R.G. (1981) Increase of peacock butterfly on island of Arran. *Western Naturalist* 1981, 27.

Emmet, A.M. and Heath, J. (1989) *The Moths and Butterflies of Great Britain and Ireland, Vol. 7, Part 1, Hesperiidae – Nymphalidae.* Harley Books, Colchester, UK.

Emmet, A.M. and Heath, J. (1990) *The Moths and Butterflies of Great Britain and Ireland, Vol. 7, Part 1, Hesperiidae – Nymphalidae* (rev. edn). Harley Books, Colchester, UK.

Emmet, A.M. and Heath, J. (1991) *The Moths and Butterflies of Great Britain and Ireland, Vol. 7, Part 2, Lasiocampidae – Thyatiridae with life history chart of the British Lepidoptera.* Harley Books, Colchester, UK.

Enfjäll, K. and Leimar, O. (2005) Density-dependent dispersal in the Glanville fritillary, *Melitaea cinxia*. *Oikos* 108, 465–472.

Engler, J.O., Balkenhol, N., Filz, K.J., Habel, J.C. and Rödder, D. (2014) Comparative landscape genetics of three closely related sympatric hesperid [*sic*] butterflies with diverging ecological traits. *PLoS ONE* 9(9), e106526, pp. 1–13. doi: 10.1371/journal.pone.0106526.

Enright, C., Maclonnreactaigh, C.Ó., Hardy, P.B. and Dennis, R.L.H. (2017) Butterflies on an Irish sea lough drumlin archipelago: the influence of survey and island geography. *Entomologist's Gazette* 68, 111–129.

Evans, D.J.A., Clark, C.D. and Mitchell, W.A. (2005) The last British ice sheet: a review of the evidence utilised in the compilation of the glacial map of Britain. *Earth-Science Reviews* 70, 253–312.

Evans, W. (1912a) The painted lady butterfly at the Isle of May. *Scottish Naturalist* 1912, 261–262.

Evans, W. (1912b) Some Lepidoptera and other insects from St. Kilda. *Scottish Naturalist* 1912, 262.

Evans, W. (1918) Insects and other terrestrial invertebrates of the Bass Rock. *Scottish Naturalist* 1918, 259–265.

Evans, W. (1922) Breeding of the painted lady in Arran. *Scottish Naturalist* 1922, 27.

Evans, W. and Grimshaw, P.H. (1916) Notes on insects captured in the Island of Raasay. *Scottish Naturalist* 1916, 299–300.

Everest, J.D., Bradwell, T., Stoker, M. and Dewey, S. (2013) New age constraints for the maximum extent of the last British–Irish Ice Sheet (NW sector). *Journal of Quaternary Science* 28, 2–7.

Eversham, B.C. and Cooper, J.M. (1998) Dragonfly species-richness and temperature: National patterns and latitude trends in Britain. *Odontologica* 27, 307–316.

Fairbanks, R.G., Mortlock, R.A., Chiu, T.-C., Cao, L., Kaplan, A. *et al.* (2005) Radiocarbon calibration curve spanning 0 to 50,000 years BP based on paired Th-230/U-234/U-238 and C-14 dates on pristine corals. *Quaternary Science Reviews* 24, 1781–1796.

Fassnidge, W. (1933) Immigrant Lepidoptera [Round Island]. *Entomologist* 66, 19–20.

Fearnehough, T.D. (1937) Single brooded *Polyommatus icarus*; Sheffield. *Entomologist's Record and Journal of Variation* 49, 83–84.

Fearnehough, T.D. (1972) Butterflies in the Isle of Wight. *Entomologist's Record and Journal of Variation* 84, 57–64, 102–109.

Feest, A., van Swaay, C. and van Hinsburg, A. (2014) Nitrogen deposition and the reduction of butterfly biodiversity quality in the Netherlands. *Ecological Indicators* 39, 115–119.

Feltwell, J. (1976) Migration of *Hipparchia semele* L. *Journal of Research on the Lepidoptera* 15, 8391.

Fenton, E.W. (1948) Miscellaneous zoological notes. *Scottish Naturalist* 1948, 223–225.

Firbank, L.G., Petit, S., Smart, S., Blain, A. and Fuller, R.J. (2008) Assessing the impacts of agricultural intensification on biodiversity: a British perspective. *Philosophical Transactions of the Royal Society B: Biological Sciences* 363, 777–787.

Fischer, K., Bot, A.N.M., Brakefield, P.M. and Zwaan, B.J. (2003) Fitness consequences of temperature-mediated egg size plasticity in a butterfly. *Functional Ecology* 17, 803–810.

Fisher, J.A.D., Frank, K.T. and Leggett, W.C. (2010) Dynamic macroecology on ecological time scales. *Global Ecology and Biogeography* 19, 1–15.

Fitter, R., Fitter, A. and Blamey, M. (1978) *The Wild Flowers of Britain and Northern Europe*. Collins, London.

Fitter, R., Fitter, A. and Farrer, A. (1984) *Grasses, Sedges, Rushes and Ferns of Britain and Northern Europe*. Collins, London.

Fletcher, T.B. (1901) The naval manoeuvres of 1900, from an entomological point of view [Isle of Sheppey]. *Entomologist* 34, 71–73.

Flux, J.E.C. and Fullagar, P.J. (1992) World distribution of the Rabbit *Oryctolagus cuniculus* on islands. *Mammal Review* 22, 151–205.

Foggit, G.T. (1981) *Danaus plexippus* in the Scilly Isles in 1981. *Entomologist's Record and Journal of Variation* 93, 202.

Forbes, W.A. (1876) Notes on the entomology of Skye. *Scottish Naturalist* 1876, 262–264.

Ford, E.B. (1945) (3rd edn 1957) *Butterflies*. Collins, London.

Ford, E.B. (1964) *Ecological Genetics*. Methuen, London.

Ford, H.D. and Ford, E.B. (1930) Fluctuation in numbers and its influence on variation in *Melitaea aurinia*. *Transactions of the Entomological Society of London* 78, 345–351.

Forrest, J.E., Waterston, A.R. and Watson, E.V. (eds) (1936) The natural history of the Isle of Barra. Outer Hebrides. *Proceedings of the Royal Physical Society of Edinburgh* 22, 241–296.

Fox, R. (2010) 2009: the year of the painted lady *Vanessa cardui* (L.). *Atropos* 40, 3–14.

Fox, R. (2014) Hot stuff. *Butterfly* 115, 18–19.

Fox, R. and Dennis, R.L.H. (2010) Winter survival of *Vanessa atalanta* (Linnaeus, 1758) (Lepidoptera: Nymphalidae): a new resident butterfly for Britain and Ireland? *Entomologist's Gazette* 61, 94–103.

Fox, R., Asher, J., Brereton, T., Roy, D. and Warren, M. (2006) *The State of Butterflies in Britain and Ireland*. Pisces Publications, Newbury, UK.

Fox, R., Warren, M.S., Brereton, T.M., Roy, D.B. and Robinson, A. (2011) A new Red List of British butterflies. *Insect Conservation and Diversity* 4, 159–172.

Fox, R., Brereton, T.M., Asher, J., August, T.A., Botham, M.S. *et al.* (2015) *The State of the UK's Butterflies 2015*. Butterfly Conservation and the Centre for Ecology and Hydrology, Wareham, UK.

Frankham, R. (2005) Genetics and extinction. *Biological Conservation* 126, 131–140.

Frazer, J.F.D. (1939) *Nymphalis io* at night. *Entomologist* 72, 20.

Freeman, R.B. (1976) The obvious insects in Alderney. (1) Butterflies. *Quarterly Bulletin of the Alderney Society* 11, 7–9.

Fremlin, H.S. (1900) Collecting in the Isle of Lewis. *Entomologist* 33, 37–39.

French, R.A. (1958) Migration records. *Entomologist* 91, 101–109.

Frohawk, F.W. (1935) *Nymphalis polychloros* in the Scilly Isles: is it a migrant? *Entomologist* 68, 262.

Frost, M.P. and Madge, S.C. (1991) *Butterflies of South-east Cornwall* [includes Looe island]. Caradon Field and Natural History Club, Cornwall, UK.

Fryer, J.C.F. (1926) Lepidoptera in the Scilly Isles. *Entomologist* 59, 82.

Fuller, R.M. (1987) The changing extent and conservation interest of lowland grasslands in England and Wales: a review of grassland surveys 1930–1984. *Biological Conservation* 40, 281–300.

Funk, V.A. and Wagner, W.L. (1995) Biogeographic patterns in the Hawaiian Islands. In: Wagner, W.L. and Funk, V.A. (eds) *Hawaiian Biogeography. Evolution on a Hotspot Archipelago*. Smithsonian, Washington, DC, pp. 379–419.

Fyfe, R.M., de Beaulieu, J.-L., Binney, H., Bradshaw, R.H.W., Brewer, S. *et al.* (2009) The European pollen database: past efforts and current activities. *Vegetation History and Archaeobotany* 18, 417–424.

Gaffney, V., Fitch, S. and Smith, D. (2009) *Europe's Lost World. The Rediscovery of Doggerland*. Research Report No 160. Council for British Archaeology, York, UK.

García-Verdugo, C. and Fay, M. (2014) Ecology and evolution on oceanic islands: broadening the botanical perspective. *Botanical Journal of the Linnean Society* 174, 271–275.

Garland, S.P. (1981) *Butterflies of the Sheffield area.* Sorby Natural History Society and Sheffield Museums, Sheffield.

Garrad, L.S. (1973) Notes on the natural history of Raasay and Scalpay. *Glasgow Naturalist* 19, 7–11.

Gaston, K.J. (1994) *Rarity.* Chapman and Hall, London.

Geiger, F., Bengtsson, J., Berendse, F., Weisser, W.W., Emmerson, M. *et al.* (2010) Persistent negative effects of pesticides on biodiversity and biological control potential on European farmland. *Basic and Applied Ecology* 11, 97–105.

Geiger, H., Descimon, H. and Scholl, A. (1988) Evidence for speciation within nominal *Pontia daplidice* (Linnaeus, 1758) in southern Europe (Lepidoptera: Pieridae). *Nota Lepidopterologica* 11, 7–20.

Genty, D., Blamart, D., Ouahdi, R., Gilmour, M., Baker, A., Jouzel, J. and Van-Exter, S. (2003) Precise dating of Dansgaard–Oeschger climate oscillations in western Europe from stalagmite data. *Nature* 421, 833–837.

Gibson, J.A. (1952) The butterflies of Ailsa Craig. *Scottish Naturalist* 1952, 112–113.

Gibson, J.A. (1976) Grayling butterflies on Ailsa Craig. *Western Naturalist* 1976, 114.

Gibson, J.A. (1982a) Notes on the butterflies of the Cumbrae islands. *Western Naturalist* 1982, 5–8.

Gibson, J.A. (1982b) Butterfly notes from Horse island, Ayrshire. *Western Naturalist* 1982, 9.

Gibson, J.A. (1982c) Butterfly notes from the Sanda island group, Kintyre. *Western Naturalist* 1982, 10.

Gibson, J.A. (1982d) Butterflies flying over the sea towards Clyde islands. *Western Naturalist* 1982, 11–12.

Gibson, J.A. (1990) The butterflies of the island of Bute. *Transactions of the Buteshire Natural History Society* 23, 41–45.

Gibson, J.A. (1992) Clouded yellow butterfly at Kilchattan bay; new record for the island of Bute. *Scottish Naturalist* 1992, 127.

Gilbertson, D.D., Kent, M. and Pyatt, F.B. (1985) *Practical Ecology for Geography and Biology.* Chapman and Hall, London.

Gilburn, A.S., Bunnefeld, N., Wilson, J.M., Botham, M.S., Brereton, T.M., Fox, R. and Goulson, D. (2015) Are neonicotinoid insecticides driving declines of widespread butterflies? *PeerJ* 3, e1402. doi: 10.7717/peerj.1402.

Gilpin, M.E. and Armstrong, R.A. (1981) On the concavity of island biogeography rate functions. *Theoretical Population Biology* 20, 209–217.

Goater, B. (1969) Entomological excursions to the Shetlands, 1966 and 1968. *Entomologist's Gazette* 20, 73–82.

Goater, B. (1973) Some further observations on Shetland Lepidoptera, 1972. *Entomologist's Gazette* 24, 7–12.

Goater, B. (1974) A remarkable year [Orkney]. *Entomologist's Record and Journal of Variation* 86, 214–219, 234–239.

Godwin, H. (1975) *The History of the British Flora* (2nd edn). Cambridge University Press, Cambridge.

Gómez-Diaz, E. and Gonzáles-Solis, J. (2007) Geographical assignment of seabirds to their origins. Combining morphological, genetic and biochemical analyses. *Ecological Applications* 17, 1484–1498.

Goslin, J., Lanoë, B.V.V., Spada, G., Bradley, S., Tarasov, L., Neill, S. and Suanez, S. (2015) A new Holocene relative sea-level curve for western Brittany (France): insights on isostatic dynamics along the Atlantic coasts of north-western Europe. *Quaternary Science Reviews* 129, 341–365.

Graham, A.G.C., Stoker, M.S., Lonergan, L., Bradwell, T. and Stewart, M.A. (2011) The Pleistocene Glaciations of the North Sea basin. In: Ehlers, J. and Gibbard, P.L. (eds) *Quaternary Glaciations 1 – Extent and Chronology. A Closer Look* (2nd edn). Elsevier, London, pp. 261–278.

Grant, M.J., Hughes, P.D.M. and Barber, K.E. (2014) Climatic influence upon early to mid-Holocene fire regimes within temperate woodlands: a multi-proxy reconstruction from the New Forest, southern England. *Journal of Quaternary Science* 29, 175–188.

Grant, P.R. and Grant, B.R. (2010) Sympatric speciation, immigration, and hybridization in island birds. In: Losos, J.B. and Ricklefs, R.E. (eds.) *The Theory of Island Biogeography Revisited.* Princeton University Press, Princeton, New Jersey, pp. 326–357.

Graves, P.P. (1930) The British and Irish *Maniola jurtina* L. (Lep., Satyridae). *Entomologist* 68, 49–54, 75–81.

Green, R.E., Krause, J., Briggs, A.W., Maricic, T., Stenzel, U. *et al.* (2010) A draft sequence of the Neandertal genome. *Science* 328, 710–722.

Gregson, C.S. (1885) Notes on the Lepidoptera taken by E.R. Curzon Esq at the island of Hoy, one of the Orkney islands, during the summer of 1885. *Young Naturalist* 6, 272–278.

Griffiths, G.C. and Bartlett, C. (1914) The natural history of Steep Holm entomology. *Proceedings of the Bristol Naturalists' Society* 4, 143–149.

Grill, A., Gkiokia, E. and Alvarez, N. (2006) Evolutionary history and patterns of differentiation among European *Maniola* butterflies (Lepidoptera: Satyrinae). *European Journal of Entomology* 103, 613–618.

Grime, J.P., Hodgson, J.G. and Hunt, R. (1988) *Comparative Plant Ecology. A Functional Approach to Common British Species.* Unwin-Hyman, London.

Grimshaw, P.H. (1906) Insects from Fair Isle. *Annals of Scottish Natural History* 1906, 118.

Grimshaw, P.H. (1908) A contribution to the insect fauna of the Isle of May. *Annals of Scottish Natural History* 1908, 89–90.

Grimshaw, P.H. (1920) Notes on the insect fauna of South Uist. *Scottish Naturalist* 1920, 85–89.

Grimshaw, P.H. (1933) Scottish insect immigration records. *Scottish Naturalist* 1933, 173–181.

Grimwood, K.W. (1965) Fritillary in a garden. *Entomologist's Record and Journal of Variation* 77, 240.

Groombridge, B. (ed.) (1992) *Global Biodiversity: Status of the Earth's Living Resources.* Chapman and Hall, London.

Gumbel, E.J. (1958) *Statistics of Extremes.* Columbia University Press, New York (reprinted 2004 by Dover Publications, New York).

Gupta, S., Collier, J.S., Palmer-Felgate, A. and Potter, G. (2007) Catastrophic flooding origin of shelf valley systems in the English Channel. *Nature* 448, 342–346.

Gupta, S., Collier, J.S., Garcia-Moreno, D., Oggioni, F., Trentesaux, A. *et al.* (2017) Two stage opening of the Dover Strait and the origin of island Britain. *Nature Communications* 8,15101. doi: 10.1038/ncomms15101.

Habel, J.C., Ulrich, W. and Assmann, T. (2013) Allele elimination recalculated: nested subset analyses for molecular biogeographical data. *Journal of Biogeography* 40, 769–777.

Haggart, G.D. (1933) Notes on Arran Lepidoptera. *Entomologist* 66, 28–30.

Hale, J.W. and Hicks, M.E. (1995) Fall of the painted lady. *Entomologist's Record and Journal of Variation* 107, 188–191.

Hall, D.W. (1950) A note on the distribution of the peacock buttorfly *Nymphalis io* in Scotland. *Scottish Naturalist* 1950, 109–110.

Hall, J., Dore, T., Smith, R., Evans, C., Rowe, E. *et al.* (2016) Defra Contract AQ0826. *Modelling and mapping of exceedance of critical loads and critical levels for acidification and eutrophication in the UK 2013–2016 Final Report: 25 July 2016*. Available at: https://uk-air.defra.gov.uk/assets/documents/reports/cat13/1611011543_AQ0826_FinalReport_25July2016.pdf (accessed 28 March 2018).

Hallatschek, O. and Nelson, D.R. (2008) Gene surfing in expanding populations. *Theoretical Population Biology* 73, 158–170.

Hallett, H.M. (1930) A contribution to the entomological fauna of Skomer island. *Transactions of the Cardiff Naturalists' Society* 1928, 72–76.

Hanbury, F.J.H. (1895) Notes on the Lepidoptera observed during a short botanical tour in west Sutherland, the Orkneys and Shetlands. *Entomologist's Monthly Magazine* 31, 1–12.

Hancock, E.G. (1978) Additional insects recorded for Puffin Island. *Proceedings of the Birmingham Natural History Society* 23, 239–241.

Hancock, E.G. (1995) Terrestrial invertebrates from Ailsa Craig. Glasgow Museum and Art Gallery (unpublished list).

Hancock, G.L.R. (1923) Contributions towards a list of the insect fauna of the South Ebudes. The Lepidoptera. *Scottish Naturalist* 1923, 125–132.

Hanski, I. (1994) Patch-occupancy dynamics in fragmented landscapes. *Trends in Ecology and Evolution* 9, 131–135.

Hanski, I.A. and Gilpin, M.E. (1997) *Metapopulation Biology. Ecology, Genetics, and Evolution.* Academic Press, London.

Hanski, I. and Thomas, C.D. (1994) Metapopulation dynamics and conservation: a spatially explicit model applied to butterflies. *Biological Conservation* 68, 167–180.

Hanski, I., Kuusaari, M. and Nieminen, M. (1994) Metapopulation structure and migration in the butterfly *Melitaea cinxia*. *Ecology* 75, 747–762.

Harding, J. and Jacob, M. (2013) Addition of Small Skipper butterfly (*Thymelicus sylvestris*) to the Irish List and notes on the Essex Skipper (*Thymelicus lineola*) (Lepidoptera: Hesperiidae). *Irish Naturalists' Journal* 32, 142–144.

Hardy, D.E. (1956) The Lepidoptera of Fair Isle. *Entomologist* 89, 261–269.

Hardy, P.B. and Dennis, R.L.H. (1999) The impact of urban development on butterflies within a city region. *Biodiversity and Conservation* 8, 1261–1279.

Hardy, P.B. and Dennis, R.L.H. (2010) A butterfly exploiting the matrix: *Pieris napi* (Linnaeus, 1758) (Lepidopetra: Pieridae) ovipositing amongst mown grass in a city park. *Entomologist's Gazette* 61, 155–158.

Hardy, P.B., Hind, S.H. and Dennis, R.L.H. (1993) Range extension and distribution-infilling among selected butterfly species in north-west England; evidence for inter-habitat movements. *Entomologist's Gazette* 44, 247–255.

Hardy, P.B., Sparks, T.H. and Dennis, R.L.H. (2014) The impact of climatic change on butterfly geography: does climatic change produce coincident trends in populations, distributions and ranges? *Biodiversity and Conservation* 23, 855–876.

Hardy, P.B., Jeffcoate, G.E. and Dennis, R.L.H. (2017) Butterflies (Papilionoidea) on the Isle of Man: losses, gains and predictions. *Entomologist's Gazette* 68, 239–254.

Hare, E.J. (1963) Shetland Macro-Lepidoptera. *Entomologist's Record and Journal of Variation* 75, 238.

Harper, M.W. (1974) *Polyommatus icarus* in Shetland. *Entomologist's Record and Journal of Variation* 86, 120.

Harrison, R.K., Stone, P., Cameron, I.B., Elliot, R.W. and Harding, R.R. (1987) *Geology, Petrology and Geochemistry of Ailsa Craig, Ayrshire.* Report of the British Geological Survey, CG16/09. HMSO, London.

Harrison, S. (1991) Local extinction in a metapopulation context: an empirical evaluation. *Biological Journal of the Linnean Society* 42, 73–88.

Harrison, S., Knight, J. and Rowan, A.V. (2015) The southernmost Quaternary niche glacier system in Great Britain. *Journal of Quaternary Science* 30, 325–334.
Harvey, P. (1990) Butterflies on Fair Isle 1978–1989. *Report of Fair Isle Bird Observatory* 42, 55–57.
Harvey, P., Riddiford, N. and Riddiford, E. (1992) Butterflies on Fair Isle 1990–92. *Report of Fair Isle Bird Observatory* 44, 73–74.
Hausdorf, H. (2011) Progress toward a general species concept. *Evolution* 65, 923–931.
Hazen, R.M., Grew, E.S., Origlieri, M.J. and Downs, R.T. (2017) On the mineralogy of the 'Anthropocene Epoch'. *American Mineralogy* 102, 595–601.
Heal, H. (1965) The Wood White and the railways. *Irish Naturalists' Journal* 15, 8–13.
Heaney, L.R. (2004) Conservation biogeography in oceanic archipelagos. In: Lomolino, M.V. and Heaney, L.R. (eds) *Frontiers of Biogeography: New Directions in the Geography of Nature*. Cambridge University Press, Cambridge, pp. 345–360.
Heath, J. (1983) Is this the earliest record of *Lycaena dispar* (Haworth) (Lepidoptera: Lycaenidae)? *Entomologist's Gazette* 34, 228.
Heath, J., Pollard, E. and Thomas, J.A. (1984) *Atlas of Butterflies in Britain and Ireland*. Viking, London.
Heath, P.M. (1975) A visit to Skomer. *Entomologist's Record and Journal of Variation* 87, 67–70.
Heath, P.M. (1976) Skomer Lepidoptera. *Nature in Wales* 15, 8–14.
Heckford, R.J. (1978) Scilly records for 1977. *Entomologist's Record and Journal of Variation* 90, 157.
Heckford, R.J. (2011) One day old larva of silver-washed fritillary *Aygynnis paphia*. *Atropos* 42, 87.
Hedges, A.V. (1947) A list of Manx Lepidoptera. *Entomologist* 80, 44–46, 62–66, 89–94.
Hedges, J. (1980) Notes on the Lepidoptera in the Isle of Man in 1980. *Entomologist's Record and Journal of Variation* 92, 303.
Hedges, J. (1981) *Danaus plexippus* in the Isle of Man. *Entomologist's Record and Journal of Variation* 93, 202.
Heikkilä, M., Kaila, L., Mutanen, M., Peña, C. and Wahlberg, N. (2012) Cretaceous origin and repeated tertiary diversification of the redefined butterflies. *Proceedings of the Royal Society B: Biological Sciences* 279, 1093–1099.
Henderson, P.A. (2003) *Practical Methods in Ecology*. Blackwell Publishing, Oxford.
Hendy, I.L., Kennett, J.P., Roark, E.B. and Ingram, B.L. (2002) Apparent synchroneity of sub-millennial scale climate events between Greenland and Santa Barbara basin. *Quaternary Science Reviews* 21, 1167–1184.
Hennig, W. (1966) *Phylogenetic Systematic*. University of Illinois Press, Urbana, Illinois.
Henrys, P., Smart, S., Barrett, G., Carter, H., Collier, R. and Hughes, P. (2016) *Defra Contract AQ0826: Modelling and Mapping of Exceedance of Critical Loads and Critical Levels for Acidification and Eutrophication in the UK 2013–2016. Final Report: 25th July 2016*. DEFRA, London.
Herbert, C. (1993) *Butterflies of the London Borough of Barnet: A Provisional Atlas*. Herts and Middlesex Wildlife Trust, East Barnet, UK.
Heron, A.M.R. (1956) Lepidoptera noted near Tenby, Pembrokeshire [Grassholm]. *Nature in Wales* 2, 350–351.
Heslop Harrison, J.W. (1937a) Rhopalocera on the Isle of Scalpay, with an account of the occurrence of *Nymphalis io* L. on Raasay. *Entomologist* 70, 1–4.
Heslop Harrison, J.W. (1937b) A contribution to our knowledge of the Lepidoptera of the islands of Eigg, Canna and Sanday. *Entomologist's Record and Journal of Variation* 49, 29–31.
Heslop Harrison, J.W. (1937c) The natural history of the Isle of Raasay and of the adjacent islands of South Rona, Scalpay, Fladday and Longay. *Proceedings of the University of Durham Philosophical Society* 9, 246–251.

Heslop Harrison, J.W. (1938a) The Rhopalocera of the islands of Coll, Canna, Sanday, Rhum, Eigg, Soay and Pabbay (Inner Hebrides) and of Barra, Mingulay and Berneray (Outer Hebrides). *Entomologist* 71, 18–20.

Heslop Harrison, J.W. (1938b) *Argynnis selene*, a butterfly new to the Isle of Rhum. *Entomologist* 71, 213.

Heslop Harrison, J.W. (1938c) The Rhopalocera of the islands of Rhum, Eigg, Muck, Eilean nan Each and Heiskeir (Inner Hebrides) and of Harris, North Uist, South Uist, Eriskay, Taransay and the Monach islands (Outer Hebrides). *Entomologist* 71, 265–267.

Heslop Harrison, J.W. (1939) *Argynnis aglaia* and *Pieris napi* flying out to sea. *Entomologist* 72, 59.

Heslop Harrison, J.W. (1940a) Lepidoptera of the Isle of Handa. *Entomologist* 73, 44–45.

Heslop Harrison, J.W. (1940b) The common blue in the Isles of Pabbay, Fladday and Fiaray (Barra Isles). *Entomologist* 73, 50.

Heslop Harrison, J.W. (1940c) *Pieris napi* flying in dusk on the Isle of Coll. *Entomologist* 73, 91.

Heslop Harrison, J.W. (1940d) More Hebridean days. 1: The Isle of Muldoanich and the Uidh peninsula of Vatersay [includes Uinessan]. *Entomologist* 73, 101–103.

Heslop Harrison, J.W. (1940e) *Aglais urticae* much earlier on the Isle of Coll than on the Isle of Tiree. *Entomologist* 73, 115.

Heslop Harrison, J.W. (1940f) *Eumenis semele* L. and *Maniola jurtina* L. on the Isles of Coll and Gunna, Inner Hebrides. *Entomologist's Record and Journal of Variation* 52, 20.

Heslop Harrison, J.W. (1940g) *Heodes phlaeas* in the Isle of Colonsay. *Entomologist's Record and Journal of Variation* 52, 129.

Heslop Harrison, J.W. (1940h) The range of *Argynnis aglaia* in the Outer Hebrides. *Entomologist's Monthly Magazine* 76, 274.

Heslop Harrison, J.W. (1941a) More Hebridean days. II. The Isle of Benbecula. *Entomologist* 74, 1–5.

Heslop Harrison, J.W. (1941b) Immigrant Lepidoptera in the Inner and Outer Hebrides in 1940. *Entomologist* 74, 19.

Heslop Harrison, J.W. (1941c) More Hebridean days. III. The Isles of Tiree and Gunna. *Entomologist* 74, 97–100.

Heslop Harrison, J.W. (1941d) Lepidoptera in the Isle of Islay. *Entomologist* 74, 238.

Heslop Harrison, J.W. (1941e) *Colias croceus* F. in the Isle of Harris. *Entomologist* 74, 278.

Heslop Harrison, J.W. (1941f) The distribution and habits of *Callophrys rubi* L. in the Isle of Rhum. *Entomologist's Record and Journal of Variation* 53, 86–88.

Heslop Harrison, J.W. (1942a) *Vanessa cardui* in the Orkney Isles. *Entomologist* 75, 16.

Heslop Harrison, J.W. (1942b) The irregularity of broods of *Pieris brassicae* L. on the Isles of Lewis, Harris and Scarp in 1941. *Entomologist* 75, 33.

Heslop Harrison, J.W. (1942c) Lepidoptera on the Isle of Scotasay. *Entomologist* 75, 36.

Heslop Harrison, J.W. (1942d) Butterflies in the Isles of Rhum and South Uist during May 1942. *Entomologist* 75, 185.

Heslop Harrison, J.W. (1942e) *Argynnis aglaia* race *scotica* Watkins in the Western Isles of Scotland. *Entomologist's Record and Journal of Variation* 54, 44–45.

Heslop Harrison, J.W. (1942f) Variation in Outer Hebridean larvae of *Aglais urticae* L. and the fate of Rhum pupae of *Pieris napi* L. reared in 1941. *Entomologist's Record and Journal of Variation* 54, 107.

Heslop Harrison, J.W. (1942g) The range of *Argynnis selene* in the Isle of Rhum. *Entomologist's Monthly Magazine* 78, 280.

Heslop Harrison, J.W. (1943a) Lepidoptera on the Isle of Baleshare. *Entomologist* 76, 58.

Heslop Harrison, J.W. (1943b) August captures of *Coenonympha tullia* Müller var. *scotica* Stdgr. in the Outer Hebrides. *Entomologist* 76, 106.

Heslop Harrison, J.W. (1943c) Other Lepidoptera noted in North Uist in 1942. *Entomologist's Record and Journal of Variation* 55, 26.
Heslop Harrison, J.W. (1943d) The range of *Euphydryas aurinia* (Lep: Nymphalidae) in the Hebrides and some possible deductions therefrom. *Entomologist's Record and Journal of Variation* 55, 27.
Heslop Harrison, J.W. (1943e) Immigrant Lepidoptera in the Inner and Outer Hebrides. *Entomologist's Record and Journal of Variation* 55, 108.
Heslop Harrison, J.W. (1944) An apparently new foodplant for *Vanessa cardui*. *Entomologist's Record and Journal of Variation* 56, 25.
Heslop Harrison, J.W. (1945a) Remarks on certain Lepidoptera from the western isles of Scotland. *Entomologist* 78, 18–21.
Heslop Harrison, J.W. (1945b) The peacock butterfly (*Nymphalis io* L.) reaches the Isle of Rhum. *Entomologist* 78, 46.
Heslop Harrison, J.W. (1945c) Lepidoptera new to, or rare in, the Outer Hebrides. *Entomologist's Record and Journal of Variation* 57, 1–3.
Heslop Harrison, J.W. (1945d) The foodplants of *Coenonympha tullia* Müller. *Entomologist's Record and Journal of Variation* 57, 125–126.
Heslop Harrison, J.W. (1946a) *Maniola jurtina* goes for a sail. *Entomologist* 79, 69–70.
Heslop Harrison, J.W. (1946b) The drinking of *Pieris napi* and the abundance of *Pieris brassicae* larvae on Rhum. *Entomologist* 79, 70.
Heslop Harrison, J.W. (1946c) Noteworthy Lepidoptera from the Isle of Rhum, with some notes on the insects captured on the adjacent islands. *Entomologist* 79, 147–151.
Heslop Harrison, J.W. (1946d) The geographical distribution of certain Hebridean insects and deductions to be made from it. *Entomologist's Record and Journal of Variation* 58, 18–21.
Heslop Harrison, J.W. (1946e) The Lepidoptera of the Isles of Coll, Tiree and Gunna, with some remarks on the biogeography of the islands. *Entomologist's Record and Journal of Variation* 58, 57–61.
Heslop Harrison, J.W. (1946f) Immigrant Lepidoptera in the Inner and Outer Hebrides in 1946. *Entomologist's Record and Journal of Variation* 58, 143.
Heslop Harrison, J.W. (1946g) The range of *Pieris napi* in the Isle of South Uist. *Entomologist's Record and Journal of Variation* 58, 144.
Heslop Harrison, J.W. (1947a) Early spring insects from the Isle of Rhum, with some remarks on the woodland fauna of the island. *Entomologist* 80, 1–4.
Heslop Harrison, J.W. (1947b) The status of the peacock butterfly in the small isles parish of Invemess-shire. *Entomologist* 80, 13.
Heslop Harrison, J.W. (1947c) *Colias hyale* L. and *C. croceus* Fourc. on the Isle of Coll, Inner Hebrides. *Entomologist's Record and Journal of Variation* 59, 139.
Heslop Harrison, J.W. (1947d) *Pieris napi* L. in the Inner and Outer Hebrides. *Entomologist's Record and Journal of Variation* 59, 139.
Heslop Harrison, J.W. (1947e) *Polyommatus icarus* race *clara* Tutt on the Isle of Ronay, Outer Hebrides. *Entomologist's Record and Journal of Variation* 59, 140.
Heslop Harrison, J.W. (1947f) Butterflies on the Isle of Tiree. *Entomologist's Record and Journal of Variation* 59, 140.
Heslop Harrison, J.W. (1947g) The Pleistocene races of certain British insects and distributional overlapping. *Entomologist's Record and Journal of Variation* 59, 141–145.
Heslop Harrison, J.W. (1948a) Lepidoptera in the Inner and Outer Hebrides during the year 1947. *Entomologist* 81, 1–6.
Heslop Harrison, J.W. (1948b) Inbreeeding in Hebridean Lepidoptera populations. *Entomologist's Record and Journal of Variation* 60, 46–50.
Heslop Harrison, J.W. (1948c) A new race of *Coenonympha pamphilus* for the Hebrides. *Entomologist's Record and Journal of Variation* 60, 111–112.

Heslop Harrison, J.W. (1949a) A contribution to our knowledge of the Lepidoptera of the Isles of Lewis and Harris. *Entomologist* 82, 16–20.

Heslop Harrison, J.W. (1949b) Further observations on the Lepidoptera of the Scottish Western Isles. *Entomologist* 82, 265–268.

Heslop Harrison, J.W. (1949c) The foodplants of *Callophrys rubi* in the Inner Hebrides. *Entomologist's Record and Journal of Variation* 61, 6.

Heslop Harrison, J.W. (1949d) Rhopalocera in the western Scottish Isles in 1948, with an account of two new forms of *Pararge aegeria* L. (Lep., Satyridae). *Entomologist's Monthly Magazine* 85, 25–28.

Heslop Harrison, J.W. (1950a) Immigrant Lepidoptera in the Scottish Western Isles in 1949. *Entomologist* 83, 70–71.

Heslop Harrison, J.W. (1950b) Lepidoptera in the Outer Hebrides. *Entomologist* 83, 241–245.

Heslop Harrison, J.W. (1950c) Observations on the ranges, habitats and variation of the Rhopalocera of the Outer Hebrides. *Entomologist's Monthly Magazine* 86, 65–70.

Heslop Harrison, J.W. (1950d) A dozen years' biogeography researches in the Inner and Outer Hebrides. *Proceedings of the University of Durham Philosophical Society* 10, 516–524.

Heslop Harrison, J.W. (1953) Lepidoptera in the Isles of Lewis and Harris in 1952. *Entomologist* 86, 53–55.

Heslop Harrison, J.W. (1954a) Lepidoptera from the Outer Hebrides in 1953. *Entomologist* 87, 83–86.

Heslop Harrison, J.W. (1954b) The return of *Pararge megera* L. to County Durham. *Entomologist* 87, 264.

Heslop Harrison, J.W. (1955a) Lepidoptera noted in the Outer Hebrides in 1954. *Entomologist* 88, 51–53.

Heslop Harrison, J.W. (1955b) Lepidoptera of the lesser Skye Isles. *Entomologist's Record and Journal of Variation* 67, 141–147, 169–177.

Heslop Harrison, J.W. (1956a) Immigrant Lepidoptera in the Outer Hebrides in 1955. *Entomologist* 89, 87.

Heslop Harrison, J.W. (1956b) Further observations on the Lepidoptera of the Outer Hebrides. *Entomologist* 89, 285–290.

Heslop Harrison, J.W. (1956c) Lepidoptera in the Isles of Lewis and Harris in 1955. *Entomologist's Record and Journal of Variation* 68, 40–43.

Heslop Harrison, J.W. (1956d) On field studies of the distribution of the plants and animals of the Scottish Western Isles. *Proceedings of the Linnean Society of London* 167, 103.

Heslop Harrison, J.W. (1957) Lepidoptera in the Isle of Longay. *Entomologist's Record and Journal of Variation* 69, 49.

Heslop Harrison, J.W. (1958a) *Argynnis aglaia* L. in Co. Durham. *Entomologist* 91, 32.

Heslop Harrison, J.W. (1958b) Lepidoptera in the Scottish Western Isles. *Entomologist* 91, 79–86.

Heslop Harrison, J.W. and Morton, J.K. (1952) Lepidoptera in the Isles of Raasay, Rhum, Lewis and Harris in 1951. *Entomologist* 85, 6–13.

Hesselbarth, G., van Oorschot, H. and Wagener, S. (1995) *Die Tagfalter der Türkei* 1–3. Selbstverlag Sigbert Wagener, Bocholt, Germany.

Hickin, N. (1992) *The butterflies of Ireland: a field guide*. Roberts Rinehart Publishers, Schull, West Cork, Ireland.

Higgs, G.E. (1986) Three unusual butterflies in Alderney. *Bulletin of the Amateur Entomologists' Society* 45, 155.

Hill, J.K., Thomas, C.D. and Huntley, B. (1999) Climate and habitat availability determine 20th century changes in a butterfly's range margin. *Proceedings of the Royal Society of London B: Biological Sciences* 266, 1197–1206.

Hill, J.K., Collingham, Y.C., Thomas, C.D., Blakeley, D.S., Fox, R., Moss, D. and Huntley, B. (2001) Impacts of landscape structure on butterfly range expansion. *Ecology Letters* 4, 313–321.

Hill, J.K., Thomas, C.D., Fox, R., Telfer, M.G., Willis, S.G., Asher, J. and Huntley, B. (2002) Responses of butterflies to twentieth century climate warming: implications for future ranges. *Proceedings of the Royal Society of London B: Biological Sciences* 269, 2163–2171.

Hill, M.O., Mountford, J.O., Roy, D.B. and Bunce, R.G.H. (1999) *Ellenberg's Indicator Values for British Plants.* ECOFACT Vol. 2 Technical Annex. Centre for Ecology and Hydrology, Huntingdon, UK.

Hirons, M.J.D. (1994) The flora of the Farne Islands. *Transactions of the Natural History Society of Northumbria* 57, 69–114.

Hockin, D.C. (1981) The environmental determinants of the insular butterfly faunas of the British Isles. *Biological Journal of the Linnean Society* 16, 63–70.

Hodgson, J.G., Tallowin, J., Dennis, R.L.H., Thompson, K., Poschlod, P. *et al.* (2014) Changing leaf nitrogen and canopy height quantify processes leading to plant and butterfly diversity loss in agricultural landscapes. *Functional Ecology* 28, 1284–1291.

Hodgson, S.B. (1935) *Euphydryas aurinia* in west Cornwall. *Entomologist* 68, 162–163.

Hodgson, S.B. (1949a) *Coenonympha tullia* at flowers [Orkney]. *Entomologist* 82, 40.

Hodgson, S.B. (1949b) Migrants in Orkney. *Entomologist* 82, 238.

Hodgson, S.B. (1951a) Unusual behaviour of *Polyommatus icarus*. *Entomologist* 84, 22.

Hodgson, S.B. (1951b) *Pieris napi* impaled on rushes. *Entomologist* 84, 177.

Hodgson, S.B. (1952a) Late flying of *Coenonympha tullia* [Orkney], *Entomologist* 85, 4.

Hodgson, S.B. (1952b) *Pieris napi* in Orkney. *Entomologist* 85, 95.

Hogan, A.R. and Haliday, A.H. (1855) Notes on various insects captured or observed in the neighbourhood of Dingle, Co. Kerry in July 1854 [Great Blasket and Beginish]. *Natural History Review* 2, 50–55.

Hogben, P. (2016) Observations of territorial holding in small copper *Lycaena phlaeas*. *Atropos* 56, 68.

Hohzaki, R. (2009) A cooperative game in search theory. *Naval Research Logistics* 56, 264–278.

Holloway, J.D. (1980) A mass-movement of *Quercusia quercus* (L.) (Lep: Lycaenidae) in 1976. *Entomologist's Gazette* 31, 150.

Horton, P. (1977) Local migrations of Lepidoptera from Salisbury Plain in 1976. *Entomologist's Gazette* 28, 281–283.

Howard, G. (1975) Lepidoptera on Hoy, Orkney. *Entomologist's Record and Journal of Variation* 87, 107–109.

Howarth, T.G. (1971a) Descriptions of a new subspecies of *Pararge aegeria* (L.) (Lep., Satyridae) and an aberration of *Cupido minimus* (F.) (Lep., Lycaenidae). *Entomologist's Gazette* 22, 112–118.

Howarth, T.G. (1971b) The status of Irish *Hipparchia semele* L. (Lep., Satyridae) with descriptions of a new subspecies and aberrations. *Entomologist's Gazette* 22, 123–129.

Howe, P.D., Bryant, S.R. and Shreeve, T.G. (2007) Predicting body temperature and activity of adult *Polyommatus icarus* using neural network models under current and projected climate scenarios. *Oecologia* 153, 857–869.

Høye, T.T., Skildsen, A., Hansen, R.R., Bowden, J.J., Schmidt, N.M. and Kissling, W.D. (2014) Phenology of high-arctic butterflies and their floral resources: species-specific responses to climate change. *Current Zoology* 60, 243–251.

Hubbard, C.E. (1968) *Grasses.* Penguin Books, Harmondsworth, UK.

Huggins, H.C. (1928) Collecting in the Aran islands. *Entomologist* 61, 13–14.

Huggins, H.C. (1956) The Burren subspecies of *Erynnis tages* L. (Lep., Hesperiidae). *Entomologist* 89, 241–242.

Huggins, H.C. (1958) Immigrants in Tresco, Isles of Scilly, 1958. *Entomologist* 91, 251–252.

Huggins, H.C. (1959) Lepidoptera in Tresco in June. *Entomologist* 92, 14–17, 246–249.

Huggins, H.C. (1972) *Euphydryas aurinia* Rott. in the Isle of Wight. *Entomologist's Record and Journal of Variation* 84, 195–196.

Hughes, H.W. (1935) An Entomologist's holiday in Aberdeen and the Shetland Isle of Unst. *Report of Cheshire Entomological Society* 1934–36, 35–44.

Hughes, P.D., Braithwaite, R., Fenton, C.R. and Schnabel, C. (2012) Two Younger Dryas glacier phases in the English Lake District: geomorphological evidence and preliminary 10Be exposure ages. *North West Geography* 12, 10–19.

Hull, J.E. and Heslop Harrison, J.W. (1937–38) The natural history of the island of Raasay. *Scottish Naturalist* 1937: 67–71, 107–113, 135–144, 145–150, 169–172; 1938: 60–64.

Hulme, D.C. (1998) Lepidoptera on Tiree. *Entomologist's Record and Journal of Variation* 110, 247.

Hungate, B.A., Kearns, D.N., Ogle, K., Caron, M., Marks, J.C. and Rogg, H.W. (2016) Hydrogen Isotopes as a sentinel of biological invasion by the Japanese Beetle, *Popillia japonica* (Newman). *PLoS ONE* 11, e0149599. doi: 10.1371/journal.pone.0149599.

Huntley, B. and Birks, H.J.B. (1983) *An Atlas of Past and Present Pollen Maps for Europe: 0–13000 Years Ago*. Cambridge University Press, Cambridge.

Hutchinson, G.E. (1957) Concluding remarks. *Cold Spring Harbor Symposia on Quantitative Biology* 22, 415–427.

Hutton, J.T. (1976) Chloride in rainwater in relation to distance from the ocean. *Search* 7, 207–208.

Innes, J.B., Chiverrell, R.C., Blackford, J.J., Davey, P.J., Gonzalez, S., Rutherford, M.M. and Tomlinson, P.R. (2004) Earliest Holocene vegetation history and island biogeography of the Isle of Man, British Isles. *Journal of Biogeography* 31, 761–772.

Institute of Biology (1992) *Safety in Biological Fieldwork Guidance Notes for Codes of Practice*, 3rd edn. Institute of Biology, London.

Isaac, N.J.B., Cruickshanks, K.L., Weddle, A.M., Rowcliffe, J.M., Brereton, T.M. *et al.* (2011a) Distance sampling and the challenge of monitoring butterfly populations. *Methods in Ecology and Evolution* 2, 585–594.

Isaac, N.J.B., Girardello, M., Brereton, T.M. and Roy, D.B. (2011b) Butterfly abundance in a warming climate: patterns in space and time are not congruent. *Journal of Insect Conservation* 15, 233–240.

Iversen, J. (1958) The bearing of glacial and interglacial epochs on the formation and extinction of plant taxa. *Uppsala Universiteit Arssk* 6, 210–215.

Ivosevic, B., Han, Y.-G. and Kwon, O. (2017) Monitoring butterflies with an unmanned aerial vehicle: current possibilities and future potentials. *Journal of Ecology and Environment* 41, 12. doi: 10.1186/s41610-017-0028-1.

Jacobson, S.L., Bliss-Ketchum, L.L., de Rivera, C.E. and Smith, W.P. (2016) A behavior-based framework for assessing barrier effects to wildlife from vehicle traffic volume. *Ecosphere* 7(4), e01345. doi: 10.1002/ecs2.1345. 15 pages.

Jeffcoate, G. (1994) An odd couple [mating between dingy skipper and burnet companion]. *Butterfly Conservation News* 58, 15.

Jefferson, R.G., Smith, S.L.N. and MacKintosh, E.J. (2014) Lowland grasslands. Chapter 3 in: *Guidelines for the Selection of Biological SSSIs. Part 2: Detailed Guidelines for Habitats and Species Groups*. Joint Nature Conservation Committee, Peterborough.

JNCC (2010) *Handbook for Phase 1 Habitat Survey – a Technique for Environmental Audit. Field Manual*. (First published by NCC, 1990; reprinted with minor additions by JNCC 2003, 2007). Joint Nature Conservation Committee, Peterborough, UK.

Johnson, C.G. (1969) *Migration and Dispersal of Insects by Flight*. Methuen, London, UK.

Johnson, E.E. (1955) Difference between adjacent colonies of *Euphydryas aurinia* Rott. *Entomologist's Record and Journal of Variation* 67, 132.

Jones, C.G. (1968) Early June in Lundy: butterflies and moths recorded. *Report of Lundy Field Society* 19, 30–33.

Jones, L., Stevens, C., Rowe, E.C., Payne, R., Caporn, S.J.M. *et al.* (2017) Can on-site management mitigate nitrogen deposition impacts in non-wooded habitats? *Biological Conservation* 212, 464–475.

Jønsson, K.A., Irestedt, M., Christidis, L., Clegg, S.M., Holt, B.G. and Fjeldså, J. (2014) Evidence of taxon cycles in an Indo-Pacific passerine bird radiation (Aves: Pachycephala). *Proceedings of the Royal Society of London B: Biological Sciences* 281, 20131727. doi: 10.1098/rspb.2013.1727.

Joyce, D.A. and Pullin, A.S. (2001) Phylogeography of the Marsh Fritillary *Euphydryas aurinia* (Lepidoptera: Nymphalidae) in the UK. *Biological Journal of the Linnean Society* 72, 129–141.

Joyce, D.A., Dennis, R.L.H., Bryant, S.R., Shreeve, T.G., Reday, J. and Pullin, A.S. (2009) Do taxonomic divisions reflect genetic differentiation? A comparison of morphological and genetic data in *Coenonympha tullia* (Müller), Satyrinae. *Biological Journal of the Linnean Society* 97, 314–327.

Kane, W.F. de V. (1907) Lepidoptera of Lambay. *Irish Naturalist* 16, 46.

Kane, W.F. de V. (1912) Clare island survey. *Proceedings of the Royal Irish Academy* 26, 1–10.

Kaye, W.J. (1922) Lepidoptera of the smaller Channel Islands. *Entomologist's Record and Journal of Variation* 34, 175–176.

Kemp, J.B. (2014) Outer Hebrides (Coastal reports). *Atropos* 51, 82–83.

Kemp, J.B. (2015) Outer Hebrides (Coastal Reports). *Atropos* [2015 Migration Review], 62–64.

Kemp, R., Hardy, P.B., Roy, D. and Dennis, R.L.H. (2008) The relative exploitation of annuals as larval host plants by phytophagous Lepidoptera. *Journal of Natural History* 42, 1079–1093.

Kennar, R.B. (1912) Small tortoiseshell butterfly hibernating in Shetland. *Scottish Naturalist* 1912, 92–93.

Kershaw, K.A. (1964) *Quantitative and Dynamic Ecology*. Edward Arnold, London.

Kertész, K., Piszter, G., Horváth, Z.E., Bálint, Z. and Biró, L.P. (2017) Changes in structural and pigmentary colours in response to cold stress in *Polyommatus icarus* butterflies. *Scientific Reports* 7, 1118. 12 pages. doi: 10.1038/s41598-017—01273-7.

Kett, S. (1993) *Satyrium w-album* (Knoch) (Lep., Lycaenidae) in a moth trap. *Entomologist's Record and Journal of Variation* 105, 282–283.

Kevan, D.K.McE. (1941) The insect fauna of the Isle of Eigg. *Entomologist* 74, 247–254.

Kier, G., Kreft, H., Lee, T.M., Jetz, W., Ibisch, P.L. *et al.* (2009) A global assessment of endemism and species richness across island and mainland regions. *Proceedings of the National Academy of Sciences* 106, 9322–9327.

Kim, M.I., Wan, X., Kim, M.J., Jeong, H.C., Ahn, N.H. *et al.* (2010) Phylogenetic relationships of true butterflies (Lepidoptera: Papilionoidea) inferred from COI, 16S rRNA and EF 1α sequences. *Molecules and Cells* 30, 409–425.

King, E.B. (1922) Lepidoptera of the smaller Channel Islands. *Entomologist's Record and Journal of Variation* 34, 217.

King, E.B. (1923) Some notes on butterflies in England and the Channel Isles in 1922. *Entomologist* 56, 18–19.

King, E.B. (1938) *D. plexippus* in the Isles of Scilly. *Entomologist* 71, 236.

King, J.J.F.X. (1901) *Vanessa antiopa* in Shetland. *Entomologist's Monthly Magazine* 37, 226–227.

Kinnear, P.K. (1976) Unusual numbers of peacocks (*Inachis io* (L.)) (Lep., Nymphalidae) in Shetland. *Entomologist's Gazette* 27, 137.

Kirkcaldy, G.W. (1899) *Erebia aethiops* (*blandina*) in the Isle of Skye. *Entomologist* 32, 236.

Kirkland, P. (2012) The enigmatic *Erebia* – the Scotch Argus in Britain. *British Wildlife* 23, 179–185.

Kisel, Y. and Barraclough, T.G. (2010) Speciation has a spatial scale that depends on levels of gene flow. *The American Naturalist* 175, 316–334.

Kitihara, M. and Fujii, K. (1994) Biodiversity and community structure of temperate species within a gradient of human disturbance. An analysis based on the concept of generalist vs. specialist strategies. *Researches on Population Ecology* 36, 187–199.

Knill-Jones, S.A. (1992) Late emergence times in the Isle of Wight during 1991. *Entomologist's Record and Journal of Variation* 104, 184.

Knill-Jones, S. (2008) Isle of Wight, Hampshire (Coastal Reports). *Atropos* 33, 53–54.

Knill-Jones, S. (2010) Isle of Wight, Hampshire (Coastal Reports) [*D. chrysippus*]. *Atropos* 39, 45–47.

Knill-Jones, S.A. and Angell, B.J. (1996) Exotic and rare migrant butterflies recorded on the Isle of Wight during 1995. *Entomologist's Record and Journal of Variation* 108, 143–144.

Kohn, A.J. (1972) *Conus miliaris* at Easter Island – ecological release of diet and habitat in an isolated population. *American Zoologist* 12, 712.

Koopman, B.O. (1980) *Search and Screening*. Pergamon, Oxford.

Koren, T., Vukotić, K. and Verovnik, R. (2015) Butterflies (Lepidoptera: Papilionoidea and Hesperioidea) of the Croatian islands: new surveys of Vir, Murter, Čiovo and Šolta. *Entomologist's Gazette* 66, 187–197.

Kotze, D.J. (2008) The occurrence and distribution of carabid beetles (Carabidae) on islands in the Baltic Sea: a review. *Journal of Insect Conservation* 12, 265–276.

Krauss, J., Schmitt, T., Seitz, A., Steffan-Dewenter, I. and Tscharntke, T. (2004) Effects of habitat fragmentation on the genetic structure of the monophagous butterfly *Polyommatus coridon* along its northern range margin. *Molecular Ecology* 13, 311–320.

Kudrna, O. (1986) *Butterflies of Europe*: 8. *Aspects of the Conservation of the Butterflies of Europe*. Aula-Verlag, Wiesbaden, Germany.

Kudrna, O. (2002) The distribution atlas of European butterflies. *Oedippus* 20, 1–342.

Kudrna, O., Pennerstorfer, J. and Lux, K. (2015) *Distribution Atlas of European Butterflies and Skippers*. Wissenschaftlicher Verlag, Schwanfeld, Germany.

Kuussaari, M., Nieminen, M. and Hanski, I. (1996) An experimental study in the Glanville fritillary butterfly *Melitaea cinxia*. *Journal of Animal Ecology* 65, 791–801.

Lack, D. (1932) Further notes on insects from St. Kilda in 1931. *Entomologist's Monthly Magazine* 68, 139–145.

Lack, D. (1976) *Island Biology Illustrated by the Land Birds of Jamaica*. Vol. 3 in *Studies in Ecology* series. University of California Press, Berkeley, California.

Laidlaw, W.B.R. (1932) Speckled wood butterfly (*Pararge aegeria*) in Mull. *Scottish Naturalist* 1932, 19.

Lake, S., Liley, D., Sill, R. and Swash, A. (2015) *Britain's Habitats. A Guide to the Wildlife Habitats of Britain and Ireland*. Princeton University Press, Woodstock, UK.

Lamb, H.H. and Johnson, A.I. (1966) *Secular Variations of the Atmospheric Circulation since 1750*. Geophysical Memoirs No. 110. HMSO, London.

Lambeck, K. (1995) Late Devensian and Holocene shorelines of the British Isles and North Sea from models of glacio-hydro-isostatic rebound. *Journal of the Geological Society* 152, 437–448.

Lambeck, K. (2004) External geophysics, climate and environment (climate) sea-level change through the last glacial cycle: geophysical, glaciological and palaeogeographic consequences. *Comptes Rendus Geoscience* 336, 677–689.

Lambeck, K., Yokoyamab, Y. and Purcella, T. (2002) Into and out of the Last Glacial Maximum: sea-level change during Oxygen Isotope Stages 3 and 2. *Quaternary Science Reviews* 21, 343–360.

Lambeck, K., Rouby, H., Purcell, A., Sun, Y. and Sambridge, M. (2014) Sea level and global ice volumes from the Last Glacial Maximum to the Holocene. *Proceedings of the National Academy of Sciences* 111, 15296–15303.

Lansbury, I. (1961) *Vanessa atalanta* in the Aran islands. *Entomologist's Gazette* 12, 35.

Lattin, G. de (1952) Two new subspecies of *Hipparchia semele* Linnaeus. *Entomologist's Record and Journal of Variation* 64, 335.

Lavery, T.A. (1989) The heath fritillary (*Mellicta athalia* Rott.): did it really occur in Ireland? *Bulletin of the Amateur Entomologists' Society* 48, 158–159.

Lawlor, M. (2011) Guernsey (Coastal Records) [*P. machaon* on Omptolle (Hommetol, Les Houmets), and Les Laurens, also *G. rhamni* at La Hougette and La Moye, both on Guernsey.] *Atropos* 42, 44.

Leamy, L.J. and Klingenberg, C.P. (2005) The genetics and evolution of fluctuating asymmetry. *Annual Review of Ecology, Evolution, and Systematics* 36, 1–21.

Lee, J.R., Rose, J., Hamblin, R.J., Moorlock, B.S., Riding, J.B. *et al.* (2011) The Glacial History of the British Isles during the Early and Middle Pleistocene: Implications for the long-term development of the British Ice Sheet. In: Ehlers, J., Gibbard, P.L. and Hughes, P.D. (eds) *Quaternary Glaciations – Extent and Chronology, A Closer Look.* Developments in Quaternary Science, 15. Elsevier, Amsterdam, pp. 59–74.

Leech, M.J. (1951) Lepidoptera in the island of South Uist, Outer Hebrides, August 17–September 4, 1950. *Entomologist* 84, 193–194.

Legendre, P. and Legendre, L. (2000) *Numerical Ecology.* Elsevier, Amsterdam.

Lekay, H.G. (1911) *Argynnis lathonia* in Guernsey. *Entomologist* 44, 300.

Lekens, W.A.H., Haflidason, H., Sejrup, H.P., Nygard, A., Richter. T., Vogt, C. and Frederichs, T. (2009) Sedimentation history of the northern North Sea Margin during the last 150ka. *Quaternary Science Reviews* 28, 469–483.

Lennon, J.J., Greenwood, J.J.D. and Turner, J.R.G. (2000) Bird diversity and environmental gradients in Britain: a test of the species-energy hypothesis. *Journal of Animal Ecology* 69, 581–598.

Le Quesne, W.J. (1945) Larvae of *Lampides boeticus* (Lep., Lycaenidae) in Jersey. *Entomologist's Monthly Magazine* 81, 280.

Le Quesne, W.J. (1946) The butterflies of Jersey. *Entomologist's Monthly Magazine* 82, 22–23.

Le Quesne, W.J. (1947) Two additions to the list of Jersey butterflies with notes on other species. *Entomologist's Monthly Magazine* 83, 134.

Le Quesne, W.J. (1973) The development of the present state of knowledge of insects and related classes in Jersey. *Annual Bulletin of Société Jersiaise* 21, 57–63.

Leslie, S., Winney, B., Hellenthal, G., Davison, D., Boumertit, A. *et al.* (2015) The fine-scale genetic structure of the British population. *Nature* 519, 309–314.

Leung, M. (2008) Butterflies [of Herschel Island, North American tundra]. Poster [Private publication received from Burn, C.]

Leverton, R. (1994) The large white, *Pieris brassicae* L. (Lep., Pieridae) apparently univoltine in Banffshire. *Entomologist's Record and Journal of Variation* 106, 190–191.

Liu, Z., Zhu, J., Rosenthal, Y., Zhang, X., Otto-Bliesner, B.L. *et al.* (2014) The Holocene temperature conundrum. *Proceedings of the National Academy of Sciences* 111, E3501–E3505.

Lomolino, M.V. (2000) A call for a new paradigm of island biogeography. *Global Ecology and Biogeography* 9, 1–6.

Lomolino, M.V. (2005) Body size evolution in insular vertebrates: generality of the island rule. *Journal of Biogeography* 32, 1683–1699.

Lomolino, M.V., Riddle, B.R. and Brown, J.H. (2005) *Biogeography* (3rd edn). Sinauer Associates, Sunderland, Massachusetts.

Lomolino, M.V., Brown, J.H. and Sax, D.F. (2010) Island biogeography theory. Reticulations and reintegration of 'a biogeography of the species'. In: Losos, J.B. and Ricklefs, R.E. (eds) *The Theory of Island Biogeography Revisited.* Princeton University Press, Princeton, New Jersey, pp. 13–51.

Long, O.M., Warren, R., Price, J., Brereton, T.M., Botham, M.S. and Franco, A.M.A. (2017) Sensitivity of UK butterflies to local climatic extremes: which life stages are most at risk? *Journal of Animal Ecology* 86, 108–116. doi: 10.1111/1365–2656.12594.

Long, R.C. (1970) Rhopalocera (Lep.) of the Channel Islands. *Entomologist's Gazette* 21, 241–251.

Long, R.C. (1987) Two butterfly records from Jersey, Channel Islands. *Entomologist's Gazette* 38, 202.

Long, R.C. (2010) European Map *Araschnia levana* in Jersey. *Atropos* 41, 63.

Long, R.C. (2015) *The State of Butterflies in Jersey – the Jersey Butterfly Monitoring Scheme: 2004 to 2013.* States of Jersey, Jersey. Available at: https://www.gov.je/SiteCollectionDocuments/Government%20and%20administration/R%20The%20State%20of%20Butterflies%20in%20Jersey%20DM%2016072015.pdf (accessed 5 March 2018).

Longstaff, G.B. (1907) First notes on the Lepidoptera of Lundy Island. *Entomologist's Monthly Magazine* 43, 241–244.

Lorimer, R.L. (1981) Orkney Lepidoptera. *Entomologist's Gazette* 32, 75–78.

Lorimer, R.L. (1983) *The Lepidoptera of the Orkney Islands.* E.W. Classey, Faringdon, UK.

Lorimer, R.L. (1988) The Lepidoptera of the Orkney Islands: supplement, 1983–87. *Entomologist's Gazette* 39, 181–186.

Lorković, Z. (1994) *Leptidea reali* Reissinger 1989 (= *lorkovicii* Réal, 1988), a new European species (Lepid., Pieridae). *Natura Croatia* 2, 1–26.

Losos, J.B. and Ricklefs, R.E. (2009) Adaptation and diversification on islands. *Nature* 457, 830–836.

Losos, J.B. and Schluter, D. (2000) An analysis of a species–area relationship. *Nature* 408, 847–850.

Louy, D., Habel, J.C., Schmitt, T., Assmann, T., Meyer, M. and Müller, P. (2007) Strongly diverging population genetic patterns of three skipper species: the role of habitat fragmentation and dispersal ability. *Conservation Genetics* 8, 671–681. doi: 10.1007/s10592-006-9213-y.

Lowe, F.E. (1904) *Lampides boeticus* in the Channel Islands. *Entomologist's Record and Journal of Variation* 16, 297.

Lowe, F.E. (1911) *Lampides boeticus* in Guernsey. *Entomologist* 44, 367.

Lowe, J.J. and Walker, M.J.C. (1997) *Reconstructing Quaternary Environments* (2nd edn). Longman, London.

Ludwig, J.A. and Reynolds, J.F. (1988) *Statistical Ecology. A Primer on Methods and Computing.* John Wiley and Sons, New York.

Luff, W.A. (1873) A list of the butterflies inhabiting Guernsey and Sark, with notes on their occurrence. *Entomologist* 6, 324–326.

Luff, W.A. (1874) Additions to the list of Macro-Lepidoptera inhabiting Guernsey and Sark. *Entomologist* 7: 42–43; 8: 29–32.

Luff, W.A. (1882) Butterflies of Guernsey and Sark. *Report and Transactions of the Guernsey Society of Natural Science* 1, 61.

Luff, W.A. (1886) *Anosia plexippus* in Guernsey. *Entomologist* 19, 278.

Luff, W.A. (1890a) *Hesperia lineola* in Jersey. *Entomologist* 23, 93.

Luff, W.A. (1890b) Captures on the Isle of Jethou. *Entomologist's Record and Journal of Variation* 1, 139–140.

Luff, W.A. (1893a) *Catocala fraxini, Colias edusa* and *hyale* in Guernsey. *Entomologist's Record and Journal of Variation* 4, 298.
Luff, W.A. (1893b) *Anthocharis cardamines* in Guernsey. *Entomologist's Monthly Magazine* 29, 139.
Luff, W.A. (1898) List of insects on Alderney. *Report and Transactions of the Guernsey Society of Natural Science* 3 (1897), 175–182.
Luff, W.A. (1900a) The insects of Alderney. *Report and Transactions of the Guernsey Society of Natural Science* 3 (1899), 388–408.
Luff, W.A. (1900b) *Papilio machaon* and *Lampides boeticus* in Guernsey. *Entomologist's Record and Journal of Variation* 12, 273.
Luff, W.A. (1905a) The insects of Herm. *Report and Transactions of the Guernsey Society of Natural Science* 4 (1904), 374–378.
Luff, W.A. (1905b) The insects of Jethou. *Report and Transactions of the Guernsey Society of Natural Science* 4 (1904), 388–390.
Luff, W.A. (1907) The insects of Sark. *Report and Transactions of the Guernsey Society of Natural Science* 5 (1906), 185–199.
Lupton, H. (1880) The Lepidoptera of Arran. *Naturalist* (Hull) 6, 72–75.
MacArthur, R.H. and Wilson, E.O. (1963) An equilibrium theory of insular zoogeography. *Evolution* 17, 373–387.
MacArthur, R.H. and Wilson, E.O. (1967) *The Theory of Island Biogeography*. Princeton University Press, New Jersey.
MacGillivray, J. (1842) An account of the island of St. Kilda, chiefly with reference to its natural history, made during a visit in 1840. *Edinburgh New Philosophical Journal* 32, 47–70.
Majerus, M.E.N. (1979) The status of *Anthocharis cardamines hibernica* Williams (Lepidoptera: Pieridae), with special reference to the Isle of Man. *Entomologist's Gazette* 30, 245–248.
Majerus, M.E.N. (1980) The brimstone butterfly in the Isle of Man. *Bulletin of the Amateur Entomologists' Society* 39, 40–41.
Mallet, J. (1995) A species definition for the modern synthesis. *Trends in Ecology and Evolution* 10, 294–299.
Mallet, J., Wynne, I.R. and Thomas, C.D. (2010) Hybridisation and climate change: brown argus butterflies in Britain (*Polyommatus* subgenus *Aricia*). *Insect Conservation and Diversity* 4, 192–199.
Manley, G. (1974) Central England temperatures: monthly means 1659–1973. *Quarterly Journal of the Royal Meteorological Society* 100, 389–405.
Marren, P. (2015) *Rainbow Dust. Three Centuries of Delight in British Butterflies.* Square Peg, Vintage (Random House Group), London.
Marsset, T., Tessier, B., Reynaud, J.-Y., De Batist, M. and Plagnol, C. (1999) The Celtic Sea banks: an example of sand body analysis from very high-resolution seismic data. *Marine Geology* 158, 89–109.
Martineau, A.H. (1894) [Exhibition of Lepidoptera collected on Lundy by R.W. Chase at Birmingham Entomological Society]. *Entomologist* 27, 303.
Marwick, J.G. (1931) Small tortoiseshell butterfly in Orkney. *Scottish Naturalist* 1931, 89.
Mason, S., Hill, J., Thomas, C., Powney, G. Fox, R., Brereton, T. and Oliver, T. (2017) Population variability of species can be deduced from citizen science records: a case study using British butterflies. *Insect Conservation and Diversity* (July 2017), 1–12 . doi: 10.1111/icad.12242.
Masters, G. (2000) Embiopteran tools. *Antenna* 24, 31–34.
Mather, A.S., Smith, J.S. and Ritchie, W. (1974) *Beaches of Northern Inner Hebrides.* Department of Geography, University of Aberdeen for Countryside Commission for Scotland. Scottish Natural Heritage, Aberdeen, UK.

Matthews, L.H. (1932) Denny Isle. *Proceedings of the Bristol Naturalists' Society* 7, 371–378.

Mayr, E. (1942) *Systematics and the Origin of Species*. Colombia University Press, New York.

Mayr, E. (1963) *Animal Species and Evolution*. Harvard University Press, Cambridge, Massachusetts.

McAllister, D.W. (1993) The spread of the speckled wood *Pararge aegeria* (L.) in Easter Ross and Sutherland. *Butterfly Conservation News* 53. 44–48.

McArthur, H. (1901) Four months collecting in the Isle of Lewis. *Entomologist* 34, 305–306.

McCormick, D. (1974a) *Islands of England and Wales*. Osprey, London.

McCormick, D. (1974b) *Islands of Scotland*. Osprey, London.

McCormick, D. (1974c) *Islands of Ireland*. Osprey, London.

McCullagh, P. and Nelder, J.A. (1982) (2nd edn 1989) *Generalized Linear Models*. Chapman and Hall, London.

McLeod, L. (1972) The distribution of insects related to railway embankments. *Entomologist's Record and Journal of Variation* 84, 69–71.

Mead-Briggs, M. and Eales, P. (2010) *Argynnis cybele* (Lepidoptera: Nymphalidae) – a 'new' record for the British Isles. *British Journal of Entomology and Natural History* 23, 167–169.

Melton, M.A. (1958) Correlation structure of morphometric properties of drainage systems and their controlling agents. *Journal of Geology* 66, 442–460.

Merckx, T. and Van Dyck, H. (2006) Landscape structure and phenotypic plasticity in flight morphology in the butterfly *Pararge aegeria*. *Oikos* 113, 226–232.

Mere, R.M. (1959) Isles of Scilly in 1958. *Entomologist's Gazette* 10, 107–110.

Mere, R.M. (1960) Isles of Scilly in 1959. *Entomologist's Gazette* 11, 118.

Metzger, C., Ursenbacher, S. and Christe, P. (2009) Testing the competitive exclusion principle using various niche parameters in a native (*Natrix maura*) and an introduced (*N. tessellata*) colubrid. *Amphibia-Reptilia* 30, 523–531.

Meyer, A. (1993) Phylogenetic relationships and evolutionary processes in East African cichlid fishes. *Trends in Ecology and Evolution* 8, 279–284.

Michener, W.K. and Brunt, J.W. (2000) *Ecological Data. Design, Management and Processing*. Blackwell Science, Oxford.

Middleton-Welling, J., Wade, R., Dennis, R.L.H., Dapporto, L. and Shreeve T.G. (2018) Optimising trait and source selection for explaining range and distribution changes: a case study using British butterflies. *Functional Ecology* doi: 10.1111/1365-2435.13082.

Miller, K.W. and Owen, J.A. (1952) A list of insects from the Isle of Ulva. *Scottish Naturalist* 1952, 31–37.

Montgomery, W.I., Provan, J., McCabe, A.M. and Yalden, D.W. (2014) Origin of British and Irish mammals: disparate post-glacial colonisation and species introductions. *Quaternary Science Reviews* 98, 144–165.

Morgan, M.J. (1969) A check list of insects from Bardsey Island, Caernarvonshire. *Entomologist's Gazette* 20, 105–117.

Morley, A.M. and Chalmers-Hunt, J.M. (1959) Some Observations on the Crimson Ringed Butterfly (*Parnassus apollo* L.) in Britain. *Entomologist's Record and Journal of Variation* 71, 273–276.

Morton, A.C.G. (1982) The effects of marking and capture on recapture frequencies of butterflies. *Oecologia* 53, 105–110.

Muir, D.A. (1954) A note on the insects from the Garvellach Islands. *Glasgow Naturalist* 17, 140–141.

Munroe, E.G. (1948) The geographical distribution of butterflies in the West Indies. PhD thesis, Cornell University, Ithaca, New York.

Murdoch, D., Lyle, T. and Holmes, M.J.G. (1987) The butterflies of London's East End: a three year survey. *London Naturalist* 66, 125–133.

Murray, S., Morgan, G. and Harris, M.P. (2015) A count of Northern Gannets on Grassholm in 2015. *British Birds* 108, 691–692.

Mutanen, M., Kivelä, S.M., Vos, R.A., Doorenweerd, C., Ratnasingham, S. *et al.* (2016) Species-level para- and polyphyly in DNA barcode gene trees: strong operational bias in European Lepidoptera. *Systematic Biology* 65, 1024–1040.

Nässig, W.A. (1995) Die Tagfalter der Bundesrepublik Deutschland: Vorschlag für ein modernes, phylogenetisch orientiertes Artenverzeichnis (kommentierte Checkliste) (Lepidoptera, Rhopalocera). *Entomologische Nachrichten und Berichte* 39, 1–28.

Nazari, V. (2014) Chasing butterflies in Medieval Europe. *Journal of the Lepidopterists' Society* 68, 223–231.

Nelson, B., Hughes, M., Nash, R. and Warren, M. (2001) *Leptidea reali* Reissinger 1989 (Lep.: Pieridae): a butterfly new to Britain and Ireland. *Entomologist's Record and Journal of Variation* 113, 97–101.

Nelson, B., Hughes, M. and Bond, K. (2011) The distribution of *Leptidea sinapis* (Linnaeus, 1758) and *L. reali* Reissinger, 1989 (Lepidoptera: Pieridae) in Ireland. *Entomologist's Gazette* 62, 213–233.

Nelson, J.M. (1980) Some invertebrates from Ailsa Craig. *Western Naturalist* 1980, 9–16.

Newman, L.H. (1931) Sark Lepidoptera. *Entomologist's Record and Journal of Variation* 43, 184–186.

Nicholson, C. (1933) Immigrant Lepidoptera [Round Island]. *Entomologist* 66, 285–286.

Nieminen, M. (1996) Migration of moth species in a network of small islands. *Oecologia* 108, 643–651.

Nonweiler, T. (1949) *Danaus plexippus* on Skye. *Entomologist* 82, 40.

Norgate, F. (1880) List of insects observed in Tresco, Scilly Isles, in August 1878. *Entomologist's Monthly Magazine* 16, 182–183.

Nye, E.R. (1957) Insect records from Bardsey. *Entomologist's Record and Journal of Variation* 69, 206–208.

Nylin, S. (1988) Host plant specialization and seasonality in a polyphagous butterfly, *Polygonia c-album* (Nymphalidae). *Oikos* 53, 381–386.

Oates, M.R. and Warren, M.S. (1990) *A review of butterfly introductions in Britain and Ireland.* World Wide Fund for Nature, London.

O'Donnell, M. and Wilson, C. (2009) *The Lepidoptera of County Wexford.* Wexford Naturalists' Field Club, Wexford, Ireland.

Oldham, C. (1931) Painted lady in Shetland. *Scottish Naturalist* 1931, 152.

Olesen, J.M., Eskildsen, L.I. and Venkatasamy, S. (2002) Invasion of pollination networks on oceanic islands: importance of invader complexes and endemic super generalists. *Diversity and Distributions* 8, 181–192.

Orme, L.C., Davies, S.J. and Duller, G.A.T. (2015) Reconstructed centennial variability of Late Holocene storminess from Cors Fochno, Wales, UK. *Journal of Quaternary Science* 30, 478–488.

OSNI (2008) OSNI 1:25,000 Activity Map. Ordnance Survey of Northern Ireland. Available at: www.osni.gov.uk (accessed 5 March 2018).

Otaki, J.M. and Yamamoto, H. (2004) Species-specific color-pattern modifications of butterfly wings. *Development, Growth and Differentiation* 46, 1–14.

Otaki, J.M., Hiyama, A., Iwata, M. and Kudo, T. (2010) Phenotypic plasticity in the range-margin population of the lycaenid butterfly *Zizeeria maha. BMC Evolutionary Biology* 10, 252 (13 pp). doi: 10.1186/1471-2148-10-252.

Owen, D.F. (1949) The Macrolepidoptera of the Moorgate, London, bombed sites. *Entomologist* 82, 59–62.

Owen, D.F. (1951) Bombed site Lepidoptera. *Entomologist* 84, 265–272.

Packer, L. (1984) The ecological genetics of the speckled wood butterfly, *Pararge aegeria* L. – a preliminary study. *Heredity* 52, 179–188.

Paine, R.T. (1966). Food web complexity and species diversity. *American Naturalist* 100, 65–75.

Pankhurst, R.J. and Allinson, J. (1985) *British grasses: a punched-card key to grasses in the vegetative state.* AIDGAP Guides, Occasional Publication 10. Field Studies Council, London.

Parker, A.G. and Goudie, A.S. (2007) Late Quaternary Environmental change in the Limestone regions of Britain. In: Goudie, A.S. and Kalvoda, J. (eds) *Geomorphological Variations.* P3K, Prague, pp. 157–182.

Parrack, J.D. (1986) Entomological investigation of the 'Snook', Holy Island, part of the Lindisfarne NNR, during 1984–86. *Vasculum* 71, 20–29.

Parsons, T. (1978) An insect island. In: Kenneth Allsop Trust and Fowles, J. (eds) *Steep Holm – A case history in the study of evolution.* Kenneth Allsop Memorial Trust, Dorset, UK, pp. 177–187.

Patton, H., Hubbard, A., Glasser, N.F., Bradwell, T. and Golledge, N.R. (2015) The last Welsh Ice Cap: Part 1 – Modelling its evolution, sensitivity and associated climate. *Boreas* 42, 471–490.

Peacock, A.D., Smith, E.P. and Davidson, C.F. (eds) (1934–1935) The natural history of South Rona. The results of a biological expedition from University College (St Andrew's, Dundee), July 1933. *Scottish Naturalist* 1934: 113–127, 149–163; 1935: 3–10, 31–34.

Pelham-Clinton, E.C. (1964) Comments on the supposed occurrence in Scotland of *Erebia ligea* (Linnaeus) (Lepidoptera, Satyridae). *Entomologist's Record and Journal of Variation* 76, 121–125.

Peltier, W.R., Shennan, I., Drummond, R. and Horton, B.P. (2002) On the postglacial isostatic adjustment of the British Isles and the shallow viscoelastic structure of the Earth. *Geophysical Journal International* 148, 443–475.

Pennington, M.G. (1993) Butterflies in Shetland. *British Butterfly Conservation Society News* 55, 45–47.

Pennington, M.G. (1996a) Camberwell Beauties *Nymphalis antiopa* (L.) (Lep. Nymphalidae) and Swallowtails *Papilio machaon* L. (Lep. Papilionidae) in Shetland. *Entomologist's Record and Journal of Variation* 108, 67–68.

Pennington, M.G. (1996b) Large White butterfly *Pieris brassicae* (L.) (Lep.: Pieridae) in Shetland. *Entomologist's Record and Journal of Variation* 108, 137–139.

Pennington, M.G. (1996c) The Peacock butterfly *Inachis io* (L.) (Lep. Nymphalidae) in Shetland. *Entomologist's Record and Journal of Variation* 108, 208–210.

Pennington, M. and Riddiford, N. (1993) Moths and butterflies in Shetland in 1992. *Shetland Entomological Group. Supplement to Newsletter* 4, 2–8.

Phelan, A. (2008) Irish butterflies. Available at: http://www.irishbutterflies.com/comma_butterfly_of_ireland.html (accessed 23 July 2017).

Philp, E.G., (1993) *The Butterflies of Kent.* Transactions of the Kent Field Club, Sittingbourne, UK.

Piquet, F.G. (1873) A list of the butterflies inhabiting Jersey, with notes of their occurrence. *Entomologist* 6, 399–401.

Plant, C. (1987) *The Butterflies of the London Area.* London Natural History Society, Passmore Edwards Museum, London.

Pocklington, C. (1872) Butterflies migrating and settling on the sea. *Entomologist* 6, 151–152.

Pollard, E. (1979) Population ecology and change in range of the white admiral butterfly *Ladoga camilla* L. in England. *Ecological Entomology* 4, 61–74.

Pollard, E. (1988) Temperature, rainfall and butterfly numbers. *Journal of Applied Ecology* 25, 819–828.

Pollard, E. and Hall, M.L. (1980) Possible movement of *Gonepteryx rhamni* (L.) (Lepidoptera: Pieridae) between hibernating and breeding areas. *Entomologist's Gazette* 31, 217–220.

Pollard, E. and Yates, T.J. (1992) The extinction and foundation of local butterfly populations in relation to population variability and other factors. *Ecological Entomology* 17, 249–254.

Pollard, E. and Yates, T.J. (1993) *Monitoring Butterflies for Ecology and Conservation.* Chapman and Hall, London.

Pollard, E., Hall, M.L. and Bibby, T.J. (1986) *Monitoring the Abundance of Butterflies 1976–85.* Research and Survey in Nature Conservation Series, no. 2. Nature Conservancy Council, Peterborough, UK.

Pollard, E., Moss, D. and Yates, T.J. (1995) Population trends of common British butterflies at monitored sites. *Journal of Applied Ecology* 32, 9–16.

Porter, S.C. and Zhisheng, A. (1995) Correlation between climate events in the North Atlantic and China during the last glaciations. *Nature* 375, 305–308.

Praeg, D., McCarron, S., Dove, D., Ó Cofaigh, C., Scott, G. *et al.* (2015) Ice sheet extension to the Celtic Sea shelf edge at the Last Glacial Maximum. *Quaternary Science Reviews* 111, 107–112.

Pratt, C. (1986–1987) A history and investigation into the fluctuations of *Polygonia c-album* L. the comma butterfly. *Entomologist's Record and Journal of Variation* 98: 197–203, 244–250; 99: 21–27, 69–80.

Pratt, C.R. and Hulme, N.A.C. (2010) The Queen of Spain Fritillary *Issoria lathonia* L. in Sussex during the twenty-first century. *Atropos* 39, 3–8.

Preston, C.D., Pearman, D.A. and Dines, T.D. (eds) (2002) *New Atlas of the British and Irish Flora: an Atlas of the Vascular Plants of Britain, Ireland, the Isle of Man and the Channel Islands.* Oxford University Press, Oxford.

Price, D. (2016) *A Field Guide to Grasses, Sedges and Rushes.* Species Recovery Trust, Salisbury, UK.

Prosser, R.S., Anderson, J.C., Hanson, M.L., Solomon, K.R. and Sibley, P.K. (2016) Indirect effects of herbicides on biota in territorial edge-of-field habitats: a critical review of the literature. *Agriculture, Ecosystems and Management* 232, 59–72.

Provan, J. and Bennett, K.D. (2008) Phylogeographic insights into cryptic glacial refugia. *Trends in Ecology and Evolution* 23, 564–571.

Prüfer, K., Racimo, F., Patterson, N., Jay, F., Sankararaman, S. *et al.* (2014) The complete genome sequence of a Neandertal from the Altai Mountains. *Nature* 505, 43–49. doi: 10.1038/nature12886.

Pulliam, H.R. (1988) Sources, sinks, and population regulation. *American Naturalist* 132, 652–661.

Pullin, A.S. and Knight, T.M. (2003) Support for decision making in conservation practice: an evidence-based approach. *Journal for Nature Conservation* 11, 83–90.

Quinette, J. and Lepertel, N. (1993) Inventaire préliminaire des Macrolépidoptères du département de la Manche. *Alexanor* 18, 157–177.

Quinn, G.P. and Keough, M.J. (2002) *Experimental Design and Data Analysis for Biologists.* Cambridge University Press, Cambridge.

Rackham, O. (1986) *The History of the Countryside.* J.M. Dent and Sons, London.

Ralston, G.S. (1959) Migrating butterflies in the northern isles of Shetland. *Entomologist* 92, 184.

Ralston, G.S. (1960) Migrating butterflies in the northern isles of Shetland. *Entomologist* 93, 194.

Ravenscroft, N.O.M. (1990) The ecology and conservation of the silver-studded blue butterfly (*Plebejus argus* L.) on the Sandlings of East Anglia, England. *Biological Conservation* 53, 21–36.

Ravenscroft, N.O.M. (1994a) The ecology of the chequered skipper butterfly *Carterocephalus palaemon* in Scotland. I. Microhabitat. *Journal of Applied Ecology* 31, 613–622.

Ravenscroft, N.O.M. (1994b) The ecology of the chequered skipper butterfly *Carterocephalus palaemon* in Scotland. II. Foodplant quality and population range. *Journal of Applied Ecology* 31, 623–630.

Ravenscroft, N.O.M. and Warren, M.S. (1996) *Species Action Plan: Chequered skipper Carterocephalus palaemon.* Butterfly Conservation, Wareham, UK.

Réal, P. (1988) Lépidoptères nouveaux principalement jurassiens. *Mémoires du Comité de Liaison pour les Recherches Ecofaunistiques dans le Jura* 4, 1–28.

Redway, D.B. (1981) Some comments on the reported occurrence of *Erebia epiphron* (Knoch) (Lepidoptera: Satyridae) in Ireland during the nineteenth century. *Entomologist's Gazette* 32, 157–159.

Reed, T.M. (1985) The number of butterfly species on British islands. *Proceedings of the 3rd Congress of European Lepidopterists*, Cambridge, 1982, pp. 146–152.

Rees, C.J. and Sutton, S.L. (1960) Entomological survey, Cape Clear Island 1960. Interim report. *Cape Clear Bird Observatory Report* 2, 29–30.

Regan, E.C., Nelson, B., Aldwell, B., Bertrand, C., Bond, K. *et al.* (2010) *Ireland Red List No. 4 – Butterflies.* National Parks and Wildlife Service, Department of the Environment, Heritage and Local Government, Wexford, Ireland.

Reimer, P.J. (2012) Atmospheric science. Refining the radiocarbon time scale. *Science* 338, 337–338.

Reissinger, E. (1989) Checkliste Pieridae Duponchel, 1835, der Westpalaearktis (Europa, Nordwestafrika, Kaukasus, Kleinasien). *Atalanta* 20, 149–185.

Richardson, A. (1960) A week in the Isle of Arran. *Entomologist's Record and Journal of Variation* 72, 112–115.

Richardson, A. (1963) Further observations on the Lepidoptera of the Isles of Scilly with particular reference to St Agnes. *Entomologist's Record and Journal of Variation* 75, 179–184.

Richardson, A. and Mere, R.M. (1958) Some preliminary observations on the Lepidoptera of the Isles of Scilly, with particular reference to Tresco. *Entomologist's Gazette* 9, 115–147.

Ricklefs, R.E. and Lovette, I.J. (1999) The roles of island area per se and habitat diversity in the species-area relationships of four Lesser Antillean faunal groups. *Journal of Animal Ecology* 68, 1142–1160.

Riley, A.M. (2007) *British and Irish Butterflies.* Brambleby Books, Taunton, UK.

Riley, N.D. (1975) Jersey Lepidoptera. *Entomologist* 58, 149–151.

Rintoul, L.J. and Baxter, E.V. (1920) Lepidoptera of the Isle of May. *Scottish Naturalist* 1920, 198.

Rintoul, L.J. and Baxter, E.V. (1937) Peacock butterflies on Islay. *Scottish Naturalist* 1937, 174.

Rintoul, L.J. and Baxter, E.V. (1950) Natural history notes from the Isle of Gigha. *Scottish Naturalist* 1950, 93–97.

Roberts, D.H., Chiverrell, R.C., Innes, J.B., Horton, B.P., Brooks, A.J. *et al.* (2006) Holocene sea levels, Last Glacial Maximum glaciomarine environments and geophysical models in the northern Irish Sea Basin, UK. *Marine Geology* 231, 113–128.

Roberts, M.J. (2006) Holocene sea-level change in North Wales: the evolution of the Menai Strait. PhD thesis, Prifysgol Bangor University. Available at: http://e.bangor.ac.uk/id/eprint/4390 (accessed 5 March 2018)

Roberts, N. (2014) *The Holocene: an Environmental History* (3rd edn). Wiley-Blackwell, Chichester, UK.

Robinson, R.A. and Sutherland, W.J. (2002) Post-war changes in arable farming and biodiversity in Great Britain. *Journal of Applied Ecology* 39, 157–176.

Rockingham, N.W. (1949) *Pieris napi* in Orkney. *Entomologist* 82, 272.

Rodríguez-Gironés, M.A. and Santamaría, L. (2006) A new algorithm to calculate the nestedness temperature of presence–absence matrices. *Journal of Biogeography* 33, 924–935.

Rodwell, J.S. (ed.) (1991–1995) *British Plant Communities*, 1–5. Cambridge University Press, Cambridge.

Roff, D.A. (1990) The evolution of flightlessness in insects. *Ecological Monographs* 60, 389–421.

Rogers, H.M. (1944) Calf Island. Some natural history notes for 1943. *Peregrine* 1, 12–14.

Rohling, E.J. and Pälike, H. (2005) Centennial-scale climate cooling with a sudden cold event around 8,200 years ago. *Nature* 434, 975–979.

Roland, J., McKinnon, G., Backhouse, C. and Taylor, P.D. (1996) Even smaller radar tags on insects. *Nature* 381, 120.

Rorison, I.H., Sutton, F. and Hunt, R. (1986) Local climate, topography and plant growth in Lathkill Dale NNR. I. A twelve-year summary of solar radiation and temperature. *Plant, Cell and Environment* 9, 49–56.

Rosenzweig, M.L. (1995) *Species Diversity in Space and Time*. Cambridge University Press, Cambridge.

Rosindell, J. and Phillimore, A.B. (2011) A unified model of island biogeography sheds light on the zone of radiation. *Ecology Letters* 14, 552–560.

Roy, D.B., Oliver, T.H., Botham, M.S., Beckmann, B., Brereton, T. *et al.* (2015) Similarities in butterfly emergence dates among populations suggest local adaptation to climate. *Global Change Biology* 21, 3313–3322.

Ruddiman, W.F., Raymo, M.E. and McIntyre, A. (1986) Matuyama 41,000-year cycles. North Atlantic ocean and north hemisphere ice sheets. *Earth and Planetary Science Letters* 80, 117–129.

Rushton, D.R.A. (1971) A peacock butterfly in Shetland. *Entomologist's Record and Journal of Variation* 83, 397.

Russell, P. (2013) Intergeneric pairings involving *Maniola jurtina* L. (Lepidoptera: Nymphalidae: Satyrinae): are sight and smell overridden by this butterfly's sex drive? *Entomologist's Gazette* 64, 11–17.

Russell, P. (2014) The probable demise of an interesting population of *Hipparchia semele* (Linnaeus, 1758) (Lepidoptera: Nymphalidae, Satyrinae) in West Sussex. *Entomologist's Gazette* 65, 3–7.

Russell, P. (2015) Review (*Britain's Butterflies. A Field Guide to the Butterflies of Britain and Ireland*. Newland, D., Still, R., Swash, A. and Tomlinson, D. (2015). Princeton University Press. Woodstock, Oxon.) *Entomologist's Gazette* 66, 284–287.

Ryrholm, N. (1988) An extralimital population in a warm climatic outpost. The case of the moth *Idaea dilutaria* in Scandinavia. *International Journal of Biometeorology* 32, 205–216.

Salmon, M.A. (2000) *The Aurelian Legacy*. Harley Books, Colchester, Essex, UK.

Sattler, K. and Tremewan, W. G. (2009) The authorship of the so-called 'Wiener Verzeichnis'. *Nota lepidopterologica* 32: 3–10.

Schär, S., Vila, R., Petrović, A., Tomanović, Z., Pierce, N.E. and Nash, D.R. (2017) Molecular substitution rate increases with latitude in butterflies: evidence for a trans-glacial latitudinal layering of populations? *Ecography* 40, 930–935. doi: 10.1111/ecog.02487.

Schermerhorn, L.J.G. (1973) What is keratophyre? *Lithos* 6, 1–11.

Schmitt, T. and Seitz, A. (2002) Influence of habitat fragmentation on the genetic structure of *Polyommatus icarus* (Lepidoptera: Lycaenidae): implications for conservation. *Biological Conservation* 107, 291–297.

Schmitt, T. and Varga, Z. (2012) Extra-Mediterranean refugia: The rule and not the exception? *Frontiers in Zoology* 9(1), 22. doi: 10.1186/1742-9994-9-22.

Schmitt, T., Giessl, A. and Seitz, A. (2002) Postglacial colonisation of western central Europe by *Polyommatus coridon* (Poda 1761) (Lepidoptera: Lycaenidae): evidence from population genetics. *Heredity* 88, 26–34.

Schoener, T.W. (2010) The MacArthur–Wilson equilibrium model. A chronicle of what it said and how it was tested. In: Losos, J.B. and Rickle, R.E. (eds) *The Theory of Island Biogeography Revisited*, Princeton University Press, Princeton, New Jersey, pp. 52–87.

Schoener, T.W., Spiller, D.A. and Losos, J.B. (2001) Predators increase the risk of catastrophic extinction of prey populations. *Nature* 412, 183–186.

Schowalter, T.D. (2016) *Insect Ecology: an Ecosystem Approach* (4th edn). Academic Press, London.

Scoble, M.J. (1992) *The Lepidoptera. Form, function and diversity.* Oxford University Press, Oxford.

Scott, J.A. (1985) The phylogeny of butterflies (Papilionoidea and Hesperioidea). *Journal of Research on the Lepidoptera* 23, 241–281.

Scottish Naturalist [Editors] (1888) The Lepidoptera of the Outer Hebrides, Orkney and Shetland. *Scottish Naturalist* 1888, 298–304.

Scottish Naturalist [Editors] (1919) The butterflies of the Isle of Coll, Inner Hebrides. *Scottish Naturalist* 1919, 64.

Sejrup, H.P., Nygård, A., Hall, A.M. and Haflidason, H. (2009) Middle and Late Weichselian (Devensian) glaciation history of south-western Norway, North Sea and eastern UK. *Quaternary Science Reviews* 28, 370–380.

Sekar, S. (2012) A meta-analysis of the traits affecting dispersal ability in butterflies: can wingspan be used as a proxy? *Journal of Animal Ecology* 81, 174–184.

Selby, K.A. and Smith, D.E. (2015) Holocene relative sea-level change on the Isle of Skye, Inner Hebrides, Scotland. *Scottish Geographical Journal* 132, 42–65.

Settele, J., Kudrna, O., Harpke, A., Kühn, I., Van Swaay, C. *et al.* (2008) *Climatic Risk Atlas of European Butterflies.* Pensoft, Sofia and Moscow.

Settele, J., Shreeve, T. G., Konvicka, M. and van Dyck, H. (eds) (2009) *Ecology of Butterflies in Europe.* Cambridge University Press, Cambridge.

Sgrò, C.M., Terblanche, J.S. and Hoffmann, A.A. (2016) What can plasticity contribute to insect responses to climate change? *Annual Review of Entomology* 61, 433–51.

Shapiro, A.M. (1970a) Notes on the spring butterflies in north-east Scotland and Orkney. *Entomologist's Record and Journal of Variation* 82, 85–86.

Shapiro, A.M. (1970b) The role of sexual behaviour in density-related dispersal of pierid butterflies. *American Naturalist* 104, 367–373.

Shaw, M.W. (1951) *Danaus plexippus* in the Hebrides. *Entomologist* 84, 279.

Shayer, C.J. (1947) *Argynnis selene* in Sark. *Entomologist* 80, 179.

Shayer, C.J. (1967) List of butterflies of the Guernsey Bailiwick. *Report and Transactions of the Guernsey Society of Natural Science* 18, 28–29.

Shennan, I., Bradley, S., Milne, G., Brooks, A., Bassett, S. and Hamilton, S. (2006) Relative sea-level changes, glacial isostatic modelling and ice-sheet reconstructions from the British Isles since the Last Glacial Maximum. *Journal of Quaternary Science* 21, 585–599.

Shreeve, T.G. (1992a) Adult behaviour. In: Dennis, R.L.H. (ed.) *The Ecology of Butterflies in Britain.* Oxford University Press, Oxford, pp. 22–45.

Shreeve, T.G. (1992b) Monitoring butterfly movements. In: Dennis, R.L.H. (ed.) *The Ecology of Butterflies in Britain.* Oxford University Press, Oxford, pp. 120–138.

Shreeve, T.G. (1995) Butterfly mobility. In: Pullin, A.S. (ed.) *Ecology and Conservation of Butterflies.* Chapman and Hall, London, pp. 37–45.

Shreeve, T.G. and Dennis, R.L.H. (2009) Cross-water transfer by common butterflies. Clear examples of ignoring the highway. *Entomologist's Gazette* 60, 107–109.

Shreeve, T.G. and Dennis, R.L.H. (2011) Landscape scale conservation: resources, behaviour, the matrix and opportunities. *Journal of Insect Conservation* 15, 179–188.

Shreeve, T.G., Dennis, R.L.H. and Williams, W.R. (1995) Uniformity of wing spotting of *Maniola jurtina* (L.) (Lepidoptera: Satyrinae) in relation to environmental heterogeneity. *Nota Lepidopterologica* 18, 77–92.

Shreeve, T.G., Dennis, R.L.H. and Pullin, A.S. (1996) Marginality: scale determined processes and the conservation of the British butterfly fauna. *Biodiversity and Conservation* 5, 1131–1141.

Shreeve, T.G., Dennis, R.L.H., Roy, D.B. and Moss, D. (2001) An ecological classification of British butterflies: ecological attributes and biotope occupancy. *Journal of Insect Conservation* 5, 145–161.

Siddall, M., Rohling, E.J., Almogi-Labin, A., Hemleben, C., Meischner, D., Schmelzer, I. and Smeed, D.A. (2003) Sea-level fluctuations during the last glacial cycle. *Nature* 423, 853–858.

Simberloff, D.S. and Wilson, E.O. (1971) Experimental zoogeography of islands: a two-year record of colonization. *Ecology* 51, 934–937.

Simmons, I.G. (1990) The mid-Holocene ecological history of the moorlands of England and Wales and its relevance for conservation. *Environmental Conservation* 17, 61–69.

Simmons, I.G. and Tooley, M.J. (eds) (1981) *The Environment in British Prehistory.* Duckworth, London.

Simpson, M.S.L. (1974) Butterflies on the Isles of Scilly, June 1973. *Bulletin of the Amateur Entomologists' Society* 33, 91–96.

Skelton, M. (2000) UK over-wintering of Clouded Yellow. *Butterfly Conservation News* 73, 26–27.

Skinner, B. (1984) *Colour Identification Guide to Moths of the British Isles.* Viking, Harmondsworth, UK.

Smee, M.R. (2011) Population ecology and genetics of the Marsh Fritillary butterfly *Euphydryas aurinia*. PhD thesis, University of Exeter, UK.

Smeed, D.A. (2003) Sea-level fluctuations during the last glacial cycle. *Nature* 423, 853–858.

Smith, A.A.D. (1933) *Danaus plexippus* in the Scilly Isles. *Entomologist* 66, 6.

Smith, D.E., Fretwell, P.T., Cullingford, R.A, and Firth, C.R. (2006) Towards improved empirical isobase models of Holocene land uplift for mainland Scotland, UK. *Philosophical Transactions of the Royal Society of London A: Mathematical, Physical and Engineering Sciences* 364, 949–72.

Smith, D.E., Hunt, N., Firth, C.R., Jordan, J.T., Fretwell, P.T. *et al.* (2012) Patterns of Holocene relative sea level change in the North of Britain and Ireland. *Quaternary Science Reviews* 54, 58–76.

Smith, K.G.V. (1958) Some records of Diptera and other insects from Lundy. *Entomologist's Monthly Magazine* 94, 94.

Smith, K.G.V. (1960) Notes on some insects from Caldey Island. *Nature in Wales* 6, 46–48.

Smith, K.G.V. and Smith, V. (1983) *A Bibliography of the Entomology of the Smaller British Offshore Islands.* E.W. Classey, Faringdon, UK.

Smith, M.J.M. (1984) Camberwell beauty in Shetland – 1982. *Bulletin of the Amateur Entomologists' Society* 43, 146–147.

Smith, R. (1993) Note on holly blue *Celastrina argiolus* (L.) (Lycaenidae) at Spurn Head in 1992. *Butterfly Conservation News* 53, 22.

Smout, A.M. and Kinnear, P. (1993) *The Butterflies of Fife.* Fife Nature, Glenrothes.

Sneath, P.H.A. and Sokal, R.R. (1973) *Numerical Taxonomy. The Principles and Practice of Numerical Classification.* W.H. Freeman and Company, San Francisco, California.

Solly, F. (2014) [Coastal Reports] Isle of Thanet, Kent. *Atropos* 51, 69–70.

Solovyev, V.I., Ilinsky, Y., and Kosterin, O.E. (2015) Genetic integrity of four species of *Leptidea* (Pieridae, Lepidoptera) as sampled in sympatry in West Siberia. *Comparative Cytogenetics* 9, 299–324.

South, R. (1888a) Lepidoptera of the Outer Hebrides. *Entomologist* 21, 25–27.

South, R. (1888b) Distribution of Lepidoptera in the Outer Hebrides, Orkney and Shetland. *Entomologist* 21, 28–30, 98–99.

South, R. (1893) Lepidoptera of the Shetland Islands. *Entomologist*, 26: 98–102.

South, R. (1941) *The Butterflies of the British Isles* (3rd edn) (revised by H.M. Edelsten from 1906 1st edn). Frederick Warne, London.

Spalding, A. (2015) *Loe Bar and the Sandhill Rustic Moth. The Biogeography, Ecology, and Cultural History of a Coastal Shingle Bar.* Brill, Leiden Netherlands.

Spalding, A. and Tremewan, W.G. (1998) Butterflies on St Michael's Mount. *Entomologist's Gazette* 49, 2.

Sparks, T.H., Roy, D.B. and Dennis, R.L.H. (2005) The influence of temperature on migration of Lepidoptera into Britain. *Global Change Biology* 11, 507–514.

Sparks, T.H., Dennis, R.L.H., Croxton, P.J. and Cade, M. (2007) Increased migration of Lepidoptera linked to climate change. *European Journal of Entomology* 104, 139–143.

Spence, B. (2008) Spurn Point, East Yorkshire (Coastal Reports) [*C. croceus, V. atalanta, V. cardui*]. *Atropos* 33, 70–71.

Spooner, G.M. (1963) On the causes of decline of *Maculinea arion* L. (Lep. Lycaenidae) in Britain. *Entomologist* 96, 199–210.

STATISTICA (1994) STATISTICA *for Windows* 1–3. Tulsa, OK: Statsoft, Inc.

Stefanescu, C., Páramo, F., Åkesson, S., Alarcón, M., Ávila, A. *et al.* (2013) Multi-generational long-distance migration of insects: studying the painted lady butterfly in the Western Palaearctic. *Ecography* 36, 474–486. doi: 10.1111/j.1600-0587.2012.07738.x.

Sterling, P. and Parsons, M. (2012) *Field Guide to the Micro Moths of Great Britain and Ireland.* British Wildlife Publishing, Milton on Stour, Gillingham, Dorset, UK.

Stevens, V.M., Turlure, C.T. and Baguette, M. (2010) A meta-analysis of dispersal in butterflies. *Biological Reviews* 85, 625–642.

Stewart, A.M. (1925) Notes from Arran. *Entomologist* 58, 272–273.

Stewart, A.M. (1933) *Colias croceus* in Arran. *Entomologist* 66, 228.

Stewart, A.M. (1939) *Nymphalis io* in Arran. *Entomologist* 72, 8.

Stewart, M. (1950) *Pieris napi* in Orkney. *Entomologist* 83, 43–44.

Stokes, M.A. and Smiley, T.L. (1996) *An Introduction to Tree-ring Dating.* University of Arizona Press, Tuscon, Arizona.

Strevens, C.M.J. and Bonsall, M.B. (2011) Density-dependent population dynamics and dispersal in heterogeneous metapopulations. *Journal of Animal Ecology* 80, 282–293.

Sturm, M. (1979) Origin and composition of clastic varves. In: Schlüchter, C. (ed.) *Moraines and Varves.* Balkema, Rotterdam, Netherlands, pp. 281–285.

Sturt, F., Garrow, D. and Bradley, S. (2013) New models of North West European Holocene palaeogeography and inundation. *Journal of Archaeological Science* 40, 3963–3976.

Sullivan, D.J. (1946) Remarkable migration of butterflies at night and during a gale in Co. Donegal. *Irish Naturalists' Journal* 8, 337.

Summers, G. (1975a) Butterflies in the Isles of Scilly. *Entomologist's Record and Journal of Variation* 87, 95–96.

Summers, G. (1975b) Butterflies on Looe (St George's) Island. *Entomologist's Record and Journal of Variation* 87, 278.

Summers, G. (1976a) Lepidoptera in the Isles of Scilly and Cornwall, June 1975. *Entomologist's Record and Journal of Variation* 88, 29–30.

Summers, G. (1976b) Late September in Scilly. *Entomologist's Record and Journal of Variation* 88, 239.

Summers, G. (1977a) October in Scilly. *Entomologist's Record and Journal of Variation* 89, 18.

Summers, G. (1977b) Treshnish Isles. *Entomologist's Record and Journal of Variation* 89, 173.

Sumner, G.N. (1978) *Mathematics for Physical Geographers*. Edward Arnold, London.

Sutcliffe, O.L. and Thomas, C.D. (1996) Open corridors appear to facilitate dispersal by ringlet butterflies (*Aphantopus hyperantus*) between woodland clearings. *Conservation Biology* 10, 1359–1365.

Suzart de Albuquerque, F., Castro-Diez, P., Rueda, M., Hawkins, B.A. and Rodriguez, M.A. (2011) Relationships of climate, residence time, and biogeographical origin with the range sizes and species richness patterns of exotic plants in Great Britain. *Plant Ecology* 212, 1901–1911.

Tait, W. (1878) *Colias edusa* in Orkney. *Scottish Naturalist* 1878, 160.

Talavera, G., Dincă, V. and Vila, R. (2013) Factors affecting species delimitations with the GMYC model: insights from a butterfly survey. *Methods in Ecology and Evolution* 4, 1101–1110.

Taylor, J.A. (1980) Environmental changes in Wales during the Holocene period. In: Taylor, J.A. (ed.) *Culture and Environment in Prehistoric Wales*. British Archaeological Report, Oxford, series 76, pp. 101–130.

Thomas, C.D. (1985) Specializations and polyphagy of *Plebejus argus* (Lepidoptera: Lycaenidae) in North Wales. *Ecological Entomology* 10, 325–340.

Thomas, C.D. (1994) Extinction, colonization and metapopulations: environmental tracking by rare species. *Conservation Biology* 8, 373–378.

Thomas, C.D. (1995) Ecology and conservation of butterfly metapopulations in the fragmented British landscape. In: Pullin, A.S. (ed.) *Ecology and Conservation of Butterflies*. Chapman and Hall, London, pp. 46–63.

Thomas, C.D. and Hanski, I. (1997) Butterfly metapopulations. In: Hanski, I.A. and Gilpin, M.E. (eds) *Metapopulation Biology: Ecology, Genetics and Evolution*. Academic Press, London, pp. 359–386.

Thomas, C.D. and Jones, T.M. (1993) Partial recovery of a skipper butterfly (*Hesperia comma*) from population refuges: lessons for conservation in a fragmented landscape. *Journal of Animal Ecology* 62, 472–481.

Thomas, C.D., Warren, M.S. and Thomas, J.A. (1992) Distributions of occupied and vacant butterfly habitats in fragmented landscapes. *Oecologia* 92, 563–567.

Thomas, C.D., Glen, S.W.T., Lewis, O.T., Hill, J.K. and Blakeley, D.S. (1999) Population differentiation and conservation of endemic races: the butterfly, *Plebejus argus*. *Animal Conservation* 2, 15–21.

Thomas, J.A. (1983a) The ecology and conservation of *Lysandra bellargus* (Lepidoptera: Lycaenidae) in Britain. *Journal of Applied Ecology* 20, 59–83.

Thomas, J.A. (1983b) A quick method for estimating butterfly numbers during surveys. *Biological Conservation* 27, 195–211.

Thomas, J.A. (1984) The conservation of butterflies in temperate countries; past efforts and lessons for the future. In: Vane-Wright, R. I. and Ackery, E R. (eds) *The Biology of Butterflies*. Academic Press, London, pp. 333–353.

Thomas, J.A. (1986) *RSNC Guide to Butterflies of the British Isles*. Country Life Books (Newnes Books), Oxford.

Thomas, J.A. (1993) Holocene climatic changes and warm man-made refugia may explain why a sixth of British butterflies possess unnatural early-successional habitats. *Ecography* 16, 278–284.

Thomas, J.A. and Lewington, R. (2010) *The Butterflies of Britain and Ireland*. British Wildlife Publishing, Gillingham, Dorset, UK.

Thomas, J.A. and Webb, N. (1984) *Butterflies of Dorset* (includes Brownsea Island). Dorset Natural History and Archaeological Society, Dorchester, UK.

Thomas, J.A., Moss, D. and Pollard, E. (1994) Increased fluctuation of butterfly populations towards the northern edges of species' ranges. *Ecography* 17, 215–220.

Thomson, G. (1970) On the nature of *Maniola jurtina splendida* B. White (Lep., Satyridae). *Entomologist's Record and Journal of Variation* 82, 261–268.

Thomson, G. (1971) The Manx race of *Maniola jurtina* (L.) (Lep., Satyridae). *Entomologist's Record and Journal of Variation* 83, 91–94.

Thomson, G. (1973) Temperature effects on *Maniola jurtina* (L.) (Lep., Satyridae). *Entomologist's Record and Journal of Variation* 85, 109–114.

Thomson, G. (1980) *Butterflies of Scotland.* Croom Helm, London.

Thomson, G. (1987) Enzyme variation at morphological boundaries in *Maniola* and related genera. PhD thesis, University of Stirling, UK.

Tilley, R.J.D. and Dennis, R.L.H. (2017) Assessing probabilities of island location and occupation: a butterfly individual's perspective. *Entomologist's Gazette* 68, 161–175.

Toghill, P. (2009) *The Geology of Britain: an Introduction.* Airlife Publishing (Crowood Press), Marlborough, UK.

Tooley, M.J. (1978) *Sea-level Changes in North-West England During the Flandrian Stage.* Clarendon Press, Oxford.

Tooley, M.J. (1991) Future sea level rise: a working paper. *Proceedings of the Institute of Civil Engineers* 90, 1101–1105.

Toucanne, S., Zaragosi, S., Bourillet, J.-F., Marieu, V., Cremer, M. *et al.* (2010) The first estimation of Fleuve Manche palaeoriver discharge during the last deglaciation: Evidence for Fennoscandian ice sheet meltwater flow in the English Channel ca 20–18 ka ago. *Earth and Planetary Science Letters* 290, 459–473.

Traill, J.W.H. (1869) Notes on the Lepidoptera of Orkney. *Entomologist* 4, 197–200.

Traill, J.W.H. (1888) The Lepidoptera of the Outer Hebrides, Orkney and Shetland. *Scottish Naturalist* 1888, 298–304.

Travis, J.M.J., Murrell, D.J. and Dytham, C. (1999) The evolution of density-dependent dispersal. *Proceedings of the Royal Society of London B* 266, 1837–1842.

Tremewan, W.G. (1953) *Aglais urticae* flying by night. *Entomologist* 86, 58.

Tremewan, W.G. (1995) The genus *Zygaena* Fabricius (Lepidoptera, Zygaenidae) on St Michael's Mount, Cornwall. *Entomologist's Gazette* 46, 256.

Trenbirth, H.E. (2010) *Lichenometry.* British Society for Geomorphology, London. *Geomorphological Techniques* (Online edition ISSN 2047-0371), Chapter 4, section 2.7. pp. 12.

Trevor, J. (1994) Butterflies of the Outer Hebrides. *Butterfly Conservation News* 58, 30–31.

Triantis, K.A., Mylonas, M., Lika, K. and Vardinoyannis, K. (2003) A model for the species–area–habitat relationship. *Journal of Biogeography* 30, 19–27.

Triantis, K.A., Vardinoyannis, K., Tsolaki, E.P., Botsaris, I., Lika, K. and Mylonas, M. (2006) Re-approaching the small island effect. *Journal of Biogeography* 33, 914–923.

Turlure, C. (2009) Habitat from a butterfly's point of view: how specialist butterflies map onto ecological resouces. PhD thesis, Biodiversity Research Centre, Université catholique de Louvain, Louvain-la-Neuve, Belgium.

Turner, H.J. (1924) Divergence of character [*Aglais urticae* on Man]. *Entomologist's Record and Journal of Variation* 36, 156.

Turner, J. (1984) Pollen diagrams from Cross Fell and their implications for former treelines. In: Haworth, E.Y. and Lund, J.W.G. (eds) *Lake Sediments and Environmental History.* University Press, Leicester, UK, pp. 317–357.

Turner, J. and Hodgson J. (1979) Studies in the vegetational history of the Northern Pennines. I. Variation in the composition of the early Flandrian forests. *Journal of Ecology* 67, 629–646.

Turner, J.R.G. (1986) Why are there so few butterflies in Liverpool? Homage to Alfred Russel Wallace. *Antenna* 10, 18–24.

Turner, J.R.G., Gatehouse, C.M. and Corey, C.A. (1987) Does solar energy control organic diversity? Butterflies, moths and the British climate. *Oikos* 48, 195–205.

Tutt, J.W. (1894) Societies [Lundy records]. *Entomologist's Record and Journal of Variation* 5, 105–111.

Tzedakis, P.C. (1993) Long-term tree populations in northwest Greece through multiple Quaternary climatic cycles. *Nature*, 364, 437–440.

van der Werf, H.M.G. (1996) Assessing the impact of pesticides on the environment. *Agricultural Ecosystems and Environment* 60, 81–96.

Van Dyck, H. and Baguette, M. (2005) Dispersal behaviour in fragmented landscapes: routine or special movements? *Basic and Applied Ecology* 6, 535–545.

Van Dyck, H., Bonte, D., Puls, R., Gotthard, K. and Maes, D. (2015) The lost generation hypothesis: could climate change drive ectotherms into a developmental trap? *Oikos* 124, 54–61. doi: 10.1111/oik.02066.

Van't Hof, A.E., Nguyen, P., Dalíkova, M., Edmonds, N., Marec, F. and Saccheri, I. (2013) Linkage map of the peppered moth, *Biston betularia* (Lepidoptera, Geometridae): a model of industrial melanism. *Heredity* 110, 283–295.

Vantieghem, P., Maes, D., Kaiser, A. and Merckx, T. (2017) Quality of citizen science data and its consequences for the conservation of skipper butterflies (Hesperiidae) in Flanders (northern Belgium). *Journal of Insect Conservation* 21, 451–463. doi: 10.1007/s10841-016-9924-4.

Vernon, P. (1966) Drumlins and Pleistocene ice flow over the Ards Peninsula/Strangford Lough area, County Down, Ireland. *Journal of Glaciology* 6, 401–409.

Vernot, B., Tucci, S., Kelso, J., Schraiber, J.C., Wolf, A.B. *et al.* (2016) Excavating Neandertal and Denisovan DNA from the genomes of Melanesian individuals. *Science* 352, 235–239. doi: 10.1126/science.aad9416.

Verovnik, R. (2011) Butterflies (Lepidoptera: Rhopalocera) of the Croatian islands: an update on published records and new surveys of Pašman and Ugljan. *Entomologogist's Gazette* 62, 251–263.

Verovnik, R. and Glogovčan, P. (2007) Morphological and molecular evidence of a possible hybrid zone of *Leptidea sinapis* and *L. reali* (Lepidoptera: Pieridae). *European Journal of Entomology* 104, 667–674.

Via, S. (2009) Natural selection in action during speciation. *Proceedings of the National Academy of Sciences* 106, 9939–9946.

Vickery, M. (1991) National garden butterfly survey. *Butterfly Conservation News* 47, 32–38.

Vickery, M. (1993) Garden butterfly survey 1992. *Butterfly Conservation News* 53, 50–59.

Vine-Hall, J.W. (1969) Some notes on Lepidoptera from the Hebrides. *Entomologist's Gazette* 20, 53–58.

Vodă, R., Dapporto, L., Dincă, V. and Vila, R. (2015) Why do cryptic species tend not to co-occur? A case study on two cryptic pairs of butterflies. *PLoS ONE* 10, e0117802 (18 pp.). doi: 10.1371/journal.pone.0117802.

Vodă, R., Dapporto, L., Dincă, V., Shreeve, T.G., Khaldi, M. *et al.* (2016) Historical and contemporary factors generate unique butterfly communities on islands. *Scientific Reports* 6, 28828. doi: 10.1038/srep28828.

Vogel, R.M., Chad, Y. and Meghan, W. (2011) Nonstationarity: flood magnification and recurrence reduction factors in the United States. *Journal of the American Water Resources Association* 47, 464–474.

Wacher, J., Worth, J. and Spalding, A., (2003) *A Cornwall Butterfly Atlas.* Pisces Publications (Nature Conservation Bureau), Newbury, UK.

Wagener, P.S. (1988) What are the valid names for the two genetically different taxa included within *Pontia daplidice* (Linnaeus, 1758)? (Lepidoptera: Pieridae). *Nota Lepidopterologica* 11, 21–38.

Wagner, W.L. and Funk, V.A. (eds) (1995) *Hawaiian Geography: Evolution on a Hot Spot Archipelago.* Smithsonian Institution Press, Washington, DC.

Wahlberg, N., Braby, M.F., Brower, A.V., de Jong, R., Lee, M.M. *et al.* (2005) Synergistic effects of combining morphological and molecular data in resolving the phylogeny of butterflies and skippers. *Proceedings of the Royal Society B: Biological Sciences* 272, 1577–1586.

Walker, D. (1970) Direction and rate in some British post-glacial hydroseres. In: Walker, D. and West, R.G. (eds) *Studies in the Vegetational History of the British Isles.* Cambridge University Press, Cambridge, pp. 117–139.

Walker, F.A. (1878) *Catalogue of Lepidoptera Rhopalocera in the Collection of the Rev. F. A. Walker, M.A., F.L.S.* T.P. Newman, London.

Walker, F.A. (1888) Lepidoptera in the Channel islands. *Entomologist* 21, 150–151.

Walker, F.A. (1892) Are Jersey insects British? *Entomologist* 25, 320.

Walker, F.A. (1897) Entomology in Alderney. *Entomologist* 30, 223.

Walker, F.A. (1932) Some records of insects from St. Kilda. *Entomologist's Monthly Magazine* 68, 146–150.

Walker, J.J. (1922) On the occasional occurrence of butterflies in Iceland; with notes on the lepidopterous fauna of the North Atlantic islands. *Entomologist's Monthly Magazine* 58, 1–7.

Walmsley, T. and Warlsley, P. (1977) [The feeding habits of butterflies]. *Report of Lundy Field Society* 28, 44.

Wang, Y., Millien, V. and Ding, P. (2016) On empty islands and the small-island effect. *Global Ecology and Biogeography* 25, 1333–1345.

Wang, Y., Wu, Q., Wang, X., Liu, C., Wu, L. *et al.* (2015) Small-island effect in snake communities on islands of an inundated lake: The need to include zeroes. *Basic and Applied Ecology* 16, 19–27.

Waring, P. and Townsend, M. (2009) *Field Guide to the Moths of Great Britain and Ireland* (ill. R. Lewington). British Wildlife Publishing, Milton on Stour, Gillingham, Dorset, UK.

Warren, M.S. (1987a) The ecology and conservation of the heath fritillary, *Mellicta athalia.* I. Host selection and phenology. *Journal of Applied Ecology* 24, 467–482.

Warren, M.S. (1987b) The ecology and conservation of the heath fritillary, *Mellicta athalia.* II. Adult population structure and mobility. *Journal of Applied Ecology* 24, 483–498.

Warren, M.S. (1987c) The ecology and conservation of the heath fritillary, *Mellicta athalia.* III. Population dynamics and the effect of habitat management. *Journal of Applied Ecology* 24, 499–513.

Warren, M.S. (1992a) Butterfly populations. In: Dennis, R.L.H. (ed.) *The Ecology of Butterflies in Britain.* Oxford University Press, Oxford, pp. 73–92.

Warren, M.S. (1992b) The conservation of British butterflies. In: Dennis, R.L.H. (ed.) *The Ecology of Butterflies in Britain.* Oxford University Press, Oxford, pp. 246–274.

Warren, M.S., Pollard, E. and Bibby, T.J. (1986) Annual and long-term changes in a population of the wood white butterfly *Leptidea sinapis. Journal of Animal Ecology* 55, 707–719.

Warren, M.S., Hill, J.K., Thomas, J.A., Asher, J., Fox, R. *et al.* (2001) Rapid responses of British butterflies to opposing forces of climate and habitat change. *Nature* 414, 65–69.

Waterston, J. (1906) On some invertebrates from St. Kilda. *Annals of Scottish Natural History* 1906, 150–153.

Waterston, J. (1911) *Vanessa cardui* in North Mavine, Shetland. *Entomologist's Monthly Magazine* 47, 217.

Waterston, J. (1913) *Pyrameis atalanta* in North Mavine (Shetland). *Entomologist's Monthly Magazine* 49, 214.

Waterston, J. (1939) Insects from Colonsay, South Ebudes. *Scottish Naturalist* 1939, 128–131.

Watkinson, A.R. and Sutherland, W.J. (1995) Sources, sinks and pseudo-sinks. *Journal of Animal Ecology* 64, 126–130.

Watson, A.B. (1893) Collecting in Arran. *Entomologist* 26, 52–54.

Wedd, D. (2004) Butterfly islands [Aran islands]. *Atropos* 21, 15–17.

Weir, J.J. (1880) The Macrolepidoptera of the Shetland Isles. *Entomologist* 13, 249–251, 289–291.

Weir, J.J. (1881a) Notes on the Lepidoptera of the Outer Hebrides. *Entomologist* 14, 218–223.

Weir, J.J. (1881b) The Macrolepidoptera of the Shetland Isles. *Entomologist* 14, 278–281.

Weir, J.J. (1882a) Notes on the Lepidoptera of the Orkney Islands. *Entomologist* 15, 1–5.

Weir, J.J. (1882b) The Macrolepidoptera of the Isle of Arran. *Entomologist* 15, 250–253.

Weir, J.J. (1884) The Macrolepidoptera of the Shetland Isles. *Entomologist* 17, 1–4.

Weninger, B., Schulting, R., Bradtmöller, M., Clare, L., Collard, M. *et al.* (2008) The catastrophic final flooding of Doggerland by the Storegga Slide tsunami. *Documenta Praehistorica* 35, 1–24.

Wheeler, G. (1931a) *Melitaea aurinia* Rott.: its fluctuation and variation. *Entomologist* 64, 137.

Wheeler, G. (1931b) The flying powers of *Melitaea aurinia*. *Entomologist* 64, 223.

White, B. (1882) The Lepidoptera of Orkney, Shetland and the Outer Hebrides. *Scottish Naturalist* 1882, 289–291, 337–344.

White, F.B.W. (1872) Insecta Scotica. *Scottish Naturalist* 1872, 162–168, 198–202, 238–241, 273–275.

White, G. (1789) *The Natural History of Selborne*. [1993 compiled by R. Davidson-Houston]. Thames and Hudson, London.

White, P.J. (1892–1893) Entomology. In: *Report for the Years 1892–93*. Puffin Island Biological Station, Anglesey, UK, pp. 10–11.

Whittaker, R.H. (1965) Dominance and diversity in land plant communities: numerical relations of species express the importance of competition in community function and evolution. *Science* 147, 250–260.

Whittaker, R.J. and Fernández-Palacios, J.M. (2007) *Island Biogeography. Ecology, Evolution, and Conservation*. Oxford University Press, Oxford.

Wigley, T.M.L., Lough, J.M. and Jones, P.D. (1984) Spatial patterns of precipitation in England and Wales and a revised, homogeneous England and Wales precipitation series. *Journal of Climatology* 4, 1–25.

Wild, O.H. (1922) Breeding of red admiral in Bute. *Scottish Naturalist* 1922, 27.

Wiley, E.O. (1981) *Phylogenetics. The Theory and Practice of Phylogenetic Systematics*. John Wiley and Sons, New York.

Wilkins, J.S. (2003) How to be a chaste species pluralist-realist: The origins of species modes and the Synapomorphic Species Concept. *Biology and Philosophy* 18, 621–638.

Wilkinson, R.S. (1975) The scarce swallow-tail: *Iphiclides podalirius* (L.) in Britain. *Entomologist's Record and Journal of Variation* 87, 289–293.

Wilkinson, R.S. (1982) The Scarce Swallow-tail: *Iphiclides podalirius* (L.) in Britain. II: Haworth's Prodromus and Lepidoptera Britannica. *Entomologist's Record and Journal of Variation* 94, 168–172.

Wilks, R.L. (1941) Lepidoptera in the Isle of Islay. *Entomologist* 74, 185–186.

Wilks, R.L. (1945a) *Argynnis cydippe* in Islay [probably error for *Argynnis aglaia*]. *Entomologist* 78, 86.

Wilks, R.L. (1945b) Some notes on the Macrolepidoptera of Islay. *Entomologist* 78, 161–165.

Williams, C.B. (1965) *Insect Migration* (2nd edn). Collins, London.

Williams, C.B. (1971) *Insect Migration* (3rd edn). Collins, London.

Williams, H.B. (1946) The Irish form of *Leptidea sinapsis* L. (Lep., Pieridae). *Entomologist* 79, 1–3.

Williams, M.A.J., Dunkerley, D.L., De Deckker, P., Kershaw, A.P. and Stokes, T. (1998) *Quaternary Environments* (2nd edn). Edward Arnold, London.

Williamson, M. (1981) *Island Populations*. Oxford University Press, Oxford.

Williamson, M. (1983) The land-bird community of Skokholm: Ordination and Turnover. *Oikos* 41, 378–384.

Williamson, M.H. (1987) Are communities ever stable? In: Gray, A.J., Crawley, M.J. and Edwards, P.J. (eds) *Colonisation, Succession and Stability*. Blackwell Scientific Publishing, Oxford, pp. 352–371.

Wilson, C.J., Goodwin, J. and Bond, K. (2007) New Irish record of the small skipper butterfly *Thymelicus sylvestris* (Poda) (Lepidoptera: Hesperiidae) (= *T. lineola*). *Irish Naturalists' Journal* 28(9), 385–386.

Wilson, E.O. (1959) Adaptive shift and dispersal in a tropical ant fauna. *Evolution* 13, 122–144.

Wilson, E.O. (1969) The species equilibrium. In: *Diversity and Stability in Ecological Systems*. Brookhaven Symposia in Biology no. 22. Brookhaven National Laboratory, Upton, New York, pp. 38–47.

Wilson, L.J., Austin, W.E.N. and Jansen, E. (2002) The last British ice sheet. Growth, maximum extent, deglaciation. *Polar Research* 21, 243–250.

Wilson, P. and Lord, T. (2014) Towards a robust deglacial chronology for the northwest England sector of the last British-Irish Ice Sheet. *North West Geography* 14, 1–11.

Windig, J.J., Rintamäki, P.T., Cassel, A. and Nylin, S. (2000) How useful is fluctuating asymmetry in conservation biology: asymmetry in rare and abundant *Coenonympha* butterflies. *Journal of Insect Conservation* 4, 253–261.

Wink, M. (2006) Use of DNA markers to study bird migration. *Journal of Ornithology* 147, 234–244.

Withrington, D.J.K and Verovnik, R. (2008) Butterflies (Rhopalocera) of the Croatian islands. *Entomologist's Gazette* 59, 3–25.

Wollaston, T.V. (1845) Note on the entomology of Lundy island. *Zoologist* 3, 897–900.

Wood, M. and Campbell, S. (1995) Flood for thought on the Great Orme. *Earth Heritage* 3, 15–19.

Woodroffe, C.D. (1986)Vascular plant species–area relationships on Nui Atoll, Tuvalu, Central Pacific: a reassessment of the small island effect. *Australian Journal of Ecology* 11, 21–31.

Woof, W.E. (1958) An ecological survey of the insects of the Farne Islands. PhD thesis, University of Durham, UK.

Wormell, P. (1982) The entomology of the Isle of Rhum NNR. *Biological Journal of the Linnean Society* 18, 291–401.

Wormell, P. (1983) Lepidoptera in the Inner Hebrides. *Proceedings of the Royal Society of Edinburgh* 83B, 531–546.

Worms, C.G.M. de (1939) An entomological trip to Shetland, July 1938. *Entomologist* 72, 60–65.

Worms, C.G.M. de (1959) Lepidoptera in the Highlands and Shetlands, August 1958. *Entomologist* 72, 1–5.

Worms, C.G.M. de (1960) *Colias croceus* (Lep.) on Tresco and near the Lizard. *Entomologist's Monthly Magazine* 93, 60.

Worms, C.G.M. de (1969) *Pyrameis cardui* and other butterflies in Orkney. *Entomologist's Record and Journal of Variation* 81, 282.

Worms, C.G.M. de (1970) Two weeks in Orkney, July–August 1969. *Entomologist's Record and Journal of Variation* 82, 67–74.

Worms, C.G.M. de (1972) Collecting Lepidoptera in Britain during 1971. *Entomologist's Record and Journal of Variation* 84, 184–193.

Wright, A.E. (1942) *Plebejus argus* Linnaeus *masseyi* (Tutt), in North Lancashire and South Westmorland. *Entomologist* LXXV, 7–13.
Wright Jr, H.E., Kutzbach, J.E., Webb III, T., Ruddiman, W.F., Street-Perrott, F.A. and Bartlein, P.J. (eds) (1993) *Global Climates Since the Last Glacial Maximum.* University of Minnesota Press, Minneapolis, Minnesota.
Wright, S. (1932) The roles of mutation, inbreeding, crossbreeding, and selection in evolution. *Proceedings of the XI International Congress of Genetics* 1, 356–366.
Xu, G., Magen, H., Tarchitzky, J. and Kafkafi, U. (2000) Advances in chloride nutrition of plants. *Advances in Agronomy* 68, 97–149.
Yang, Y.H., Yao, J., Hu, S. and Qi, Y. (2000) Effects of agricultural chemicals on DNA sequence diversity of soil microbial community: a study with RAPD marker. *Microbial Ecology* 39, 72–79.
Young, M.R. and Harper, M.W. (1986) *Periclepis cinctana* (D & S) and other Lepidoptera on Tiree in 1984. *Entomologist's Gazette* 39, 199–205.

Websites

BBC News (2012) Rabbits overrun the Isle of Canna. http://www.evergreenrabbitcontrol.co.uk/Isle-of-Canna/c-1-32.htm (accessed 28 July 2017).
BBC News (2016) Rare bog butterfly flutters back from brink [Webb, J.]. http://www.bbc.co.uk/news/science-environment-36865426 (accessed 23 July 2017).
BMS (2006) Biological Records Centre. http://www.ukbms.org/SpeciesList.aspx, http://www.ukbms.org/reportsAndPublications.aspx and http://www.ukbms.org/docs/reports/2013/Country-level%20Summary%20of%20changes%20Table%202013.pdf (accessed 24July 2017).
British Geological Survey, NERC (undated). http://mapapps.bgs.ac.uk/geologyofbritain/home.html (accessed 22 July 2017).
Butterfly Conservation (2010a) Code on collecting, breeding and photography. http://butterfly-conservation.org/files/bc-policy_collecting-breeding--photography-2010.pdf (accessed 23 July 2017).
Butterfly Conservation (2010b) Policy on Introductions and Re-introductions. http://butterfly-conservation.org/files/bc-policy_introductions--re-introductions_2010.pdf (accessed 23 July 2017).
Butterfly Conservation (2015a) http://butterfly-conservation.org/files/1.bnm-newsletter-winter15.pdf (accessed 23 July 2017).
Butterfly Conservation (2015b) Wider Countryside Butterfly Survey [Randle, Z., Brereton, T., Harris, S., Noble, D., Botham, M. and Roy, D.]. http://www.ukbms.org/Downloads/Wider_Countryside/WCBS2015_hires.pdf (accessed 23 July 2017).
Butterfly Conservation (2017a) Big Butterfly Count. www.bigbutterflycount.org/ (accessed 23 July 2017).
Butterfly Conservation (2017b) Migrant Watch. http://butterfly-conservation.org/612/migrant-watch.html (accessed 24 July 2017).
Butterfly Conservation (2017c) [for list of current regional recorders]. https://butterfly-conservation.org/2390/recording-contacts.html (accessed 23 July 2017).
Irish National Biodiversity Data Centre (2017). http://www.biodiversityireland.ie/ (accessed 23 July 2017).
Manx Biodiversity (2016). http://www.manxbiodiversity.org/ (accessed 23 July 2017).
Ordnance Survey of Northern Ireland (OSNI) (2008) OSNI 1:25,000 Activity Map. https://www.osni.gov.uk (accessed 5 March 2018).

Phelan, A. (2008) Irish butterflies. http://www.irishbutterflies.com/comma_butterfly_of_ireland.html (accessed 23 July 2017).

United Kingdom Butterfly Monitoring scheme (2017). http://www.ukbms.org/ (accessed 23 July 2017).

Wikipedia (2017) The 100 year flood. https://en.wikipedia.org/wiki/100-year_flood (accessed 31 July 2017).

Wikipedia (2017) List of islands of the British Isles. https://en.wikipedia.org/wiki/List_of_islands_of_the_British_Isles (accessed 20 July 2017).

Wikipedia (2017) List of butterflies of Ireland. https://en.wikipedia.org/wiki/List_of_butterflies_of_Ireland (accessed 25 July 2017).

Wilkins (2006) (1 October). http://scienceblogs.com/evolvingthoughts/2006/10/01/a-list-of-26-species-concepts/ (accessed 16 March 2017).

Index

Note: Page numbers in **bold** type refer to **figures**
Page numbers in *italic* type refer to *tables*
Page numbers followed by 'P' refer to plates